Algal Technologies for Wastewater Treatment and Resource Recovery

In Focus – Special Book Series

Algal Technologies for Wastewater Treatment and Resource Recovery

Editors

Raul Muñoz, Hardy Temmink, Anthony M. Verschoor and Peter van der Steen

IWA PUBLISHING

Published by

IWA Publishing
Alliance House
12 Caxton Street
London SW1H 0QS, UK
Telephone: +44 (0)20 7654 5500
Fax: +44 (0)20 7654 5555
Email: publications@iwap.co.uk
Web: www.iwapublishing.com

First published 2019

British Library Cataloguing in Publication Data
A CIP catalogue record for this book is available from the British Library

ISBN: 9781789060928 (paperback)
ISBN: 9781789060935 (eBook)
ISBN: 9781789060942 (ePUB)

Contents

Editorial: Algal technologies for wastewater treatment and resource recovery 1
Raul Muñoz, Hardy Temmink, Anthony M. Verschoor and Peter van der Steen

Comparison of the treatment performance of a high rate algal pond and a facultative waste stabilisation pond operating in rural South Australia 3
Neil Buchanan, Paul Young, Nancy J. Cromar and Howard J. Fallowfield

Energy recovery in high rate algal pond used for domestic wastewater treatment 12
Paula Peixoto Assemany, Maria Lúcia Calijuri, Eduardo de Aguiar do Couto, Fernanda Pereira da Silva and Mauro Henrique Batalha de Souza

Investigation and modelling of high rate algal ponds utilising secondary effluent at Western Water, Bacchus Marsh Recycled Water Plant 20
Digby Wrede, Syed U. Hussainy, William Rajendram and Stephen Gray

Pond walls: inclined planes to improve pathogen removal in pond systems for wastewater treatment? 31
A. L. Hawley and H. J. Fallowfield

Exploring the influence of meteorological conditions on the performance of a waste stabilization pond at high altitude with structural equation modeling 37
Long T. Ho, Duy T. Pham, Wout Van Echelpoel, Andres Alvarado, Juan E. Espinoza-Palacios, Maria B. Arevalo-Durazno and Peter L. M. Goethals

Tilapia rearing with high rate algal pond effluent: ammonia surface loading rates and stocking densities effects 49
I. A. Sánchez, R. K. X. Bastos and E. A. T. Lana

Microalgae cultivation in agro-industrial effluents for biodiesel application: effects of the availability of nutrients 57
Mariana Daniel Tango, Maria Lúcia Calijuri, Paula Peixoto Assemany and Eduardo de Aguiar do Couto

Microalgae cultivation for wastewater treatment and biogas production at Moscow wastewater treatment plant 69
N. Shchegolkova, K. Shurshin, S. Pogosyan, E. Voronova, D. Matorin and D. Karyakin

Improved methane yield from wastewater grown algal biomass 81
Mohit Thawani, Nidhi Hans, Saurabh Samuchiwal and Sanjeev Kumar Prajapati

Enhanced biogas production potential of microalgae and swine wastewater using co-digestion and alkaline pretreatment 92
K. Panyaping, R. Khiewwijit and P. Wongpankamol

Biogas from mono- and co-digestion of microalgal biomass grown on piggery wastewater 103
P. Carminati, D. Gusmini, A. Pizzera, A. Catenacci, K. Parati and E. Ficara

Start-up of a microalgae-based treatment system within the biorefinery concept: from wastewater to bioproducts 114
Enrica Uggetti, Joan García, Juan Antonio Álvarez and María Jesús García-Galán

Co-digestion of microalga-bacteria biomass with papaya waste for methane production 125
Glenda Cea-Barcia, Jaime Pérez and Germán Buitrón

Valorization of waste micro-algal biomass – collected from coke oven effluent treatment plant and evaluation of sorption potential for fluoride removal 132
Gargi Biswas, Philips Prince Pokkatt, Aratrika Ghosh, Biswajit Kamila, Kalyan Adhikari and Susmita Dutta

Removal of copper from an electroplating industrial effluent using the native and modified *spirogyra* 147
Nimra Ilyas, Sadia Ilyas, Sajjad-ur-Rahman, Sidra Yousaf, Aqsa Zia and Sidra Sattar

Effective adsorption of nickel (II) with *Ulva lactuca* dried biomass: isotherms, kinetics and mechanisms 156
Jianyou Long, Xiaona Huang, Xiaoli Fan, Yan Peng and Jianrong Xia

Microalgal luxury uptake of phosphorus in waste stabilization ponds – frequency of occurrence and high performing genera 165
A. Crimp, N. Brown and A. Shilton

Enhanced nitrogen removal and energy saving in a microalgal–bacterial consortium treating real municipal wastewater 174
P. Foladori, S. Petrini, M. Nessenzia and G. Andreottola

Photosynthetic oxygenation for urine nitrification 183
Maarten Muys, Joeri Coppens, Nico Boon and Siegfried E. Vlaeminck

Outdoor flat-panel membrane photobioreactor to treat the effluent of an anaerobic membrane bioreactor. Influence of operating, design, and environmental conditions 195
J. González-Camejo, R. Barat, M. V. Ruano, A. Seco and J. Ferrer

Activity assessment of microalgal-bacterial consortia based on respirometric tests 207
S. Rossi, M. Bellucci, F. Marazzi, V. Mezzanotte and E. Ficara

Exploiting symbiotic interactions between *Chlorella protothecoides* and *Brevundimonas diminuta* for an efficient single-step urban wastewater treatment 216
Martina Pastore and Eleonora Sforza

Treatment of food waste digestate using microalgae-based systems with low-intensity light-emitting diodes 225
Andrés Felipe Torres Franco, Scarlet da Encarnação Araújo, Fabiana Passos, Carlos Augusto de Lemos Chernicharo, César Rossas Mota Filho and Cleber Cunha Figueredo

Synthesis, characterization and application of green seaweed mediated silver nanoparticles (AgNPs) as antibacterial agents for water disinfection 235
D. Dixit, D. Gangadharan, K. M. Popat, C. R. K. Reddy, M. Trivedi and D. K. Gadhavi

Editorial: Algal technologies for wastewater treatment and resource recovery

PROMISING FEATURES OF ALGAL TECHNOLOGIES

Over 80% of the globally produced wastewater receives none or hardly any treatment before it is disposed into the environment (WWAP 2017). Therefore, it is urgent to develop new wastewater treatment technologies that are sustainable in the broad sense of the word, i.e. not only produce high quality effluents, but also minimise energy expenses, recover energy and nutrients, and apply technology that is appropriate in relation to the availability of skilled personnel.

The requirements of such a technology could be summarised by a number of characteristics. Firstly, the effluent quality should be better than commonly achieved by e.g. waste stabilisation ponds, especially if nutrient removal is required. With natural systems such as ponds or wetlands extremely large areas are required, which in many circumstances makes these technologies not practically feasible. If the local market conditions are favourable for nutrient recovery, nitrogen and particularly phosphorus recovery are important. Rapidly growing algae cultures are very efficient to capture nutrients from water. However, the areal footprint of wastewater treatment systems is an important cost factor, especially in urbanising areas where land is scarce and/or expensive. As the required area for mechanised systems such as activated sludge is typically around $0.5\ m^2$ per capita, algae-based technologies preferably should be only marginally more area demanding.

The low global coverage of wastewater treatment indicates that many treatment technologies are too complex, too expensive and/or lack robustness. The technological complexity therefore should be kept to a minimum. A thorough understanding of the biological processes and ecological engineering of microbial communities is required, and relatively simple reactors are likely to be preferred over multi-step processes that encompass many different unit operations.

'Wastewater treatment' is increasingly transformed into 'resource recovery', including recovery of the energy and nutrients contained in wastewaters. Algal-based systems may have strong advantages in this respect, as no artificial aeration is required to provide oxygen, and the algal biomass could be used to produce biofuels, biogas, biofertilisers or in special cases even valuable biopharmaceuticals. Further process improvements are necessary to make some of these products feasible in terms of economics and net energy production or consumption.

As water scarcity in the world is increasing, so will the need for reuse of treated wastewater in agriculture or for other non-potable uses. Disinfection of the wastewater is therefore of great importance. Many research papers have been published about pathogen removal in natural systems such as ponds and wetlands, but the more controlled conditions in a photo-bioreactor potentially could be used to achieve a much more rapid removal of pathogens. Such a characteristic probably is a condition for widespread application of modern algal technologies for production of irrigation water. The global fight against climate change also requires wastewater treatment technologies with a minimum CO_2 footprint. The low energy demand of microalgae-based wastewater treatment, along with the ability of microalgae to capture CO_2 from the aerobic heterotrophic oxidation of organic matter, rank algal systems among the most sustainable technologies for water reclamation.

THE INTERDISCIPLINARY BASIS OF ALGAL REACTOR TECHNOLOGY

In order to exchange and discuss the latest achievements towards the realisation of the above described technology, scientists, algaeneers and practitioners gathered for the 1st IWA International Conference on Algal Technologies for Wastewater Treatment and Resource Recovery at the premises of IHE Delft in the Netherlands, on 16–17 March 2017. The objective of the meeting was to bring two perspectives together, i.e. the perspective of treating wastewater (with algae) and the perspective of growing algae (on wastewater). Representatives from both fields benefited from the interdisciplinary discussions. Wastewater treatment engineers benefited from knowledge about the use of algae to produce biofuels, food/fodder supplements or green pharmaceuticals. Likewise, wastewater as a cheap source of carbon and nutrients was

doi: 10.2166/wst.2018.323

shown to be promising for the production of algae-based commodities.

THIS SPECIAL ISSUE

This special issue compiles the main outcomes of the recent efforts to improve the design of waste stabilisation ponds and confirms the superior performance of high rate algal ponds as a result of process intensification. Anaerobic digestion devoted to biogas production continues to be the preferred strategy for the energy valorisation of the algal biomass, co-digestion with multiple high C/N ratio substrates gathering significant attention over the past years. The potential of algal biomass as a biosorbent for heavy metal removal (Cu, Ni, F) maintains its share in the research field of water bioremediation, while research on nutrient removal has focused on providing new insights on the mechanism of nitrogen and phosphorus removal from wastewater in algal–bacterial systems. Finally, it is worth noticing that breakthroughs in complementary fields of research such as nanotechnology or lighting technology are gradually being implemented in algal biotechnology, with new products such as nanoparticles for water disinfection or photobioreactors illuminated by low intensity LED panels.

Guest Editors
Raul Muñoz
Department of Chemical Engineering and Environmental Technology, School of Industrial Engineering, Universidad de Valladolid, Dr. Mergelina s/n, 47011, Valladolid, Spain

Hardy Temmink
Sub-department of Environmental Technology, Wageningen University, P.O. Box 17, 6700 AA Wageningen, The Netherlands

Anthony M. Verschoor
KWR Watercycle Research Institute, P.O. Box 1072, 3430 BB Nieuwegein, The Netherlands

Peter van der Steen
Department of Environmental Engineering and Water Technology, IHE Delft Institute for Water Education, Delft, The Netherlands
E-mail: *p.vandersteen@un-ihe.org*

REFERENCE

WWAP 2017 *The United Nations World Water Development Report 2017: Wastewater, The Untapped Resource*. United Nations World Water Assessment Programme, UNESCO, Paris.

Comparison of the treatment performance of a high rate algal pond and a facultative waste stabilisation pond operating in rural South Australia

Neil Buchanan, Paul Young, Nancy J. Cromar and Howard J. Fallowfield

ABSTRACT

South Australian community wastewater management schemes (CWMS) treat wastewater using waste stabilisation ponds before disposal or reuse. This study compared the performance of a facultative pond, 6,300 m^2, 27.5 d theoretical hydraulic retention time (THRT), with a high rate algal pond (HRAP) operated at depths of 0.32, 0.43 and 0.55 m with THRT equivalent to 4.5, 6.4 and 9.1 d respectively. Both ponds received influents of identical quality, differing only in quantity, and were operated in similar climatic conditions. The depth of HRAP operation had only a minor influence on treatment performance. The study showed that the quality of the treated effluent from the HRAP was equivalent to that of the facultative pond, 5-day biochemical oxygen demand removal >89%, NH_4-N removal 59.09–74.45%. Significantly, *Escherichia coli* log_{10} reduction values by the HRAP, 1.74–2.10, were equivalent to those of the facultative pond. Consequently, HRAPs could replace facultative ponds within CWMS while maintaining treated effluent quality. The benefit would be halving the surface area requirement from 4.2 m^2 $capita^{-1}$ for the facultative pond to between 2.0 and 2.3 m^2 $capita^{-1}$, depth dependent, for an HRAP, with significant attendant reductions in the capital costs for construction.

Key words | faecal indicator organisms, high rate algal ponds, log_{10} reduction values, waste stabilisation pond, wastewater treatment

Neil Buchanan[†]
Paul Young
Nancy J. Cromar
Howard J. Fallowfield (corresponding author)
Health and Environment Group, College of Science and Engineering,
Flinders University,
GPO Box 2100, Adelaide 5001,
Australia
E-mail: *howard.fallowfield@flinders.edu.au*

[†]Deceased 2 July 2015.

INTRODUCTION

Community wastewater management schemes (CWMS), formerly known as septic tank effluent disposal schemes, are adopted in rural South Australia when environmental and public health concerns arise, associated with on-site disposal of treated effluents from residential septic tanks (Palmer *et al.* 2001). Septic tank effluents are reticulated to central treatment facilities that frequently include waste stabilisation ponds (WSPs). Current design guidelines (LGA SA not dated) require a five-cell WSP system, operated in series, comprising a facultative pond, 36 d theoretical hydraulic retention time (THRT), and four maturation ponds, combined 30 d THRT, all with a recommended operational depth of 1.2 m. To assist estimates of dry weather flow, the guidelines assume 3.5 persons per household and a daily flow of 140 L $capita^{-1}$ d^{-1}, which equates to a total WSP surface area equivalent to 7.7 m^2 $capita^{-1}$. CWMS with WSPs, therefore, occupy relatively large surface areas.

WSPs and high rate algal ponds (HRAPs) depend upon the same treatment mechanisms for their efficacy. Algal growth and photosynthesis assimilates inorganic nitrogen and phosphorus and provides oxygen to heterotrophic bacteria for growth and the mineralisation of organic carbon, yielding CO_2 for photosynthesis (Young *et al.* 2017). WSPs are large open ponds, prone to thermal, pH and dissolved oxygen vertical stratification leading to physically heterogeneous reaction environments throughout their depth (Sweeney *et al.* 2005). These conditions result in spatial variations in treatment performance (Sweeney *et al.* 2007). Mixing is effected only by inflowing wastewater and wind (Sweeney *et al.* 2003). Wind speed and direction influence treatment performance and potentially the outcomes of performance monitoring (Sweeney

doi: 10.2166/wst.2018.201

et al. 2007). In contrast, HRAPs are shallow, 0.2–0.5 m, intentionally mixed, most commonly by a slowly rotating paddlewheel, and consequently do not experience vertical stratification. Disinfecting UVB and UVA irradiances are rapidly attenuated with depth in turbid wastewaters (Bolton *et al.* 2010). Intentional mixing increases exposure of the wastewater to these disinfecting irradiances at and near the surface of an HRAP, with the potential to increase pathogen inactivation and $\log_{10}$ reduction value – the difference ($\log_{10}$) between the concentration of the organism ($\log_{10}$ 100 mL^{-1}) in inlet wastewater and that of the treated effluent (Fallowfield *et al.* 1996).

Fallowfield & Garrett (1985) using data from Azov & Shelef (1982) and Arthur (1983), primarily 5-day biochemical oxygen demand (BOD_5) reduction, predicted significantly reduced retention times and area requirements if HRAPs were employed to treat domestic wastewaters rather than WSPs. Real world comparisons were congruent with these predictions. Picot *et al.* (1992) compared two HRAP systems with a standard three-cell WSP at Meze in the South of France. They concluded that HRAPs are comparable to WSPs regarding wastewater treatment performance but require a fifth of the surface area. In another study, El Hamouri *et al.* (2003) compared the wastewater treatment performance of an HRAP and a three-cell WSP in Ouarzazate, Morocco. They showed that the adoption of the HRAP in place of a series of three facultative ponds reduces the net land area requirement by at least 40%.

The reduced depth and surface area of HRAPs compared to WSPs results in a reduction in earthworks and consequently construction costs (Young *et al.* 2017). Additionally, the reduced surface area of HRAPs also results in reduced evaporative losses. Young *et al.* (2017) estimated that compared to a five-cell facultative-maturation WSP system an HRAP treating the same volume of wastewater, 100 m^{-3} d^{-1}, would reduce evaporative losses from 30% to 12.3–15.3%. This increase in treated wastewater for reuse is especially significant for communities in water-scarce regions like those in rural South Australia (Cox *et al.* 2016). Over the past decade, South Australia has experienced a decline in surface water flows and groundwater levels compared to long-term means, with this expected to worsen due to climate change (Cox *et al.* 2016).

This study is, to the authors' knowledge, the first to compare the wastewater treatment performance of a facultative WSP and an HRAP both receiving treated septic tank effluents in Australia. The objective of the research was to determine if an HRAP could replace the facultative pond in a five-cell WSP CWMS.

METHODS

The community wastewater management systems

Lyndoch WSP

In the 2011 census, the population of Lyndoch (34.6000° S, 138.8833° E) was recorded as 1,909. The daily inflow from on-site septic tanks to the three-cell WSP system was 165 m^3 d^{-1}, which was delivered in 5-minute pumping events every 15 minutes throughout the day. The Lyndoch WSP comprises three ponds with gravity flow between each pond. The facultative pond (WSP, 180 m × 35 m) is a 6,300 m^2 pond with a design depth of 1.2 m (Figure 1). However, the pond had not been desludged since construction in 1979. A bathymetric survey of the WSPs in 2010 determined a mean facultative pond depth of 0.68 m. The effective volume was 4,533 m^3 resulting in a THRT of 27.5 d.

Kingston on Murray HRAP

The HRAP (Figure 2) was built in 2008 at Kingston on Murray (KoM; E 140 20′, S 34 14′) for a population of approximately 300 permanent residents. The HRAP consisted of two channels (base dimensions 30 m × 5 m) and was operated at depths of 0.32, 0.43 and 0.55 m. The central baffle dividing the channels was a floating curtain constructed from the same high density polyethylene material as used to line the HRAP. The 3:1 wall batter increased the effective surface area from 192 m^2 (0.32 m) to 208 m^2 (0.43 m) and 226 m^2 (0.55 m). An eight-bladed, stainless

Figure 1 | Facultative waste stabilisation pond at Lyndoch, SA, wastewater treatment plant.

Figure 2 | High rate algal pond at the Kingston on Murray, SA, wastewater treatment plant.

steel paddlewheel powered by a 750 W electric motor directly coupled to a 1:100 reduction gearbox provided mixing, with a mean surface water velocity of $0.2\ m\ s^{-1}$. The HRAP received $12\ m^3\ d^{-1}$ of wastewater pre-treated in on-site septic tanks (THRT 24 h) delivered in six daily pumping operations. A consequence of the constant daily inflow was that the THRT of the HRAP increased with increasing operational depth to 4.5, 6.4 and 9.1 d. Following a change in operational depth, a minimum of three retention times elapsed, to enable equilibration before sampling was conducted over a minimum of three retention times. The median temperature of the wastewater over the sampling period was 17.6 °C. Individual depths were operated and evaluated over periods greater and less than the median temperature. The data from both periods for each depth were aggregated and presented in the comparison.

Climate data

Both sites were provided with a weather station, Environdata WeatherMaster 2000™ (Warwick, Queensland, Australia), which continuously logged photosynthetically active radiation (400–700 nm), total UV irradiation and air temperature over the study period May (winter) 2010 to April (Autumn) 2011.

In situ pond measurements

The mid-pond *in situ* wastewater temperature (T-TEC A™, Temperature Technology, Adelaide), dissolved oxygen (DO; Danfoss Oxy1100 partnered with Danfoss EMCO transmitter-controller) and pH (Hach pH probe and GLI Pro transmitter-controller) were continuously monitored (0.2 m below the surface) at all three WSPs at Lyndoch and the data logged (T-TEC A™, Temperature Technology, Adelaide) every 30 minutes. Identical equipment was used at the KoM Murray HRAP, to measure and log mid-stream and mid-depth parameters similarly.

Wastewater sample collection

Wastewater samples (1 L) were manually collected, following pumping events, from the inlets to the Lyndoch WSP 1 and the KoM HRAP. At Lyndoch, wastewater samples (1 L) were collected manually from the outlet weir points for WSP 1, refrigerated, transported to the laboratory and analysed within 24 h of collection. At the HRAP samples were collected into the same sample bottle at 01:00 and 13:00, each 400 mL, providing a daily 800 mL composite sample, using a refrigerated auto-sampler (ISCO Avalanche). The samples were held at 1 °C in the sampler until retrieval. They were then transported and stored (1 °C) until analysis within 24 h of retrieval.

Wastewater analysis

The methods described in *Standard Methods for the Examination of Water and Wastewater* (Greenberg *et al.* 1992) were used to analyse influent and treated effluent wastewater samples for turbidity (NTU), suspended solids (SS) and chlorophyll *a*. Filtered water (GFC Whatman, exclusion size 1.2 μm) was used for the analysis of BOD_5, and the nutrients NH_4-N and combined NO_2-N and NO_3-N (expressed as NO_x-N) and PO_4-P. *Escherichia coli* per 100 mL was enumerated using the Colilert® chromogenic MPN system (IDDEX Ltd) as described by Young *et al.* (2016).

Log_{10} reduction value calculations

The $\log_{10}$ reduction values (LRVs) of the indicator organisms for both ponds were equal to the difference between the $\log_{10}$ concentration of the organisms entering each pond and the $\log_{10}$ concentration of the organisms leaving each pond.

Statistical analysis

Statistical analysis and graphical presentations were prepared using Microsoft Excel and IBM SPSS Statistics, version 23 (Armonk, NY, USA).

The combined data from the operation of the HRAP at all depths were compared with all data from the WSP. Additionally, for *E. coli* concentrations and LRVs the values collected for each HRAP depth were separately compared to the values collected from the facultative WSP during each depth. Data sets were tested for normality using the Shapiro–Wilk test for normality (Supplementary Material, available with the online version of this paper). When found to be normally distributed, data were analysed using independent-samples t-test for equality of means; if found to violate normality, data were compared using the independent-samples Mann–Whitney U-test. Significance was tested to the 0.05 level for all statistical comparisons.

RESULTS AND DISCUSSION

Climate at the two study sites

The comparison between Lyndoch WSP and the KoM HRAP was conducted under similar climatic conditions (Figure 3), suggesting that any climatic impact on the relative performance of the WSP and the HRAP was negligible.

Composition of the inlet wastewaters

This comparison between the performance of the HRAP and the facultative WSP was made on almost identical wastewaters since independent-samples Mann–Whitney U-test showed there were no statistically significant differences in the concentrations of BOD_5 ($p = 0.395$), PO_4-P ($p = 0.055$), SS ($p = 0.055$) and *E. coli* ($p = 0.068$) in the inlet wastewaters entering the HRAP or facultative WSP (Table 1). As expected in treated septic tank effluent the concentration of oxidised nitrogen was low for both inlets and, while the difference in concentrations was statistically significant ($p < 0.001$) according to an independent-samples Mann–Whitney U-test comparison, practically the difference was negligible. The HRAP inlet NH_4-N was, however, slightly higher than that going into the facultative WSP; the difference in means was statistically significant ($p < 0.001$) according to independent-samples Mann–Whitney

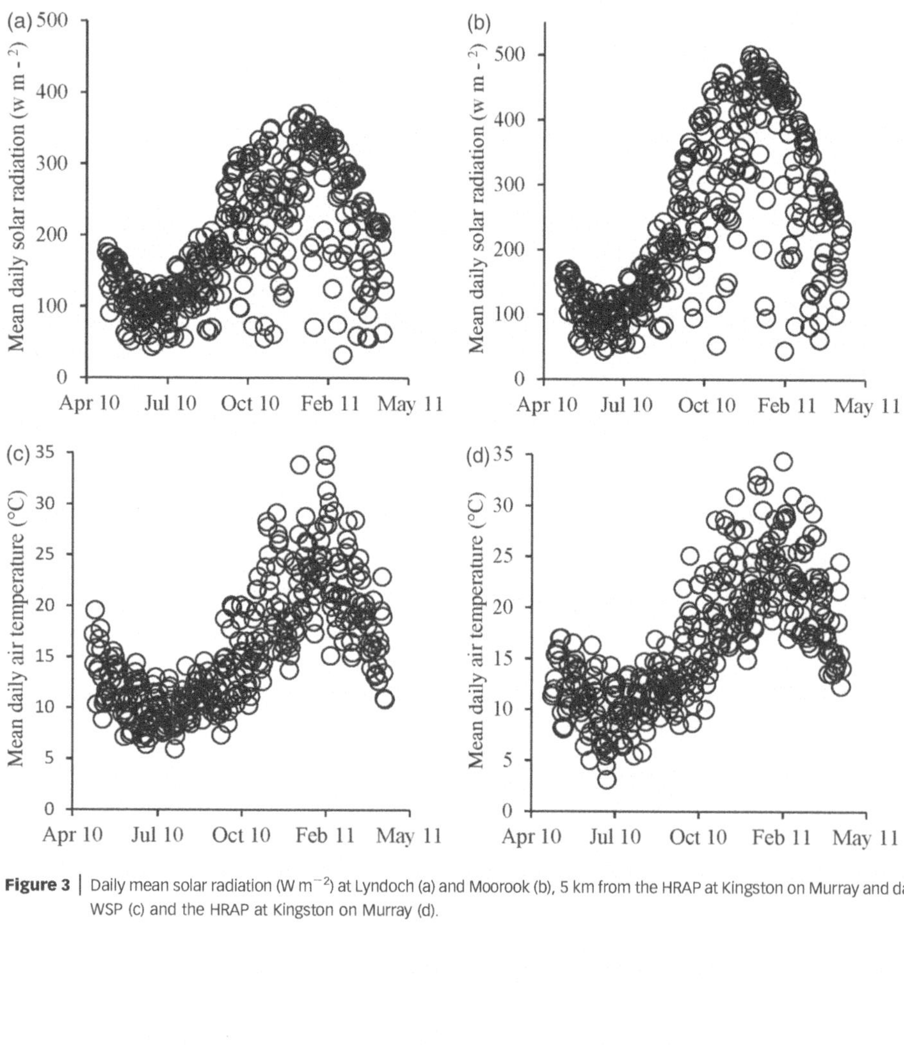

Figure 3 | Daily mean solar radiation (W m^{-2}) at Lyndoch (a) and Moorook (b), 5 km from the HRAP at Kingston on Murray and daily mean air temperature (°C), recorded on site at the Lyndoch WSP (c) and the HRAP at Kingston on Murray (d).

Table 1 | The composition of the inlet wastewater (mean ± standard deviation) from on-site septic tanks to the Kingston on Murray HRAP at and the Lyndoch facultative WSP between 1 May 2010 and 28 March 2017, where n = number of samples analysed

	WSP	HRAP
n	36	>121
BOD_5 (mg L^{-1})	216.86 ± 36.43	203.99 ± 39.60
BOD_5 areal loading rate (kg ha^{-1} d^{-1})	52.80 ± 10.51	135.28 ± 43.11
Suspended solids (mg L^{-1})	102.75 ± 15.82	107.29 ± 27.41
E. coli (log_{10} 100 mL^{-1})	6.29 ± 0.16	6.38 ± 0.33
NH_4-N (mg L^{-1})	76.27 ± 10.77	90.06 ± 11.92
NO_x-N (mg L^{-1})	0.03 ± 0.06	0.40 ± 0.47
PO_4-P (mg L^{-1})	12.44 ± 3.41	13.79 ± 4.16

U-test comparison. The mean BOD_5 areal loading rate for the HRAP was also 2.5 times higher, 135 kg BOD_5 kg ha^{-1} d^{-1}, than the Lyndoch facultative WSP, 54 kg BOD_5 kg ha^{-1} d^{-1}; the difference was statistically significant ($p < 0.001$) reported by independent-samples Mann–Whitney U-test comparison. This is consistent with, or higher than, other reported HRAP loading rates. El Hamouri *et al.* (1995) reported organic loading rates of 86 and 97 kg BOD_5 kg ha^{-1} d^{-1}, respectively, in the hot and the cold season in Morocco.

Temperature, pH and DO regimes of the WSP and the HRAP

These parameters were measured 0.2 m below the water surface in the facultative WSP so do not account for potential variations throughout the pond depth (Sweeney *et al.* 2005). The values shown (mean ± one standard deviation) for the HRAP are metrics produced by analysing values for all operational pond depths. The daily mean temperatures of the WSP was in the range 9.73–31.94 °C with a mean of 18.83 ± 6.07 °C. The HRAP recorded slightly lower temperatures with a minimum of 7.94 °C, a maximum of 29.00 °C and a mean of 17.95 ± 5.40 °C. The wastewater temperatures within the WSP and the HRAP at all depths were remarkably similar and an independent-samples Mann–Whitney U-test found these values not to be significantly different ($p = 0.072$). The wastewater mean DO concentration was lower in the HRAP than the facultative WSP, 6.43 ± 3.31 mg DO L^{-1} compared with 7.34 ± 5.00 mg DO L^{-1}, although the difference was not statistically significant ($p = 0.454$) according to an independent-samples Mann–Whitney U-test comparison. The mean pH and range were lower in the HRAP (7.40–8.73; mean of 7.84 ± 0.21) than the facultative WSP (8.07–10.56; mean of 8.76 ± 0.51) and an independent-samples Mann–Whitney U-test showed this difference was significant ($p < 0.001$). The ranges of pH and DO values recorded for the HRAP are lower than normally reported for these systems. In general, the pH was not as high as reported by some other authors in HRAPs (Azov & Shelef 1982, 1987; Picot *et al.* 1993; El Hamouri *et al.* 1995; El Hamouri *et al.* 2003). It was, however, consistent with the peak range found by other authors (Fallowfield & Garrett 1985; Craggs *et al.* 2003).

Wastewater treatment

The composition of the respective wastewaters following treatment in the HRAP operated at three depths and in the facultative WSP at Lyndoch are shown in Table 2. There were subtle differences identified in the performance of the HRAP associated with operational depth and moderated by ambient temperature, which are beyond the scope of this paper. The data from an HRAP operated at a specific pond depth collected over periods above and below the median temperature were aggregated. This facilitated comparison of HRAP performance data with the mean performance data for the facultative WSP over the duration of the study, which ranged from winter to early autumn.

It can be inferred (Table 2) that irrespective of the depth of operation the percentage of BOD_5 removed by the HRAP, from effluents of virtually identical composition, was almost identical to the facultative WSP, with an independent-samples Mann–Whitney U-test indicating that statistically there was no significant difference between the systems ($p = 0.948$). Most notably, the HRAP at all depths achieved this similar level of removal with a significantly shorter THRT compared to the facultative WSP operating at a THRT of 27.5 d. There was, however, poor phosphorus removal by both the HRAP at all depths of

Table 2 | The composition of the treated wastewater effluent (mean ± standard deviation) from the Kingston on Murray HRAP operated at three depths and the Lyndoch facultative WSP between 1 May 2010 and 28 March 2017, where n = number of samples analysed

	HRAP			WSP
Depth (m)	0.32	0.43	0.55	0.68
Surface area (m^2)	192	208	226	6,300
THRT (d)	4.5	6.4	9.1	27.5
n	58	35	31	36
BOD_5 (mg L^{-1})	13.34 ± 8.48	16.09 ± 9.44	21.79 ± 13.45	28.76 ± 33.40
NH_4-N (mg L^{-1})	25.88 ± 17.79	22.95 ± 12.29	36.92 ± 19.36	39.30 ± 16.24
NO_2-N (mg L^{-1})	5.29 ± 8.19	1.96 ± 1.73	2.21 ± 2.49	0.14 ± 0.24
NO_3-N (mg L^{-1})	7.66 ± 4.29	15.39 ± 4.54	6.96 ± 6.71	0.17 ± 0.26
PO_4-P (mg L^{-1})	11.50 ± 2.77	12.41 ± 2.04	15.19 ± 5.54	9.17 ± 2.93
Chlorophyll *a* (mg L^{-1})	3.86 ± 3.54	3.81 ± 4.32	0.90 ± 0.75	2.56 ± 2.73
Log_{10} *E. coli* 100 mL^{-1}	4.62 ± 0.38	4.19 ± 0.75	4.52 ± 0.51	4.07 ± 0.60

operation and the facultative WSP (Tables 2 and 3). Even though the removal of phosphorus from both ponds was poor, the facultative WSP did outperform the HRAP at all depths and this difference was found to be statistically significant from the WSP ($p = 0.010$) as reported by independent-samples Mann–Whitney U-test.

The concentration of NH_4-N in the treated effluent increased with increasing depth of HRAP operation. The concentration in the treated effluent from the 0.55 m deep HRAP was equivalent to that of the 0.68 m deep facultative WSP (Table 2). At all depths the HRAP outperformed the facultative WSP in NH_4-N removal with this difference found to be statistically significant ($p < 0.001$) according to an independent-samples Mann–Whitney U-test comparison. This increased removal is particularly impressive given the shorter hydraulic retention time of the HRAP compared to the facultative WSP (Table 3).

The greatest difference in the nitrogen speciation of the systems was that the HRAP demonstrated high rates of nitrification compared with the facultative WSP (Table 2). The mean inorganic, oxidised nitrogen (NO_x) concentration, calculated for the facultative WSP as the sum of NO_2-N and NO_3-N concentrations, was 0.31 mg NO_x L^{-1}. The equivalent oxidised nitrogen concentrations in the HRAP, operated at 0.32, 0.43 and 0.55 m were considerably higher at 12.95, 17.35 and 9.17 mg NO_x L^{-1} respectively, indicating increased nitrification of ammonium in the HRAP. An independent-samples Mann–Whitney U-test showed this difference in NO_2-N and NO_3-N between the ponds was statistically significant, $p < 0.001$ and $p < 0.001$ respectively. Oxidation of inorganic nitrogen within the treatment system has environmental benefits since it decreases subsequent oxygen demand associated with nitrification of ammonium by the biota in natural aquatic systems receiving the effluent. Furthermore, although not an inorganic nitrogen removal mechanism *per se*, conversion of ammonium to nitrate may facilitate subsequent effluent treatment via denitrification (Craggs 2005; Pearson 2005).

The difference in the *E. coli* count between the inlet and treated effluent wastewater is used as one indicator

Table 3 | Percentage removal from the respective inlet wastewater of chemical constituents and mean log_{10} removal value for *E. coli* following treatment in the Kingston on Murray HRAP operated at three depths and the Lyndoch facultative WSP between 1 May 2010 and 28 March 2017

	HRAP			WSP
Depth (m)	0.32 m	0.43 m	0.55 m	0.68 m
THRT (d)	4.5	6.4	9.1	27.5
BOD_5	93.03 ± 4.89	92.48 ± 4.32	89.19 ± 5.53	87.7 ± 14.73
NH_4-N	69.70 ± 19.68	74.45 ± 14.87	61.1 ± 23.8	48.75 ± 18.67
PO_4-P	6.14 ± 7.83	15.53 ± 9.02	9.31 ± 10.58	20.95 ± 20.89
E. coli log_{10} removal	1.74 ± 0.28	2.10 ± 0.68	2.00 ± 0.46	2.22 ± 0.68

of the disinfection performance and reported as LRV. Table 2 shows that septic tank effluent treated in the HRAP over all depths returned mean *E. coli* concentrations of the same order of magnitude as those of the facultative WSP. Despite the small difference in absolute values, an independent-samples Mann–Whitney U-test reported this difference as being significantly different ($p = 0.001$). Further investigation compared the *E. coli* concentrations in the HRAP effluents from ponds operated at the three depths with contemporaneous WSP effluent concentrations. The *E. coli* concentration in the medium depth (0.43 m) HRAP was found not to be significantly different according to an independent-samples t-test for equality of means ($p = 0.002$) from that of the WSP operated concurrently. The *E. coli* concentrations of the shallow (0.32 m) and deep (0.55 m) HRAP were, however, shown to be significantly different from the WSP concentrations according to an independent-samples Mann–Whitney U-test ($p = 0.001$) and an independent-samples t-test for equality of means ($p = 0.002$).

The LRVs achieved by each system were also found to be within the same order of magnitude but differed significantly when comparing the aggregated HRAP data with that from the WSP (Mann–Whitney U-test comparison, $p = 0.001$). Comparing the LRVs for individual depths with those achieved by the WSP over the same period showed that this statistically significant difference was only apparent for the shallow (0.32 m) HRAP as reported by an independent-samples Mann–Whitney U-test ($p = 0.004$). The mean *E. coli* LRVs of the HRAPs operated at 0.43 m and 0.55 m did not differ significantly from those of the WSP according to an independent-samples Mann–Whitney U-test ($p =$ 0853) and an independent-samples t-test for equality of means ($p = 0.002$). The LRVs achieved by the HRAP at all depths fall within the range reported by other studies (Young *et al.* 2017).

Furthermore, the 5th percentile *E. coli* LRVs were remarkably similar: 1.34 for the Lyndoch facultative WSP and 1.36 for the HRAPs, determined by combing LRV data from all HRAP operational depths. These values were obtained following exposure to UV irradiances that differed significantly ($p < 0.001$); the daily mean UV irradiation (W m^{-2}) was 5.73 ± 3.74 at Lyndoch and slightly less at KoM, 3.95 ± 1.86. Significantly, while the 5th percentile values were similar for both pond systems the LRV was achieved at much shorter THRTs within the HRAP, between 4.5 and 9.1 d, compared with 27.5 d for the Lyndoch facultative WSP.

Treated effluents from CWMS are discharged for beneficial reuse, typically for irrigation of non-food crops, commonly woodlots, or amenity spaces. These disposal pathways must be consistent with the *Australian National Guidelines for Water Recycling: Managing Health and Environmental Risks (Phase 1)* (NRMMC 2006) to minimise potential exposure of the population to pathogenic microorganisms. These guidelines set the water quality objective for non-food crop irrigation as an *E. coli* median concentration of $<10^4$ 100 mL^{-1} in the final effluent. This guideline would not normally be applied to effluent from a facultative WSP since there is a requirement that it be followed by four maturation ponds operating in series. This guideline was marginally exceeded by the facultative WSP operating at a THRT of 27.5 d and the HRAP operating between 4.5 and 9.1 d. The interpretation of the relative influence of HRAP depth and retention time on disinfection is confounded by the fact that at constant daily influent flow increasing HRAP operation depth also increases retention time. Furthermore, the study design attempted to account for seasonal variation by operating the single HRAP twice at each depth, above and below the annual median air temperature, over three hydraulic retention times. Notwithstanding, the indicative performance and potential benefit associated with the adoption of HRAPs for CWMS was such that subsequently an independent, winter validation of the HRAP was conducted (Fallowfield *et al.* 2018) consistent with that required by the *Australian National Guidelines for Water Recycling: Managing Health and Environmental Risks (Phase 1)* (NRMMC 2006). This clearly showed that HRAPs operated at a depth of 0.3 m at a THRT of 10 d achieved a mean LRV of 3.30 producing a mean concentration of 2.89 *E. coli* 100 mL^{-1} in the final treated effluent (Fallowfield *et al.* 2018). A design guideline for the incorporation of HRAPs into CWMS was subsequently promulgated.

It is widely recognised that HRAPs treating wastewater have the potential to yield large quantities of biosolids which comprise algae, bacteria, zooplankton and detritus and are potentially a source of biomass energy, e.g. via anaerobic digestion (Pearson 2005; Christenson & Sims 2011). Chlorophyll *a* is frequently used as a surrogate measure of algae biomass. Chlorophyll *a* concentrations were high in effluents treated in the 0.32 and 0.43 m deep HRAP and decreased when the operation depth was increased to 0.55 m, which had a mean chlorophyll *a* content lower than that of the facultative WSP at Lyndoch. It is interesting to note that the *E. coli* LRV measured for the 0.55 m deep pond was similar to that for other depths.

This may be as a consequence of its lower chlorophyll *a* content increasing light penetration and compensating for any adverse effect of its greater depth (Curtis *et al.* 1994).

The results of this research clearly demonstrate that an HRAP could replace the facultative WSP in a community wastewater management scheme while maintaining effluent quality. It is likely that the performance of the HRAP is related to its shallow depth (compared to a WSP) and improved hydrodynamics, a consequence of intentional mixing via incorporation of a low energy paddlewheel. Mechanical mixing has been perceived as a disadvantage of HRAP technology for application in rural and remote communities since it requires electrical energy and connection to the power grid. Recently, 5.3 kW h^{-1} of solar voltaic cells coupled with battery storage has been installed at the KoM HRAP demonstrating operation independent of the electricity grid (Figure 4).

Application of HRAP technology to replace only the facultative WSP would halve the pond surface area required to treat domestic wastewater to an equivalent quality. Using KoM data presented here the per capita area requirement would reduce from 4.2 m^2 for a facultative WSP to 2.0 m^2 (0.32 m, 4.5 d); 2.1 m^2 (0.43 m, 6.4 d) and 2.3 m^2 (0.55 m, 9.1 d) for the HRAP. These reductions are consistent with those predicted by Fallowfield & Garrett (1985) and those observed by Picot *et al.* (1992) and El Hamouri *et al.* (2003). A consequence of the reduced area requirements is significant savings in the capital costs for construction. They also enable application of a low energy, low greenhouse gas emission pond technology in locations where the land area was insufficient for larger facultative WSPs (Shilton *et al.* 2008). The finding that depth of HRAP operation had little overall influence on treatment performance offers more flexible HRAP operation. Townships that experience marked seasonal fluctuations in population, e.g. vacation destinations, could increase the depth of HRAP operation during that period without decreasing treated effluent quality or incurring increased capital costs to manage such.

Figure 4 | The 5.3 kW h^{-1} solar voltaic cells installed at the Kingston on Murray, SA, wastewater treatment plant.

CONCLUSIONS

- The wastewater treatment performance of the HRAP operated at all depths and THRT equalled or exceeded that of the facultative WSP in regards to BOD_5 and NH_4-N.
- Although little PO_4-P was removed by either the WSP or the HRAP, the former outperformed the latter at all depths.
- The facultative WSP and the HRAP achieved remarkably similar *E. coli* $\log_{10}$ reduction values and both produced effluent of similar quality with regard to *E. coli* concentration.
- The depth of operation had little effect on the treatment performance of the HRAP.
- Replacing the facultative pond with an HRAP in CWMS would reduce the surface area requirement by 50% with concomitant savings in the capital cost of construction.
- The reduced THRT required for effective treatment by the HRAP reduces evaporative losses compared with the facultative WSP operated at an extended THRT comparatively. Consequently, approximately twice as much effluent is available for beneficial reuse following HRAP treatment.

ACKNOWLEDGEMENTS

The authors acknowledge the Local Government Association of South Australia for construction of the HRAP at Kingston on Murray and to the Flinders Centre for Coastal and Catchment Environments for co-funding the study; Loxton Waikerie District Council and Light District Council for access to Kingston on Murray and Lyndoch wastewater treatment plants respectively; Mr Richard Gayler of Gayler Professional Services, for engineering and management support; and the Department of Health and Ageing, South Australia (Dr David Cunliffe, Tony Farror and Michelle

Wittholz), for advice on the application of the national recycling guidelines.

REFERENCES

Arthur, J. 1983 *Notes on the Design and Operation of Waste Stabilization Ponds in Warm Climates of Developing Countries*. The World Bank, Washington, DC.

Azov, Y. & Shelef, G. 1982 Operation of high-rate oxidation ponds: theory and experiments. *Water Research* **16** (7), 1153–1160.

Azov, Y. & Shelef, G. 1987 The effect of pH on the performance of high-rate oxidation ponds. *Water Science and Technology* **19** (12), 381–383.

Bolton, N., Cromar, N., Hallsworth, P. & Fallowfield, H. 2010 A review of the factors affecting sunlight inactivation of micro-organisms in waste stabilisation ponds: preliminary results for enterococci. *Water Science and Technology* **61** (4), 885–890.

Christenson, L. & Sims, R. 2011 Production and harvesting of microalgae for wastewater treatment, biofuels, and bioproducts. *Biotechnology Advances* **29** (6), 686–702.

Cox, J. W., Akeroyd, M. & Oliver, D. P. 2016 Integrated water resource assessment for the Adelaide region, South Australia. *Proceedings of the International Association of Hydrological Sciences* **374**, 69.

Craggs, R. J. 2005 Nutrients. In: *Pond Treatment Technology* (A. Shilton, ed.). IWA Publishing, London.

Craggs, R., Davies-Colley, R., Tanner, C. & Sukias, J. 2003 Advanced pond system: performance with high rate ponds of different depths and areas. *Water Science and Technology* **48** (2), 259–267.

Curtis, T., Mara, D. D., Dixo, N. & Silva, S. A. 1994 Light penetration in waste stabilization ponds. *Water Research* **28** (5), 1031–1038.

El Hamouri, B., Jellal, J., Outabiht, H., Nebri, B., Khallayoune, K., Benkerroum, A., Hajli, A. & Firadi, R. 1995 The performance of a high-rate algal pond in the Moroccan climate. *Water Science and Technology* **31** (12), 67–74.

El Hamouri, B., Rami, A. & Vasel, J. 2003 The reasons behind the performance superiority of a high rate algal pond over three facultative ponds in series. *Water Science and Technology* **48** (2), 269–276.

Fallowfield, H. & Garrett, M. 1985 The treatment of wastes by algal culture. *Journal of Applied Bacteriology* **59** (s14), 187S–205S.

Fallowfield, H. J., Cromar, N. & Evison, L. 1996 Coliform die-off rate constants in a high rate algal pond and the effect of operational and environmental variables. *Water Science and Technology* **34** (11), 141–147.

Fallowfield, H. J., Young, P., Taylor, M. J., Buchanan, N., Cromar, N., Keegan, A. & Monis, P. 2018 Independent validation and regulatory agency approval for high rate algal ponds to treat wastewater from rural communities. *Environmental Science: Water Research & Technology* **4**, 195–205.

Greenberg, A. E., Clesceri, L. S. & Eaton, A. D. 1992 *Standard Methods for the Examination of Water and Wastewater*. American Water Works Association, American Public Health Association, Washington, DC.

LGA SA (Local Government Association of South Australia) not dated Septic tank effluent drainage scheme: Design criteria. In: Department of Health GoSA (ed.).

NRMMC 2006 *AHMC, Australian Guidelines for Water Recycling: Managing Health and Environmental Risks (Phase 1)*. Natural Resource Ministerial Management Council, Environment Protection and Heritage Council and Australian Health Ministers.

Palmer, N., Lightbody, P., Fallowfield, H. & Harvey, B. 2001 *Australia's Most Successful Alternative to Sewerage: South Australia's Septic Tank Effluent Disposal Schemes*. Local Government of South Australia.

Pearson, H. 2005 *Microbiology of Waste Stabilization Ponds*. IWA Publishing, London.

Picot, B., Bahlaoui, A., Moersidik, S., Baleux, B. & Bontoux, J. 1992 Comparison of the purifying efficiency of high rate algal pond with stabilization pond. *Water Science and Technology* **25** (12), 197–206.

Picot, B., Moersidik, S., Casellas, C. & Bontoux, J. 1993 Using diurnal variations in a high rate algal pond for management pattern. *Water Science and Technology* **28** (10), 169–175.

Shilton, A., Mara, D., Craggs, R. & Powell, N. 2008 Solar-powered aeration and disinfection, anaerobic co-digestion, biological CO_2 scrubbing and biofuel production: the energy and carbon management opportunities of waste stabilisation ponds. *Water Science and Technology* **58** (1), 253–258.

Sweeney, D., Cromar, N., Nixon, J., Ta, C. & Fallowfield, H. 2003 The spatial significance of water quality indicators in waste stabilization ponds – limitations of residence time distribution analysis in predicting treatment efficiency. *Water Science and Technology* **48** (2), 211–218.

Sweeney, D., Nixon, J., Cromar, N. & Fallowfield, H. 2005 Profiling and modelling of thermal changes in a large waste stabilisation pond. *Water Science and Technology* **51** (12), 163–172.

Sweeney, D., Nixon, J., Cromar, N. J. & Fallowfield, H. J. 2007 Temporal and spatial variation of physical, biological, and chemical parameters in a large waste stabilisation pond, and the implications for WSP modelling. *Water Science and Technology* **55** (11), 1–9.

Young, P., Buchanan, N. & Fallowfield, H. 2016 Inactivation of indicator organisms in wastewater treated by a high rate algal pond system. *Journal of Applied Microbiology* **121** (2), 577–586.

Young, P., Taylor, M. & Fallowfield, H. 2017 Mini-review: high rate algal ponds, flexible systems for sustainable wastewater treatment. *World Journal of Microbiology and Biotechnology* **33** (6), 117.

First received 1 November 2017; accepted in revised form 20 April 2018. Available online 2 May 2018

Energy recovery in high rate algal pond used for domestic wastewater treatment

Paula Peixoto Assemany, Maria Lúcia Calijuri, Eduardo de Aguiar do Couto, Fernanda Pereira da Silva and Mauro Henrique Batalha de Souza

ABSTRACT

High rate algal pond (HRAP) was evaluated according to its energy potential and productivity by two rates, net energy ratio (NER) and specific biomass productivity. All energy inputs were calculated according to one HRAP with pre-ultraviolet disinfection treating anaerobic domestic sewage. The outputs were calculated for two energetic pathways: lipid and biogas production for the raw biomass (RB) and biomass after lipid extraction. The non-polar lipid content in dry biomass was 7.6%, reaching a daily lipid productivity of 0.2 g/m^2·day and the biogas production potential was 0.20 m^3/kg solids. For the biomass after lipid extraction, the biogas production reached 2.6 m^3/kg solids. NER values of 10^{-3} for the RB were similar for lipids and biogas routes. The specific biomass productivity was 0.7 mg/kJ. For the residual biomass, after lipid extraction, NER value was 10^{-2} for the integrated route (lipids + biogas) and the specific biomass productivity of the extracted biomass was 0.4 mg/kJ. The best energetic pathway was to integrate both lipids and biogas route.
Key words | bioenergy, biogas, high rate algal pond, lipids, microalgae

Paula Peixoto Assemany (corresponding author)
Maria Lúcia Calijuri
Eduardo de Aguiar do Couto
Fernanda Pereira da Silva
Mauro Henrique Batalha de Souza
Department of Civil Engineering,
Universidade Federal de Viçosa,
Núcleo de Pesquisas Ambientais Avançadas – nPA,
Viçosa, MG,
Brazil
E-mail: *paula_assemany@hotmail.com*

Paula Peixoto Assemany
Av. P.H. Rolfs – Centro de Ciências Exatas e Tecnológicas,
Campus da Universidade Federal de Viçosa,
Viçosa, MG 36570-900,
Brazil

INTRODUCTION

The search for renewable energy sources is a world reality across the depletion of fossil fuels, and also due to the environmental impacts caused by them. Among renewable energy sources like sun, wind and hydroelectric, biomass is a promising source. Energy from biomass is regarded as one of the most important future renewable energy sources. It can provide a continuous power generation, compared with solar and wind energies, and it plays an important role in the current CO_2 mitigation policy (Appels *et al.* 2011). Non-food alternatives and alternatives more efficient than vegetable crops used as biomass for biofuels production have been studied, and among them, microalgae come with a great potential.

As a way to make microalgae production process sustainable from both economic and environmental aspects, the use of effluents and exploration of the concept of biorefinery are viable alternatives. Within this context, the high rate algal ponds (HRAPs) are one of the main alternatives for wastewater treatment and algae biomass production. Despite the greater area requirement and lower productivity yields compared with closed reactors, they have low installation costs and energy consumption, which reduces operating costs (de Godos *et al.* 2010).

Even though there is a lot of research dealing with microalgae bacteria systems for wastewater bioremediation, the investigation devoted to the use of this generated biomass is scarce (González-Fernández *et al.* 2015). Nowadays, the most targeted energy use for algal biomass is the production of biodiesel, due to large capacity of these microorganisms to accumulate lipids. However, microalgae with low lipid content are common in effluent cultivation (Mehrabadi *et al.* 2015). Faced with effluent cultivation and low lipid accumulation, anaerobic digestion of algae biomass could be a potentially attractive alternative for energetic purposes, besides the advantage of using humid biomass, minimizing costs of harvesting and concentration of this biomass. Biogas can be used as a primary source of energy in a great part of the algal biomass production process, thus reducing the costs of biodiesel production and other products with high aggregate value (Chisti 2007). More than that, Sialve *et al.* (2009) estimated that the anaerobic digestion of microalgae biomass after

doi: 10.2166/wst.2017.570

lipid extraction for methane generation can generate as much energy as that obtained with the extraction of lipids.

Collet *et al.* (2011) have concluded that the combination of lipid extraction with biogas production would be optimum from both environmental and economic points of view. In contrast, Quinn *et al.* (2014) illustrated the importance of considering the effects of the removal of lipids from microalgae for anaerobic digestion performance in a life cycle analysis of microalgae biofuels, since these components are energy rich and extracted biomass (EB) would provide lower methane yield than raw biomass (RB). If, on one hand, the previous lipid extraction can be considered as pre-treatment, increasing the bioavailability of microalgae intracellular content; on the other hand, such process includes dehydration, destruction of the cell wall and use of solvents, steps that influence the quality of the biomass for digestion.

Therefore, in this study, the potential for biogas production and the anaerobic biodegradability of the biomass, before and after lipid extraction, were assessed. The main objective was to apply energy analysis in an integrated context of biorefinery, in order to define the best use of biomass cultivated in an HRAP using domestic sewage as culture medium.

MATERIAL AND METHODS

Biomass production unit

The experiment was carried out outdoors in the city of Viçosa, Minas Gerais, Brazil (20°45′14″S and 42°52′54″W). Viçosa is located at approximately 648 m above sea level; the annual average precipitation is 1,221 mm and the annual average temperature ranges between 19 °C and 20 °C. The average relative humidity is 81%. The local climate is characterized as tropical altitude, with hot and rainy summers, and cold and dry winters.

The experimental HRAP was operated with domestic sewage pre-treated by a full-scale upflow anaerobic sludge blanket (UASB) reactor and pre-disinfected by ultra-violet (UV) disinfection. The HRAP was operated in batch mode (four batch operations: July and September 2014; and February and March 2015), until the decay phase of algal growth was reached, measured every day by the variable chlorophyll-*a* (according to NEN 6520 1981; APHA 2005). The duration of batch operations was on average 8 days. Chemical and physical composition of HRAP input and output is presented in Table 1.

Table 1 | Chemical and physical composition of HRAP input and output

Variable	Unit	Input	Output
pH	–	7.7	8.6
Temperature	°C	23.5	25.9
Dissolved oxygen	mg L^{-1}	0.3	11.9
Total suspended solids	mg L^{-1}	93	100
Volatile suspended solids	mg L^{-1}	56	62
Ammoniacal nitrogen	mg L^{-1}	50.3	29.6
Nitrate	mg L^{-1}	0.3	0.1
TKN	mg L^{-1}	55.8	37.5
Dissolved phosphorus	mg L^{-1}	3.9	4.5
Soluble COD	mg L^{-1}	142.0	67.9
Dissolved organic carbon	mg L^{-1}	32.8	9.5
Escherichia coli	MPN 100 mL^{-1}	5.81×10^5	1.78×10^3

The HRAP had the following characteristics: width = 1.28 m, length = 2.86 m, total depth = 0.5 m, culture depth = 0.4 m, surface area = 3.3 m^2, culture volume = 1.32 m^3. The disinfection system was designed to achieve a final concentration of 10^3 MPN (100 mL)$^{-1}$ of *Escherichia coli*, with an adopted effective dose of 21 mJ/cm^2 and absorbance of 42%. The disinfection phase had the objective of removal microalgae predator's organisms and competitors for nutrients, helping the algae biomass growth, as demonstrated by Santiago *et al.* (2013).

Biomass characterization

After production, the biomass was separated and concentrated by sedimentation for posterior energy analysis. A proportion of the biomass was subjected to the processes of lipid extraction and anaerobic digestion, whereas another part of the biomass was submitted to anaerobic digestion only. The biomass obtained after lipid extraction and submitted to anaerobic tests was named EB. Similarly, the biomass sent directly for anaerobic digestion was named RB.

Both biomasses, RB and EB, were characterized in terms of chemical oxygen demand (COD), total solids (TS), total volatile solids (TVS), water content, pH, total Kjeldahl nitrogen (TKN) and total phosphorus (TP), according to APHA (2005). All analysis were done in duplicate. The phytoplankton community was also assessed at the end of each batch operation. For qualitative analysis, the samples were preserved with formaldehyde (4%). The identification and cell counting was performed with an inverted optical microscope (Olympus CK2). The morphologic and morphometric characteristics of the vegetative and reproductive cycles of significant taxonomic value for the species were analyzed according to specialized literature (Bourrelly 1970; Parra *et al.* 1982). The cells were counted by using 2 mL sedimentation chambers.

Lipids

The lipid content was assessed by solvent extraction, as described by Assemany *et al.* (2015). The biomass was oven dried at 50 °C for 12 h and the cells disrupted with hydrochloric acid (HCl) 3 M. For 0.2 g of dry biomass, 5 mL of HCl were added under stirring during 10 min. Then, 25 mL of methanol and 5 mL of petroleum ether were added under heat at 30 °C during 15 min. After disruption, petroleum ether was added to the dry biomass, totalizing three cycles of extraction (10 mL in the first and third cycles, 20 mL in the second cycle). The non-polar lipid content (% of lipids in the dry biomass) was determined gravimetrically after the evaporation of the solvent for 2 h at 105 °C.

Biogas

The Biogas Production Potential (BPP) tests followed the procedures described by Jawed & Tare (1999), with modifications described by Viana *et al.* (2012). Tests were carried out using 250 mL Erlenmeyer's reagent (200 mL of useful volume), filled with sludge from an UASB reactor (12.5 mL) as anaerobic inoculum, substrate (RB and EB) (100 mL), buffer solution (1 g of sodium bicarbonate) and distilled water (67.5 mL), with a food to microorganism (F/M) ratio of eight to one (v/v). The best F/M was previously chosen after vary specific methanogenic activity (SMA) tests, during which others' ratios were tested (four and six) and eight achieved the best performance. Micro- and macronutrients were added (20 mL) to prevent deficiency during the tests, according Viana *et al.* (2012). The test was also performed on the control without the addition of substrate to measure the endogenous respiration of the microorganisms. After closure of the bottles, N_2 was injected for 4 min to purge the oxygen. The tests were performed at 35 ± 2 °C under continuous agitation (120 rpm) in an incubator (TECNAL, TE-420). The volume of biogas was monitored daily in a Mariotte bottle filled with 25 g NaCl/L (pH = 2) solution for biogas measurement. The calculation of the BPP was based on the cumulative production of biogas after 30 days of incubation, and on the mass of substrate used in the tests. Anaerobic biodegradability was also assessed, converting the biogas production to organic load, considering that at 35 °C, 0.395 L of methane correspond to 1 g of COD and that the biogas is composed of 62.5% of methane (mean value of the interval reported by European Commission 2005).

Energy analysis

The energy performance of the production system was evaluated in terms of specific productivity of biomass, P_B/E_C (mg/kJ), defined as the productivity of biomass (P_B, mg/L·d) per input energy (E_C, kJ/L·d) (Pegallapati *et al.* 2014) and the net energy ratio (NER), which relates the total energy produced and that consumed by the system. Equation (1) presents NER calculation:

$$\mathrm{NER} = \frac{\sum \text{Energy produced}}{\sum \text{Energy consumed}} \tag{1}$$

The produced energy can be with respect to its lipid and/or BPP. The energy consumption for the agitation of the culture medium by the paddlewheels was considered, and also for the UV lamps in the disinfection unit.

Quantification of the energy consumption

The energy consumption of each operation (COP, kWh/day) was determined with Equation (2):

$$C_{OP} = \frac{24 \times Po_t}{1000} \tag{2}$$

where Pot is the power in W. The annual energy consumption (C_T, kJ/year) can be determined by Equation (3):

$$C_T = C_{OP} \times \text{days of operation in the year} \times 3600 \tag{3}$$

Considering the continuous operation of the HRAP during the year, Equation (3) can be substituted by Equation (4):

$$C_T = C_{OP} \times 1,314,000 \tag{4}$$

Quantification of the energy produced from lipids and biogas production

The annual energy production from the anaerobic digestion of the biomass (E_G, kJ/year) and lipids (E_L, kJ/year) was determined from the annual lipid and biogas productions, considering that: (1) the energy content of 1 L of lipids is equivalent to 35,133.33 kJ (Jorquera *et al.* 2010), (2) lipid density is approximately 0.9 kg/L (Jorquera *et al.* 2010), (3) 1 m^3 of biogas equals 23,400 kJ (mean value of the interval reported by Chisti 2007), and (4) the duration of each operation as being equal to 8 days, according to Equations (5) and (6):

$$E_L = \frac{P_{L,annual}}{0.9} \times 35,133.33 \quad (5)$$

where P_L, annual is the annual lipid production (kg/year) and

$$E_G = P_{G,\,annual} \times 23400 \quad (6)$$

where P_G, annual is the annual biogas production (m^3/year).

RESULTS AND DISCUSSION

Biomass characterization

The phytoplanktonic community was dominated by Chlorophyceae during all the batch operations. The algae *Chlorella vulgaris* was the predominant specie, with an average of 9.4×10^5 individuals/mL, followed by the gender *Scenedesmus* sp. with 6.5×10^5 individuals/mL.

Table 2 presents the physical and chemical characterization of the biomass used in the energetic tests.

The pH of RB presented a value close to neutrality; on the other hand, EB had an acid pH. The average value below 2, can be explained by the addition of hydrochloric acid for cell wall disruption during lipid extraction. COD concentration of EB was two times higher than RB. This may be related to the use of solvents, such as methanol and petroleum ether, in the extraction procedure, with the possibility of incorporation of the carbon of such reagents into the biomass. For phosphorus and nitrogen (NTK), we highlighted lower concentrations in EB compared to RB, caused by the extraction of phosphorus elements and complex lipids, such as phospholipids, that have besides phosphorus, nitrogen on its composition. Specifically, for nitrogen reduction in EB, membrane proteins could also be extracted due to cell wall disruption.

We highlight greater concentration of solids and lower value of water content in RB, due to the dilution of EB. Specifically, for EB the high water content, 99%, was due to the addition of water after the separation of the solvents and hydrochloric acid. The separation, carried out by sedimentation of the biomass, was necessary since the reagents used in the extraction are extremely toxic to the anaerobic microorganisms, potentially inhibiting anaerobic digestion. This inhibition was confirmed in BPP preliminary tests, during which no methanogenic activity was detected in the bottles with biomass, additional to the reagents after extraction.

Moreover, the high content of ash in both biomasses (45 and 43% of TS) should be pointed out as an intrinsic characteristic of the culture medium used to grow the biomass. In this study, the domestic sewage was used in the HRAP after a prior treatment in an UASB reactor. However, the UASB reactor is not properly operated, since anaerobic sludge is discarded with no time control, consequently with a large proportion of material already stabilized (inorganic) to the detriment of volatile material.

Energetic output

Lipids

The biomass presented a non-polar lipid content of 7.6 ± 0.6%, reaching an average lipid productivity of 0.2 $g/m^2 \cdot d$. The cultivation in effluents under less appropriate conditions, and in competition with other microorganisms, may have been the cause of the low lipid accumulation. According to Mehrabadi *et al.* (2015), the main reason for the low lipid content of biomass grown in HRAPs with wastewater is the composition of the biomass, since biomass contains a mixture of algae and bacteria. As the lipid content of bacteria is typically below 10% (Brown *et al.* 1996), this

Table 2 | Characterization of the algal biomass used as substrate for the anaerobic tests (mean and standard deviation values, $n = 2$)

Biomass	pH	Water content (%)	TS (g/L)	Ash (%TS)	Volatile (%TS)	TKN (g/L)	N-NH_4^+ (g/L)	TP (g/L)	COD (g/L)
RB	7.5 (1)	98 (1)	21 (7)	45	55	3 (3)	ND	0.65 (0.1)	53 (27)
EB	<2 (1)	99 (1)	7 (2)	43	57	0.1 (0.06)	–	0.3 (0.1)	100 (25)

ND: not detected; –: not performed.

reduces the overall lipid content and hence energy content of the biomass.

Biogas

Figure 1 presents the cumulative biogas production for the biomass before (RB) and after (EB) lipid extraction.

The incubation time of EB was over the 30 days. The objective was to maintain incubation until the stabilization of biogas production, which was observed next to the 70th day. Despite the high biogas production, reaching 2,114 mL at the end of incubation, during the first 10 days of incubation there was no production observed. The exponential phase of production was only observed after the 20th day of incubation of EB. This time lag was probably due to the adaptation of the anaerobic microorganisms to the substrate EB. A factor that may have contributed to this adaptation was that the pH in the EB was extremely acid (lower than 2), as discussed earlier (Table 2). Thus, an alternative to decrease the adaptation time would be the correction of the pH of the biomass prior to incubation, following a pH target of 6.8–7.2 or an adoption of a period of acclimation before the anaerobic digestion. A long lag period of more than 18 days was also observed by Yun *et al.* (2014) when digesting microalgae biomass after lipid extraction with a solvent concentration of 200 mL chloroform/L.

The results of the anaerobic biodegradability and BPP tests for the biomass, before and after lipid extraction, are presented in Table 3.

We can observe the low anaerobic biodegradability of the RB. The cellular wall is considered the main characteristic of the difficult biodegradability of algal biomass (Zamalloa *et al.* 2012). According to Golueke *et al.* (1957), microscopic analysis showed that a great amount of algal cells were intact, but not viable, after digestion, probably indicating that the cellular wall inhibited contact of the anaerobic microorganisms with substrate. The ultrastructure of the cellular wall of the genera found in this study, *Chlorella* and *Scenedesmus* (*Chlorococcales*), consists of two shells, the internal one consisting of microfibers of cellulose and amorphous matrix; and the external, which in its trilaminar shape can contain substances such as carotenoids and lignin (sporopollenin) (Weng & Chapple 2010). Therefore, components such as cellulose and lignin could be the cause of the low biodegradability indices observed for RB.

We highlight the high BPP value for EB, reaching 2.6 m^3 biogas/kg TVS, a yield approximately 10 times higher than the results of the RB and those reported in the literature. Studies for pre-treated algal biomass also cultivated in HRAPs with wastewater presented results of: ~0.3 m^3 biogas/kg TVS after 45 days of incubation and application of microwaves (Passos *et al.* 2013) and 0.155 m^3 CH_4/kg TVS for pre-treatment of freezing and thawing and 0.136 m^3 CH_4/kg TVS for thermal pre-treatment (Kinnunen *et al.* 2014).

This result reinforces the use of lipid extraction as a pre-treatment of algal biomass. The process of lipid release using solvents probably increased the bioavailability of the intracellular content of the microalgae, making the substrate

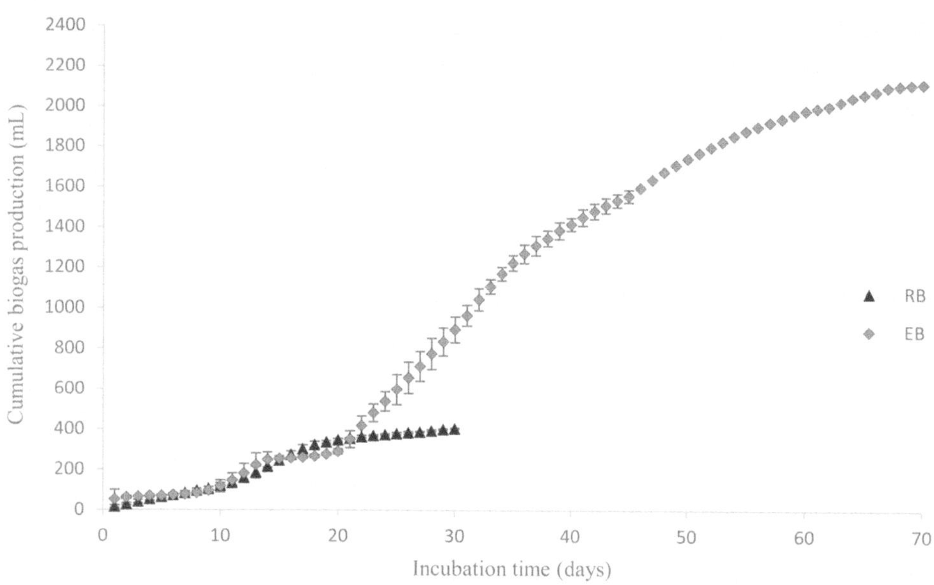

Figure 1 | Cumulative biogas production for the biomass before (RB, $n = 4$) and after (EB, $n = 2$) lipid extraction (vertical bars represent standard deviation).

Table 3 | Anaerobic biodegradability, biogas production potential (BPP) tests and energetic output for the biomasses from the HRAP, before and after lipid extraction ($n = 4$ for RB and $n = 2$ for EB, mean and standard deviation values)

Biomass	Anaerobic biodegradability (%)	BPP (m^3 biogas/kg TVS)
RB	10.9 (7.8)	0.20 (0.1)
EB	28.5 (15.4)	2.6 (0.5)

more available for the anaerobic microorganisms to perform digestion. Also, it is possible to affirm that the best yield of the digestion using EB was due to the shortening of the hydrolysis phase, leaving the hydrolytic bacteria very little to do, since they received the substrate almost or completely hydrolyzed. As already mentioned by González-Fernández *et al.* (2012), when working with particulate substrates, as microalgae biomass, the hydrolysis step determines the successful production of methane. Otherwise, in easily degradable substrates, methanogenesis is considered the limiting phase.

Besides the pre-treatment effect, lipid extraction also increased by two times the COD content in the biomass (see Table 2), contributing for better biogas performance of EB, compared with RB. An external source of organic carbon from the solvents used in the extraction was incorporated into the EB, consequently increasing biogas production.

ENERGETIC ANALYSIS

In order to operate paddlewheels and UV disinfection, the HRAP production system had an annual energy consumption of 5.4 GJ. Its biomass productivity of RB was 7.75 mg/L·d, with a volatile suspended solids (VSS) concentration of 62 mg/L after 8 days of batch operation (Table 4). Productivities reported for HRAP treating domestic sewage are usually greater than those obtained in this study (Assemany *et al.* 2015; Mehrabadi *et al.* 2015). Batch mode operation and the 40 cm of HRAP depth could be considered the main factors for low biomass yield. For the EB, the productivity was 62.7% of RB, discounting lipid content of 7.6% and 29.7% of losses in the lipid extraction process, totalizing 4.86 mg/L·d. Due to the small scale of the extraction process, loss was potentiated, since part of the biomass was retained in the walls of the glass used for the process, and during reagents separation. However, such weaknesses can be minimized by increasing the scale of the process. The low values of the specific productivity of biomass (P_B/E_C) of both biomasses (Figure 2) are also a consequence of the low HRAP biomass productivity.

Annual energy outputs of the EB were 0.01 GJ for lipids and 0.02 GJ for biogas routes. The EB presented a biogas energy gain of 0.14 GJ/year, contributing in large proportion

Table 4 | Energy analysis data

Energy input						
Energy Consumption	**Power (W)**	**KW**	**KWh/day**	**Daily energy consumption (KJ/day)**	**Daily energy consumption (GJ/day)**	**Annual energy consumption (GJ/year)**
HRAP	170	0.17	4.08	14,688	0.014688	5.36112
Biomass productivity						
	HRAP VSS (mg/L)	**Operation days**	**Volume (L)**	**Productivity (mg/L·d)**	**Productivity mg/day**	**Productivity kg/day**
RB	62	8	1,320	7.75	10,230	0.01023
EB	38.9	8	1,320	4.86	6,419	0.006419
Energy Output						
	Content (%)	**Lipid productivity (mg/day)**	**Daily energy yield (KJ/day)**	**Daily energy yield (GJ/day)**	**Annual energy yield (GJ/year)**	
Lipids	7.6	0.00077748	30.35	0.000030	0.01	
		BPP (m^3/kg SV)	**Biogas productivity (m^3/day)**	**Daily energy yield (KJ/day)**	**Daily energy yield (GJ/day)**	**Annual energy yield (GJ/year)**
Biogas	RB	0.2	0.002046	47.8764	0.000048	0.02
	EB	2.6	0.016690	390.552	0.000391	0.14

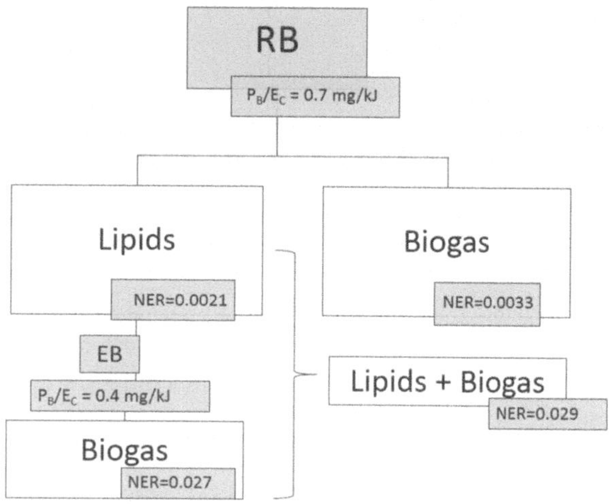

Figure 2 | NER and specific biomass productivity for different energetic routes.

to the energy output of the integrated (lipids + biogas) route of 0.15 GJ/year.

NER values were below one for all the energetic routes (Figure 2), indicating that energy gain was not high enough to compensate the energetic input. For RB, both lipids and biogas presented similar NER values of 10^{-3}, associated with low lipid content and low BPP of this biomass. For the EB, NER value of biogas route was 10^{-2}, indicating the improvement of energy efficiency due to the greater value of BPP, although the lower EB productivity. Because of the RB NER value, the integrated route presented NER of 2.9×10^{-2}, demonstrating that energy gain can be obtained through route integration.

Although a favorable result was obtained in the present study for the integrated route, a same energy gain could be obtained through a biomass treatment prior the anaerobic digestion in order to disrupt microalgae cell wall. Data of energy consumption for concentration and dewatering of the biomass for lipid extraction, as well for the extraction process itself, could be determinant for choosing either a pre-treatment of the whole biomass for anaerobic digestion or the biogas production after lipid extraction.

According to Sialve *et al.* (2009), lipid extraction of biomass containing less than 40% of lipids combined with anaerobic digestion of the residual biomass is not effective in terms of energy nor in terms of costs. For these authors, the anaerobic digestion of the whole biomass appeared to be the optimal strategy on an energy balance basis.

The scarcity of energy analyses on cultivation using effluent makes comparison difficult; however, these results show the necessity of intervention in the productive system in order to improve its energy performance, by using low-energy equipment and improve biomass productivity.

CONCLUSIONS

Lipid extraction could be considered as a biomass pre-treatment for anaerobic digestion tests. According to NER values, the best energetic pathway was the integrated route (lipids + biogas), having the biogas production of the EB major contribution for this result. However, for all the studied energetic routes, the energy input was higher than the output, indicating that the HRAP needs intervention in order to improve its energy performance and increase its biomass productivity.

Moreover, other stages of the production system should be improved in order to scale-up the technology and consequently get more precision in measuring energy efficiency. Biomass harvesting and concentration, lipids extraction and a specific anaerobic reactor for algal biomass digestion are processes that should be improved. For instance, in a larger scale, chemicals should be included in the economic balance, and also pumping energy demand in the energetic analysis.

The integration between energetic, economic and environmental spheres should be included in a major context analysis, as a life cycle analysis, for example.

ACKNOWLEDGEMENTS

The authors gratefully acknowledge the Brazilian National Council for Scientific and Technological Development, CNPq (grant numbers 403013/2013-5 and 405230/2015-0) for partial financial support.

REFERENCES

APHA 2005 *Standard Methods for the Examination of Water and Wastewater*, 20th edn. American Public Health Association/American Water Works Association/Water Environment Federation, Washington, DC, USA.

Appels, L., Laumwers, J., Degréve, J., Helsen, L., Lievens, B. & Willems, K. 2011 Anaerobic digestion in global bio-energy production: potential and research challenges. *Renewable Sustainable Energy Reviews* **15**, 4295–4301.

Assemany, P. P., Calijuri, M. L., Couto, E. A., Santiago, A. F. & Reis, A. J. D. 2015 Biodiesel from wastewater: lipid production in high rate algal pond receiving disinfected effluent. *Water Science and Technology* **71**, 1229–1234.

Bourrelly, P. 1970 Les alges d'eau douce. Initiation à la sistematique. III Les algues bleues et rouges. Les Eugléniens, Peridiniens et Cryptomonadines. Nouvelle Boubée & Cie, Paris, France [in French].

Brown, M. R., Barret, S. M., Volkman, J. K., Nearhos, S. P., Nell, J. A. & Allan, G. L. 1996 Biochemical composition of new yeast and bacteria evaluated as food for bivalve aquaculture. *Aquaculture* **143**, 341–360.

Chisti, Y. 2007 Biodiesel from microalgae beats bioethanol. *Trends in Biotechnology* **26**, 126–131.

Collet, P., Arnaud, H., Laurent, L., Monique, R., Romy-Alice, G. & Steyer, J. P. 2011 Lifecycle assessment of microalgae culture coupled to biogas production. *Bioresource Technology* **102**, 207–214.

de Godos, I., Blanco, S., Gárcis-Encina, P. A., Becares, E. & Muñoz, R. 2010 Influence of flue gas sparging on the performance of high rate algae ponds treating agro-industrial wastewaters. *Journal of Hazardous Materials* **179**, 1049–1054.

European Commission 2005 *Reference Document on Best Available Techniques for the Waste Treatments Industries. Integrated Pollution Prevention and Control*. The European IPPC Bureau, Seville, Spain.

Golueke, C. G., Oswald, W. J. & Gotaas, H. 1957 Anaerobic digestion of algae. *Applied Microbiology* **5**, 47–55.

González-Fernández, C., Sialve, B., Bernet, N. & Steyer, J. P. 2012 Impact of microalgae characteristics on their conversion to biofuel. Part II: focus on biomethane production. *Biofuels Bioproducts and Biorefinering* **6**, 205–218.

González-Fernández, C., Sialve, B. & Molinuevo-Salces, B. 2015 Anaerobic digestion of microalgal biomass: challenges, opportunities and research needs. *Bioresource Technology* **198**, 896–906.

Jawed, M. & Tare, V. 1999 Microbial composition assessment of anaerobic biomass through methanogenic activity tests. *Water SA* **25**, 345–350.

Jorquera, O., Kiperstok, A., Sales, E. A., Embiruçu, M. & Ghirardi, M. L. 2010 Comparative energy life-cycle analyses of microalgal biomass production in open ponds and photobioreactors. *Bioresource Technology* **101**, 1406–1413.

Kinnunen, V., Craggs, R. & Rintala, J. 2014 Influence of temperature and pretreatments on the anaerobic digestion of wastewater grown microalgae in a laboratory-scale accumulating volume reactor. *Water Research*. **57**, 247–257.

Mehrabadi, A., Craggs, R. & Farid, M. M. 2015 Wastewater treatment high rate algal ponds (WWT HRAP) for low-cost biofuel production. *Bioresource Technology* **184**, 202–214.

Nederlands Norm, NEN 6520 1981 *Water: Spectrophotometric Determination of Chlorophyll-a Content*. Nederlands Normalisatie- instituut, Delft, The Netherlands.

Parra, O. O., González, M. & Dellarosa, V. 1982 *Manual Taxonômico del Fitoplancton de água continentales*. V- Chlorophyceae, Concepcion, Chile [in Spanish].

Passos, F., Solé, M., García, J. & Ferrer, I. 2013 Biogas production from microalgae grown in wastewater: effect of microwave pre-treatment. *Applied Energy* **108**, 168–175.

Pegallapati, A. K., Arudchelvam, Y. & Nirmalakhandan, N. 2014 Energetic performance of photobioreactors for algal cultivation. *Brief Review. Environmental Science Technology Letters* **1**, 2–7.

Quinn, J. C., Hanif, A., Sharvelle, S. & Bradley, T. H. 2014 Microalgae to biofuels: life cycle impacts of methane production of anaerobically digested lipid extracted algae. *Bioresource Technology* **171**, 37–43.

Santiago, A. F., Calijuri, M. L., Assemany, P. P., Calijuri, M. C. & Reis, A. J. 2013 Algal biomass production and wastewater treatment in high rate algal ponds receiving disinfected effluent. *Environmental Technology* **34**, 1877–1885.

Sialve, B., Bernet, N. & Bernard, O. 2009 Anaerobic digestion of microalgae as a necessary step to make microalgal biodiesel sustainable. *Biotechnology Advances* **27**, 409–416.

Viana, M. B., Freitas, A. V., Leitão, R. C. & Santaella, S. T. 2012 Biodegradability and methane production potential of glycerol generated by biodiesel industry. *Water Science and Technology* **66**, 2217–2222.

Weng, J. K. & Chapple, C. 2010 The origin and evolution of lignin biosynthesis. *New Phytologist* **187**, 273–285.

Yun, Y., Cho, S., Jung, K., Kim, M., Shin, H. & Kim, D. 2014 Inhibitory effect of chloroform on fermentative hydrogen and methane production form lipid-extracted microalgae. *Hydrogen Energy* **39**, 19256–19261.

Zamalloa, C., Boon, N. & Verstraete, W. 2012 Anaerobic digestibility of *Scenedesmus obliquus* and *Phaeodactylum tricornutum* under mesophilic and thermophilic conditions. *Applied Energy* **92**, 733–738.

First received 3 July 2017; accepted in revised form 26 October 2017. Available online 14 November 2017

Investigation and modelling of high rate algal ponds utilising secondary effluent at Western Water, Bacchus Marsh Recycled Water Plant

Digby Wrede, Syed U. Hussainy, William Rajendram and Stephen Gray

ABSTRACT

There is growing interest in the ability of high rate algal ponds (HRAP) to treat wastewater. This method reduces the costs of algal production while treating the wastewater quicker and more efficiently than standard lagoon practices. Two parallel HRAPs were used in this study to treat secondary effluent. Nitrogen levels were significantly reduced with a mean reduction of 71% for ammonia and 64% for total nitrogen. The use of the HRAPs significantly increased the algal biomass levels compared to the algal growth in the storage lagoons, with a mean increase of 274%. Beneficial use of algae can be used to reduce treatment costs; so being able to predict and optimise the amount of algal biomass produced in HRAPs is vital. However, most models are complicated and require specific, detailed information. In this study, a predictive microalgal growth model was developed for HRAP by adapting two previously established models: the Steele and Monod models. The model could predict algal growth based on temperatures and solar radiation and account for limiting ammonia concentrations in an elevated pH environment with natural variations in the algal community. This model used experimental data that would be readily available to any established HRAP study.

Key words | algae, algal biomass model, high rate algal ponds, secondary effluent, wastewater treatment

Digby Wrede (corresponding author)
Syed U. Hussainy
Stephen Gray
Institute for Sustainability and Innovation, School of Engineering and Science,
Victoria University,
3030, Werribee, VIC,
Australia
E-mail: *digby.wrede@live.vu.edu.au*

William Rajendram
Western Water,
3429, Sunbury, VIC,
Australia

INTRODUCTION

High rate algal ponds (HRAPs) are open shallow raceway ponds that are utilised for wastewater treatment and the production of microalgae. HRAPs have been used to treat wastewater for over half a century, and there have been many studies investigating HRAPs for wastewater treatment and biofuel production (Oswald *et al.* 1957; Craggs *et al.* 2012; Chisti 2016). For large-scale algal production, the use of HRAPs is one of the cheapest methods as they have a relatively low operating and construction cost. The cost of using microalgae to treat wastewater could potentially be offset by making biofuels with the algal biomass produced (Christenson & Sims 2011; Craggs *et al.* 2012). This method treats the wastewater and provides microalgae with nutrients, substantially improving the economics of the algal production (Rawat *et al.* 2013). These nutrients include nitrogen, phosphorus, carbon and other micronutrients that are required for growth (Abdel-Raouf *et al.* 2012; Craggs *et al.* 2012).

Microalgae are simple, acellular or colonial organisms which require nutrients to grow. These nutrients can become limiting quite quickly especially in wastewater, where the carbon to nitrogen ratio is generally 3:1 instead of the required 6:1 (Benemann 2003; Sutherland *et al.* 2016). Modelling algal growth and being able to predict how well the microalgae will grow can help determine if and when the water may require extra nutrients or if other factors are affecting the microalgal growth. A predictive model can also be used to optimise processes and ensure harvesting is done at the ideal time to maximise biomass production. Furthermore, prediction of algal growth in wastewater will also help determine the discharge times or retention times required.

Due to substantial variation in different types of microalgae such as red microalgae, green microalgae and diatoms, prediction of algal growth in open HRAPs can be complex. It is known, however, that green algae are the most abundant

doi: 10.2166/wst.2018.202

algae found in wastewater treatment facilities, and that the majority of microalgae require a few essential elements for growth: nutrients, a specific temperature and a particular solar radiation range (Abdel-Raouf *et al.* 2012). The nutrients, most commonly needed are carbon, nitrogen and phosphorus. Ammonia is algae's preferred form of nitrogen and is assimilated into the biomass first (Glibert *et al.* 2016).

Temperature has a significant impact on the production of algae, and different species of algae prefer different temperature regimes. A large number of microalgae species prefer temperatures in the range of 20–30°C (Undurraga *et al.* 2016). Solar radiation is the primary form of energy for microalgae, as they are photosynthetic organisms. Most algae have an optimal solar radiation range in which they prefer to grow, and the shallow design of HRAPs maximises the amount of light for algal growth (Young *et al.* 2017). Regular harvesting and mixing can help minimise self-shading that occurs at high concentrations. The effect of temperature and solar radiation on algal growth are strongly linked given their impact on photosynthesis and cellular respiration (Béchet *et al.* 2015). Nutrient availability in wastewater if not adequately controlled can severely diminish algal growth. Carbon is the primary nutrient which is limited in the secondary effluent. Algal cells contain 50% carbon by weight and, therefore, carbon drives the growth of the cell (Putt *et al.* 2011). Being able to incorporate nutrient concentrations, temperature and solar radiation into a growth model, and determine how fluctuations of each of them can affect the algal growth is essential for the design of HRAPs and optimisation of their operation.

Numerous algal growth models have developed over the years, and currently the models we use for predicting algal growth are complicated. These models tend to be species-specific and less reliable for ambient open HRAPs that contain numerous species. In general, most models focus on one or two growth aspects, such as light and temperature, and the models are proven to be useful in a controlled laboratory setting (Jayaraman & Rhinehart 2015; Undurraga *et al.* 2016). However, they are less applicable to large-scale open ponds due to fluctuations in weather, nutrient concentrations and microalgal species. Most models require large amounts of pre-determined values such as optimal and minimal growth rates in certain conditions, and at specific nutrient concentrations. They are usually too specific to a particular set of conditions to be utilised outside of the controlled laboratory environment. Most of the models assume that either the nutrients are not limiting or that the temperature and light are kept constant. Some previous models are species specific and would not be able to predict the biomass production in a naturally occurring culture (Wu *et al.* 2013; Béchet *et al.* 2015; Undurraga *et al.* 2016). A few studies have considered systems with numerous species, but tend to be very complicated and require data that is not commonly available (Huesemann *et al.* 2016). A simple model with readily available data would be beneficial for design and operation of large-scale HRAPs.

The Steele model has been previously used to predict algal cell growth based on light intensity and is shown in Equation (1) (Wu *et al.* 2013).

$$\mu = \mu_{\max} \cdot \frac{l}{l_{opt}} \cdot e^{1(l/l_{opt})} \quad (1)$$

where μ (d^{-1}) is the specific growth rate under light intensity of l (lx); $\mu_{\max}$ (d^{-1}) is the maximum specific growth rate when light intensity is optimal; l (lx) is the light intensity, and l_{opt} (lx) is the optimal light intensity (Wu *et al.* 2013). Béchet *et al.* (2011) developed a universal temperature model for shallow algal ponds, but it required vast amounts of information and site-specific data. A simple algal growth model which incorporates both temperature and solar radiation and their effects on each other would probably find broader application.

The Monod model has previously been used to describe the algal growth rate based on nutrient levels and is described in Equation (2) (Wu *et al.* 2013; Béchet *et al.* 2015; Jayaraman & Rhinehart 2015).

$$\mu = \mu_{\max(N)} \cdot \frac{S_N}{K_N + S_N} \quad (2)$$

where μ (d^{-1}) is the specific growth rate under the nutrient concentration of S_N (mg/L), $\mu_{\max(N)}$ (d^{-1}) is the maximal specific growth rate when nutrient concentrations are saturated, S_N (mg/L) is the nutrient concentration in the medium; and K_N is the half-saturation constant of the nutrient (Wu *et al.* 2013).

Here the use of microalgae to treat secondary effluent is examined, and the development of a simple model for biomass production in an open pond environment with elevated pH was investigated. The paper considered the following variables for prediction of algal growth: temperature, solar radiation and ammonia concentration.

EXPERIMENTAL DESIGN/ METHODS

Bacchus Marsh Recycled Water Plant (BMRWP)

The two HRAPs used in this experiment were located in Victoria, Australia, at Western Waters' Bacchus Marsh

Recycled Water Plant (BMRWP) (37°72′44.09″S, 144°47′61.20″E). The HRAPs received water from the third secondary treatment lagoon at the BMRWP, which treats municipal wastewater in a series of lagoons, see Figure 1. Wastewater is first fed into an aeration lagoon (AL) that sparges air into the raw sewage to ensure an aerobic environment which enhances treatment and reduces odour. After 3 days, the aerobic effluent is fed into three primary settling lagoons operated in parallel, where the larger solids settle out and that operate as facultative lagoons. Subsequently, the primary effluent from all three primary ponds is combined and fed into a series of three secondary lagoons. The secondary lagoons further treat the water and remove nutrients, mainly using algae. The effluent from the third secondary lagoon feeds into the winter storage lagoon. The winter storage lagoon is a final polishing lagoon. The effluent from the winter storage lagoon is 'Class A' water, and is discharged onto the surrounding farmland. Typical effluent quality from secondary lagoon 3 is shown in Table 1.

High rate algal ponds

The two HRAPs were single loop raceway ponds with a central baffle, had working depths of 0.3 m, surface areas of 2.8 m^2 and a total volume of 850 L (see Figure 2). The HRAPs were continuously mixed with paddle wheels. Effluent from secondary lagoon 3 was pumped into the HRAPs via pipework and float valves. The HRAPs were cleaned and filled at the start of each run, and the water level was kept constant with the float valves. The HRAPs were operated in batch mode with a retention time of 7 days during most of the year and a shorter retention time of 4 days for four summer runs.

Methods and investigation

Chemical methodology

Each HRAP had a YSI Quatro probe installed with sensors for temperature, conductivity, pH, ammonia and dissolved oxygen. The YSI Quatro probe recorded measurements on a half-hourly basis. Waters samples were taken at the start and end of each experiment, and additional samples were taken during each run to observe the changes in the nutrient concentrations, species diversity (not shown) and biomass concentrations. An initial sample was taken and passed through a 0.45 µm syringe filter immediately, and the filtrate was tested for nitrite (NO_2-N) and orthophosphate (P-PO_4) concentrations using Hach kits and analysed on a handheld Hach DR890 colorimeter (Hach 2008). A second sample was taken and placed on ice and returned to the laboratory. It was filtered through a glass fibre grade C Whatman filter and tested for ammonia (NH_3-N), nitrate (NO_3-N), total nitrogen, and total phosphorus concentrations using Hach test kits and analysed on a bench top Hach spectrophotometer (Hach 2008).

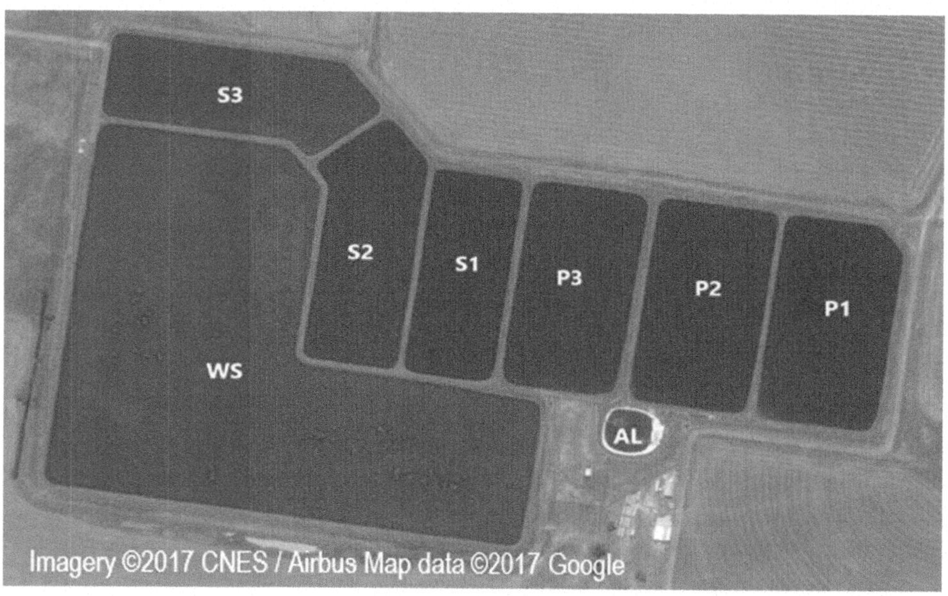

Figure 1 | Western Waters' Bacchus Marsh Recycled Water Plant. Aerobic lagoon (AL), primary lagoons (P1–P3), secondary lagoons (S1–3) and winter storage lagoon (WS) (Google Maps 2017).

Table 1 | Minimum, maximum, mean and standard deviation of the nutrient concentrations of HRAP influent measured during the 12 months of operation (May 2016–May 2017)

	Nitrate and nitrite (mg/L)	Ammonia (mg/L)	Total N (mg/L)	Total P (mg/L)	Biomass (mg/L)	pH
	Annual					
Min	0.45	0.41	2.89	4.28	5.00	8.03
Max	8.70	25.50	29.53	12.31	129.00	10.10
Mean	2.93	10.31	13.30	8.11	41.93	8.68
Std dev	2.21	9.13	9.76	2.23	2.77	0.61
	Autumn					
Min	0.90	0.82	3.80	5.82	12.00	8.10
Max	2.90	19.70	23.64	7.97	92.00	9.63
Mean	1.52	6.56	9.70	6.76	50.40	8.74
Std dev	0.81	7.85	8.36	0.78	37.45	0.70
	Winter					
Min	1.10	12.95	15.27	6.26	5.00	8.03
Max	8.70	25.50	29.53	10.76	29.00	8.57
Mean	3.99	20.29	23.87	8.26	15.50	8.23
Std dev	3.11	4.59	5.32	1.68	9.41	0.25
	Spring					
Min	2.10	4.48	7.05	9.23	8.50	8.14
Max	6.80	23.20	26.67	12.31	50.00	8.89
Mean	4.34	13.10	15.53	10.93	30.70	8.47
Std dev	1.77	7.65	9.17	1.36	18.27	0.28
	Summer					
Min	0.45	0.41	2.89	4.28	28.50	8.21
Max	4.00	2.64	6.93	8.70	129.00	10.10
Mean	1.85	1.28	4.10	6.49	71.10	9.18
Std dev	1.36	0.84	1.63	1.69	36.73	0.68

Biological methods

Dry weight and absorbance were used to measured algal biomass. Dried biomass was determined using the method described in APHA/AWWA/WEF (1998). Absorbance was measured at 440, 680 and 750 nm (results not shown) using a Biochrom Libra S22 spectrophotometer (APHA/AWWA/WEF 1998; Wang *et al.* 2010; Huesemann *et al.* 2013). Growth rates were determined using the dried biomass results.

$$G = \frac{fb - ib}{t} \tag{3}$$

where G is growth rate (mg/(L·D)), fb is the final biomass (mg/L), ib is the initial biomass (mg/L), and t is time D (days). Growth rates were taken from the growth phase of the algal growth, and this was assumed to be the case for all runs unless the ammonia concentration was less than 1.3 mg/L. For ammonia concentrations <1.3 mg/L the biomass production rates dropped, and the algae were in the stationary or death phase of their growth.

Model

Development of a growth model for algal biomass focused on the main factors required for algal growth and utilised data that was easily accessible and available. The model has considered three factors: solar radiation, temperature and ammonia concentration. Temperature and solar radiation were the main drivers for algal growth, with solar radiation providing energy and temperature regulating the kinetics of enzyme activation for growth (Singh & Singh 2015). Ammonia concentration was included to identify when the system was

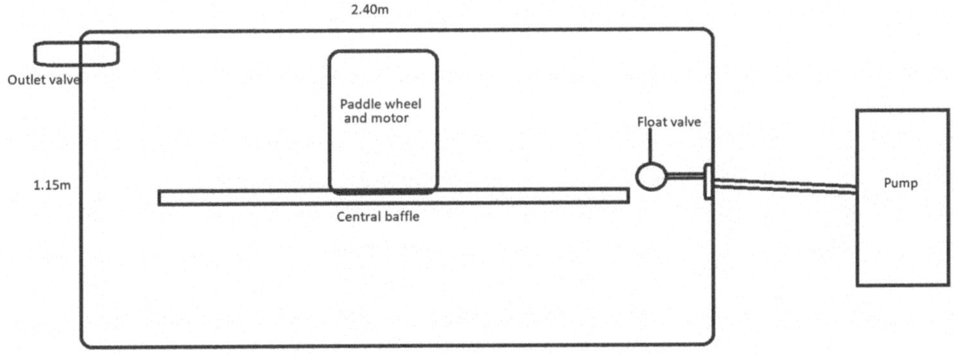

Figure 2 | Schematic of HRAP and pump.

nitrogen-limited, and a concentration of 0.3 mg/L was used as the minimum ammonia concentration below which the algal growth rate was in the stationary or death phase based on experimental data. Algal growth is believed to occur exponentially if there are no limiting factors. Limiting factors can include physical, chemical and biological elements. Physical factors include solar radiation, temperature, self-shading, the design of the HRAP and mixing. Chemical factors that can limit growth include nutrient concentrations, nutrient availability, toxicants and pH. Biological factors that can limit growth are predators, inter-species competition, parasites and microbiological contaminants such as bacteria, fungi or viruses (Collos & Harrison 2014; Carney *et al.* 2016; Chisti 2016; Mehrabadi *et al.* 2016).

Modified versions of Steele's and Monod's equations were utilised in this study. The modified Steele's equation was used to determine the effect of solar radiation and temperature on algal growth, and the modified Monod's equation was employed to investigate the impact limiting ammonia concentrations had on the algal growth. Steele's equation was modified to include both temperature and solar radiation. The optimum value used in the equation was determined from results observed in the studies' experimental period. The temperature and solar radiation were multiplied together and replaced the light intensity value used in the Steele model (see Equation (4)).

$$\mu = \mu_{\max(tl)} \cdot \frac{tl}{tl_{opt}} \tag{4}$$

where μ (mg/(L·D)) is the specific growth rate under conditions of tl (°C/(W·m^2)); $\mu_{max(tl)}$ (g/(L·D)) is the maximum specific growth rate when temperature and solar radiation were optimal; tl (°C/(W·m^2)) is the temperature (°C) multiplied by solar radiation (W/m^2). The use of both temperature and solar radiation incorporated effects associated with the kinetics of enzyme activation as well as the amount of energy available for growth. Mean solar radiation and mean temperature results were utilised based on data over the trial period. The temperature was recorded on a half-hourly basis from the YSI Quatro probe in the HRAPs. Solar radiation was recorded on a half-hourly basis by Western Water's weather station.

A modified Monod equation (Equation (5)) was used to incorporate the negative effect limited ammonia concentration would have on biomass productivity.

$$\mu N = \frac{N}{n+N} \cdot \mu \tag{5}$$

where μN (mg/(L·D)) is the biomass growth rate under N at μ, N (mg/(L·D)) is the ammonia concentration, n is the ammonia half-saturation constant, and μ is the biomass growth rate predicted by the Steele equation. The ammonia half-saturation constant was determined using the ammonia concentration when the growth rate was optimal in the experimental data. The Monod equation was only utilised when the initial ammonia concentration was below 1.3 mg/L. This allowed for less than 1 mg/L of ammonia to be utilised per day, as 0.3 mg/L ammonia was determined from the HRAP experiments to be the minimum ammonia concentration required for growth. Above the concentration of 1.3 mg/L ammonia algal growth was believed not to be limited by ammonia concentrations.

STATISTICS

Mean, minimum and maximum values for the influent water quality results were analysed. Percentages values were used to show the changes in nutrient concentrations and biomass concentrations during HRAP trials, and the results were separated into annual and seasonal values to highlight the significant variations that occur throughout the year. R^2 values were used to signify the accuracy of the correlation between the recorded growth rates and predicted growth rates. R^2 values for exponential and linear growth rates for each HRAP trial were compared using the using the t-test assuming equal variances, and a *p*-value of 0.05 used. Root mean square error (RMSE) was used to determine the amount of error there was between the recorded and predicted daily biomass productivity data sets. Graphs and trendlines were compiled using Microsoft Excel 365.

RESULTS AND DISCUSSION

Water quality

Influent water quality to the HRAPs varied throughout the year with seasonal fluctuations. The most substantial variations were between summer and winter. Table 1 indicates that in the warmer summer months the ammonia concentrations were reduced and the biomass concentrations were elevated, the reverse of which was found in the colder winter months with high ammonia and low biomass concentrations. The concentration of ammonia, microalga's preferred nitrogen source, was highest in winter with a mean ammonia concentration of 20.29 mg/L, and lowest in summer with a mean concentration of 1.28 mg/L (Glibert *et al.* 2016). This indicates that in the lagoons before the final secondary lagoon, large amounts of ammonia were removed during summer; however, during winter ammonia removal was not as effective. Total nitrogen followed the same pattern with the highest mean concentration recorded in winter and the lowest in summer with 23.87 mg/L and 4.10 mg/L respectively. This was inversely reflected in the influent's starting concentration of biomass when summer had the highest biomass concentration, and winter had the lowest with 129 mg/L and 5 mg/L respectively.

High rate algal ponds

The two HRAPs were operated under the same conditions during the 12 months of operation and acted as duplicate experiments. Similar nutrient removal and biomass concentrations were recorded in both HRAPs during the operating period (May 2016–May 2017). The HRAPs had an annual mean removal of 71% of ammonia and 64% of total nitrogen, and an annual mean increase in biomass concentrations of 274% (see Table 2). The highest seasonal percentage removal of ammonia and total nitrogen were recorded in spring with a mean removal of 97% and 84% respectively. Spring also had the highest seasonal increase in biomass, with a mean increase of 399%. Winter had the lowest seasonal ammonia and total nitrogen removal and smallest biomass increase with means of 59%, 34% and 160% respectively. Based on this research and previous research (Wu *et al.* 2013; Mehrabadi *et al.* 2016), it is clear there is a relationship between nitrogen removal and biomass production. This relationship is displayed in Figure 3, which shows the ammonia concentration versus biomass concentrations from a spring HRAP run with a 7-day retention time (R^2 value of 0.99). As the biomass increased, the ammonia decreased in a linear fashion. The low biomass production rate and nutrient removal in winter were due to the lower temperatures and are reflected in the low biomass concentrations and high ammonia concentrations in the colder winter and high biomass concentrations and low ammonia concentrations in the warmer summer (see Table 2). Low temperatures hinder microalgal growth due to slower enzyme reactions and less available energy (Singh & Singh 2015).

According to Redfield (1958) the ratio of carbon:nitrogen:phosphorus in algae is approximately 106:16:1. A mean HRAP influent N:P ratio of 1.64:1 and a maximum N:P ratio of 4.16:1 were recorded during this study. This confirms that there was no phosphorus limitation during the experiments and microalgae were believed to have taken up extra phosphorus as a form of luxury uptake (Wang *et al.* 2010; Beuckels *et al.* 2015). Nitrogen was thought to be a limiting factor in some runs, other than carbon, and carbon limitation was highlighted by the elevated pH values.

Table 3 shows an annual mean pH value of 9.35, a maximum of 11.02 and a minimum of 7.93 during the algal growth runs. The highest mean pH results were recorded in summer with a mean of 9.98, the lowest in winter with a mean of 8.71, and autumn and spring means were 9.15 and 9.30 respectively. These elevated pH levels negatively affect the growth of the microalgae. When pH exceeds 8.3, the amount of free/dissolved CO_2 reduces to virtually zero. Algae are still able to grow in elevated pH conditions, but the growth rates are diminished by the algae's need to break down bicarbonate to access the carbon. High pH values cause ammonia volatilisation, and this would further hinder the growth of algae due to a lack of nitrogen (Cai *et al.* 2013).

A large variety of microalgae inhabited the lagoons and the HRAPs. These included *Scenedesmus* sp., *Dictyosphaerium* sp., *Chlamydomonas* sp., *Euglena* sp., *Micractinium* sp.,

Table 2 | Mean percentage nutrient removal results and mean percentage algal biomass increases during HRAP trials throughout the experimental period

	Autumn	Winter	Spring	Summer	Annual
Nitrate and nitrite (mg/L)	4%	−83%	3%	−11%	−6%
Orthophosphate (mg/L)	15%	3%	53%	57%	32%
Ammonia (mg/L)	68%	56%	97%	65%	71%
Total nitrogen (mg/L)	45%	34%	84%	26%	64%
Total phosphorus (mg/L)	21%	5%	60%	61%	37%
Dried biomass (mg/L)	215%	160%	399%	299%	274%
750 nm	269%	237%	513%	267%	330%
680 nm	245%	198%	483%	256%	302%
440 nm	216%	149%	397%	223%	252%

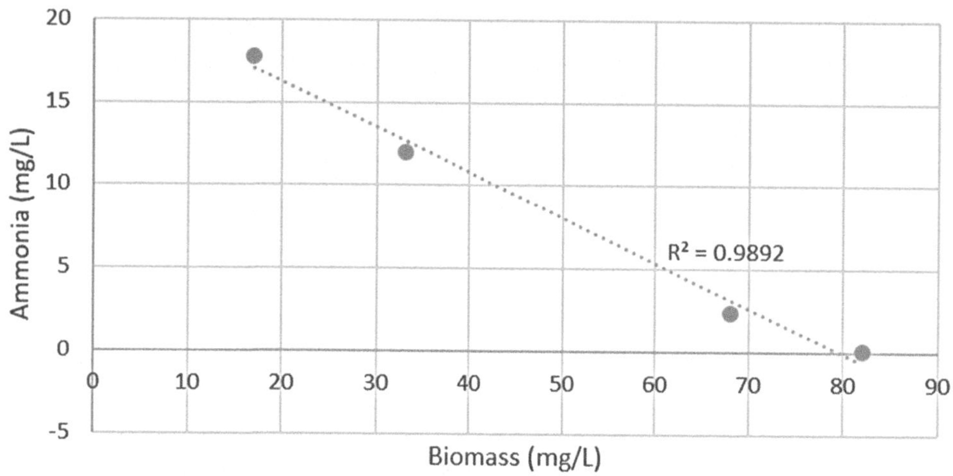

Figure 3 | Ammonia (mg/L) versus biomass (mg/L) for an HRAP run commencing on 5 October 2016 with a retention time of 7 days (sampling on days 0, 2, 5 and 7).

Table 3 | Mean, standard deviation, maximum and minimum pH results in each season and throughout the experimental period (May 2016–May 2017)

	Autumn	Winter	Spring	Summer	Annual
Mean	9.15	8.71	9.30	9.98	9.35
Max	9.63	8.82	10.18	10.49	11.02
Min	8.69	8.41	8.29	9.10	7.93
Mean std dev	0.22	0.11	0.43	0.31	0.27

Golenkinia sp. and *Oocystis* sp. The species diversity and population density varied throughout the year responding to fluctuations in temperature and nutrients. Large amounts of zooplankton were observed in the third secondary lagoon. These included rotifers (*Brachionus* sp.), cladocerans (*Daphnia* sp.) and copepods (*Cyclops* sp.) but these zooplankton were not found in significant numbers in the HRAPs. Montemezzani *et al.* (2015) stated that zooplankton might be killed by the generation of shear forces exerted by a water pump. It is believed that during the process of transferring the water from the lagoon to the HRAPs the shear forces exerted on the zooplankton in the pipework and pump killed large amounts of zooplankton due to powerful impacts with the pipe walls, causing the zooplankton to be damaged. Elevated pH levels and free ammonia toxicity may also have a lethal effect on the zooplankton population (Montemezzani *et al.* 2015).

Environmental factors

Previous studies showed the significant impact solar radiation and temperature have on algal growth (Singh & Singh 2015; Huesemann *et al.* 2016). The concentration of algal biomass recorded during each run was plotted against the number of days passed in that run, linear and exponential trendlines were plotted, and the R^2 values are recorded in Table 4. Analysis of variance and t-test analysis on the two sets of R^2 values showed no difference. Algae are expected to grow exponentially, and a linear growth relationship indicates that the algal growth was limited. The use of a linear

Table 4 | R^2 values based on time versus biomass concentration for each separate HRAP run

	Linear R^2	Exponential R^2
25/05/2016	0.9767	0.9757
3/06/2016	0.9483	0.9934
18/07/2016	0.9786	0.99
31/08/2016	0.0112	0.0206
19/09/2016	0.5595	0.1317
5/10/2016	0.9917	0.9629
19/10/2016	0.9713	0.9713
16/11/2016	0.7068	0.6712
2/12/2016	0.9924	0.9676
12/12/2016	0.8846	0.7394
6/01/2017	0.5967	0.6318
16/01/2017	0.8957	0.8581
13/02/2017	0.8935	0.8563
6/03/2017	0.882	0.9083
13/03/2017	0.7923	0.7835
10/04/2017	0.8313	0.8193
1/05/2017	0.982	0.9875
29/05/2017	0.5578	0.5152
Mean	0.8029	0.7657

relationship in the predictive growth rate modelling was determined to be more accurate than an exponential relationship for these data due to the slightly higher mean R^2 value: the linear and exponential mean R^2 values were 0.80 and 0.77 respectively.

The growth rate was correlated to solar radiation, ammonia concentration and temperature. Linear regression of the mean temperature and growth rate resulted in an R^2 value of 0.52, while linear regression of mean solar radiation and growth rate achieved an R^2 value of 0.65. The effect of differing photoperiods was investigated and gave the same R^2 value as solar radiation. While these R^2 values are low, the results do indicate a relationship between daily production and mean temperature and solar radiation during the growth period. A simple combination of mean temperature and mean solar radiation was tested, and when measured against growth rate an R^2 value of 0.66 was achieved as shown in Figure 4.

Model validation

The use of modified versions of the Steele and Monod equations was shown to enhance the predictive capabilities of this work. Steele's equation was modified by removing the exponential function from the equations as it was determined that the algal growth rates in the HRAP followed a linear trend rather than the expected exponential trend. This result is caused by the various ecological and chemical limitations the HRAPs contain. For example, most models developed for use in a laboratory do not incorporate light limitation factors such as diurnal variation and self-shading, which are common in an HRAP. Inter-species competition between algae can also affect the rate of growth. The modified Steele equation (Equation (4)) was used to model the effect of light and temperature. When comparing the predicted growth rate using the modified Steele equation against the recorded growth rate an R^2 value of 0.66 resulted. The main outliers were HRAPs trials in which high temperatures and solar radiations were observed, such as during the summer months. In these instances, the model overpredicted the observed biomass growth, and this was believed to be caused by an ammonia limitation during these trials. Removing the results in which the initial ammonia concentrations were less than 1.3 mg/L increased the R^2 value to 0.83. Overprediction of algal biomass was also observed under cold temperatures conditions where growth measurements were inaccurate due to the low algal biomass concentrations.

To correct for the ammonia limitation, the modified Monod model was used to predict the biomass production based on the amount of ammonia that was available. An ammonia concentration of 2 mg/L was determined to be sufficient for normal growth in these experiments, and below 1.3 mg/L of available ammonia in the system was limiting. The model typically requires a maximum growth rate based on the optimal growth rate determined by saturated nitrogen conditions. However, the maximum growth rate used was the predicted value calculated using the modified Steele equation using solar radiation and temperature.

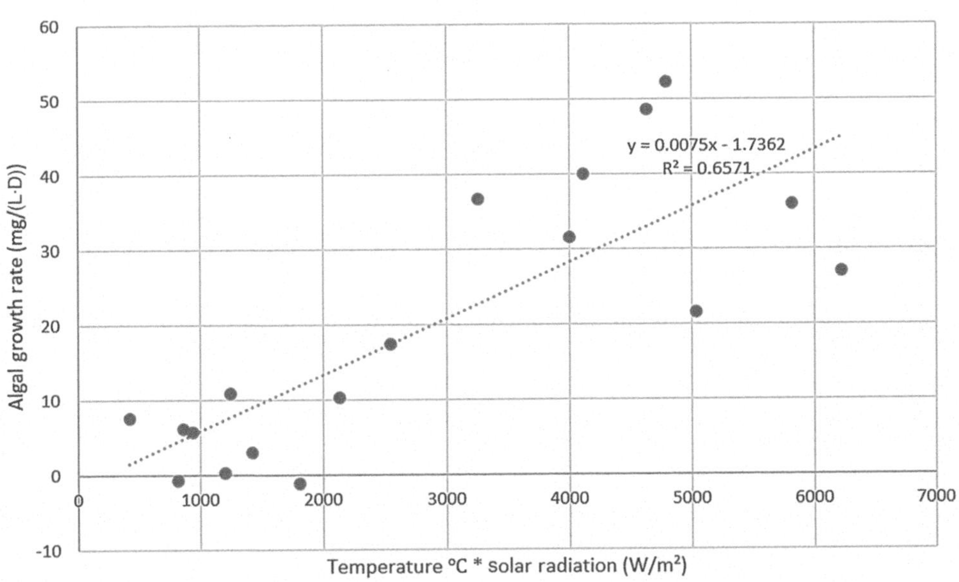

Figure 4 | Algal growth rate (mg/(L·D)) versus mean temperature (°C) x mean solar radiation (W/m²).

A combination of the two equations achieved an R^2 value of 0.82 when comparing the predicted growth rates and the recorded growth rates (see Figure 5). This result is promising, and while more complex models could predict the algae growth more accurately, this would require more detailed characterisation of the growth conditions. Wu *et al.* (2013) used an integrated growth model on nine different wastewater samples and obtained R^2 values ranging from 0.64 to 0.99; however, each wastewater sample had different model parameters, thus adding to the complexity of the modelling. An RMSE value of 8.82 mg/(L·D) was obtained when comparing all the predicted daily biomass productivity to the recorded daily biomass productivity. The daily biomass productivity data were separated into two sets, above and below the average temperature to highlight the differences in RMSE. The RMSE of the above-average temperature (15°C) data is 5.47 mg/(L·D), and the RMSE of the below-average temperature data is 10.77 mg/(L·D). These results are displayed in Figure 5, which shows which data correspond to which RMSE set. This highlights that there is a more significant error for model prediction at colder temperatures than at warmer temperatures. Therefore, estimating growth times and lagoon areas for operation of an HRAP during cold weather is prone to greater risk than for higher temperatures, while better estimates can be made for higher temperatures (>15 °C) that correspond to the usual conditions for industrial use of HRAPs. The discrepancies in the current model could be attributed to a few factors experienced during the study. Firstly, elevated pH levels reduce the amount of available free carbon in the system, and this negatively affects growth. Secondly, the difference in algal species and cyanobacteria composition throughout the year would affect the growth rate. Thirdly, there were two results obtained in winter 2016 in which the biomass productivity rates were slightly negative, and while these results were expected to be low, negative values were not expected. The negative values could have been produced by higher respiration rates compared to their photosynthetic rates or experimental error. Fourthly, evaporation could have contributed to the higher than predicted biomass concentrations in summer. The water in the HRAPs was kept constant via the addition of effluent from the secondary lagoon 3; this additional water would have contained unaccounted-for ammonia and consequently enhanced biomass growth rates. Lastly, zooplankton may have consumed algae and skewed the results, although zooplankton were killed by the pump when the lagoons were filled with secondary effluent.

From the proposed relationship and initial ammonia concentration, we can estimate how long it would take for microalgae to reach a stationary growth phase. The relationship will predict the growth rate and consumption of ammonia during growth. This will assist with identifying the optimum harvesting period and retention times for HRAPs. Additionally, a significant deviation of growth rates from this relationship might also identify if there are limiting conditions for growth or if microalgae growth can be enhanced by the addition of nutrients or a change in pH.

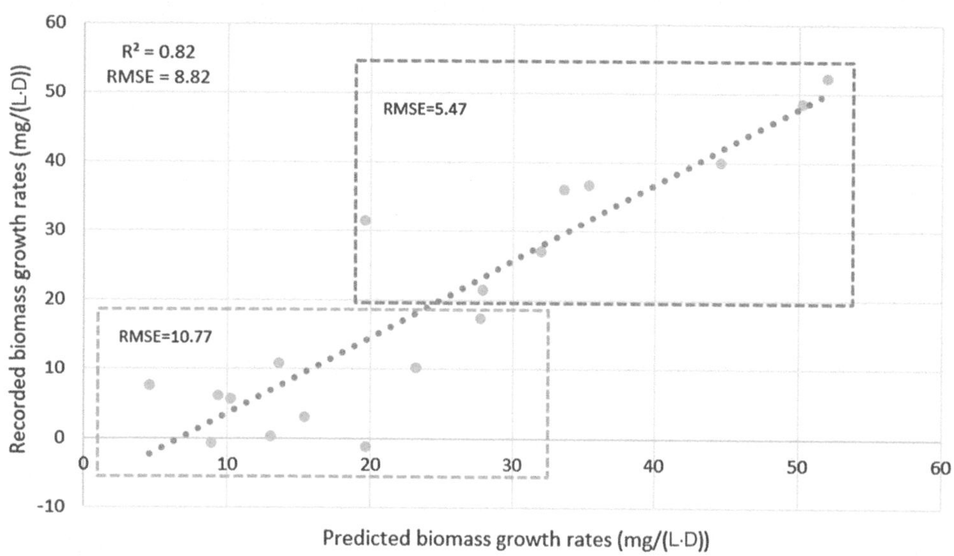

Figure 5 | The values from the combined modified Steele and Monod models. Predicted algal growth rates (mg/(L·D)) versus recorded algal growth rates (mg/(L·D)).

CONCLUSION

This study investigated the use of batch HRAPs to remove nutrients from secondary effluent by means of growing algae. The two HRAPs were operated in parallel for 12 months with no control of pH, nutrient concentration or the algal community. The HRAPs significantly reduced the nutrient concentrations by means of incorporation of nutrients into algal biomass. This further reduced the nutrient level from the secondary effluent before its discharge for future use. A simple predictive algal growth model was developed from modified versions of the Steele and Monod models. The model incorporated temperature, solar radiation and ammonia concentrations which varied significantly throughout the year, and an R^2 correlation of 0.82 was obtained with an RMSE value of 8.82 mg/(L·D). A predictive algal biomass model for HRAPs that requires readily available data is of key importance for the development of more extensive wastewater treatment HRAPs. This model can help predict if effluent quality and environmental conditions are suitable for a large-scale HRAP from limited data and if HRAPs would be viable for biomass cultivation.

ACKNOWLEDGEMENTS

The authors would like to thank Western Water for providing facilities in the field for the conduct of the experiment and for the provision of the weather data. DW would like to thank Western Water and Victoria University for providing a postgraduate scholarship that allowed him to undertake this research.

REFERENCES

Abdel-Raouf, N., Al-Homaidan, A. A. & Ibraheem, I. B. M. 2012 Microalgae and wastewater treatment. *Saudi Journal of Biological Sciences* **19**, 257–275.

APHA/AWWA/WEF 1998 *Standard Methods for the Examination of Water and Wastewater*, 20th edn. American Public Health Association/American Water Works Association/Water Environment Federation, Washington, DC, USA.

Béchet, Q., Shilton, A., Park, J. B., Craggs, R. J. & Guieysse, B. 2011 Universal temperature model for shallow algal ponds provides improved accuracy. *Environmental Science & Technology* **45**, 3702–3709.

Béchet, Q., Chambonnière, P., Shilton, A., Guizard, G. & Guieysse, B. 2015 Algal productivity modeling: a step toward accurate assessments of full-scale algal cultivation. *Biotechnology and Bioengineering* **112**, 987–996.

Benemann, J. R. 2003 *Bio-fixation of CO_2 and Greenhouse Gas Abatement with Microalgae-Technology Roadmap*. Final Report to the US Department of Energy, National Energy Technology Laboratory, No. 7010000926.

Beuckels, A., Smolders, E. & Muylaert, K. 2015 Nitrogen availability influences phosphorus removal in microalgae-based wastewater treatment. *Water Research* **77**, 98–106.

Cai, T., Park, S. Y. & Li, Y. 2013 Nutrient recovery from wastewater streams by microalgae: status and prospects. *Renewable and Sustainable Energy Reviews* **19**, 360–369.

Carney, L. T., Wilkenfeld, J. S., Lane, P. D., Solberg, O. D., Fuqua, Z. B., Cornelius, N. G., Gillespie, S., Williams, K. P., Samocha, T. M. & Lane, T. W. 2016 Pond crash forensics: presumptive identification of pond crash agents by next generation sequencing in replicate raceway mass cultures of *Nannochloropsis salina*. *Algal Research* **17**, 341–347.

Chisti, Y. 2016 Large-scale production of algal biomass: raceway ponds. In: *Algae Biotechnology: Products and Processes* (F. Bux & Y. Chisti, eds). Springer International Publishing, Cham, Switzerland, pp. 21–40.

Christenson, L. & Sims, R. 2011 Production and harvesting of microalgae for wastewater treatment, biofuels, and bioproducts. *Biotechnology Advances* **29**, 686–702.

Collos, Y. & Harrison, P. J. 2014 Acclimation and toxicity of high ammonium concentrations to unicellular algae. *Marine Pollution Bulletin* **80**, 8–23.

Craggs, R., Sutherland, D. & Campbell, H. 2012 Hectare-scale demonstration of high rate algal ponds for enhanced wastewater treatment and biofuel production. *Journal of Applied Phycology* **24**, 329–337.

Glibert, P. M., Wilkerson, F. P., Dugdale, R. C., Raven, J. A., Dupont, C. L., Leavitt, P. R., Parker, A. E., Burkholder, J. M. & Kana, T. M. 2016 Pluses and minuses of ammonium and nitrate uptake and assimilation by phytoplankton and implications for productivity and community composition, with emphasis on nitrogen-enriched conditions. *Limnology and Oceanography* **61**, 165–197.

Google Maps 2017 Parwan, Victoria 3340 Map.

Hach 2008 *Procedure Manual*. Hach, Loveland, CO, USA.

Huesemann, M. H., Van Wagenen, J., Miller, T., Chavis, A., Hobbs, S. & Crowe, B. 2013 A screening model to predict microalgae biomass growth in photobioreactors and raceway ponds. *Biotechnology and Bioengineering* **110**, 1583–1594.

Huesemann, M., Crowe, B., Waller, P., Chavis, A., Hobbs, S., Edmundson, S. & Wigmosta, M. 2016 A validated model to predict microalgae growth in outdoor pond cultures subjected to fluctuating light intensities and water temperatures. *Algal Research* **13**, 195–206.

Jayaraman, S. K. & Rhinehart, R. R. 2015 Modeling and optimization of algae growth. *Industrial & Engineering Chemistry Research* **54**, 8063–8071.

Mehrabadi, A., Farid, M. M. & Craggs, R. 2016 Variation of biomass energy yield in wastewater treatment high rate algal ponds. *Algal Research* **15**, 143–151.

Montemezzani, V., Duggan, I. C., Hogg, I. D. & Craggs, R. J. 2015 A review of potential methods for zooplankton control in wastewater treatment high rate algal ponds and algal production raceways. *Algal Research* **11**, 211–226.

Oswald, W., Gotaas, H., Golueke, C., Kellen, W., Gloyna, E. & Hermann, E. 1957 Algae in waste treatment [with discussion]. *Sewage and Industrial Wastes* **29**, 437–457.

Putt, R., Singh, M., Chinnasamy, S. & Das, K. C. 2011 An efficient system for carbonation of high-rate algae pond water to enhance CO_2 mass transfer. *Bioresource Technology* **102**, 3240–3245.

Rawat, I., Ranjith Kumar, R., Mutanda, T. & Bux, F. 2013 Biodiesel from microalgae: a critical evaluation from laboratory to large scale production. *Applied Energy* **103**, 444–467.

Redfield, A. C. 1958 The biological control of chemical factors in the environment. *American Scientist* **46**, 230A-221.

Singh, S. P. & Singh, P. 2015 Effect of temperature and light on the growth of algae species: a review. *Renewable and Sustainable Energy Reviews* **50**, 431–444.

Sutherland, D. L., Montemezzani, V., Mehrabadi, A. & Craggs, R. J. 2016 Winter-time CO_2 addition in high rate algal mesocosms for enhanced microalgal performance. *Water Research* **89**, 301–308.

Undurraga, D., Poirrier, P. & Chamy, R. 2016 Microalgae growth kinetic model based on the PSII quantum yield and its utilization in the operational curves construction. *Algal Research* **17**, 330–340.

Wang, L., Min, M., Li, Y., Chen, P., Chen, Y., Liu, Y., Wang, Y. & Ruan, R. 2010 Cultivation of green algae *Chlorella* sp. in different wastewaters from municipal wastewater treatment plant. *Applied Biochemistry and Biotechnology* **162**, 1174–1186.

Wu, Y.-H., Li, X., Yu, Y., Hu, H.-Y., Zhang, T.-Y. & Li, F.-M. 2013 An integrated microalgal growth model and its application to optimize the biomass production of *Scenedesmus* sp. LX1 in open pond under the nutrient level of domestic secondary effluent. *Bioresource Technology* **144**, 445–451.

Young, P., Taylor, M. & Fallowfield, H. J. 2017 Mini-review: high rate algal ponds, flexible systems for sustainable wastewater treatment. *World Journal of Microbiology and Biotechnology* **33**, 117.

First received 1 November 2017; accepted in revised form 20 April 2018. Available online 2 May 2018

Pond walls: inclined planes to improve pathogen removal in pond systems for wastewater treatment?

A. L. Hawley and H. J. Fallowfield

ABSTRACT

Attenuation of sunlight in wastewater treatment ponds reduces the depth of the water exposed to disinfecting irradiances. Shallow pond depth with paddlewheel rotation increases exposure of pathogens to sunlight in high rate algal ponds. Generation of thin films, using pond walls as inclined planes, may increase inactivation of pathogens by increasing sunlight exposure. The performance of a laboratory based model system incorporating an inclined plane (IP) was evaluated. F-RNA bacteriophage, in tap water or wastewater, was exposed to sunlight only on the IP with the bulk water incubated in the dark. MS2 inactivation was significantly higher when the IP was present ($P < 0.05$) with a 63% increase observed. Prolonged exposure increased MS2 die-off irrespective of IP presence. Versatility of the IP was also demonstrated with faster inactivation observed in both optically clear tap water and wastewaters. IPs of different surface areas produced similar inactivation rates when operated at similar hydraulic loading rates regardless of slope length.

Key words | dark inactivation, high rate algal ponds, inclined plane, pathogen removal, solar exposure

A. L. Hawley (corresponding author)
H. J. Fallowfield
Health and Environment Group, College of Science and Engineering,
Flinders University,
G.P.O. Box 2100, Adelaide 5001,
South Australia,
Australia
E-mail: *amy.hawley@flinders.edu.au*

INTRODUCTION

Wastewater treatment ponds utilise the germicidal properties of sunlight for the disinfection of pathogens (Clancy *et al.* 2000). Ultraviolet (UV) and visible (Vis) light when absorbed, directly (photo-inactivation) or indirectly (photo-oxidation), cause the genetic material or membranes of the organism to be damaged (Muela *et al.* 2002). In wastewater, however, attenuation reduces the depth of light penetration through the water column, particularly of the more germicidal, shorter wavelength UVB spectrum, (Caslake *et al.* 2004). This light decay is more prominent in turbid waters, and increases with pond depth (Kirk 1994; Fallowfield *et al.* 1996). In turbid water, the majority of the light involved in inactivation is absorbed in the first 1 m of water (Haag & Hoigne 1986) and UVB in the first 0.03 m (Kohn & Nelson 2007). More specifically, Bolton (2012) identified the extinction depths of 0.03 m (UVB; 280–315 nm), 0.07 m (UVA; 315–400 nm) and 0.14 m (Vis; 400–700 nm), in waste stabilisation pond (WSP) effluent, respectively. For improved disinfection, it is essential that the availability of light within the water column is increased and the effects of attenuation reduced.

High rate algal ponds (HRAPs) are intentionally mixed, shallow treatment ponds (0.2–0.5 m) arranged in a raceway configuration (Park *et al.* 2011). The hydraulic retention time of these ponds is between 2 and 8 days (Shilton 2005). Increased exposure to sunlight is achieved in these ponds through large surface area to volume ratios and continual mixing, most commonly by paddlewheel (Fallowfield & Garrett 1985). Elevated removal rates are achieved in these ponds with disinfection up to six times faster than other WSP systems (Buchanan *et al.* 2011). The removal of helminth (El Hamouri *et al.* 1994), protozoa (Araki *et al.* 2001) and bacteria (El Hamouri *et al.* 1994; Fallowfield *et al.* 1996) have been reported with a focus on the reduction of faecal coliforms and *Escherichia coli*. The removal of enteric viruses is a priority to protect human health; however, limited information exists regarding the removal of these viruses and their bacteriophage surrogates. To help bridge the gap this study will focus on the removal of MS2, an F-RNA bacteriophage.

The existing pond walls surrounding both HRAPs and WSPs provide a natural inclined plane (IP) (45°), formed during construction. At present, these embankments serve no purpose other than to contain the pond water. However, they may provide an opportunity for a cost-effective means of increasing solar exposure, and consequently virus inactivation within the systems. Generation of a thin film of wastewater flowing down the pond wall will increase

doi: 10.2166/wst.2018.269

exposure to disinfecting wavelengths of light. The solar exposure (UV energy) experienced by an IP has been characterised. An increase of 3–4% was reported when the incline was 37° compared to a horizontal surface (Navntoft *et al.* 2012). A 10% increase was reported when the plane was in the direction of the sun and the incline equal to the degree of latitude (Iqbal 1983; Duffie & Beckman 2013).

The objective of this research was to establish if increasing the area available for solar exposure, through the addition of an IP and the generation of a thin film, improved the inactivation of the F-RNA coliphage MS2. To achieve this, the study examined and compared removal rates achieved in model HRAPs in the presence and absence of an IP.

METHODS

Inclined planes (IP)

IPs were constructed from black Perspex sheet (width 0.67 m) fixed to a steel frame base. Two sizes were used, a small IP (SIP; length 0.55 m, 0.37 m^2) and a large IP (LIP; length 1.10 m, 0.75 m^2). A valved manifold was attached to the top of the plane through which water was pumped (Aqua PRO AP950) and the flow rates controlled to generate the thin film on the plane. Figure 1 provides a schematic diagram of the IPs used.

Model HRAP systems

Model HRAPs were constructed from 100 L plastic vessels, filled to a depth of 0.30 m (87.0 L) with either optically clear (tap) water or wastewater collected from a treatment plant comprising an aerated lagoon and maturation pond (Mt Barker, South Australia). The bulk water in the plastic vessels was continuously mixed using aquarium pumps (Aqua One 102). In order to determine the inactivation potential of the IP the bulk water in the vessels was covered so that only the IP was exposed to sunlight. A similarly mixed and covered HRAP without an IP ($HRAP_d$) was used for comparison (Figure 1). Operating conditions are presented in Table 1.

Sample collection

Triplicate 10 mL samples were collected in 1.5 h intervals over a 24 h period. For water analysis, 120 mL samples were collected every 3 hours with an additional 1 L collected prior to and at completion of each experimental run.

Environmental conditions

In situ water measurements were recorded at each collection interval. Parameters monitored included water temperature (YSI model 55, Xylem), pH (370 pH meter,

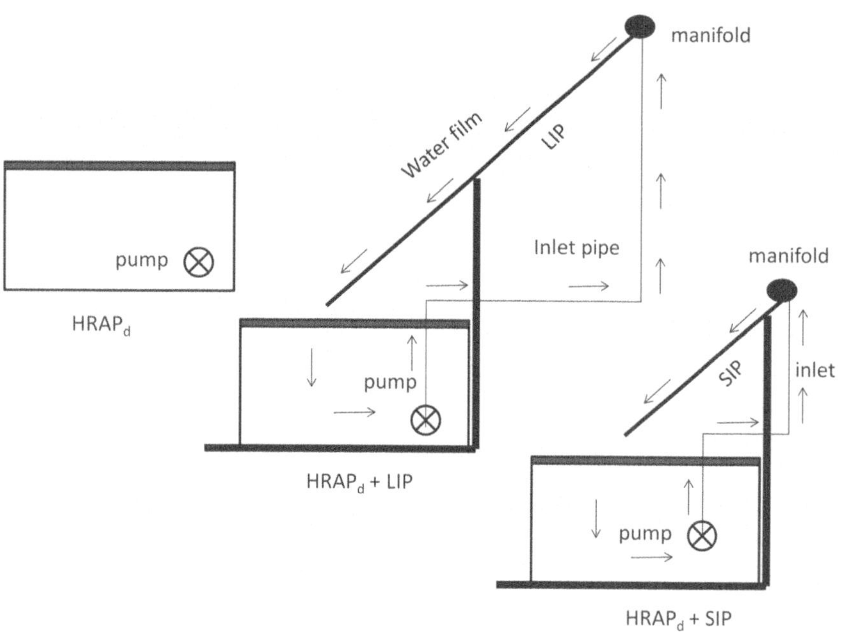

Figure 1 | Schematic diagram of the model $HRAP_d$ and $HRAP_d$ + IP used. Two sizes of IPs were used: a short (SIP, 0.37 m^2, 0.55 m) and a long (LIP, 0.75 m^2, 1.10 m) IP. The diagram also demonstrates the function of the IP where pond water is pumped to the top of the IP via an inlet pipe connected to both pump and manifold. The water is converted into a thin film by holes in the manifold and exposed to sunlight as it travels down the IP back to the pond. Systems were operated concurrently.

Table 1 | Model HRAP operating characteristics for two slopes of differing lengths and surface area; small-IP (SIP; 0.37 m^2) and large-IP (LIP; 0.75 m^2), where Q: flow rate ($L h^{-1}$), HLR: hydraulic loading rate ($L h^{-1} m^{-2}$)

			1		2		3		4	
Model HRAP	**Area (m^2)**	**Initial Q**	**HLR**	**Q**	**HLR**	**Q**	**HLR**	**Q**	**HLR**	**Q**
SIP	0.37	129.6	350.3	129.6	167.8	62.1	232.7	87.7	300.0	111.0
LIP	0.75	131.1	350.3	262.7	174.4	130.8	232.8	174.6	300.0	225.0

Jenway) and dissolved oxygen (DO, YSI model 55, Xylem). Onsite UVA and UVB radiation was recorded using a Solar Light PMA2100 data logging radiometer with UVA (PMA 2110-WP, Solar Light) and UVB (PMA 2106-WP, Solar Light) sensors. Daily solar exposure ($MJ m^{-2}$) and sunshine duration (h) were monitored from the Australian Bureau of Meteorology. Data were obtained from Adelaide airport (Station number 23034; 9.58 km from site).

Water quality analysis

Water quality was assessed, using Greenberg *et al.* (1992) standard methods for the analysis of wastewater methods, for turbidity (NTU), suspended solid (SS, $mg L^{-1}$), and chlorophyll *a* (chl *a*, $mg L^{-1}$).

Microbial quantification

F-RNA bacteriophage MS2 (ATTC#15597-B1) was spiked into all model systems (3.09×10^{10} PFU 100 mL^{-1}) and inactivation determined using the quantitative double layer agar plaque assay adapted from Debartolomeis & Cabelli (1991) and Noble *et al.* (2004). Briefly, *E. coli* Famp (ATTC# 700891) was used as the bacterial host. Both host and MS2 stock for HRAP inoculation were grown with ampicillin sodium salt (Sigma) and streptomycin sulphate (Sigma) antibiotics. Samples were serially diluted (10×) with 0.5% tryptone water (Oxoid) for analysis when required. Counts were reported as plaque forming units (PFU) per 100 mL for both systems.

Statistical analysis

Log_{10} reduction values (LRV) were calculated using Equation (1); where N_0 represents initial concentration, N_t is concentration at time t and LRV_t is the log_{10} reduction obtained after t hours.

$$LRV_t = \log_{10}N_0 - \log_{10}N_t \quad (1)$$

Inactivation rate constants (K) were determined using GInaFiT; a Microsoft Excel add-in (Geeraerd *et al.* 2005). Statistical analyses were carried out using statistical software packages: R version 3.1.2 (Vienna, Austria) and SPSS version 20.0 (Armonk, NY). Normal distribution of data was determined from quantile-comparison (Q-Q) plots and Shapiro–Wilk normality tests. Statistical analyses included linear regression (multiple with stepwise regression), independent samples t-tests, one-way analysis of variance with Tukey's post hoc comparison and Pearson's product correlation. Statistical significance was inferred at $P < 0.05$.

RESULTS AND DISCUSSION

MS2 inactivation with IP inclusion

The global solar irradiation received during this experiment ranged from 26.0 to 28.2 $MJ m^{-1}$ with a mean of 26.9 ± 1.1 $MJ m^{-1}$. Water temperatures ranged from 17.4 to 32.6 °C in the model systems with means of 26.0 ± 3.7 °C ($HRAP_d$) and 25.9 ± 5.2 °C ($HRAP_d$ + SIP). Elevated MS2 inactivation was achieved by the incorporation of the SIP (Table 2). The inclusion of the SIP resulted in an inactivation 1.6 times higher than the $HRAP_d$ operated in the absence of the IP. The difference between the mean LRV was statistically significant ($P < 0.05$).

Table 2 shows the inactivation rates obtained after 24 h incubation. The log_{10} linear + tail model described by Geeraerd *et al.* (2000) was identified as the best representation of the data. Tailing of the data suggests a lag in die-off for the presence of a mixed or sub culture where one of the populations exhibits a greater resistance to disinfection (Bevilacqua *et al.* 2015). In Table 2, the K_{max} values are derived from the log_{10} linear + tail inactivation curves.

Since the bulk water of both HRAPs was in the dark, the elevated MS2 inactivation seen in the $HRAP_d$ + SIP can be attributed to the exposure of water to sunlight received

Table 2 | MS2 inactivation in model systems: $\log_{10}$ reduction values (LRV) and inactivation rate constant (K_{max}, mean ± standard deviation for $HRAP_d$ and $HRAP_d$ + SIP, after 24.0 h incubation

		Inactivation				
Time (h)	**System**	***n***	**LRV_t**	**K_{max}**	**R^2**	***P*-value**
24.0	$HRAP_d$	21	1.445	0.240 ± 0.067	0.7795	2.2×10^{-12}
	$HRAP_d$ + SIP	21	2.354	0.297 ± 0.078	0.9110	2.0×10^{-12}

R^2 and *P*-values relate to the strength of the statistics associated with the determination of K_{max}.

whilst on the slope. The inclusion of the IP increased the LRV_{24} by 63% when compared with the $HRAP_d$.

A direct comparison could not be made with the current literature due to the novel nature of this research. Comparison was made with studies examining F-RNA phage inactivation in other pond systems, for example WSPs, under both dark and solar exposures. The values reported here were higher than the inactivation rates reported by Sinton *et al.* (2002), who examined the inactivation of F-RNA phage in WSP effluent exposed to sunlight (summer; 0.070 h^{-1}, winter; 0.050 h^{-1}) and in the absence of light (0.014 h^{-1}).

The role of hydraulic loading rate

Further experiments comparing the performance of SIP and LIP at different flow rates which maintained different but constant hydraulic loading rate (HLR) on the two IPs within each experiment resulted in MS2 $LRV_{49.5}$ that were statistically similar ($P \geq 0.05$), an observation consistent across the four HLRs examined (Figure 2). The similar inactivation rates for slopes of different lengths and surface areas operated over a range of HLRs suggest that HLR is the factor which influences inactivation by the IP operated under the same climatic conditions. The higher inactivation exhibited for HLR 1 (Figure 2) was attributed to the higher solar irradiance received throughout the experimental period. Corresponding solar irradiances for the HLR experiments were 18.5 ± 4.3 (HLR 1), 8.9 ± 10.7 (HLR 2), 4.4 ± 3.5 (HLR 3) and 5.5 ± 6.2 MJ m^{-2} (HLR 4).

Incubation time

Solar and dark disinfection are affected by the duration of incubation and exposure to sunlight (Kirk 1994). Figure 3 shows that inactivation was related to incubation time both in the dark-incubated $HRAP_d$ and in the $HRAP_d$ + IP exposed to sunlight. The LRV_{5d} were 1.882 ± 0.282, 2.852 ± 0.627 and 3.046 ± 0.322 for the HRAP, HRAP + SIP and HRAP + LIP, respectively.

In situ water conditions

Inactivation can vary depending on different environmental parameters. Higher pH, DO and water temperature were identified in the $HRAP_d$ + LIP (pH 7.73 ± 0.42, 6.08 ± 2.01 mg DO L^{-1}, 21.9 ± 6.0 °C) and HRAP + SIP (pH

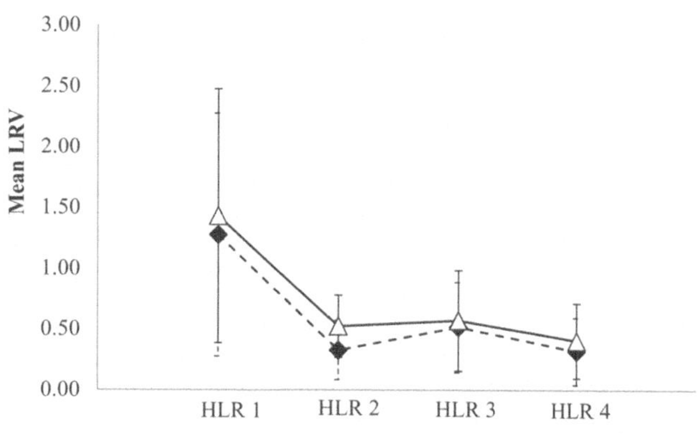

Figure 2 | MS2 $\log_{10}$ reduction values (mean ± standard deviation) obtained after an incubation time of 49.5 h for small (♦) and large (▲) IP operated at different HLRs. Mean HLRs were HLR 1, 350.3 L m^{-2} h^{-1}; HLR 2, 171.1 L m^{-2} h^{-1}; HLR 3, 232.75 L m^{-2} h^{-1} and HLR 4, 300 L m^{-2} h^{-1}.

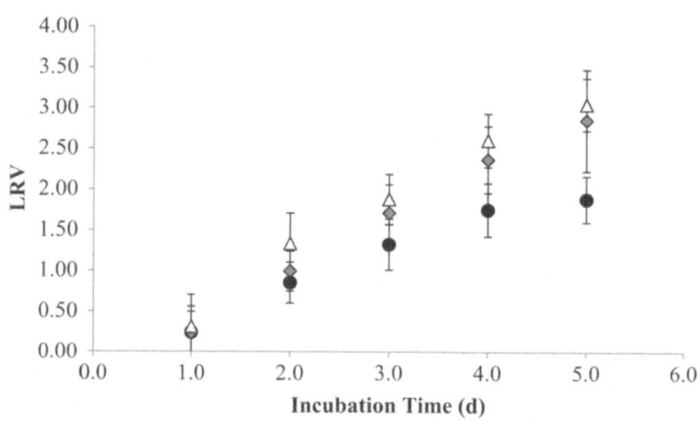

Figure 3 | The relationship between LRV (mean ± 1 standard deviation) and the incubation time for the dark-incubated $HRAP_d$ (●), $HRAP_d$ + SIP (♦) and $HRAP_d$ + LIP (▲). The inclined planes were operated at an HLR of 300.0 L m^{-2} h^{-1}.

7.64 ± 0.41, 6.58 ± 2.11 mg DO L^{-1}, 20.7 ± 5.9 °C) compared to the $HRAP_d$ (pH 7.31 ± 0.43, 5.51 ± 2.26 mg DO L^{-1}, 20.4 ± 6.2 °C). However, pH was the only parameter identified as being significantly higher in the $HRAP_d + IP$. This is probably a response to the elevated chlorophyll *a* levels identified in the systems incorporating IPs. The chl *a* levels were 0.76 ± 0.53 ($HRAP_d + LIP$), 0.71 ± 0.38 ($HRAP_d + SIP$) and 0.49 ± 0.21 mg L^{-1} ($HRAP_d$), respectively.

Summer incubations exhibited greater MS2 inactivation (Figure 4) indicative of the effect of higher solar irradiances received throughout summer. During the experimental period the mean summer irradiance, 27.9 ± 1.1 MJ m^{-2}, was 2.6 and 5.5 times higher than the mean irradiance received during autumn (11.2 ± 8.1 MJ m^{-2}) and winter (9.9 ± 0.0 MJ m^{-2}), respectively.

These results support those of Sinton *et al.* (1999) who identified slower inactivation of F-RNA phage throughout winter (both in sunlight and the dark).

Effect of water type

Figure 5 shows the variation of MS2 inactivation in clear (tap) waters and turbid wastewater. Inactivation was found to be 0.358 $\log_{10} h^{-1}$ slower in wastewater ($0.186 \pm 0.019 \log_{10} h^{-1}$) than in tap water ($0.544 \pm 0.214 \log_{10} h^{-1}$), suggesting inactivation efficiency was affected by water type. High turbidity and the presence of algae and particulate matter were probably responsible for the slower inactivation rates, with all capable of affecting light dispersion in the water column (Curtis *et al.* 1994). Davies-Colley *et al.* (1999) and Kohn & Nelson (2007) also reported lower F-RNA phage inactivation in WSP effluent than in reverse osmosis water. Inclusion of the IP continued to produce elevated inactivation in wastewater and tap water, with the $HRAP_d + SIP$ yielding an inactivation 1.5 and 1.2 times higher than the tap water and wastewater $HRAP_d$, respectively.

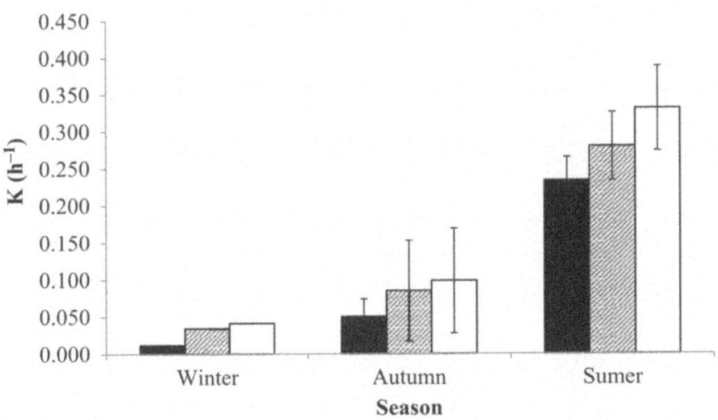

Figure 4 | Seasonal MS2 inactivation rates (K) recorded in the model systems; $HRAP_d$ (■), $HRAP_d + SIP$ (▨) and $HRAP_d + LIP$ (□). Solar radiation range was 9.9 ± 0.0 MJ m^{-2} (winter), 11.2 ± 8.1 MJ m^{-2} (autumn) and 27.9 ± 1.1 MJ m^{-2} (summer).

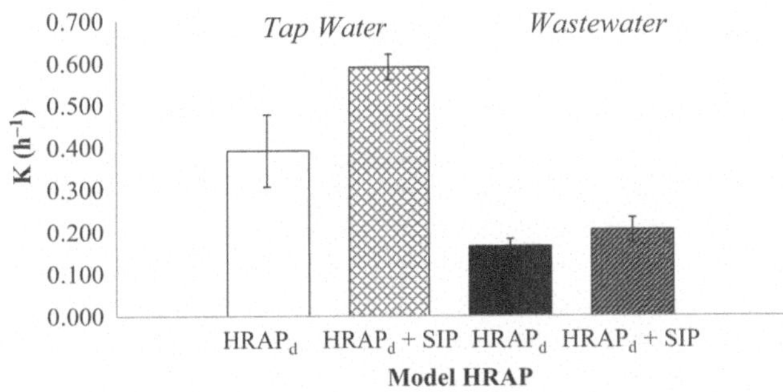

Figure 5 | MS2 inactivation rates (K; mean ± 1 standard deviation) for systems operated with either tap water or wastewater over 24 h. Solar irradiation ranged from 5.6 to 10.8 MJ m^{-2}. Mean water temperatures ranged from 10.8 to 28.1 °C in the tap water and 10.3 to 27.3 °C in the wastewater.

CONCLUSION

It is clear from the results that pathogen inactivation can be improved by the inclusion of an IP to a model HRAP. Inactivation increased by 63% following inclusion of the IP compared to the dark-incubated pond. HLR was identified as an important factor influencing inactivation rates. The improvement in inactivation rates following incorporation of an IP was less for turbid wastewaters compared to operation with optically clear tap waters. Additional work is still required with the aim of transferring the concept to a fully functioning field system. It would be beneficial to gain an understanding of the inactivation efficiency when both pond and IP are solar exposed and exposed to different seasonal variations. Multiple disciplines within the water industry would benefit from this research with higher quality effluent produced, safer for reuse via the management of the risk associated with exposure to pathogens.

ACKNOWLEDGEMENTS

We would like to acknowledge Flinders Medical Centre biomedical engineers for the construction of the model HRAPs and the Community Wastewater Management System operators at Mt Barker Wastewater Treatment Plant for their assistance and permission to collect wastewater samples when required.

REFERENCES

Araki, S., Martín-Gomez, S., Bécares, E., De Luis-Calabuig, E. & Rojo-Vazquez, F. 2001 Effect of high-rate algal ponds on viability of *Cryptosporidium parvum* oocysts. *Applied and Environmental Microbiology* **67** (7), 3322–3324.

Bevilacqua, A., Speranza, B., Sinigaglia, M. & Corbo, M. 2015 A focus on the death kinetics in predictive microbiology: benefits and limits of the most important models and some tools dealing with their application in foods. *Foods* **4** (4), 565–580.

Bolton, N. 2012 *Photoinactivation of Indicator Micro-Organisms and Adenovirus in Sunlit Waters*. PhD thesis, Health and the Environment Group, Flinders University, Adelaide, South Australia.

Buchanan, N., Cromar, N., Bolton, N. & Fallowfield, H. 2011 Comparison of a high rate algal pond with a standard secondary facultative waste stabilisation pond in rural South Australia. In: *10th Specialised Conference on Small Water and Wastewater Treatment Systems*, pp. 292–299.

Caslake, L. F., Connolly, D. J., Menon, V., Duncanson, C. M., Rojas, R. & Tavakoli, J. 2004 Disinfection of contaminated water by using solar irradiation. *Applied and Environmental Microbiology* **70** (2), 1145–1151.

Clancy, J., Bukhari, Z., Hargy, T., Bolton, J., Dussert, B. W. & Marshall, M. M. 2000 Using UV to inactivate *Cryptosporidium*. *Journal of the American Water Works Association* **92** (9), 97–104.

Curtis, T., Mara, D. D., Dixo, N. & Silva, S. A. 1994 Light penetration in waste stabilization ponds. *Water Research* **28** (5), 1031–1038.

Davies-Colley, R., Donnison, A., Speed, D., Ross, C. & Nagels, J. 1999 Inactivation of faecal indicator micro-organisms in waste stabilisation ponds: interactions of environmental factors with sunlight. *Water Research* **33** (5), 1220–1230.

Debartolomeis, J. & Cabelli, V. J. 1991 Evaluation of an *Escherichia coli* host strain for enumeration of F male-specific bacteriophages. *Applied and Environmental Microbiology* **57** (5), 1301–1305.

Duffie, J. A. & Beckman, W. A. 2013 *Solar Engineering of Thermal Processes*. Wiley, Hoboken, NJ, USA.

El Hamouri, B., Khallayoune, K., Bouzoubaa, K., Rhallabi, N. & Chalabi, M. 1994 High-rate algal pond performances in faecal coliforms and helminth egg removals. *Water Research* **28** (1), 171–174.

Fallowfield, H. & Garrett, M. 1985 The treatment of wastes by algal culture. *Journal of Applied Microbiology* **59** (s14), 187S–205S.

Fallowfield, H. J., Cromar, N. & Evison, L. 1996 Coliform die-off rate constants in a high rate algal pond and the effect of operational and environmental variables. *Water Science and Technology* **34** (11), 141–147.

Geeraerd, A., Herremans, C. & Van Impe, J. 2000 Structural model requirements to describe microbial inactivation during a mild heat treatment. *International Journal of Food Microbiology* **59** (3), 185–209.

Geeraerd, A., Valdramidis, V. & Van Impe, J. 2005 GInafit, a freeware tool to assess non-log-linear microbial survivor curves. *International Journal of Food Microbiology* **102** (1), 95–105.

Greenberg, A., Clesceri, L. & Eaton, A. 1992 *Standard Methods for the Examination of Water and Wastewater*. American Public Health Association/American Water Works Association/ Water Environment Federation, Washington, DC, USA.

Haag, W. R. & Hoigne, J. 1986 Singlet oxygen in surface waters. 3. Photochemical formation and steady-state concentrations in various types of waters. *Environmental Science & Technology* **20** (4), 341–348.

Iqbal, M. 1983 *An Introduction to Solar Radiation*. Academic Press, Toronto, ON, Canada.

Kirk, J. T. O. 1994 *Light and Photosynthesis in Aquatic Ecosystems*. Cambridge University Press, Cambridge, UK.

Kohn, T. & Nelson, K. L. 2007 Sunlight-mediated inactivation of MS2 coliphage via exogenous singlet oxygen produced by sensitizers in natural waters. *Environmental Science & Technology* **41** (1), 192–197.

Muela, A., García-Bringas, J., Seco, C., Arana, I. & Barcina, I. 2002 Participation of oxygen and role of exogenous and endogenous sensitizers in the photoinactivation of *Escherichia coli* by photosynthetically active radiation, UV-A and UV-B. *Microbial Ecology* **44** (4), 354–364.

Navntoft, L., Fernandez-Ibañez, P. & Garreta, F. 2012 UV solar radiation on a tilted and horizontal plane: analysis and comparison of 4 years of measurements. *Solar Energy* **86** (1), 307–318.

Noble, R. T., Lee, I. M. & Schiff, K. C. 2004 Inactivation of indicator micro-organisms from various sources of faecal contamination in seawater and freshwater. *Journal of Applied Microbiology* **96** (3), 464–472.

Park, J., Craggs, R. & Shilton, A. 2011 Wastewater treatment high rate algal ponds for biofuel production. *Bioresource Technology* **102** (1), 35–42.

Shilton, A. N. 2005 *Pond Treatment Technology*. IWA Publishing, London, UK.

Sinton, L. W., Finlay, R. K. & Lynch, P. A. 1999 Sunlight inactivation of fecal bacteriophages and bacteria in sewage-polluted seawater. *Applied and Environmental Microbiology* **65** (8), 3605–3613.

Sinton, L. W., Hall, C. H., Lynch, P. A. & Davies-Colley, R. J. 2002 Sunlight inactivation of fecal indicator bacteria and bacteriophages from waste stabilization pond effluent in fresh and saline waters. *Applied and Environmental Microbiology* **68** (3), 1122–1131.

First received 1 November 2017; accepted in revised form 1 June 2018. Available online 14 June 2018

Exploring the influence of meteorological conditions on the performance of a waste stabilization pond at high altitude with structural equation modeling

Long T. Ho, Duy T. Pham, Wout Van Echelpoel, Andres Alvarado, Juan E. Espinoza-Palacios, Maria B. Arevalo-Durazno and Peter L. M. Goethals

ABSTRACT

Algal photosynthesis plays a key role in the removal mechanisms of waste stabilization ponds (WSPs), which is indicated in the variations of three parameters, dissolved oxygen, pH, and chlorophyll *a*. These variations can be considerably affected by extreme climatic conditions at high altitude. To investigate these effects, three sampling campaigns were conducted in a high-altitude WSP in Cuenca (Ecuador). From the collected data, the first application of structure equation modeling (SEM) on a pond system was fitted to analyze the influence of high-altitude characteristics on pond performance, especially on the three indicators. Noticeably, air temperature appeared as the highest influencing factors as low temperature at high altitude can greatly decrease the growth rate of microorganisms. Strong wind and large diurnal variations of temperature, 7–20 °C, enhanced flow efficiency by improving mixing inside the ponds. Intense solar radiation brought both advantages and disadvantages as it boosted oxygen level during the day but promoted algal overgrowth causing oxygen depletion during the night. From these findings, the authors proposed insightful recommendations for future design, monitoring, and operation of high-altitude WSPs. Moreover, we also recommended SEM to pond engineers as an effective tool for better simulation of such complex systems like WSPs.

Key words | algal photosynthesis, high altitude, spatiotemporal variation, structure equation modeling, waste stabilization ponds

Long T. Ho (corresponding author)
Duy T. Pham
Wout Van Echelpoel
Juan E. Espinoza-Palacios
Maria B. Arevalo-Durazno
Peter L. M. Goethals
Department of Animal Sciences and Aquatic Ecology,
Ghent University,
Coupure Links 653, Ghent 9000,
Belgium
E-mail: *long.tuanho@ugent.be*

Andres Alvarado
Departamento de Recursos Hídricos y Ciencias Ambientales,
Universidad de Cuenca,
Av. 12 de Abril s/n, Cuenca,
Ecuador
and
Facultad de Ingeniería,
Universidad de Cuenca,
Av. 12 de Abril s/n, Cuenca,
Ecuador

INTRODUCTION

Waste stabilization ponds (WSPs) are large shallow lagoons surrounded by earth embankments, in which wastewater is purified by completely natural processes (Hosetti & Frost 1995; von Sperling 1996; Babu *et al.* 2010). Taking advantages of the photosynthetic aeration to reduce operation costs and potential risks from the emission of volatile organic compounds, WSPs are increasingly favored in numerous countries (Munoz *et al.* 2004; Dias *et al.* 2014). Indeed, due to their simplicity and low cost, WSPs are desirable in many developing countries (Mara 2004). However, contrary to their simple design and operation, processes happening inside the pond systems are complicated. These processes are characterized by complex interactions of bio-geochemical cycles, biochemical processes and reactions which are strongly dependent on climatic conditions and hydraulic patterns (Ho *et al.* 2017). Therefore, it is expected that the performance of WSPs can be greatly influenced by severe high-altitude meteorology, i.e. strong light intensity, low air temperature, great variation of temperature and low oxygen pressure (Juanico *et al.* 2000).

Until now, the number of studies concerning the efficiency of WSPs at high altitude remains low despite some reports on their poor effluent quality, e.g. Pearson *et al.* (1987b), Juanico *et al.* (2000) and Lloyd *et al.* (2002). More

doi: 10.2166/wst.2018.254

specifically, according to a study of Lloyd *et al.* (2002) on fourteen WSPs at the altitude of 2,675 m to 3,827 m in Mexico, adverse environmental conditions, such as large daily temperature variation from −4 °C to 30 °C and strong wind speeds, caused night-time thermal stratification and short-circuiting leading to the under-performance of these ponds. Indeed, these climatic conditions are also concluded as predominant controlling factors on the pond removal efficiency in the other two studies. However, since these studies only take into account the data from pond influent and effluent, their conclusions are relatively common and it is still unclear how high-altitude characteristics affect the pond performance. From that perspective, there is a lack of proper guidelines for pond designers and operators to deal with the distinctive features of high-altitude meteorology in order to ensure compliance with effluent quality standards (Ho *et al.* 2018). This paper aims at filling these gaps of knowledge. From three sampling campaigns thoroughly conducted in Ucubamba WSP system in Cuenca, Ecuador, an advanced multivariate data analysis, structural equation modeling (SEM), is applied to investigate the effects of climatic conditions on the efficiency of the pond system. More specifically, we focus on a distinguishing element of WSPs which differentiates it from conventional WWTPs, i.e. the involvement of algal photosynthesis. As such, the interaction between the high-altitude characteristics and the spatio-temporal variations of three indicators of algal photosynthesis (dissolved oxygen (DO), pH, and chlorophyll *a*) is meticulously analyzed. At the end, insightful recommendations on pond design, monitoring, and operation at high altitude are drawn.

MATERIALS AND METHODS

Study area

The WSP is located at Ucubamba (2°52′21″ S 78°56′30″ W) at altitude of 2,400 m a.s.l. and operated by the Municipal Company ETAPA. The largest WWTP in Ecuador is designed to treat the domestic effluent coming from more than 400,000 inhabitants in Cuenca. The average influent concentrations of organic matters and nutrients are biochemical oxygen demand (BOD_5) of 94 mg $O_2.L^{-1}$, chemical oxygen demand (COD) of 240.73 mg $O_2.L^{-1}$, total phosphorus (TP) of 5.18 mg $P.L^{-1}$, total Kjeldahl nitrogen (TKN) of 8.15 mg $N.L^{-1}$ (Alvarado 2013). The dry season is between June and December and the rainy season is between January and May with temperatures between 7 and 20 °C and 12 and 25 °C, respectively (Alvarado 2013). The WSP receives wastewater through a pre-treatment step including screening and grit chamber and, then, the wastewater is divided into two identical flow lines including an aerated pond (AP), a facultative pond (FP) and a maturation pond (MP) before being discharged into Cuenca river (see Figure 1). The total surface of the WSP is 45 ha with 11.5 days of theoretical hydraulic retention time (HRT), in which APs occupy 6 ha, FPs 26 ha, and MPs (Alvarado *et al.* 2012b).

Sampling campaigns

Three sampling events were implemented in the FPs and MPs of the two parallel lines on July 25–26, 2013, August 14–15, 2013 and August 26–27, 2013. To obtain

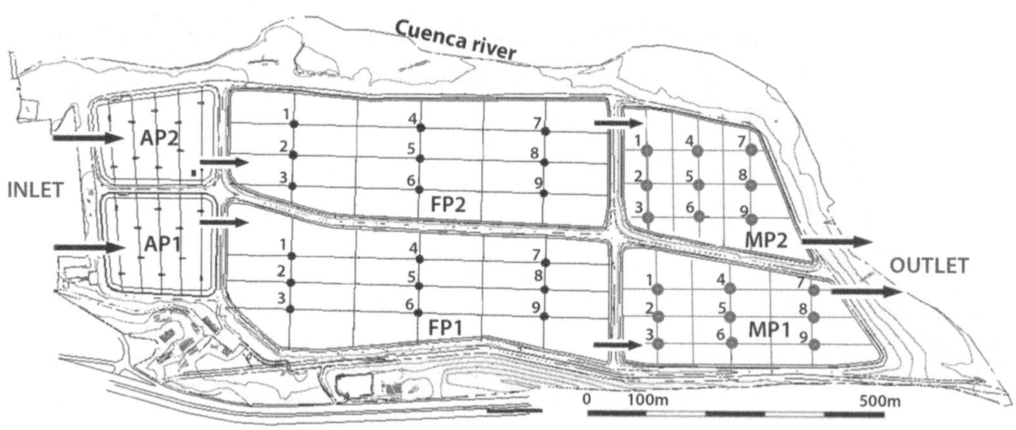

Figure 1 | Map of Ucubamba waste stabilization pond in Cuenca, Ecuador. Two parallel lines of WSP contain aerated ponds, facultative ponds, and maturation ponds. Grids were formed within the last two ponds in order to ensure that the collected samples are representative of variations along the ponds.

comprehensive insights into the spatial variation of water quality parameters, grids were formed by dividing FPs and MPs into six sections along the longitudinal direction and four sections breadthways. From the grids, samples were taken at nine different locations representing three zones of these ponds: influent (locations 1, 2, and 3), middle (locations 4, 5, and 6), and effluent (locations 7, 8, and 9) (Figure 1). In each location, samples were collected at two depths, 30 cm below the water surface and 15 cm above the sediment layer. Due to the sludge accumulation at location 1 and 2 in both FPs, only samples at 30 cm below the water surface were collected. Each sampling campaign lasted for two days as the sampling of one line required the period of one day, i.e. from 08.00 until 18.00. To investigate the diurnal dynamics of wastewater constituents, the samples of the bottom line were collected within a different period in the third campaign.

At each sampling location, temperature (°C), total dissolved solids (TSS, $g.L^{-1}$), pH (–), ammonium (mg NH_4^+-N. L^{-1}), chlorophyll *a* (μg *a*.L^{-1}), and DO (mg $O_2.L^{-1}$) were determined with two manual multi-probes, YSI 6600 V2 and YSI 6920 V1. These probes were carefully calibrated every three days by following their manual in order to ensure their accuracy. At the same time, the samples at three locations within one zone of the ponds were collected by Teledyne ISCO 6712 and then rigorously mixed. These homogeneous samples were analyzed for BOD_5 (mg O_2. L^{-1}), COD (mg $O_2.L^{-1}$), TP (mg $P.L^{-1}$), TKN (mg $N.L^{-1}$) and total solids (TS) (mg.L^{-1}) using the American Public Health Association methods (APHA 2005). Additionally, meteorological data, including air temperature (°C), solar radiation ($W.m^{-2}$) and wind speed ($m.s^{-1}$) were obtained from the Meteorological Station of CELEC Hidropaute located around 600 m away from the WSP.

Structural equation modeling and Kruskal-Wallis tests

One main objective of our research is to determine the key factors influencing pond performance, especially on the three indicators of algal photosynthesis, DO, pH, and chlorophyll *a*. To this end, SEM as an advanced statistical analysis tool for multivariate data was applied in R (Team 2010) using the lavaan package (Rosseel 2012). Principally, a hypothesized model was composed, including a theoretical covariance structure between random variables ($\sum^{(\theta)}$), which, subsequently, was compared to the covariance matrix of the actual data ($\sum$) (Sutton-Grier *et al.* 2010; Kline 2011). The agreement between these two covariance matrices was evaluated via the fundamental null hypothesis in SEM as follows (Equation (1)).

$$H_O: \sum = \sum(\theta) \tag{1}$$

where θ is a vector of all free parameters. Therefore, in the first step as model specification, we constructed a conceptual model or a path diagram as a pictorial demonstration of a system of simultaneous equations, based on the existing knowledge on the mechanisms of WSPs and our research questions. After model specification, the collected data were applied for model estimation by the most widely used fitting function, maximum likelihood (ML). After that, we followed three key aspects to evaluate the developed model, i.e. existing knowledge and research questions on the system, model fit indices, and model parsimony to evaluate the fitted model (Musil *et al.* 1998). Particularly, various fit indices were used to assess the goodness-of-fit and the parameter estimation of the hypothesized model. According to Hu & Bentler (1999), besides chi-square (χ^2) which was highly sensitive to sample size, other indicators should be applied to avoid the problems of sample size, i.e. comparative fit index (CFI), Tucker-Lewis index (TLI), root mean square error of approximation (RMSEA), and standardized root mean square residual (SRMR). Particularly, the recommended values were CFI and TLI >0.95, RMSEA between 0.05 and 0.08, and SRMR <0.08 to indicate a good model fit (Schumacker & Lomax 2015). The detailed formulas of these fit indices can be found in Hu & Bentler (1999). To achieve a parsimonious model which should not be more complicated than necessary for the description of the data (Omlin *et al.* 2001), the significance of simulated pathways among latent and observed variables was assessed via z statistic. Non-significant variables were removed in the final model, hence, allowing to identify the significant impacts of high-altitude characteristics on pond performance, especially on DO, pH, and chlorophyll *a*.

Together with SEM, a non-parametric statistical tool, Kruskal-Wallis test, was applied to determine the spatiotemporal differences of the measured wastewater constituents and climate conditions regarding the sampling time and the locations within a pond and among the ponds.

RESULTS

Spatial variations among and within the ponds

The spatial variations between different depths and ponds of the three parameters (DO, pH, and chlorophyll *a*) are demonstrated in Figure 2. The variations of other biochemical and meteorological parameters are shown in appendix 1 (available with the online version of this paper). Similar to the fluctuation of meteorological parameters, a wide variation of wastewater constituents was found among four ponds, which corresponded to very low *p*-values of Kruskal-Wallis tests in Table 1. Interestingly, even though the two flow lines were designed to be identical, their performances were highly diverse. The first line received around 20% higher pollutant concentrations, i.e. BOD, COD, TKN, TP, and TS, compared to the second line as demonstrated via low *p*-values of these pollutants between two FPs and two MPs in Table 1. Similarly, we also found a higher level of oxygen and pH in the first line compared to those in the second line, which was contrary to comparable concentrations of chlorophyll *a* between two lines. Between the surface and the bottom, the values of most biochemical parameters were relatively homogeneous, while DO, pH, and chlorophyll *a* were characterized by a high variability (*p*-values <0.01). As demonstrated in Figure 2, higher values in terms of both concentration and variation were found near the water surface. In contrast to the high variability of water quality parameters between two depths, pond zone had only a minor impact on the variations of biochemical parameters, except for oxygen level. Regarding the meteorological parameters, the climatic conditions intensively changed within a short period of time, which was confirmed by low *p*-values in all the cases in Table 1. Two exceptions were the values between two depths which were measured subsequently after each other and among the three campaigns where their similar ranges were observed.

For a further investigation on the spatial variation of DO, pH, and chlorophyll *a*, their concentrations and deviations in three different zones of the pond system are

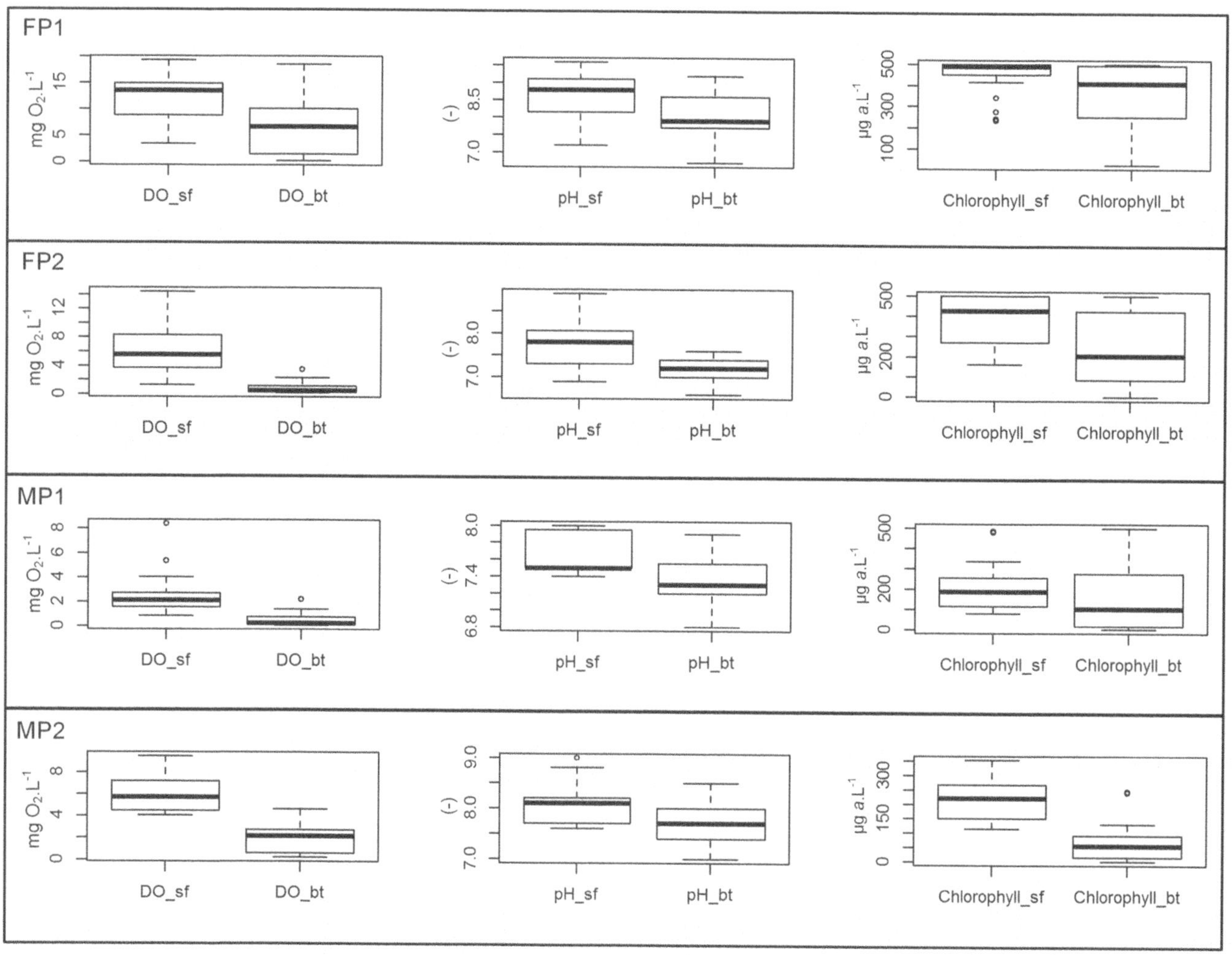

Figure 2 | Variations of DO, pH, and chlorophyll *a* between different depths and ponds. Sf: surface; Bt: bottom.

Table 1 | Summary of the *p*-values of the Kruskal-Wallis tests

Parameters	Different ponds	Between two FPs	Between two MPs	Between depths	Different zones	Different daytimes	Different campaigns
DO	<0.001*	<0.001*	<0.001*	<0.001*	<0.001*	<0.001*	<0.001*
Chlorophyll *a*	<0.001*	0.0531	0.1824	<0.001*	0.7840	0.2049	<0.001*
pH	<0.001*	<0.001*	<0.001*	<0.001*	0.0914	<0.001*	<0.001*
BOD	<0.001*	<0.001*	<0.001*	0.0381*	0.6457	<0.001*	<0.001*
COD	<0.001*	<0.001*	<0.001*	0.0139*	0.7384	<0.001*	0.1642
TKN	<0.001*	<0.001*	<0.001*	0.3259	0.6191	<0.001*	<0.001*
TP	<0.001*	0.0107*	0.1353	0.7974	0.1315	<0.001*	<0.001*
TS	<0.001*	0.0003*	0.203	0.2006	0.3253	<0.001*	<0.001*
Solar radiation	<0.001*	0.5286	<0.001*	0.9064	<0.001*	<0.001*	0.3846
Air temperature	<0.001*	<0.001*	<0.001*	0.6345	<0.001*	<0.001*	0.8809
Wind speed	<0.001*	<0.001*	<0.001*	0.8030	<0.001*	<0.001*	0.7822

*: *p*-value ≤0.05 (threshold of illustrating a significant difference).

illustrated in Figure 3. As agreed with the results for the Kruskal-Wallis tests, the concentrations of the three parameters were homogeneous within one pond, but very different among them and between two depths. More specifically, despite having higher level of DO and pH in FP1 compared to FP2, these values of its consecutive pond (MP1) were lower than MP2. In fact, the oxygen level at the water surface dropped from 10 mg $O_2.L^{-1}$ in FP1 to around 3 mg $O_2.L^{-1}$ in MP1 while there was a slight increase in the upper line. This trend also occurred at the bottom layers of the ponds. Interestingly, the concentrations of chlorophyll *a* were relatively comparable between two lines and these values decreased by around 50% from the FPs to the MPs in both lines. This finding is in line with the study of Pham *et al.* (2014) on algal biodiversity of this pond system, which suggested that the algal abundance in the FPs was lower than in the MPs. It is also noteworthy that the standard deviations of DO and pH were much higher than those of chlorophyll *a*, which could be associated with stronger impact of climatic conditions on DO and pH leading to more fluctuation of their values.

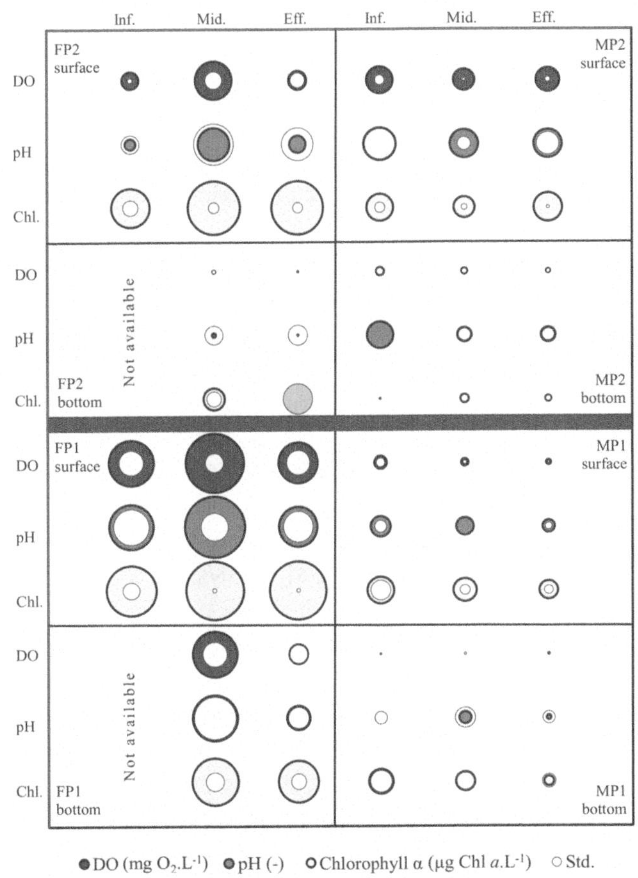

Figure 3 | DO, pH, and chlorophyll *a* (chl.) with their standard deviation (Std.) at the three zones of the ponds, influent (inf.), middle (mid.), and effluent (eff.). The order of the ponds in the graph is analogous to the real system, where the top four boxes correspond to flow line two and the bottom four boxes to flow line one. The normalized values of DO, pH, and chlorophyll *a* were calculated by subtracting the minimum value and then dividing by the difference between the maximum and minimum values (Aksoy & Haralick 2001). Due to the sludge accumulation, the values of the influent zone at the bottom of FPs were not available.

Temporal variations

Figure 4 demonstrates relatively analogous patterns of diurnal fluctuation of DO and pH, in which they were lowest at the early time of the day, increased during the day, and peaked from 13.00 to 15.00 when light intensity reached its maximum. These temporal fluctuations were in contrast with the static level of chlorophyll *a*. This difference was relevant with the obtained *p*-values of the three parameters regarding different daytimes in Table 1. During high

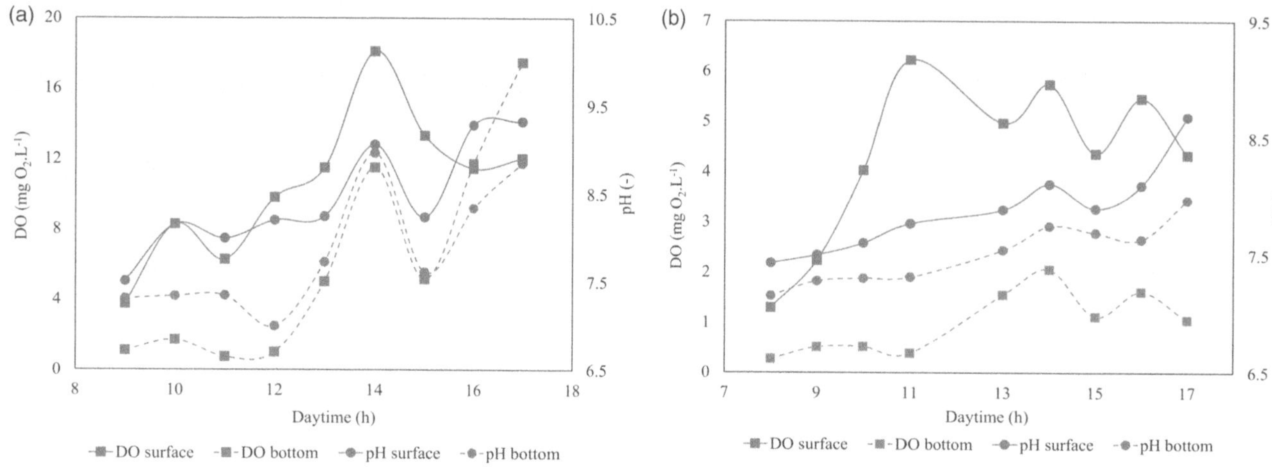

Figure 4 | Diurnal variations of DO and pH between two depths of FPs (a) and MPs (b).

light-intensity period, average oxygen concentrations and pH could reach above 20 mg $O_2.L^{-1}$ and 9 at the surface of the FPs while such high peaks were not found in the MPs. Much less expected is the extremely high concentration of oxygen at the bottom of FP1 around 16 mg $O_2.L^{-1}$ during the last 2 h of the afternoon in the first sampling campaign. pH at the surface of the ponds during the late afternoon remained at high values, around 9 for FPs and 8.5 for MPs, which was generally regarded as beneficial for the inactivation of fecal microorganisms and ammonia volatilization (Davies-Colley *et al.*, 1999). Regarding other water quality parameters and meteorological conditions, as indicated in the results of the Kruskal-Wallis tests, daytime also contributed significantly on their variations, which proved the crucial role of temporal aspects on the variations of the performance of the pond system.

SEM conceptual model and path analysis

The specification of the conceptual SEM model was composed, based on existing knowledge in literature and our research questions about the meteorological effects on pond performance, especially on the three parameters, DO, pH, and chlorophyll *a*. Four latent variables were chosen, i.e. meteorology, water pollution, time, and space, which included 14 observed variables from the sampling campaigns. The effects of time and space as two latent exogenous variables (ξ) was further investigated on not only the three parameters but also other two latent endogenous variables (η), meteorological conditions and water pollutants. Moreover, we also evaluated the direct correlations among the three parameters as well as the indirect effects of daytime on DO via meteorological conditions, i.e. wind speed, air temperature, and solar radiation. By allowing the estimation of these direct and indirect effects of the latent and manifest variables, the SEM model was able to provide a comprehensive overview of the interconnected web of biochemical processes and reactions happening inside the pond system. Indeed, thanks to its ability to simultaneously test theoretical relations among multiple variables, the complex interactions of the in-pond bio-geochemical cycles and meteorological characteristics were illustrated in Figure 5. This conceptual model can hence be applied to other WSPs systems as the first step of model specification.

Subsequently, the conceptual SEM model was fitted on the collected data, leading to a simpler model in which all non-significant pathways were removed (Figure 6). Particularly, the effect of water pollution on the three parameters was not significant (p-values >0.05), hence, none of its observed parameters is present in the fitted SEM model. The reason could be because of high oxygen concentrations as a result of the enhanced algal photosynthesis which was accelerated by strong solar radiation in this meridional high-altitude WSP system. Compared to this high production of DO, the amount of oxygen, which was consumed for bacterial mineralization and nitrification process, appeared to be inconsiderable. Moreover, nutrient concentrations in the influent also exceeded their required amount for the growth of algae, hence, their impact on chlorophyll *a* is not significant. Conversely, the latent variable, space, including depth and pond type, appeared to be greatly important for the variation of the three parameters. Pond location, however, was evaluated as a non-significant parameter which was relevant to the high p-values of Kruskal-Wallis tests (Table 1). Sampling campaign was also removed as a non-significant

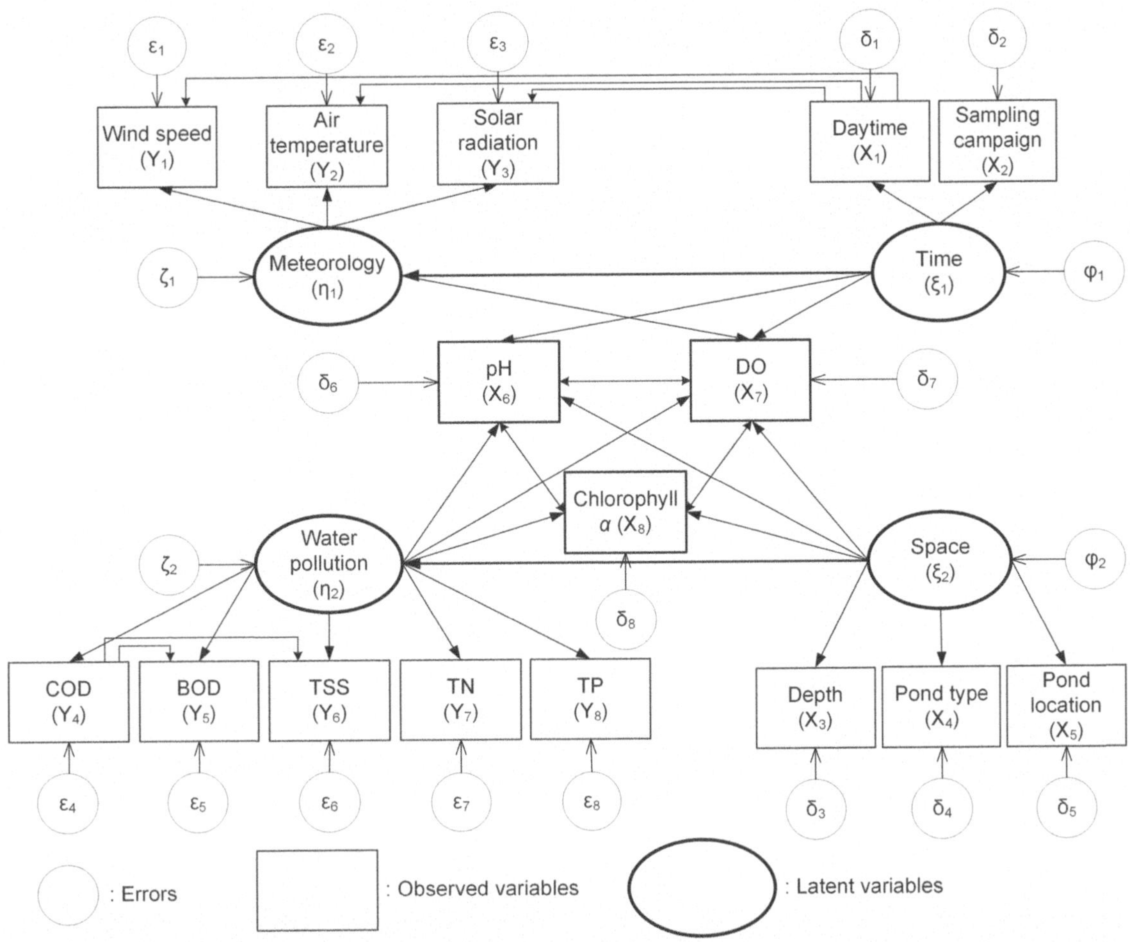

Figure 5 | Conceptual structural equation model (SEM) of a WSP. This SEM focuses on the influence of high-altitude characteristics on pond behavior, especially on the three parameters linked with algal photosynthesis, i.e. DO, pH, and chlorophyll *a*.

variable (p-value $= 0.136$), while daytime significantly affected the oxygen level via its impacts on meteorological conditions in both direct and indirect pathways.

From the path analysis of SEM, the space exerted the major impact on DO and pH with its high path coefficients of -14.898 and 0.196, respectively. Daytime was one of the most significant contributors to DO variations with its direct path coefficient of 2.509 and its significant indirect effect via meteorological parameters. These substantial impacts of space and daytime on the oxygen level were contradicted with non-significant effects of the water pollutants (p-values >0.05). Among three observed meteorological parameters, air temperature had the greatest influence on the oxygen level while solar radiation played the least important role, which is relevant with the degree of the influence of daytime on these meteorological parameters.

Regarding the performance of the SEM, a good agreement was found between the model-implied covariance matrix and the sample covariance matrix of the collected data. According to the guide of Schumacker & Lomax (2015) for model-fit criteria, our SEM model has most of its fit indices in the acceptable range for good fit model, i.e. CFI: 0.982, TLI: 0.969, RMSEA: 0.074, and SRMR: 0.087. With respect to model parsimony, after fitting on the collected data, the final model became simpler than the conceptual structure, which clearly highlighted the crucial roles of meteorology, as well as time and space in the variations of the three parameters in the pond system.

DISCUSSION

Spatial variations among and within the ponds

WSPs are normally subdivided into two parallel lines so that both lines are able to have similar treatment performance.

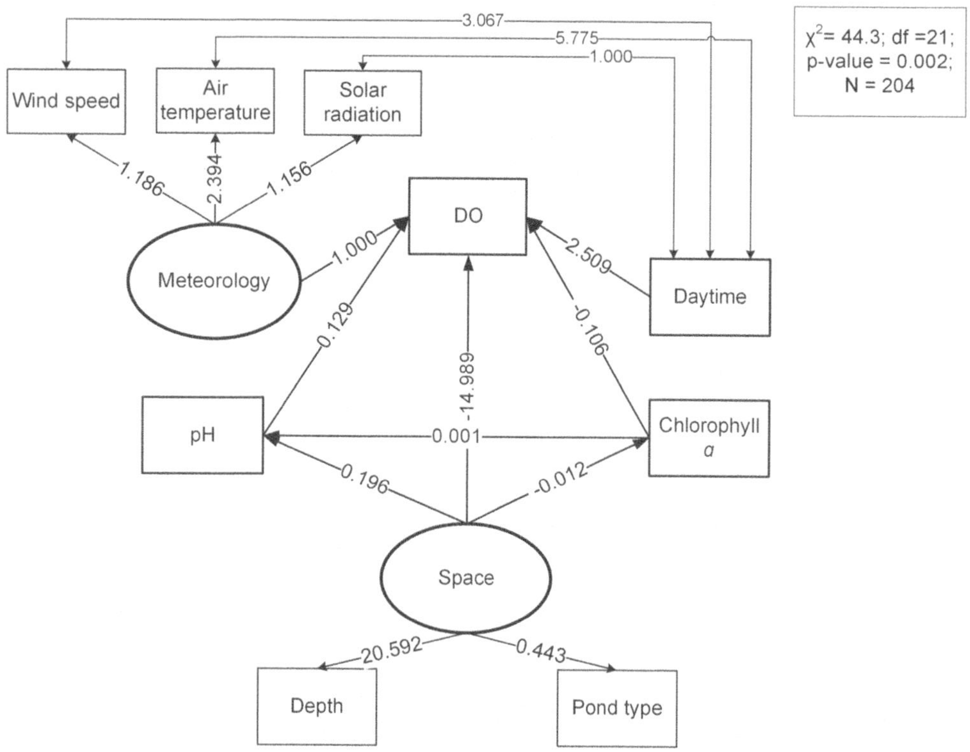

Figure 6 | Simple SEM model for Ucubamba WSP with only significant pathways among the latent and observed variables. χ^2: chi-squared test values, df: the degree of freedom, N: sample size.

However, the two identical flow lines of Ucubamba WSP performed differently, in which FP1 received higher pollutant loadings than FP2, especially organic matter. Indeed, their average surface organic loadings were up to 250 and 185 $kg.ha^{-1}.d^{-1}$, respectively, while the recommended limitations were 240 $kg.ha^{-1}.d^{-1}$ for WSPs at tropical and subtropical regions and only 200 $kg.ha^{-1}.d^{-1}$ for WSPs at altitudes above 2,400 m a.s.l. (Juanico *et al.* 2000; Ho *et al.* 2017). These high loadings were associated with the sludge accumulation, which was also the reason for unavailable data at the bottom of location 1 and 2 in FPs. According to the study of Alvarado *et al.* (2012a) in this pond system, the sludge volume of the FPs reached up to 34% of the pond volume which substantially reduced its active volume. More importantly, low temperature at high altitude from 7 to 20°C during our sampling campaigns could significantly reduce the growth rate of microorganisms involved in biological wastewater treatment (Zimmo *et al.* 2004). A much higher temperature was recommended for the optimum growth of nitrifying bacteria, i.e. about 35 °C, being similar to the recommended value for heterotrophic bacteria collected from WSPs (Mayo & Noike 1996). Indeed, air temperature was weighed as the most influential meteorological parameter on the fluctuation of the oxygen level in the SEM model.

Between the two depths, there was a wide disparity in the concentrations of DO, pH, and chlorophyll *a* caused by the algal stratification while pollutant concentrations were relatively homogeneous. This difference is also demonstrated in the fitted SEM as depth is the most influential factor with respect to space on DO, pH, and chlorophyll *a* while pollutants appear to be unaffected. This homogeneity of wastewater constituents was a result of sufficient mixing which can be influenced by the intensive temperature variation at high altitude, from 7 to 20°C, and strong wind. Wind speed also appeared in the fitted SEM model as the second most influential meteorological parameter affecting pond performance, especially the oxygen level. In the model, its positive coefficients, 1.186 and 3.067 for the direct and indirect effects, respectively, reveal its beneficial role on enhancing the oxygen level in the ponds via promoting the mixing efficiency, hence increasing oxygen reaeration rate. Note that the location of these ponds with negligible protection of surrounding tree encourages more exposure to wind turbulence from downstream to upstream Cuenca river, facilitating the mixing conditions. Indeed, the interaction between this external factors and pond configurations, such as design of input and output, can affect greatly the hydraulic performance of WSP systems (Passos *et al.* 2016). The vertical stratification of DO and pH were, however, induced by the movement of algal layer approaching the water surface in response to the light limitation caused by the water turbidity. This turbid condition was also demonstrated via a high TSS/BOD ratio in the FPs, around 5, while 1.5 is the recommended value from EPA (2011) for properly operated FPs. This high concentration of suspended solid can be caused by the overgrowth of algae and the loss of old sludge particles (EPA 2011). Strong solar radiation at high altitude and abundant nutrients in these ponds can promote algal growth as high algal biomass was recorded in our sampling campaigns, above 420 μg $a.L^{-1}$ near the surface of the FPs. Old sludge age can be another reason for the high concentration of TSS.

Temporal variations

In Cuenca, daylight lasted from 06.00 until 18.00 with the peak of solar radiation between 13.00 and 15.00 when the maximum rate of algal photosynthesis and the highest DO and pH values occurred. The comparable variation of DO and pH in Figure 4 can be explained by the carbonate bicarbonate buffering system in WSPs. More specifically, CO_2 consumption via algal photosynthetic activity causes the increase of pH during the day while during the night or light limited conditions, respiratory processes generate CO_2, causing the drop of pH and oxygen level. More importantly, during this peak period, an extremely high oxygen level was recorded, more than 20 mg $O_2.L^{-1}$ in our sampling campaigns and up to 39 mg $O_2.L^{-1}$ in the sampling campaign of Alvarado (2013) on the same system. These abnormally high oxygen levels can be induced by immense light intensity, up to 1,500 $W.m^{-2}$, high algal biomass promoting the photosynthesis rate and low air temperature increasing gas solubility in water. Compared to Lake Van (Turkey) located at 1,648 m a.s.l. with 907 $W.m^{-2}$ of the maximum solar radiation, the maximum oxygen concentration was merely above 10 mg $O_2.L^{-1}$ in (Duzen & Aydin 2012; Stockhecke *et al.* 2012). Other climate conditions, such as rainfall, can considerably affect the oxygen level as a heavy rain with high wind speed, above 5 $m.s^{-1}$, was recorded in the first sampling campaign causing abnormally high oxygen concentrations at the bottom layers, up to 16 mg $O_2.L^{-1}$, in the last 2 h of the afternoon of the first sampling campaign. As such, additional sampling campaigns in rainy season are required for further study on the impact of rainfall or seasonal changes in other environmental conditions on WSP systems.

SEM evaluation

In contrast to the popularity of SEM in many research areas, such as sociology, psychology, medicine, political science, and economics, the application of SEM on ecological and environmental sciences remains limited (Arhonditsis *et al.* 2006). To the authors' knowledge, this is the first time SEM models are applied on a pond treatment system. Normally, when encountering with the complex interactions of multifaceted factors in ecological systems, ecological engineers are inclined to give away some degrees of generality for better realism and accuracy (Vepsalainen & Spence 2000). As such, regression models, as a simple description of collected data from laboratory experiments or sampling campaigns, are frequently a favorite choice of pond engineers. For instance, two well-known regression equations, Mara (1987) and Gloyna *et al.* (1976), are widely applied for designing and operating numerous pond systems. While useful for having a general idea of the performance of WSPs, these equations offer little insight on the complex interconnections between abiotic and biotic parameters within the pond system, which can cause many difficulties for engineers to extrapolate them on a global scale. As such, fourteen high altitude WSPs in Mexico experienced poor quality effluent because large daily temperature variation from −4 °C to 30 °C and strong wind speed caused night-time thermal stratification and short-circuiting at high altitude, which were probably omitted in their designs (Lloyd *et al.* 2002).

On the other hand, a SEM model of a complex pathway structure is capable of reflecting both direct and indirect interrelationships between observed variables and hypothesized variables in the presence of measurement errors and unexplained variations (Malaeb *et al.* 2000). Thanks to its ability to simultaneously evaluate multiple causal relationships, the mechanistic understanding of hierarchical ecological structure in different time and space scales can be illustrated in SEM models. Indeed, in our case, the conceptual SEM model accurately reflects the interactions among not only multiple inputs but also many kinds of inputs, i.e. biochemical, microbial, meteorological and spatiotemporal data. As such, unlike other descriptive models with limited ability for global extrapolation, pond engineers can apply this conceptual SEM model to describe the causal mechanisms inside other pond systems. After fitting the data, this SEM demonstrates the essential effects of meteorology on the performance of the pond treatment system via the spatiotemporal variations of the three parameters. The fitted SEM also proves its high goodness-of-fit via different evaluation criteria in the acceptable range for a good fit and parsimonious model. However, few limitations of the SEM model should be noticed. For instance, the effects of configuration characteristics related to hydraulic performance, such as the position of pond inlet and outlet or dead zones, on the flow efficiency as well as the pond performance, are not able to be simulated in SEM models. From this perspective, mechanistic models, which can include both hydraulic processes and biochemical reactions, can be a viable alternative.

Insights for pond design, monitoring, and operation at high altitude

The results from the fitted SEM model indicate the predominant role of climatic conditions at high altitude on the performance of the pond system, in which air temperature contributes the most among meteorological parameters. This confirms our expectations, as low temperature constrains the growth of biological community in the pond system which may reduce the removal efficiencies of treatment plants. Hence, in order to ensure compliance with the wastewater discharge standards, an extra safety factor of pond designs at high altitude should be considered. As such, longer HRT or a buffer system may be required at high altitude. Besides, wind speed is also evaluated as an essential affecting factor on pond efficiency, suggesting that engineers and plant owners should keep in mind its interactions with other parameters, such as inlet/outlet configurations, pond geometry, and hydraulic regimes in their designs. For this, a spatially explicit dynamic model, such as computational fluid dynamics (CFD) model, and more detailed hydraulic assessment are needed (Alvarado *et al.* 2012a; Passos *et al.* 2016).

Due to the strong solar radiation, the possibility of algal overgrowth at high altitude is much higher, especially in dry season as in our case. High algal biomass, above 420 μg *a*.L^{-1} near the surface of the FPs, was recorded in our sampling campaigns. Similarly, Pearson *et al.* (1987b) also found an extremely high level of chlorophyll *a*, up to 1,500 μg *a*.L^{-1}, in a high-altitude WSP in Mexico. This algal overgrowth generates the supersaturated DO condition during the day but, on the other hand, depletes the oxygen level due to their respiration during the night or light-limited conditions (EPA 2011). As a result, there can be a vast fluctuation of the pond performance at high altitude between early morning and mid-afternoon. Indeed, from the results of the Kruskal-Wallis tests and the SEM model, daytime proves its significant impacts, both direct and indirect, on the behavior of the pond system, especially on the three parameters related to algal metabolism, DO, pH, and chlorophyll *a*. To minimize these temporal influences and maintain a stable effluent quality, an extra aeration can be applied during the night for the oxygen consumption of aerobic heterotrophs to mineralize soluble organic matters and nutrients. However, in order to ensure a cost effective trade-off of this promising solution, optimal mixing performance with minimum number of aerators should be investigated via applying CFD models which were successfully employed by Alvarado *et al.* (2013) in the aerated ponds where the number of aerators reduced by factors of 2.5, without affecting the pond hydraulic behavior.

Pond operators should also take extra care on monitoring the three parameters and TSS/BOD ratio to control the algal biomass. More specifically, pH should be larger than 6.8 and chlorophyll *a* should be larger than 300 μg *a*.L^{-1} to avoid turning anoxic while measures should be taken to avoid the oxygen level dropping to zero during the night (König 1984; EPA 2011). Furthermore, it is important to keep in mind the spatial effect on the three parameters, which is clearly demonstrated in the high path coefficients between space and the three parameters in the final SEM model and extremely low *p*-values in the Kruskal-Wallis tests. These spatiotemporal variations suggest a need for different monitoring strategy for the three parameters which, according to previous guidelines for pond performance evaluation, i.e. Pearson *et al.* (1987a), Mara & Pearson (1998), and Shilton (2005), can be the same as other parameters. Particularly, instead of monitoring only the influent of pond system and the effluent of each pond, the determination of the three parameters should be measured daily at atleast at two depths, i.e. (i) 20–30 cm from the surface where the photosynthesis occurs and (ii) near the bottom. To evaluate the hydraulic efficiency, their samples should be collected monthly in different locations, as indicated in our sampling campaigns. Furthermore, we recommend that the samples of DO and pH should be collected daily at sunrise and again in midafternoon, to monitor algal respiration causing the oxygen depletion. From that, the necessity of extra aeration for compensation of oxygen depletion can be evaluated. Overgrowth of algae can be indicated by the high ratio between TSS/BOD which also suggests the problem of old sludge age as a result of high sludge accumulation. Since excessive solids buildup in the bottoms of ponds considerably reduce the active volume of the FPs, a regular schedule for physical solid removal is required. It is also important to keep in mind the seasonal effect on pond performance as we observed the great influence of heavy rain in combination with strong wind on the hydraulic performance of the two FPs.

CONCLUSIONS

An advanced multivariate data analysis, SEM model, is applied on the data collected from the three meticulous sampling campaigns in Ucubamba WSP to investigate the effect of severe meteorological conditions at high altitude on the pond performance. From the output of this first application of SEM on WSPs, it is concluded that low temperature is the main reason for an extra safety factor that should be taken into account in the designs of high-altitude ponds. Similarly, high wind speed can create a significant impact on flow efficiency; however, a further investigation of its interactions with other parameters, such as inlet/outlet configurations, pond geometry, and hydraulic regimes, is needed. Strong solar radiation can be both beneficial and detrimental to pond removal efficiency as it boosts the oxygen level during the day but also promotes algal overgrowth causing oxygen depletion during the night. From that perspective, extra care should be given to controlling the algal biomass via the three indicators, DO, pH, and chlorophyll *a*. More specifically, daily sampling at two different depths in early morning and mid-afternoon and monthly sampling at different pond locations as indicated in our campaigns are necessary for better evaluation of pond performance. Pond operators at high altitude should also record daily the variations of climatic conditions in terms of light intensity, cloudiness, precipitation and especially air temperature since extra measures are highly likely to be needed in unusual conditions at high altitude. More importantly, being able to illustrate the complex mechanisms of WSPs in different time and space scales, SEM models are recommended as an effective tool for better understanding and simulation to pond engineers and researchers.

ACKNOWLEDGEMENTS

This research was performed in the context of the VLIR Ecuador Biodiversity Network project. This project was funded by the Vlaamse Interuniversitaire Raad-Universitaire Ontwikkelingssamenwerking (VLIR-UOS), which supports partnerships between universities and university colleges in Flanders and the South. We are grateful to ETAPA for allowing us to use their facilities and the WSP to perform this research. We thank Mariska Barendse from Department of Data Analysis, Ghent University, for her support on the development of the SEM model. We thank three anonymous reviewers for their careful reading of our manuscript and their many insightful comments and suggestions. Long Ho is supported by the special research fund (BOF) of Ghent University. Duy Tan Pham is supported by a PhD grant of the Vietnamese government.

DECLARATION OF INTEREST

Conflicts of interest: none

REFERENCES

Aksoy, S. & Haralick, R. M. 2001 Feature normalization and likelihood-based similarity measures for image retrieval. *Pattern Recognition Letters* **22**, 563–582.

Alvarado, A. 2013 *Advanced Dynamic Modelling of Wastewater Treatment Ponds*. PhD Thesis, Ghent University, Ghent, Belgium.

Alvarado, A., Sanchez, E., Durazno, G., Vesvikar, M. & Nopens, I. 2012a CFD analysis of sludge accumulation and hydraulic performance of a waste stabilization pond. *Water Science and Technology* **66**, 2370–2377.

Alvarado, A., Vedantam, S., Goethals, P. & Nopens, I. 2012b A compartmental model to describe hydraulics in a full-scale waste stabilization pond. *Water Research* **46**, 521–530.

Alvarado, A., Vesvikar, M., Cisneros, J. F., Maere, T., Goethals, P. & Nopens, I. 2013 CFD study to determine the optimal configuration of aerators in a full-scale waste stabilization pond. *Water Research* **47**, 4528–4537.

APHA 2005 *Standard Methods for the Examination of Water and Wastewater*. American Public Health Association (APHA), Washington, DC, USA.

Arhonditsis, G. B., Stow, C. A., Steinberg, L. J., Kenney, M. A., Lathrop, R. C., McBride, S. J. & Reckhow, K. H. 2006 Exploring ecological patterns with structural equation modeling and Bayesian analysis. *Ecological Modelling* **192**, 385–409.

Babu, M. A., Hes, E. M. A., Van Der Steen, N. P., Hooijmans, C. M. & Gijzen, H. J. 2010 Nitrification rates of algal-bacterial biofilms in wastewater stabilization ponds under light and dark conditions. *Ecological Engineering* **36**, 1741–1746.

Davies-Colley, R. J., Donnison, A. M., Speed, D. J., Ross, C. M. & Nagels, J. W. Inactivation of faecal indicator microorganisms in waste stabilisation ponds: Interactions of environmental factors with sunlight. *Water Res.* **1999** (33), 1220–1230.

Dias, D. F. C., Possmoser-Nascimento, T. E., Rodrigues, V. a. J. & Von Sperling, M. 2014 Overall performance evaluation of shallow maturation ponds in series treating UASB reactor effluent: ten years of intensive monitoring of a system in Brazil. *Ecological Engineering* **71**, 206–214.

Duzen, H. & Aydin, H. 2012 Sunshine-based estimation of global solar radiation on horizontal surface at Lake Van region (Turkey). *Energy Conversion and Management* **58**, 35–46.

Epa, U. 2011 *Principles of Design and Operations of Wastewater Treatment Pond Systems for Plant Operators, Engineers, and Managers*. United States Environmental Protection Agency, Office of Research and Development, Cincinnati, OH, USA.

Gloyna, E. F., Malina, J. F. & Davis, E. M. 1976 *Ponds as A Wastewater Treatment Alternative*. Center for Research in Water Resources, College of Engineering, University of Texas at Austin, University of Texas Press, Austin, TX, USA.

Ho, L. T., Van Echelpoel, W. & Goethals, P. L. M. 2017 Design of waste stabilization pond systems: a review. *Water Research* **123**, 236–248.

Ho, L., Pham, D., Van Echelpoel, W., Muchene, L., Shkedy, Z., Alvarado, A., Espinoza-Palacios, J., Arevalo-Durazno, M., Thas, O. & Goethals, P. 2018 A closer look on spatiotemporal variations of dissolved oxygen in waste stabilization ponds using mixed models. *Water* **10**, 201.

Hosetti, B. B. & Frost, S. 1995 A review of the sustainable value of effluents and sludges from wastewater stabilization ponds. *Ecological Engineering* **5**, 421–431.

Hu, L. T. & Bentler, P. M. 1999 Cutoff criteria for fit indexes in covariance structure analysis: conventional criteria versus new alternatives. *Structural Equation Modeling-A Multidisciplinary Journal* **6**, 1–55.

Juanico, M., Weinberg, H. & Soto, N. 2000 Process design of waste stabilization ponds at high altitude in Bolivia. *Water Science and Technology* **42**, 307–313.

Kline, R. B. 2011 *Principles and Practice of Structural Equation Modeling*. Guilford Press Guilford Publications, New York, NY, USA.

König, A. 1984 *Ecophysiological Studies on Some Algae and Bacteria of Waste Stabilization Ponds. Doctoral dissertation*, University of Liverpool, Liverpool, UK.

Lloyd, B. J., Leitner, A. R., Vorkas, C. A. & Guganesharajah, R. K. 2002 Under-performance evaluation and rehabilitation strategy for waste stabilization ponds in Mexico. *Water Science and Technology* **48**, 35–43.

Malaeb, Z. A., Summers, J. K. & Pugesek, B. H. 2000 Using structural equation modeling to investigate relationships among ecological variables. *Environmental and Ecological Statistics* **7**, 93–111.

Mara, D. D. 1987 Waste stabilization ponds: problems and controversies. *Water Quality International.*

Mara, D. D. 2004 *Domestic Wastewater Treatment in Developing Countries*. Earthscan Publications, Routledge, Abingdon, UK.

Mara, D. D. & Pearson, H. W. 1998 Waste stabilization ponds: design manual for Mediterranean Europe. In: D. D. Mara & H. W. Pearson (eds), *Waste Stabilization Ponds: Design Manual for Mediterranean Europe*. World Health Organization. Regional Office for Europe, Copenhagen, Denmark.

Mayo, A. W. & Noike, T. 1996 Effects of temperature and pH on the growth of heterotrophic bacteria in waste stabilization ponds. *Water Research* **30**, 447–455.

Munoz, R., Kollner, C., Guieysse, B. & Mattiasson, B. 2004 Photosynthetically oxygenated salicylate biodegradation in a continuous stirred tank photobioreactor. *Biotechnology and Bioengineering* **87**, 797–803.

Musil, C. M., Jones, S. L. & Warner, C. D. 1998 Structural equation modeling and its relationship to multiple regression and factor analysis. *Research in Nursing & Health* **21**, 271–281.

Omlin, M., Brun, R. & Reichert, P. 2001 Biogeochemical model of Lake Zurich: sensitivity, identifiability and uncertainty analysis. *Ecological Modelling* **141**, 105–123.

Passos, R. G., Dias, D. F. C. & Von Sperling, M. 2016 Review of practical aspects for modelling of stabilization ponds using Computational Fluid Dynamics. *Environmental Technology Reviews* **5**, 78–102.

Pearson, H. W., Mara, D. D. & Bartone, C. R. 1987a Guidelines for the minimum evaluation of the performance of full-scale waste stabilization pond systems. *Water Research* **21**, 1067–1075.

Pearson, H. W., Mara, D. D., Thompson, W. & Maber, S. P. 1987b Studies on high-altitude waste stabilization ponds in Peru. *Water Science and Technology* **19**, 349–353.

Pham, D. T., Everaert, G., Janssens, N., Alvarado, A., Nopens, I. & Goethals, P. L. M. 2014 Algal community analysis in a waste stabilisation pond. *Ecological Engineering* **73**, 302–306.

Rosseel, Y. 2012 Lavaan: an R package for structural equation modeling. *Journal of Statistical Software* **48**, 1–36.

Schumacker, R. E. & Lomax, R. G. 2015 *A Beginner's Guide to Structural Equation Modeling: Fourth Edition*. Taylor & Francis, Routledge, New York, NY, USA.

Shilton, A. 2005 *Pond Treatment Technology*. IWA Publishing, London, UK.

Stockhecke, M., Anselmetti, F. S., Meydan, A. F., Odermatt, D. & Sturm, M. 2012 The annual particle cycle in Lake Van (Turkey). *Palaeogeography Palaeoclimatology Palaeoecology* **333**, 148–159.

Sutton-Grier, A. E., Kenney, M. A. & Richardson, C. J. 2010 Examining the relationship between ecosystem structure and function using structural equation modelling: a case study examining denitrification potential in restored wetland soils. *Ecological Modelling* **221**, 761–768.

Team, R. C. 2010 *R: A Language and Environment for Statistical Computing* (R. D. C. Team, ed.). Vienna, Austria.

Vepsalainen, K. & Spence, J. R. 2000 Generalization in ecology and evolutionary biology: from hypothesis to paradigm. *Biology & Philosophy* **15**, 211–238.

Von Sperling, M. 1996 Comparison among the most frequently used systems for wastewater treatment in developing countries. *Water Science and Technology* **33**, 59–72.

Zimmo, O. R., Van Der Steen, N. P. & Gijzen, H. J. 2004 Nitrogen mass balance across pilot-scale algae and duckweed-based wastewater stabilisation ponds. *Water Research* **38**, 913–920.

First received 28 November 2017; accepted in revised form 24 May 2018. Available online 7 June 2018

Tilapia rearing with high rate algal pond effluent: ammonia surface loading rates and stocking densities effects

I. A. Sánchez, R. K. X. Bastos and E. A. T. Lana

ABSTRACT

In two pilot-scale experiments, fingerlings and juvenile of tilapia were reared in high rate algal pond (HRAP) effluent. The combination of three different total ammonia nitrogen (TAN) surface loading rates (SLR1 = 0.6, SLR2 = 1.2; SLR3 = 2.4 kg TAN·ha^{-1}·d^{-1}) and two fish stocking densities (D1 = 4 and D2 = 8 fish per tank) was evaluated during two 12-week experiments. Fingerlings total weight gain varied from 4.9 to 18.9 g, with the highest value (equivalent to 0.225 g·d^{-1}) being recorded in SLR2-D1 treatment; however, high mortality (up to 67%) was recorded, probably due to sensitivity to ammonia and wide daily temperature variations. At lower water temperatures, juvenile tilapia showed no mortality, but very low weight gain. The fish rearing tanks worked as wastewater polishing units, adding the following approximate average removal figures on top of those achieved at the HRAP: 63% of total Kjeldahl nitrogen; 54% of ammonia nitrogen; 42% of total phosphorus; 37% of chemical oxygen demand; 1.1 log units of *Escherichia coli*.

Key words | high rate algal ponds, tilapia rearing, wastewater treatment

I. A. Sánchez
Departamento de Recursos Hidrobiológicos, Universidad de Nariño, Ciudad Universitaria, Barrio Torobajo, Pasto-Nariño, Colombia

R. K. X. Bastos (corresponding author)
Departamento de Engenharia Civil, Universidade Federal de Viçosa, Viçosa-Minas Gerais CEP: 36570-000, Brazil
E-mail: *rkxb@ufv.br*

E. A. T. Lana
Departamento de Zootecnia, Universidade Federal de Viçosa, Viçosa-Minas Gerais CEP: 36570-000, Brazil

INTRODUCTION

It is estimated that in 2015 only 39 per cent of the global population (2.9 billion people) used a safely managed sanitation service (i.e., excreta safely disposed of *in situ* or treated off-site), and 2.3 billion people still lacked even a basic sanitation service (WHO & UNICEF 2017). In turn, the vast majority of the hungry lives in the developing regions, where an estimated 780 million people were undernourished in 2014–16, with an undernourishment prevalence of 12.9% (FAO *et al.* 2015). As a whole, this points out the need for low-cost sanitation and sewage treatment systems in low-income countries, preferably combined with wastewater use in agriculture or aquaculture, which is seen as method of water and nutrient recycling as well as improving household food and nutrition security (WHO 2006).

Under favorable climate conditions and where land is available, waste stabilization ponds (WSP) are one of the most appropriate methods of wastewater treatment in developing countries, as they are low-cost, low-maintenance, highly efficient, entirely natural and highly sustainable systems (Peña & Mara 2004). WSP are also one of the treatment options most compatible and used with aquaculture, since they provide a highly enriched environment that supports the growth of plankton, which serves as fish food (WHO 2006). High rate algal ponds (HRAP) are an advanced type of pond – paddlewheel-mixed, shallow, open raceway – developed for the treatment of wastewater and resource recovery. According to Craggs *et al.* (2012), HRAPs retain the advantages of conventional ponds (simplicity and economy), overcome some of their drawbacks, and have the additional benefit of recovering wastewater nutrients as harvestable algal/bacterial biomass for beneficial use as biofuel fertilizer or feed.

The literature reports several studies on wastewater use in fish culture in a variety of situations, looking both at toxic effects/fish survival and at fish productivity. For instance, there are quite a few reports on carp polyculture in sewage-fed ponds in India, either in large or experimental scale (e.g., Raychaudhuri *et al.* 2008; Mandal *et al.* 2018). As stated by Watanabe *et al.* (2002), a fundamental advantage of tilapia for aquaculture is that it feeds on a low trophic level; members of the genus *Oreochromis* are all omnivores, feeding on algae, aquatic plants, small invertebrates, detritus and associated bacterial films, as well as

doi: 10.2166/wst.2018.285

on a variety of feeds of animal origin. Thus, there also are a quite a few works on wastewater-fed tilapia culture, either in WSP (e.g., Silva *et al.* 2000 in Brazil; Abdul-Rahaman *et al.* 2012 in Ghana), or in fish rearing tanks/ponds fed with WSP effluents. In the latter case, some works are based on batch experiments (i.e., tanks intermittently fed with effluents), e.g., Gaigher & Krause (1983) in South Africa and Mota *et al.* (2009) in Brazil. Others, more closely resembling the experimental arrangement of the present work, used tanks/ponds continuously fed with effluents of domestic sewage treated in WSP or HRAP. For example, Edwards *et al.* (1981), in Thailand, reared *Tilapia nilotica* in 4 m^3 concrete tanks and 200 m^2 earth ponds using HRAP effluent; Moscoso (1996), in Peru, used the effluent of a tertiary WSP for rearing Nile tilapia in 370 m^2 earth ponds; in Brazil, Mota *et al.* (2009) used 50 m^3 concrete tanks fed with an effluent of anaerobic-facultative-maturation pond series for rearing Nile tilapia. In the same location of the present work, Bastos *et al.* (2003) reared Nile tilapia in two parallel 1 m^3 plastic tank systems: (i) one receiving the effluent of a wastewater treatment system composed of an upflow anaerobic sludge blanket reactor followed by a series of three pilot-scale polishing ponds, where the plankton produced in the ponds was the only available food source; (ii) and another, of conventional aquaculture in freshwater and provision of artificial feed. Briefly, at low stocking densities fingerlings showed similar weight gains both in the wastewater-fed and in the conventional systems; however, during the more advanced fish growth stages higher weight gains were obtained in the conventional system. The authors concluded that the WSP phytoplankton proved to be of excellent nutritional value for the fingerlings but may have been insufficient to sustain high production of tilapia juvenile. Overall, these above-mentioned studies recommended using up to 10 fish per m^3 or 2 fish per m^2, and reported high fish survival as well as fish productivity estimates up to 6 $t{\cdot}ha^{-1}{\cdot}y^{-1}$.

One of the major concerns regarding wastewater use in fish culture is ammonia toxicity. Un-ionized ammonia (NH_3-N) is the most toxic form of ammonia because of its ability to move across cell membranes; thus the toxicity of total ammonia nitrogen ($TAN = NH_4^+ + NH_3$) is dependent on the percentage present in the un-ionized form, which, in turn, is pH-dependent. In spite of having no shortage of studies on ammonia toxicity to fish, and also more specifically to tilapia (e.g., Wrigley *et al.* 1988; Benli & Köksal 2005; Hanna *et al.* 2013), engineers still lack information on design criteria for wastewater-fed fish culture systems based on the very important variable of ammonia surface loading rates (kg $TAN{\cdot}ha^{-1}{\cdot}d^{-1}$). Actually, there are only a few reports on general design criteria, and not so recent. Mara *et al.* (1993), quoted by WHO (2006), based on the concept of minimal wastewater treatment in WSP for maximal production of microbially safe fish, suggested that a wastewater-fed fish pond receiving an effluent of a facultative pond should be designed on the basis of a surface loading of total nitrogen of 4 kg $N{\cdot}ha^{-1}{\cdot}d^{-1}$ (according to the authors, 'too little nitrogen results in a low algal biomass in the fish pond and consequently small fish yields; too much nitrogen gives rise to high concentration of algae, with the resultant high risk of severe dissolved oxygen depletion at night and consequently fish kills'). Regarding ammonia toxicity the authors only stated that free ammonia in the fish pond should be <0.5 mg L^{-1} in order to avoid acute toxicity to fish. Edwards (1992), thinking of maintaining good dissolved oxygen conditions, indicated that fish ponds should receive maximum loading rates of 25 kg $BOD_5{\cdot}ha^{-1}{\cdot}d^{-1}$ (BOD_5: 5-day biochemical oxygen demand).

Aiming at filling some of these information gaps, in this study we evaluated the effect of three ammonia surface loading rates (SLR) and two fish stocking densities on the growth and mortality of fingerlings and juvenile of tilapia cultured in tanks fed with an effluent of HRAP, using plankton as the only food source

METHODS

The experiment was carried out in Viçosa, State of Minas Gerais, Brazil (20°45′14″S, 42°52′54″W; average altitude = 648 m), at an experimental site in the University of Viçosa. This Brazilian town presents a humid subtropical climate (Köppen classification), with average, maximum and minimum temperatures of 19.8, 32.4 and 7.2 °C, respectively; annual average relative humidity and precipitation of, respectively, 81%, and 1,221.4 mm – spread over rainy (spring–summer) and dry (autumn–winter) seasons.

Domestic sewage was collected fortnightly, using a submersible pump, from a septic tank of a local small wastewater treatment plant. This anaerobically primarily treated wastewater was transported in tanks of 500 L and transferred to 3,000 and 5,000 L reservoirs at the experimental site, which then fed fiberglass HRAP with the following characteristics: length = 2.86 m, width = 1.28 m, free board = 0.2 m, pond depth = 0.3 m, surface area = 3.3 m^2, volume = 1 m^3 (Figure 1 left). The ponds' stainless steel paddlewheels were driven by a 1 hp electric motor, whose

Figure 1 | High rate algal ponds and sewage reservoirs (left); and fish rearing tanks (right): (1) system inlet; (2) water level control and system outlet.

rotation was controlled by a frequency inverter (WEG, series CFW-10) to provide a mean horizontal water velocity between 0.10 and 0.15 m·s^{-1}. The HRAP were operated at an average hydraulic retention time (HRT) of 7 days with an inlet continuous flow regulated two or three times a day.

Under natural climatic conditions, fingerlings and juvenile of genetically improved farmed tilapia (GIFT) were reared in 18 plastic tanks with 210 L as useful volume (Figure 1 right). Three groups of six tanks received, each, a continuous flow rate of treated effluent, so that three different ammonia SLR were tested: SLR1 = 0.6, SLR2 = 1.2, and SLR3 = 2.4 kg of TAN·ha^{-1}·d^{-1}. The loading rates were chosen based on values published in the literature of sewage-fed fish ponds; the lowest value was similar to those reported by Bastos *et al.* (2003) and Moscoso (1996), and in order to evaluate the effect of higher levels of ammonia two and four times the lowest value were applied. Based on the values of wastewater flow applied to the tanks, the average HRT in the rearing tanks was: SLR1 = 55 d, SLR2 = 27.5 d, and SLR3 = 13.8 d. For each group of tanks, two fish stocking densities (D1 = 4, D2 = 8 fish per tank, corresponding respectively to 6 and 12 fish per m^2) were randomly located. Taking into account the high phytoplankton concentration usually found in HRAP effluents, D1 was selected to be twice the values recommended by Edwards *et al.* (1981) and Bastos *et al.* (2003), which were based on conventional ponds.

The rearing units were filled with treated wastewater and left therein for at least 10 days to reduce the levels of ammonia and its toxicity. In order to avoid breeding grounds for *Aedes aegypti* mosquitos in the stored wastewater during that period of time, at the beginning of the experiments the HRAP effluent was diluted as follows: 60% of the tank's useful volume was filled with tap water free of chlorine and 40% with the effluent.

A first experiment was carried out for 12 weeks during the end of the summer and the first 3 weeks of the autumn. Tanks were stocked with GIFT fingerlings with an average initial weight of 2.60 ± 0.15 g in a randomized distribution of the three treatment replicates: T1: SLR1-D1; T2: SLR1-D2; T3: SLR2-D1; T4: SLR2-D2; T5: SLR3-D1 and T6: SLR3-D2. A second experiment was conducted for 12 weeks during autumn and the first week of winter. Tanks were stocked with juvenile GIFT with an average initial weight of 58.24 ± 7.61 g, according to the replicate's randomized distribution for the six treatments, as previously mentioned. Before starting the experiment, the juveniles were reared in a recirculating aquaculture system and were fed with commercial fish feed with 34% dietary protein.

Fish were weighed at the beginning and at the end of the experiments in order to calculate the average total weight gain for each treatment, as well as the daily weight gain (total weight gain/duration of the experiment). Fish tanks were checked twice a day in order to monitor fish mortality, as well as to remove and weigh dead fish. Analysis of variance was applied to look at differences between treatments (fish weight gain and fish mortality) using the software R 3.1.0. Correlation between fish weight gain and water quality parameters was evaluated by the Spearman's rank test using the software Excel.

During the experiments the following physical and chemical variables were monitored according to the recommendations of APHA *et al.* (2012) (the methods used for the respective analyses are within brackets): chemical oxygen demand (COD: 5220D), total phosphorus (TP: 4500-P C), ammonia nitrogen (AN: 4500 – NH_3D), total Kjeldahl nitrogen (TKN: 4500-N D), dissolved oxygen – membrane electrode method (DO: 4500-O G), pH – electrometric method (4500-H^+ B), temperature (2550 B). The pH, DO and temperature were monitored using a Hach portable meter, model HQ40d; the chromogenic-fluorogenic method (Colilert®) was used to measure *Escherichia coli* (enzyme substrate coliform test: 9323). Chlorophyll-*a* was extracted with 80% ethanol and measured by spectrophotometry (APHA *et al.* 2012). COD, TP, AN, TKN, DO, pH, temperature and *E. coli* were monitored in the septic tank and HRAP effluents once a month during the first experiment, and fortnightly during the second experiment; in the second experiment composite samples were also analyzed from the fish tanks of each SLR line. Chlorophyll-*a* was measured in the same frequency for the HRAP effluent during the two experiments, and in the fish tanks at the end of the two experiments. The pH, DO and temperature were measured every 5 days in the septic tank effluent, in the HRAP and in the 18 fish culture tanks.

RESULTS AND DISCUSSION

Experiment with tilapia fingerlings

The average concentrations of TKN, AN, COD and TP recorded in the effluent samples were, respectively: (i) 77; 91.8; 1,454.7; and 16.2 mg L^{-1} in the septic tank (ST) effluent; and (ii) 31.7; 49.7; 295.2; and 7.6 mg L^{-1} in the HRAP effluent. Thus, average removal efficiencies of 50.8% for TKN, 49.1% for AN, 80.9% for COD and 53% for TP were recorded in the HRAP during the experiments The high values of COD in the septic tank effluent were mainly due to sludge accumulation in the treatment unit and the transport of sludge particles during the pumping of sewage samples. The average chlorophyll-*a* concentrations recorded in the treatment–fish culture system were: 1,736 μg L^{-1} in the HRAP effluent; 161.53 μg L^{-1} in SLR3; 660 μg L^{-1} in SLR1 and 680.4 μg L^{-1} in SLR2.

The average water temperature remained rather stable, usually around 23–24 °C. According to a local database, during the same period, the minimum, average and maximum values of air temperature were, respectively, 11.4, 22.3 and 33.5 °C; often, the air temperature varied as widely as 11.4 °C in 24 hours. The pH varied from 5 to 10.2 in the SLR treatments, with average values higher than 8.3; in the HRAP pH varied from 6.9 to 9.1. The DO varied widely, from 2.3 to 21.1 mg L^{-1} in the SLR treatments, and from 2.2 to 15.9 mg L^{-1} L in the HRAP, with lower values found at dawn; after midday, supersaturated concentrations of DO were recorded due to intense photosynthetic activity. Table 1 shows the wastewater DO, pH and temperature mean and standard deviation values found in the HRAP and in the SLR treatments during the first experiment.

Table 2 presents the results of the total fish weight gain and total mortality during the experiment – average values of the three replicates of each stocking density in each SLR treatment. In general, the daily weight gain was lower than those reported by other authors, e.g., Edwards *et al.* (1981) and Mota *et al.* (2009). Marked differences in fish weight gains were noticed between treatments, although not statistically significant, probably due to the large data variability (high standard deviation values). In fact, differences in growth within the same fish group are not unexpected. Johnsson *et al.* (2006) suggested that appetite suppression associated with stress-induced anorexia and decreased nutrient digestibility in subordinate fish can lead to decreased growth in the short- and long-term metabolism. Fernandes & Volpato (1993), studying Nile tilapia, registered a higher metabolic cost for the subordinates compared to the dominants.

With the exception of SLR1 treatments (T1 and T2), the lower the fish stocking densities the higher was the weight gain (T3 and T5, compared to T4 and T6, respectively). The highest weight gain (18.9 g, equivalent to 0.225 g per day) was obtained in T3 treatment, maybe due to the intermediate ammonia SLR level (thus less toxic ammonia effects) coupled with the also intermediate chlorophyll-*a* concentration (hence reasonable availability of plankton as food source). In SLR3 treatments, fish growth may have

Table 1 | Average and standard deviation (within brackets) values of DO, temperature and pH in the HRAP and in the rearing tanks during the tilapia fingerlings experiment

Treatment or rearing unit	DO (mg·L^{-1})	Temperature (°C)	pH
HRAP	6.5 (2.86)	22.4 (2.84)	7.9 (0.66)
T1 (SLR1-D1)	12.4 (5.03)	23.7 (2.94)	8.7 (1.29)
T2 (SLR2-D2)	9.4 (3.64)	23.6 (2.95)	8.8 (1.03)
T3 (SLR2-D1)	9.2 (3.10)	23.7 (3.03)	9.1 (0.39)
T4 (SLR2-D2)	9.2 (3.72)	23.7 (3.01)	8.6 (1.17)
T5 (SLR3-D1)	10.6 (4.29)	23.8 (2.96)	8.5 (0.94)
T6 (SLR3-D2)	11 (4.64)	23.8 (3.06)	8.4 (1.43)

Table 2 | Average and standard deviation (within brackets) values of total fish weight gain and total mortality during the tilapia fingerlings experiment

Treatment	Average total weight gain (g)	Total mortality (%)
T1 (SLR1-D1)	4.86 (0.33)	66.7 (23.57)
T2 (SLR1-D2)	7.76 (1.77)	41.7 (35.84)
T3 (SLR2-D1)	18.90 (14.50)	41.7 (42.49)
T4 (SLR2-D2)	10.07 (0.27)	62.5 (44.49)
T5 (SLR3-D1)	9.96 (4.82)	16.7 (11.79)
T6 (SLR3-D2)	5.91 (1.66)	54.2 (41.25)

been inhibited due to the high ammonia concentrations, despite the highest chlorophyll-*a* concentration. El-Sherif & El-Feky (2008) also observed that tilapia weight gain decreased with increasing levels of ammonia. In turn, the low weight gain in SLR1 may have been associated with limited availability of plankton (in spite of the weight gain in T2 being higher than in T6). Overall, these results suggest the potential for rearing fingerlings at low stocking densities, even at moderate or high ammonia loading rates.

A strong negative correlation was found between fish weight gain and DO (Spearman's rank correlation coefficient $r_s = -0.90$). Surprisingly at first sight, such finding may be explained by the larger variability of DO data in those treatments with higher DO concentrations. As suggested by Wang *et al.* (2009), it is likely that dynamic changes in oxygen levels would increase the routine metabolic rate and reduce the amount of energy available for growth. There was no correlation between fish weight gain and water temperature ($r_s = -0.014$), certainly because the water temperature did not vary much between treatments. However, correlation was found between fish weight gain and both water pH ($r_s = 0.371$) and chlorophyll-*a* ($r_s = 0.486$). Notwithstanding these associations being only weak to moderate they suggest that higher plankton productivity may have promoted fish growth.

Fish mortality varied widely, and this certainly contributed to no statistically significant differences being found between the treatments' mortality values. Again, this may be in part explained by the behavioral and physiological consequences of the social order imposed between dominant and subordinate fish (Johnsson *et al.* 2006). Full mortality was registered in 7 out of the 18 rearing tanks. Mortality events occurred mostly, or more intensively, following wide daily temperature variations, which may have exerted a synergic effect with the ammonia concentrations. Fish mortality may also help to explain the above-mentioned exceptions in the weight gain results (T1 vs. T2 and T2 vs. T6), in as much as, in general, the lower the fish mortality the higher was the weight gain (T2, T3 and T5, compared to T1, T4 and T6, respectively).

Experiment with tilapia juveniles

Table 3 shows the average and the standard deviation values of the wastewater characteristics found in the septic tank and the HRAP effluents and in the fish rearing tanks during the second experiment.

The maximum sewage constituent values recorded in the wastewater treatment–fish rearing system were: (i) ST – 131.9 mg·L^{-1} TKN; 156.1 mg·L^{-1} AN; 18.7 mg·L^{-1} TP; 1,485 mg·L^{-1} COD; (ii) HRAP – 70.8 mg·L^{-1} TKN;

Table 3 | Average and standard deviation (within brackets) values of TKN, AN, TP, COD, *E. coli* and chlorophyll-*a* in the septic tank and HRAP effluents and in the rearing tanks during the tilapia juvenile experiment

Treatment or rearing unit	TKN (mg·L^{-1})	AN (mg·L^{-1})	TP (mg·L^{-1})
Septic tank	113.2 (20.65)	147.3 (7.38)	16.4 (1.40)
HRAP	40.1 (14.63)	33.9 (21.44)	12.7 (0.68)
SLR3	20.3 (7.50)	24.1 (4.37)	9.8 (2.00)
SLR1	10.5 (3.81)	10.7 (3.53)	5.5 (1.87)
SLR2	14.1 (5.95)	11.9 (8.55)	6.9 (2.13)
Treatment or rearing unit	**COD (mg·L^{-1})**	***E. coli* (MPN.100 mL^{-1})**	**Chlorophyll-*a* (μg·L^{-1})**
Septic tank	1,064 (396.96)	6.6×10^5 (1.01×10^6)	–
HRAP	231.5 (57.79)	3.2×10^4 (3.2×10^4)	475.3 (147.11)
SLR3	167.8 (45.40)	3.7×10^3 (6.3×10^3)	234.1 (83.35)
SLR1	124.7 (34.54)	1.8×10^3 (1.45×10^3)	178.4 (70.53)
SLR2	145.5 (51.06)	2.8×10^3 (2.29×10^3)	174.7 (71.61)

54.7 mg·L^{-1} AN; 13.3 mg·L^{-1} TP; 343 mg·L^{-1} COD; (iii) SLR3 – 33.1 mg·L^{-1} TKN; 30.6 mg·L^{-1} AN; 11.9 mg·L^{-1} TP; 248 mg·L^{-1} COD; (iv) SLR1 – 17.4 mg·L^{-1} TKN; 14.6 mg·L^{-1} AN; 8.2 mg·L^{-1} TP; 191 mg·L^{-1} COD; (v) SLR2 – 24.8 mg·L^{-1} TKN; 20.5 mg·L^{-1} AN; 9.4 mg·L^{-1} TP; 202 mg·L^{-1} COD. Again, the concentrations of chlorophyll-*a* varied widely in the HRAP effluent (from 280.3 to 723.6 μg·L^{-1}) as well as in the SLR lines: 125.5 to 392.5 μg·L^{-1} in SLR3; 109.5 to 320.4 μg·L^{-1} in SLR1, and 101.5 to 325.7 μg·L^{-1} in SLR2.

Based on the values shown in Table 3, the average removal efficiencies in the HRAP and in the fish rearing system were: (i) HRAP – 64.6% TKN; 77.0% AN; 22.9% TP; 78.2% COD; 1.31 log units *E. coli*; (ii) SLR3 – 49% TKN; 29% AN; 23% TP; 28% COD; 0.94 log units *E. coli*; (iii) SLR1 – 74% TKN; 68% AN; 57% TP; 46% COD; 1.25 log units *E. coli*; (iv) SLR2 – 65% TKN; 65% AN; 46% TP; 37% COD; 1.1 log units *E. coli*. The removal results in the HRAP (from both the experiments with the fingerlings and the juvenile) were in accordance with previous reports, like in Young *et al.* (2017). The reduction of chlorophyll-*a* concentration from the HRAP effluent to the fish rearing tanks (again, in both experiments) may be a result of phytoplankton consumption by the fish. It is noticeable that the chlorophyll-*a* values found in the experiment with juveniles were lower than those from the experiments with fingerlings, which may be explained by the higher plankton consumption by bigger fish in the second experiment, as well as by the lesser average radiation recorded during the second study (less 17%). Clearly, the final wastewater quality was improved by rearing juvenile tilapias, which fed on particulate organic matter, plankton and detritus present in the HRAP effluent; fish also controlled the breeding grounds for mosquitos by eating the mosquito larvae.

Table 4 shows the mean and standard deviation values for DO, pH and temperature of the wastewater in the wastewater treatment–fish rearing system during the second experiment. The water temperature was lower than in the first experiment, usually under 20 °C. According to the local database the minimum, average and maximum air temperature values during the experiment were 7.3, 17.5 and 31 °C respectively. DO and pH values were also lower than those from the experiment with fingerlings, probably as a result of the lower temperatures and lesser photosynthetic activity. The pH varied from 5.5 to 8 in the rearing tanks, with average values below 7; in the HRAP pH varied from 6 to 7.9. DO varied from 0.7 to 11.8 mg·L^{-1} in the rearing tanks, and from 4.4 to 10.7 mg·L^{-1} in the HRAP, with lower values occurring early morning whereas supersaturated concentrations were found in the afternoon.

Table 4 | Average and standard deviation (within brackets) values of DO, temperature and pH in the HRAP and in the rearing tanks during the experiment with tilapia juvenile

Treatment or rearing unit	DO (mg·L^{-1})	Temperature (°C)	pH
HRAP	6.8 (2.07)	18.2 (3.39)	7.1 (0.53)
T1 (SLR1-D1)	6.0 (2.38)	19.8 (2.87)	6.6 (0.63)
T2 (SLR2-D2)	4.7 (2.32)	19.7 (2.82)	6.7 (0.53)
T3 (SLR2-D1)	6.1 (2.49)	19.8 (2.85)	6.7 (0.64)
T4 (SLR2-D2)	5.0 (2.50)	19.8 (2.82)	6.8 (0.64)
T5 (SLR3-D1)	4.9 (3.37)	19.9 (2.78)	6.9 (0.53)
T6 (SLR3-D2)	4.9 (3.00)	19.9 (2.89)	6.9 (0.64)

There were no significant differences of weight gain between treatments. Very low weight gain was recorded in treatments SLR2-D1 and SLR3-D1 – up to 3 g over 12 weeks; no weight gain was recorded in SLR2-D2 and SLR3-D2, whereas there were even weight losses in SLR1 treatments. These findings may be explained by the low water temperatures which prevailed in these second experiments since, as stated by El-Sayed (2006), tilapia feeding is sharply reduced below 20 °C, and stops at about 16 °C, while severe mortality occurs at 12 °C. In an experiment using a four pilot-scale WSP series in South Africa, Wrigley *et al.* (1988) also reported that low winter temperatures and low DO concentrations during spring restrict production. Small African sharptooth catfish (*Clarias gariepinus*) stocked in the first three ponds grew rapidly in summer (relative daily growth rate of 5.9%), but during winter growth rate was reduced to 0.2% d^{-1}. The fourth pond was stocked with a polyculture of *Oreochromis mossambicus*; *Hypopthamichthys molitrix*, *Cyprinus carpio* and *Labeo umbratus*. Again, during summer, *O. mossambicus* and *C. carpio* had average relative daily growth rates of 1.4 and 2.1%, respectively, while in winter both *H. molitrix* and *L. umbratus* lost weight (respectively, 0.2 and 0.3% of body weight·d^{-1}) and *C. carpio* grew slower.

Even though during this experiment water temperature and pH values varied only narrowly, and fish weight gain was very low, plankton productivity still seems to have somehow favored fish growth: strong positive correlations were found between fish weight gain, water temperature ($r_s = 0.786$), chlorophyll-*a* ($r_s = 0.843$) and pH ($r_s = 0.629$). No fish mortality occurred during the second experiment, probably because the sensitivity to ammonia is reduced at the more advanced growing stages of the fish; as observed by Benli & Köksal (2005) in ammonia toxicity tests with tilapia, larvae were less tolerant than fingerlings to the same levels of ammonia.

CONCLUSIONS

Tilapia fingerlings were highly sensitive to ammonia and short term temperature variation, which led to high mortality. On the other hand, tilapia juveniles withstood the ammonia effects and, although no mortality was recorded, the low temperature affected drastically the fish weight gain. The rearing tanks worked as polishing units and contributed to improving the effluent water quality in terms of nutrients, organic matter and *E. coli*. The results obtained herein suggest that, under the specific circumstances of these experiments (for instance, the prevailing climatic conditions), tilapia fingerlings culture using high rate algal pond effluent is feasible, during summer–autumn, with fish stocking density of 6 fish per m^2 and ammonia surface loading rate of 1.2 $kg{\cdot}ha^{-1}{\cdot}d^{-1}$. The feasibility of tilapia juvenile rearing is still to be confirmed over warmer temperatures than those tested here.

ACKNOWLEDGEMENTS

The authors would like to thank the Brazilian agency CAPES for providing a postgraduate scholarship and research financial support.

REFERENCES

Abdul-Rahaman, I., Owusu-Frimpong, M. & Ofori-Danson, P. K. 2012 Sewage fish culture as an alternative to address the conflict between hunters and hunting communities in Northern Region. *Journal of Agriculture and Sustainability* **1** (1), 1–22.

APHA (American Public Health Association), AWWA (American Water Works Association) & WEF (Water Environment Federation) 2012 *Standard Methods for the Examination of Water and Wastewater*, 22nd edn. APHA, AWWA and WEF, Washington, DC.

Bastos, R. K. X., Pereira, C. M., Pivelli, R. P., Lapolli, F. R. & Lanna, E. A. T. 2003 Utilização de esgotos sanitários em piscicultura (Use of domestic sewage in fish culture). In: *Utilização de Esgotos Tratados em Fertirrigação, Hidroponia e Piscicultura – Projeto PROSAB (Use of Treated Sewage in Fertirrigation, Hydroponics and Fish Culture – PROSAB Project)* (R. K. X. Bastos, coord.). ABES, RiMa, Rio de Janeiro, pp. 193–221.

Benli, A. Ç. K. & Köksal, G. 2005 The acute toxicity of ammonia on tilapia (*Oreochromis niloticus* L.) larvae and fingerlings. *Turkish Journal of Veterinary and Animal Sciences* **29**, 339–344.

Craggs, R., Sutherland, D. & Campbell, H. 2012 Hectare-scale demonstration of high rate algal ponds for enhanced wastewater treatment and biofuel production. *Journal of Applied Phycology* **24**, 329–337.

Edwards, P. 1992 *Reuse of Human Wastes in Aquaculture, a Technical Review*. The World Bank, UNDP-World Bank Water and Sanitation Program, Washington, DC, USA.

Edwards, P., Sinchumpasak, O. A. & Tabucanon, M. 1981 The harvest of microalgae from the effluent of a sewage fed high rate stabilization pond by *Tilapia nilotica*: Part 2: studies of the fish ponds. *Aquaculture* **23** (1–4), 107–147.

El-Sayed, A.-F. M. 2006 *Tilapia Culture*. CABI Publishing, Wallingford, Oxfordshire, UK.

El-Sherif, M. S. & El-Feky, A. 2008 Effect of ammonia on Nile tilapia (*O. niloticus*) performance and some hematological and histological measures. In: *8th International Symposium on Tilapia in Aquaculture, 12–14 October 2008, Cairo, Egypt*, pp. 513–530.

FAO, IFAD & WFP 2015 The state of food insecurity in the world 2015. In: *Meeting the 2015 International Hunger Targets: Taking Stock of Uneven Progress*. FAO, Rome.

Fernandes, M. O. & Volpato, G. L. 1993 Heterogeneous growth in the Nile tilapia: social stress and carbohydrate metabolism. *Physiology & Behavior* **54** (2), 319–323.

Gaigher, I. G. & Krause, J. B. 1983 Growth rates of Mozambique tilapia (*Oreochromis mossambicus*) and silver carp (*Hypophthalmichthys molitrix*) without artificial feeding in floating cages in plankton-rich wastewater. *Aquaculture* **31** (2–4), 361–367.

Hanna, M. I., El-Maedawy, S. A., Kenawy, A. M. & Girgis, S. M. 2013 Sublethal effects of acute ammonia exposure on *Oreochromis niloticus*. *Global Veterinaria* **11** (5), 592–603.

Johnsson, J. I., Winberg, S. & Sloman, K. A. 2006 Social interactions. In: *Behaviour and Physiology of Fish: Volume 24 (Fish Physiology)* (K. A. Sloman, R. W. Wilson & S. Balshine, eds). Academic Press Elsevier B.V., Amsterdam, pp. 151–196.

Mandal, R. N., Chakrabarti, P. P., Paul, B. N., Chattopadhyay, D. D., Das, A., Hussan, A. & Jayasankar, P. 2018 Recycling of sewage in aquaculture: decadal technical advancement. In: *Wastewater Management through Aquaculture* (B. Jana, R. Mandal & P. Jayasankar, eds). Springer, Singapore, pp. 95–118.

Mara, D. D., Edwards, P., Clask, D. & Mills, S. W. 1993 A rational approach to the design of wastewater-fed fishponds. *Water Research* **27** (12), 1797–1799.

Moscoso, J. 1996 *Aquaculture Using Treated Effluents from the San Juan Stabilization Ponds, Lima, Peru – Executive Summary*. OPS/CEPIS, Lima, Peru.

Mota, S., Santos, E. S. & Santos, A. B. 2009 Water and fish quality during the reuse in aquaculture of sewage treated in stabilization ponds: experience of Ceará state, Brazil. In: *8th IWA Specialist Group Conference on Waste Stabilization Ponds, 26–30 April 2009, Belo Horizonte, Brazil*.

Peña, M. & Mara, D. 2004 *Waste Stabilisation Ponds*. IRC International Water and Sanitation Centre, Delft, The Netherlands.

Raychaudhuri, S., Mishra, M., Nandy, P. & Thakur, A. R. 2008 Waste management: a case study of ongoing traditional practices at East Calcutta Wetland. *American Journal of Agricultural and Biological Sciences.* **3** (1), 315–320.

Silva, F. J. A., Mara, D. D., Pearson, H. W. & Mota, S. E. 2000 Informal fish culture in the Maracanaú waste stabilization ponds in Fortaleza, Brazil. *Water Science and Technology* **42** (10–11), 393–398.

Wang, T., Lefevre, S., Huong, D. T. T., Cong, N. V. & Bayley, M. 2009 The effects of hypoxia on growth and digestion. In: *Hypoxia: Volume 27 (Fish Physiology)* (J. G. Richards, A. P. Farrell & C. J. Braunes, eds). Academic Press Elsevier B.V., London, pp. 361–396.

Watanabe, W. O., Losordo, T. M., Fitzsimmons, K. & Hanley, F. 2002 Tilapia production systems in the Americas: technological advances, trends, and challenges. *Reviews in Fisheries Science* **10** (3–4), 465–498.

WHO (World Health Organization) 2006 *WHO Guidelines for the Safe use of Wastewater, Excreta and Greywater. Volume 3. Wastewater and Excreta Use in Aquaculture.* WHO, Geneva.

WHO (World Health Organization) & UNICEF (United Nations Children's Fund) 2017 *Progress on Drinking Water, Sanitation and Hygiene: 2017 Update and SDG Baselines.* WHO and UNICEF, Geneva.

Wrigley, T. J., Toerien, D. F. & Gaigher, I. G. 1988 Fish production in small oxidation ponds. *Water Research* **22** (10), 1279–1285.

Young, P., Taylor, M. & Fallowfield, H. J. 2017 Mini-review: high rate algal ponds, flexible systems for sustainable wastewater treatment. *World Journal of Microbiology and Biotechnology* **33** (6), 117.

First received 1 November 2017; accepted in revised form 13 June 2018. Available online 28 June 2018

Microalgae cultivation in agro-industrial effluents for biodiesel application: effects of the availability of nutrients

Mariana Daniel Tango, Maria Lúcia Calijuri, Paula Peixoto Assemany and Eduardo de Aguiar do Couto

ABSTRACT

The present study evaluated the cultivation of microalgae in a photobioreactor using effluents from the meat-processing industry, which had been previously treated at the primary and secondary levels. *Scenedesmus* sp. was the dominant genus in the phytoplankton community in both of the evaluated effluents. The different nutritional conditions affected the production of biomass, which reached 1,160 mg/L of volatile suspended solids (VSS) and 371 mg/L of VSS with cultivation in the primary (PE) and secondary effluents (SE), respectively. In both effluents, great removal efficiencies close to quantification limits were observed for ammoniacal nitrogen and soluble phosphorus. Regarding the accumulation of lipids, there were no considerable differences between the effluents. The highest lipid productivity that was observed in the PE, which reached 3.7 g/m^2·d, was attributed to its larger production of biomass as a consequence of its better nutritional condition in relation to the SE.

Key words | biomass, bubble column photobioreactor, effluent treatment, lipids, meat-processing industry

Mariana Daniel Tango (corresponding author)
Maria Lúcia Calijuri
Paula Peixoto Assemany
Eduardo de Aguiar do Couto
Federal University of Viçosa (Universidade Federal de Viçosa/UFV),
Department of Civil Engineering, Environmental Engineering Group – nPA, Campus Viçosa, Viçosa, MG 36570-000,
Brazil
E-mail: *tango.mariana@gmail.com*

INTRODUCTION

Globally, livestock contributes 15% of total food energy and 25% of dietary protein and is one of the fastest growing parts of the agricultural economy. In order to meet rising demand, global annual meat production is expected to expand from 228 currently to 463 million tonnes by 2050 with the cattle population estimated to grow from 1.5 billion to 2.6 billion and that of goats and sheep from 1.7 billion to 2.7 billion (FAO 2009). Brazil is one of the leading countries in world meat exportation for having the largest commercial cattle herd in the world, with approximately 198 million head in 2014 (ANUALPEC 2015). Brazil is also being recognized with one of the main producers and exporters of pork and poultry. Each year, Brazilian participation in international trade is growing, particularly in the production of beef, pork, and chicken. By 2020, it is expected that the national production of meat will supply 44.5% of the world market (MAPA 2015).

Inherent to the production process and as consequence of the promising economic scenario, the meat-processing industry generates a large volume of effluents, which represent serious environmental and health problems primarily due to their biological instability and pathogen development potential (De Sena *et al.* 2009). Issues such as increasing demand for animal protein, environmental and health restrictions imposed by national legislation and international agreements (De Sena *et al.* 2009), and the search for viability and sustainability of industrial processes highlight the importance of the development of innovative effluent treatment technologies. These technologies must combine efficient and inexpensive treatments, recyclable nutrients, and the production of added-value items.

In meat production processes, during the various stages (from slaughtering to food processing and animal products), there is an immense production potential for added-value items, such as pigments and biofuels. Previous research has proven the efficiency of microalgae in the removal of nutrients and metals, in agro-industrial effluents, such as previously digested dairy effluents, slaughterhouse, piggery

doi: 10.2166/wst.2018.180

(Qi *et al.* 2017) and tannery effluents. Despite the global importance of meat production and the great potential use of generated effluents, it was not found in the literature studies that assessed the utilization of effluents from the meat-processing stage in the cultivation of microalgae.

Microalgae can be cultivated in suspended form, in open photobioreactors (such as a high rate pond – HRP), or in enclosed photobioreactors (PBRs). Considering the use of effluents as culture media, HRPs represent a technology that is consolidated to a certain extent and has been extensively evaluated. Alternatively, PBRs have a higher productivity of biomass in comparison to HRPs due to a better control of the cultivation conditions (illumination, pH, and mixture), which reduces the area and lowers the costs associated with biomass collection.

However, this optimization also generates negative consequences, such as an increase in power consumption, which is primarily associated with the mixture and still restricts its application on a full scale. For example, a PBR, depending on its type, consumes 10–50 W/m^3 (Xu *et al.* 2018) for its functioning versus 3.72 W/m^3 for raceways ponds (Jorquera *et al.* 2010). However, together with airlift reactors, they are easy to scale up and have a low power consumption compared with that of other types of PBRs. These are essential characteristics when considering the treatment of large volumes of effluents and intensive production of algal biomass. However, most of the studies that have investigated the simultaneous treatment of effluents and biomass production in PBRs use production systems on a laboratory scale (Wang *et al.* 2010) and with synthetic or diluted culture media (Qi *et al.* 2017), which poorly depicts reality and makes it difficult to replicate the results on a large scale.

Considering the previous discussion, the goal of the present study is to evaluate the cultivation of microalgae in real effluents of the meat-processing industry with a focus on both the biomass production and the accumulation of lipids as well as the treatment of effluents in a pilot-scale bubble column PBR. The objective is to identify the nutritional factors that increase the production of algal biomass in primary and secondary pre-treated effluents, which would also provide support in the production of biodiesel.

METHODS

Experimental unit

The experiment was conducted in an outside area of the Laboratory of Sanitary and Environmental Engineering at the campus of the Federal University of Viçosa – UFV, Brazil (20°45′14″S, 42°52′54″W) from September to December of 2014.

As shown in Figure 1, the bubble column PBR that was used consisted of three independent acrylic tubes with diameters of 15 cm and a total volume of 16 liters each. The culture medium was mixed continuously by bubbling air (0.625 vvm, volume of injected air per minute divided by the volume of the culture medium) enriched with CO_2 (6.5%, v/v).

The air for the mixture was supplied by a diaphragm air compressor with a power of 0.25 kW and driven to each acrylic tube through a pneumatic hose followed by a PVC pipe connected to a cylindrical porous stone disperser (22 × 12 mm). To control the flows, valves and flowmeters with an accuracy of 0–15 L/min were installed.

The CO_2 supply was controlled by the pH variation in the units, which was maintained at between 6 and 8. For this automated system, a probe was used for real-time measurement of the pH and temperature (controller sc200, brand HACH and pH differential analog sensor for effluents) with an electrical signal transmission system compatible with the solenoid valve to control the addition of CO_2.

The main components of the system are presented in Figure 1(a), and the dimensions of the acrylic tubes that were used are shown in Figure 1(b).

Culture medium and inoculum

To test different nutritional conditions, primary (PE) and secondary effluents (SE) evaluated as culture media were collected from a wastewater treatment plant (WWTP) installed in a meat-processing industry after flotation treatment (PE) and after an activated sludge unit (SE). For each operation, 10% (v/v) of inoculum collected from HRP was added to the evaluated culture medium (Table 1). This HRP was preceded by an upflow anaerobic sludge blanket (UASB) reactor, which was used to treat domestic effluent, as previously studied by Santiago *et al.* (2013).

Monitoring of the photobioreactor

Each culture medium was evaluated during three outdoor operations of the PBR, which were performed in batches: OPR 1, OPR 2, and OPR 3 (PE) and OPR 4, OPR 5, and OPR 6 (SE). The three acrylic tubes had identical dimensions and were simultaneously operated with the same culture medium, mixture, supplement of CO_2, and percentage of inoculum to represent a single-treatment system.

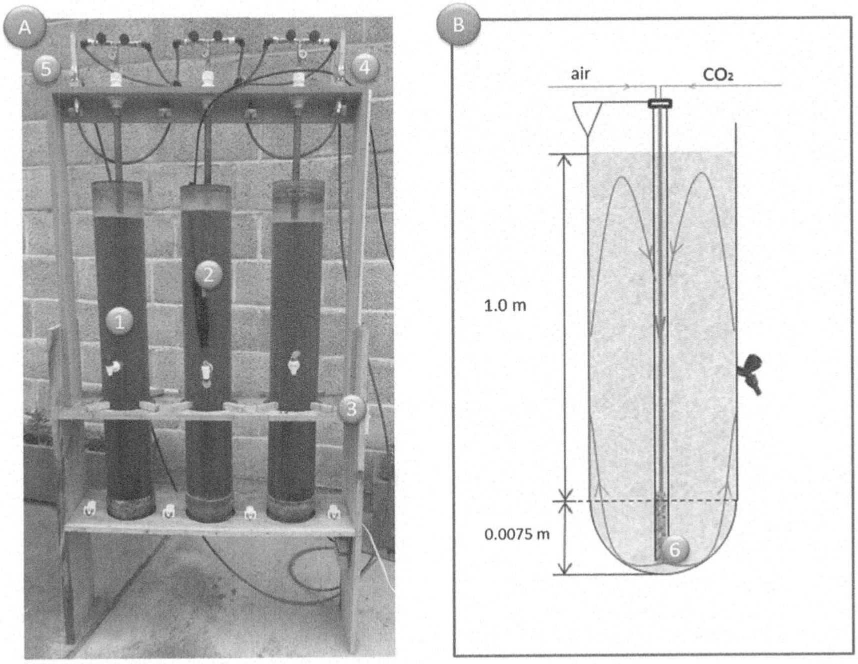

Figure 1 | (a) Bubble column PBR. Numbering indicates the constituent parts: (1) acrylic tube; (2) pH sensor; (3) wooden structure for support; (4) and (5) flowmeters; (b) (6) disperser and dimensions of each tube.

The operation was interrupted when the concentration of phaeophytin exceeded the concentration of chlorophyll-*a* for two consecutive days. In the next operation, the culture medium was replaced, and new inoculum was added.

The incident photosynthetically active radiation (PAR, 400–700 nm) was measured at 12:00 pm by means of a radiometer (LICOR – LI-193 *Underwater Spherical Quantum Sensor*). The dissolved oxygen (DO) was measured three times per day at 6:30 am, 12:00 pm, and 5:30 pm using a probe (Hach, model HQ40d, luminescent dissolved oxygen).

The culture medium was collected in composite form daily at 5:30 pm. For this purpose, simple samples of 300 mL collected from each PBR tube were mixed, forming a composite sample for analyses. The physical and chemical analyses essentially followed the provisions of *Standard Methods for the Examination of Water and Wastewater* (APHA 2012) and the methods used for analysis of each variable are between parentheses: total alkalinity (Alk) (2330–Alk B), chemical oxygen demand (COD) (5220–COD D) and filtered chemical oxygen demand (CODf) (5220–CODf D – sample filtered at 0.45 μm), total suspended solids (TSS) (2540–TSS D) and volatile suspended solids (VSS) (2540–VSS E), nitrate ($N\text{-}NO_3^-$) (4500–N-NO3 E), ($N\text{-}NO_2^-$) nitrite (4500–NO2 B), ammoniacal nitrogen ($N\text{-}NH_4^+$) (4500–NH3 C), total Kjeldahl nitrogen (N-TKN) (4500–Norg C), total phosphorus (Pt) (4500–FT D), and soluble phosphorus (P_S) (4500–FS D – sample filtered at 0.45 μm). The N-Norg was determined by the difference between the N-TKN and the $N\text{-}NH_4^+$. The forms of total

Table 1 | Inoculum characteristics

Sample	Genus/species	Total density (individuals/mL)	Observations
OPR 1 and 2	*Scenedesmus* sp. (*Scenedesmus accuminatus* and *Scenedesmus arvensis*)	3.5×10^6	High density of yeast and bacteria
OPR 3	*Scenedesmus* sp.	1.2×10^6	Presence of yeasts and bacteria
OPR 4		1.9×10^6	
OPR 5		6.6×10^5	
OPR 6	*Scenedesmus* sp. (free cells)	1.6×10^6	High density of bacteria

dissolved carbon (TDC) and dissolved organic carbon (DOC) were determined by TOC 5000 Shimadzu equipment.

The variable VSS was used as an indirect measurement of the concentration of total biomass at the end of the operations because, as it is not an axenic culture, other microorganisms develop in consortium with the microalgae. The development of microalgae was accompanied by monitoring chlorophyll-*a* and phaeophytin according to Nush (1980) which was modified according to NEN 6520 (1981).

Phytoplankton community

Samples (200 mL) of the inocula and the culture medium at the end of each operation were collected to characterize the phytoplankton community, which was performed at the genus level, and for the dominant genus, the species present were identified.

Qualitative analysis

The samples were preserved with a 4% formalin solution. The identification was conducted on an inverted microscope (Olympus CK2) by the method of Uthermöhl (1958). The morphological and morphometric characteristics of the vegetative and reproductive life of the taxonomic value for the species were analyzed according to the specialized literature.

Quantitative analysis

Due to the high density of organisms in the ponds, a 1:50 dilution of the samples was performed before quantification. The counting of the organisms was performed in sedimentation chambers with a volume of 2 mL, as described in Uthermöhl (1958), with an inverted microscope OLYMPUS CK2. The total density of the phytoplankton (individuals/mL) was calculated according to Equation (1):

$$D = \frac{C_{org} \times A_t}{A_f \times F \times V_s} \tag{1}$$

where D: total density (individuals/mL); C_{org}: number of counted organisms; A_t: total area of the bottom of the sedimentation chamber (mm^2); A_f: area of the counting field (mm^2); F: number of counted fields; and V_s: volume of the sedimented sample (mL).

Lipid quantification

The lipid content was evaluated at the end of each batch by solvent extraction, as described by Assemany *et al.* (2014).

The lipid productivity was obtained by the product of the total biomass productivity (mg VSS/L·d and/or g VSS/m^2·d) by the lipid content obtained in each operation.

RESULTS AND DISCUSSION

Characterization of the culture medium

Table 2 shows the physical and chemical characterization of the effluents used as culture media.

There was a wide variability in the physical and chemical composition registered in the operations of each effluent, which is typically observed in effluents from multiproduct factories due to the variations in the production system. Still, characteristics such as the concentrations of nitrogen and phosphorus were different in relation to the origin of the effluents and assessable for microalgal cultivation.

Table 2 | Characterization of the effluents used as culture media

	PE			SE		
	OPR 1	**OPR 2**	**OPR 3**	**OPR 4**	**OPR 5**	**OPR 6**
pH	5.2	5.6	6.5	7.6	7.6	7.4
Alk (mg $CaCO_3$/L)	207.1	152.3	221.9	262.2	174.2	90.4
DO (mg O_2/L)	3.3	2.7	2.2	4.7	3.1	5.4
Temperature (°C)	29.0	29.0	26.2	25.7	26.3	20.2
COD total (mg/L)	346.0	347.1	350.9	99.3	99.4	88.0
COD filtered (mg/L)	345.4	342.2	274.6	63.1	48.9	78.0
N-NH_4^+ (mg/L)	22.8	23.1	3.7	62.5	18.3	13.5
N-Norg (mg/L)	83.7	73.5	44.9	6.4	3.5	1.7
N-NO_3^- (mg/L)	6.1	2.4	1.2	1.2	0.2	2.1
N-NO_2^- (mg/L)	UD	0.2	UD	UD	UD	UD
Pt (mg/L)	8.8	2.5	12.0	1.7	1.8	1.1
P_S (mg/L)	6.0	1.6	5.5	0.7	0.6	0.3
TSS (mg/L)	588.3	381.3	668.3	67.0	77.5	44.3
VSS (mg/L)	560.3	333.8	588.3	43.0	67.5	32.9
DOC (mg/L)	236.6	125.8	170.0	28.1	18.1	14.4
TDC (mg/L)	245.3	132.1	177.0	78.1	50.9	31.6
N/P*	4.8	16.1	0.9	91.0	30.8	52.0

*N/P: nitrogen was considered in the forms of N-NH_4^+, N-NO_3^-, N-NO_2^-, and phosphorus in the form of P_S in the medium.
UD = undetected values.

As expected, due to the higher depuration provided by the activated sludge system, there was a lower concentration of organic material, phosphorus, and suspended solids in the SE in relation to the PE. The soluble fraction of organic matter was represented by the concentration of DOC and filtered COD. It was observed that the DOC fraction in the TDC was lower in the SE (35–46%, vs. 78–99% in the PE) due to the degradation of the organic matter in the treatment of the activated sludge. The high concentration of organic carbon is particularly important in the absence of light. Genera such as *Chlorella* sp. and *Scenedesmus* sp., which are recognized as preferentially photoautotrophic, can survive in heterotrophic conditions using some forms of organic carbon in the absence of light. It was found that most of the N-TKN present in the PE corresponded to the organic fraction (N-Norg). The opposite situation occurs with the forms of nitrogen that exist in the SE, which much of the N-TKN that is present corresponds to ammoniacal nitrogen (N-NH_4^+). Two of the three records of N-NH_4^+ from the SE (OPR 5 and OPR 6) are suited for the national effluent discharge requirements (limit of 20 mg N-NH_4^+/L) (Brasil 2011). The N/P ratio was considerably high primarily because of the limitation of soluble phosphorus in the SE. The optimal N/P ratio for algae growth in fresh water is 10 (Martin *et al.* 1985). Only the primary effluent used in OPR 1 (N/P: 4.8) and OPR 2 (N/P: 16.1) presented a relation close to the optimum suggested by the authors. The concentration of the TSS of the PE was high and showed a high variability (545.9 ± 121.0 mg TSS/L). High concentration of suspended solids can reduce the penetration of solar radiation. However, the low concentration of suspended solids may cause photoinhibition (Zhu *et al.* 2013), which is very frequent in culture systems with a high light surface. Moreover, associated with the solids particles, microorganisms such as bacteria may be adhered, which may result in increased or reduced microalgae production, depending on the culture conditions or related species.

Characterization of the environmental conditions

Figure 2 shows the environmental conditions that were recorded during the operations with the evaluated effluents. To better visualize the pH, a representative operation of each effluent was selected to illustrate the behavior of that variable throughout the culture time.

With the exception of operation 4, which showed mean PAR values of 2,254 ± 172 μE/m²·s, wide variations of PAR were recorded during the remaining operations: 1,902 ± 529 (OPR 1), 1,797 ± 777 (OPR 2), 2,101 ± 342 (OPR 3), 1,269 ± 743 (OPR 5), and 1,987 ± 578 μE/m²·s (OPR 6). The following are the recorded temperatures: 11 to 48 °C (OPR 1), 14 to 43 °C (OPR 2), 19 to 44 °C (OPR 3), 15 to 45 °C (OPR 4), 19 to 37 °C (OPR 5), and 14 to 37 °C (OPR 6). As a consequence of the high temperatures, on very hot days, *biofouling* was observed, in which cells adhere to the walls of the PBR, as reported by Arbib *et al.* (2013) in an airlift type PBR. The maximum values of DO were 16 mg/L (OPR 1), 18 mg/L (OPR 2), 14.5 mg/L (OPR 3), 17 mg/L (OPR 4), 12 mg/L (OPR 5), and 12.5 (OPR 6). The lowest values recorded throughout the operations, i.e., 3.34 mg/L (OPR 1), 1.81 mg/L (OPR 2), 2.17 mg/L (OPR 3), 6.38 mg/L (OPR 4), 5.69 mg/L (OPR 5) and 6.49 mg/L (OPR 6), were measured at sunrise. This fact can result from the microorganism's respiration and the absence of photosynthesis during the night.

From the pH behavior, it is possible to note the relationship between the carbon addition cycles and the photosynthetic activity, which is illustrated by the difference between the higher frequency of gas addition during the day compared with that during the night. This behavior was also observed by Arbib *et al.* (2013) and Valdés *et al.* (2012) who operated a tubular airlift and a bubble column PBR, respectively. These authors stated that during the day, microalgae use light and nutrients to assimilate CO_2, which causes the pH of the medium to increase. In a period of intense solar radiation, the balance of photosynthesis in relation to respiration is great and the withdrawal of carbon dioxide causes a rise of pH. During the night, the photosynthetic activity is null, and due to microorganism's respiration, CO_2 is released. However, the pH still increases slightly during the night. According to Valdés *et al.* (2012), this fact is attributed to the release of CO_2 from the liquid medium to the air, which is promoted by the continuous injection of compressed air into the reactor.

During the initial culture period, the pH of the PE took a longer time to reach 8 (t > 50 hours) in contrast with the SE, which took a maximum of 4 minutes. The slow increase in the pH represents an adaptation period of the inoculum to the cultivation conditions. This fact can be associated with factors such as the lower initial value of pH (5.8 ± 0.5) and the higher concentration of VSS and organic matter (Table 2) that were observed in the PE in relation to the SE. Initially, because there was a higher concentration of other microorganisms in the PE, the respiration process may have been more intense than photosynthesis during the first 50 hours. Moreover, due to high availability of

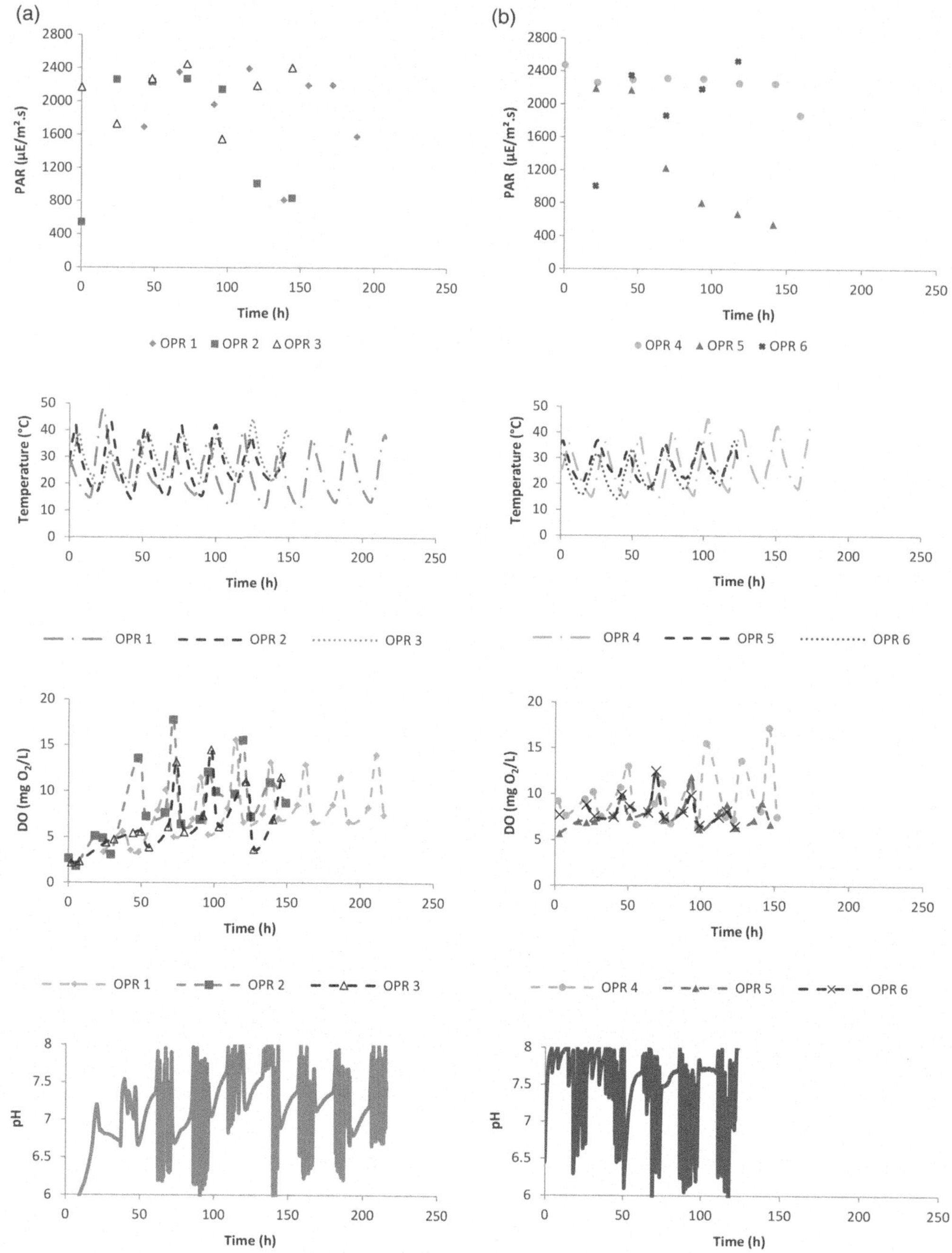

Figure 2 | Environmental conditions recorded during the operations: PAR; temperature; DO; and pH throughout the monitored period for both effluents (a) PE and (b) SE.

organic matter, CO_2 from heterotrophic bacteria respiration may have been enough to allow algae growth, resulting in a slow increase in pH. Similarly, early in the mornings, the pH of the PE also took a longer time to reach 8, which is also attributed to the release of CO_2 as a result from respiration and organic matter degradation. This fact highlights the symbiotic relationship that is established between algae and heterotrophic microorganisms in the PE.

Characterization of the biomass and phytoplankton community

It is important to note that both the inocula and the post-PBR effluents were not fully known cultures. Identifying the genera and, primarily, determining their counts were made difficult due to the high concentration of microorganisms, such as yeasts and bacteria, present in the samples.

The genus *Scenedesmus* sp. was dominant in the phytoplankton community with densities of 6.6×10^5 to 3.5×10^6 individuals/mL (inoculum) and 2.1×10^6 to 3.7×10^6 individuals/mL (final effluents, without considerable differences regarding the origin of the culture medium). Table 3 shows the characteristics of the phytoplankton community.

The dominance of the genus *Scenedesmus* (mixotrophic) in all the operations showed the ability of that genus to survive in extreme environments. As expected, in conditions of restricted availability of nutrients and consisting of DOC and P_S, which occurred in OPR 5 and 6, there was a limited growth of heterotrophic organisms, such as bacteria and yeasts. However, when the objective is to exploit the energy of the total biomass, characteristics of the effluents that favor the presence of other microorganisms, apart from microalgae, can be advantageous in certain cases. Cheirsilp *et al.* (2011) evaluated the lipid productivity of mixed cultures of yeasts (*Rhodotorula glutinis*) and microalgae (*Chlorella vulgaris*) in industrial effluents from the processing of seafood and sugarcane molasses. These authors reported that lipid productivity was higher in the mixed culture than in pure cultures, which was possibly due to the symbiotic nutritional circuit that is established between algae and yeast – similar to that between algae and bacteria.

The concentrations of VSS and chlorophyll-*a* recorded during the microalgal cultivation in the evaluated effluents are shown in Figures 3 and 4, respectively.

Wang *et al.* (2010) evaluated the cultivation of *Chlorella* sp. in raw, primary, and secondary municipal effluents as well as concentrated effluent after centrifugation. Although studies such as that by Martin *et al.* (1985) have suggested that the optimum N/P ratio for the growth of algae in fresh water is 10/1, the concentrate evaluated by Wang *et al.* (2010) (N/P of 0.36) showed best performance, which was assessed in terms of specific growth rate. According to the authors, despite the low N/P of the concentrate, the growth of microalgae was higher in this effluent due to the higher concentration of nitrogen, phosphorus, and DOC (131.5 mg/L of NT, 201.5 mg/L of Pt, and 2250.0 mg/L of DOC) in relation to the other effluents. This may explain the higher growth of biomass in the PE that was evaluated in the present study. The highest concentration of VSS (1,160.0 mg VSS/L, after 101 hours of operation) and the highest concentration of chlorophyll-*a* (10.9 mg chlorophyll *a*/L, after 53 hours of operation) were observed in the cultivation in PE (OPR 2). The cultivation of biomass in SE reached concentrations of 225 to 371 mg VSS/L in OPR 6 and 4, respectively, and from 2.3 to 3.5 mg/L of chlorophyll-*a* in OPR 4 and 5, respectively.

Table 3 | Phytoplankton community observed in the OPR 1-6

Sample	Genus/species	Total density (individuals/mL)	Observations
OPR 1	*Scenedesmus* sp.	–	High density of yeast and bacteria
OPR 2	*Scenedesmus* sp.	–	High density of yeast and bacteria
OPR 3	*Scenedesmus* sp.	2.1×10^6	Presence of yeasts and bacteria
OPR 4	*Scenedesmus* sp.	–	High density of yeast and bacteria
OPR 5	*Scenedesmus Staurastrum*	2.3×10^6	Presence of bacteria
OPR 6	*Scenedesmus Staurastrum*	3.7×10^6	Absence of bacteria

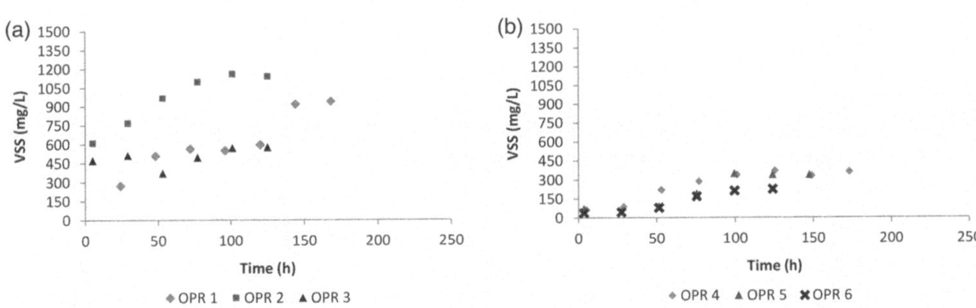

Figure 3 | Concentrations of VSS recorded during microalgal cultivation in the evaluated effluents (a) PE and (b) SE.

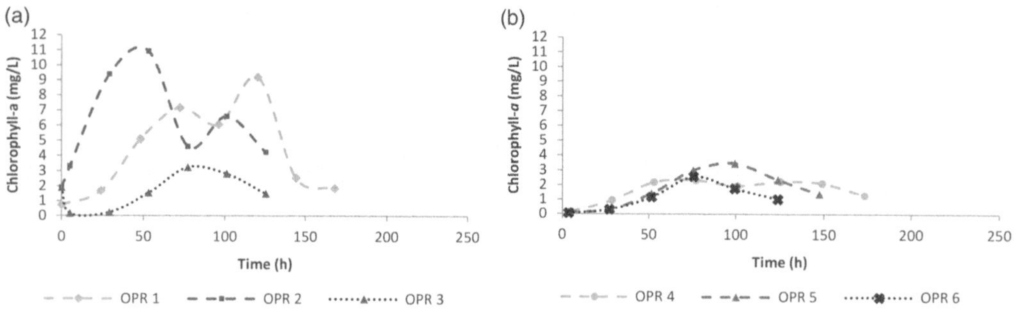

Figure 4 | Concentrations of chlorophyll-*a* during microalgal cultivation in the evaluated effluents (a) PE and (b) SE.

When the objective is the utilization of biomass, lower performances resulting from cultivation in secondary or very diluted effluents are frequently reported in the literature. Wang *et al.* (2010) observed that the lower production of microalgae in the SE was attributed primarily to the recalcitrant nature of the organic matter. Xin *et al.* (2010), using autoclaved secondary municipal effluent as the culture medium (0.5 mg/L of Pt, similar to the concentration observed in the present study for the SE, of 1.6 ± 0.31 mg/L), observed a production of 0.11 g *Scenedesmus* sp. LX1 per liter. These authors attributed the limited production of algal biomass to the reduced concentration of nutrients, primarily phosphorus, in the culture medium. In the present study, the lower performance of SE in terms of algae biomass production is related both to the recalcitrant nature of the existing organic matter and the lower concentration of phosphorus.

Table 4 | Removal of organic matter during microalgal cultivation in the evaluated effluents

	PE			SE		
Variables	**OPR 1**	**OPR 2**	**OPR 3**	**OPR 4**	**OPR 5**	**OPR 6**
[CODs]i (mg/L)	345.4	342.2	274.6	63.1	48.9	78.0
[CODs]f (mg/L)	155.8	156.5	190.0	86.0	44.0	39.0
Removal (%)	54.9	54.3	30.8	−36.3	10.1	50.0
[TDC]i (mg/L)	245.3	132.1	177.0	78.1	50.9	31.6
[TDC]f (mg/L)	72.8	79.8	98.8	54.8	49.1	30.8
Removal (%)	70.3	39.6	44.2	29.8	3.5	2.3
[DOC]i (mg/L)	236.6	125.8	170.0	28.1	18.1	14.4
[DOC]f (mg/L)	60.4	59.6	66.4	49.2	26.7	14.6
Removal (%)	74.5	52.7	61.0	−75.1	−48.1	−1.6

[...]i: initial concentration of the variable (in the raw effluent); [...]f: final concentration of the variable (after the cultivation).

Removal of pollutants

Organic matter

Table 4 shows the removal of organic matter by the cultivation of biomass in the evaluated effluents.

A lower removal of filtered COD was expected in the SE due to the higher recalcitrance of the organic matter. However, there was a removal of 50% (OPR 6) – similar to the PE (OPR 1 and 2) – a removal of 10% (OPR 5), and an increase of 36.3% (OPR 4) of filtered COD. These results reflect the wide variation in the effluent composition. The increase of CODs in OPR 4 can be related to the release of organic compounds by the microalgae. According to Wang *et al.* (2014), a high nutrient load and highly unbalanced N/P ratio may cause unfavorable conditions for algae, which can lead to the production of a large amount of exocellular protein. OPR 4 was the operation in which the PBR worked with the higher concentration of ammoniacal nitrogen, and consequently, a higher N/P ratio (Table 2).

During all of the operations, TDC was removed. However, the DOC behavior was different between the effluents. While a fast removal of DOC was observed at the start of cultivation in the PE, there was an increase in DOC during the cultivation in the SE. This behavior can be related to the recalcitrance of the existing organic matter in the SE.

Nitrogen and phosphorus

Figure 5 shows the concentrations of N-NH_4^+ and P_S recorded during the cultivation in the primary and SE.

The removal of N-NH_4^+ in all the operations was higher than those obtained by Wang *et al.* (2010) through the cultivation of *Chlorella* sp. in raw municipal effluent (82.4%), primary municipal effluent (74.7%), and concentrated municipal effluent after centrifugation (78.3%). The required

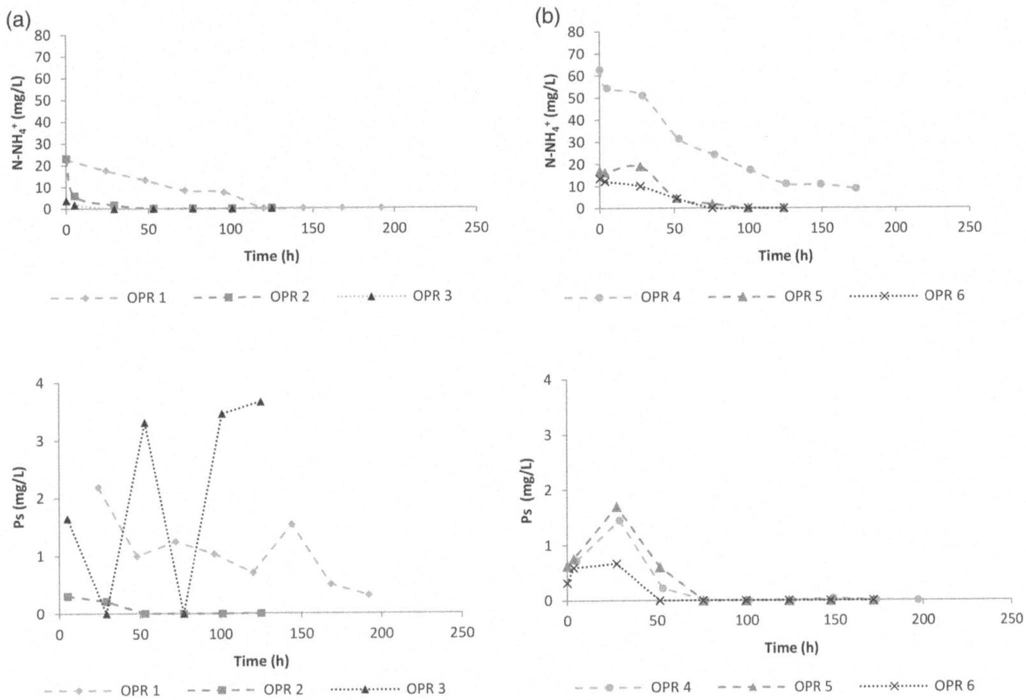

Figure 5 | Concentration of N-NH_4^+ and P_S recorded during cultivation in both effluents (a) PE and (b) SE.

time for the maximum removal of N-NH_4^+ was similar only between OPR 2 and 3 (50 hours) and OPR 5 and 6 (75.5 hours). As also reported by Ruiz-Martinez *et al.* (2012), the pH control allowed the assumption that much of the removal of N-NH_4^+ is due to its assimilation by the biomass because values that would favor volatilization were not achieved.

After achieving complete removal of N-NH_4^+, N-NO_3^- was expected to be the preferably consumed form of nitrogen. However, there was a wide variation in the concentration of N-NO_3^- during all operations, which even occurred in increments of this component in the final effluent (OPR 2, 5, and 6). Among the effluents, higher removal of N-NO_3^- was observed in OPR 1 (63.9%) and OPR 4 (66.7%). The increases in N-NO_3^- were possible due to the characteristics of the culture medium, such as the presence of N-NH_4^+ and DO. Evaluating the consortium cultivation of mixotrophic algae in the primary municipal effluent that had been previously sterilized and filtered, Mahapatra *et al.* (2014) attributed the increase in the nitrate concentration (from 1.07 ± 0.15 to 1.54 ± 0.28 mg/L) to nitrification, a two-stage process performed by nitrifying bacteria – autotrophic organisms that are extremely common in mixotrophic cultivation systems. Furthermore, the permanence of N-NO_3^- in the culture medium can be associated with the absence of another nutrient, such as phosphorus, which can limit the biomass growth and subsequently the assimilation of this nutrient.

With the PE, a significant presence of N-NO_2^- was not identified, whereas the reduced concentration of N-NO_2^- in the SE was completely removed in the first 75 hours (OPR 5 and 6) and in the first 120 hours (OPR 4). The removal of N-NO_2^- recorded during the cultivation in SE is possibly associated with the occurrence of nitrification because the presence of oxidizing bacteria of N-NH_4^+ and N-NO_2^- occurs frequently in activated sludge systems.

As shown in Figure 5(b), different from the PE, the behavior of the SE regarding the removal of P_S was similar between OPR 4, 5, and 6, which resulted in complete removal in up to 75 hours. A similar performance was obtained by Ruiz-Martinez *et al.* (2012), who identified removal of 97.8% of P-PO_4 through the treatment of secondary domestic effluent by the cultivation of microalgae and cyanobacteria. Both in the PE and in the SE, the P_S removal was attributed to the assimilation by the biomass because the pH was controlled between 6 and 8.

With the exception of OPR 3, which showed a high variation in the P_S concentration in the medium, alternating between increases and complete removal of this nutrient, all of the other PE operations showed high removal of P_S, i.e., 95% (OPR 1, after 192 hours) and 100% (OPR 2, after 53 hours), which are higher than what was reported by

Wang *et al.* (2010) in the cultivation of *Chlorella* sp. in PE (91%, in terms of P-PO_4).

Similarly to OPR 3, fluctuations of P-PO_4 were reported by Kumar *et al.* (2010) in all the feeding frequencies that were evaluated for the cultivation of *Chlorella* sp. in pig-farming effluent. The authors stated that the alterations of P-PO_4 in the culture medium can be associated with various factors, such as the concentration of organic particles and sediments. In certain cases, the removal of P-PO_4 can be the result of its adsorption by the sediments and organic particles that exist in the culture medium. When saturation of the sediments is reached, the phosphorus content is released, which increases its concentration in the medium.

According to Wang *et al.* (2010), a limitation that hinders the spread of effluent treatment technologies based on algae is their long hydraulic retention time (HRT) in relation to conventional treatment systems, such as activated sludge. However, the full removal of N-NH_4 and P_S in the first 53 hours of OPR 2 allowed the recovery of nutrients and a high production of biomass with a wide range of application, which again indicates the potential use of the system at the secondary level of treatment.

Quantification of the accumulated lipid and lipid productivity

Table 5 shows the results of the total lipid content, total biomass productivity, and lipid productivity that were determined for each operation.

Both the evaluated effluents had a lipid content of less than 10%. Therefore, the highest lipid productivity observed in the PE, i.e., 3.7 g/m²·d (considering PBR surface area) or 15.6 mg/L·day (considering its volume), was attributed to its higher biomass productivity in relation to the SE due to its nutritional condition, which is more favorable for the growth of biomass.

In this study, the nutritional condition influenced only the production of biomass, showing no difference in the accumulation of lipids by the algal cell, which is primarily due to the abundant availability of (ammoniacal) nitrogen in both effluents. This fact may explain the low lipid content of the cultivated biomass, which did not need to store fat at any moment but needed to grow. Although, OPR1 and OPR3 were exceptions, with a low N/P ratio and indicating nitrogen limitation, lipid content for these operations was also low. The biomass cultivation during a longer period of time could result in a larger accumulation of lipids because the culture would be subjected to longer periods of nutritional restriction. However, under strict nutrient conditions, despite the larger accumulation of lipids, a loss of algal biomass can occur.

Even with low lipid content, we can estimate a biodiesel productivity of 12,333.33 L/ha·year. This performance is similar to that observed by Woertz *et al.* (2009) with 9,000 L/ha·year, which is 25 times greater than the productivity of soybean oil (490 L/ha·year) (USDA 2005). However, it must be highlighted that biodiesel from microalgae, up to date, is economically unfeasible, due to the overall cost of downstream processing.

Therefore, the accumulated energy in the biomass could alternatively be used in other energy routes. According to Sialve *et al.* (2009), a more energetically favorable use of the biomass with a lipid content of less than 40% is anaerobic digestion. For these authors, the anaerobic digestion of this substrate for methane generation, considering the control of the several parameters involved, can generate as much energy as that obtained with the use of lipids. Moreover, the digestate after anaerobic digestion can be reused for microalgae culture medium as it still contains considerable amounts of nutrients. Besides energetic purposes, microalgae biomass can be used as an organic fertilizer in the agriculture. In the present study, the complete phosphorus and nitrogen assimilation into the biomass has to be outlined and microalgae cultivation can be pointed out as a system for nutrient recovery from water and effluents and use as fertilizer.

Table 5 | Total lipid content (% in the dry biomass), total biomass productivity (g/m².d), and lipid productivity (g/m²·d) for each evaluated effluent

		PE			SE		
	Unit	**OPR 1**	**OPR 2**	**OPR 3**	**OPR 4**	**OPR 5**	**OPR 6**
Total lipids	%	7.2	7.1	6.8	5.5	6.7	3.2
Biomass productivity[a]	g/m²·d	32.2	52.5	26.5	12.1	11.4	10.5
Lipid productivity[a]	g/m²·d	2.3	3.7	1.8	0.7	0.8	0.3

[a]Calculations based on the PBR surface area.

CONCLUSIONS

The results demonstrate the potential of cultivating microalgae in agro-industrial effluents previously treated at the primary and secondary levels. Distinct nutritional conditions differently affected the production of biomass, which reached the highest value in the PE, mainly due to its higher availability of phosphorus, nitrogen and organic matter. Complete removal of ammoniacal nitrogen and soluble phosphorus were observed in both effluents. The nutritional condition, although crucial for the growth of biomass, did not influence lipid accumulation. Moreover, it is important to highlight the potential of the proposed microalgae system to recover valuable products from wastewater, whether lipids or nutrients, which can be valued for several purposes.

ACKNOWLEDGEMENTS

The authors acknowledge the financial assistance provided by the Technology and the Research Support Foundation of Minas Gerais – FAPEMIG (grant number TEC-APQ-01892-16).

REFERENCES

ANUALPEC – ANUÁRIO DA PECUÁRIA BRASILEIRA 2015 *Pecuária de corte (estatísticas) [Brazilian livestock breeding Statistics]*. FNP Consultoria & Comércio, São Paulo (in Portuguese).

APHA 2012 *Standard Methods for the Examination of Water and Wastewater*, 22nd edn, American Public Health Association/American Water Works Association/Water Environment Federation, Washington, DC, USA.

Arbib, Z., Ruiz, J., Álvarez-Días, P., Garrido-Pérez, C., Barragan, J. & Perales, J. A. 2013 Long term outdoor operation of a tubular airlift pilot photobioreactor and a high rate algal pond as tertiary treatment of urban wastewater. *Ecological Engineering* **52**, 143–153.

Assemany, P. P., Calijuri, M. L., Santiago, A. F., Couto, E. A., Leite, M. O. & Sierra, J. J. 2014 Effect of solar radiation on the lipid characterization of biomass cultivated in high-rate algal ponds using domestic sewage. *Environmental Technology* **35** (17–20), 2296–2305.

Brasil. Conselho Nacional do Meio Ambiente. Resolução n° 430, 13 de Maio de 2011. Dispõe sobre as condições e padrões de lançamento de efluentes, complementa e altera a Resolução no. 357, de 17 de março de 2005, do Conselho Nacional do Meio Ambiente-CONAMA [Brazilian Guideline for wastewater conditions and standard for discharge] (in Portuguese).

Cheirsilp, B., Suwannarat, W. & Niyomdecha, R. 2011 Mixed culture of oleaginous yeast *Rhodotorula glutinis* and microalga *Chlorella vulgaris* for lipid production from industrial wastes and its use as biodiesel feedstock. *New Biotechnology* **28** (4), 362–368.

De Sena, R. F., Tambosi, J. L., Genena, A. K., Moreira, R. F. P. M., Schröder, H. F. & José, H. J. 2009 Treatment of meat industry wastewater using dissolved air flotation and advanced oxidation processes monitored by GC–MS and LC–MS. *Chemical Engineering Journal* **152** (1), 151–157.

FAO – Food and Agriculture Organization 2009 *The State of Food and Agriculture 2009*. Food and Agriculture Organization of the United Nations, Rome.

Jorquera, O., Kiperstok, A., Sales, E. A., Embirucu, M. & Ghirardi, M. L. 2010 Comparative energy life-cycle analyses of microalgal biomass production in open ponds and photobioreactors. *Bioresource Technology* **101** (4), 1406–1413.

Kumar, M. S., Miao, Z. H. & Wyatt, S. K. 2010 Influence of nutrient loads, feeding frequency and inoculum source on growth of *Chlorella vulgaris* in digested piggery effluent culture medium. *Bioresource Technology* **101**, 6012–6018.

Mahapatra, M. D., Chanakya, H. N. & Ramachandra, T. V. 2014 Bioresource technology bioremediation and lipid synthesis through mixotrophic algal consortia in municipal wastewater. *Bioresource Technology* **168**, 142–150.

MAPA. Ministério da Agricultura, Pecuária e Abastecimento 2015 www.agricultura.gov.br/animal/exportacao (accessed 10 June 2015).

Martin, C., De La Noun, J. & Picard, G. 1985 Intensive cultivation of freshwater microalgae on aerated pig manure. *Biomass* **7**, 245–259.

Nederlands Norm (NEN). NEN 6520 1981 *De spectrofotometrische bepaling van chlorophyll-a in oppervlaktewater kritisch bekeken (Water: Spectrophotometric Determination of Chlorophyll-a Content)*. Nederlands Normalisatie-Instituut, Delft, The Netherlands.

Nush, E. 1980 Comparison of different methods for chlorophyll and phaeopigments determination. *Archiv für Hydrobiologie* **4**, 14–36.

Qi, F., Xu, Y., Yu, Y., Liang, X., Zhang, L., Zhao, H. & Wang, H. 2017 Enhancing growth of *Chlamydomonas reinhardtii* and nutrient removal in diluted primary piggery wastewater by elevated CO2 supply. *Water Science and Technology* **75** (10), 2281–2290.

Ruiz-Martinez, A., Garcia, N. M., Romero, I., Seco, A. & Ferrer, J. 2012 Microalgae cultivation in wastewater: nutrient removal from anaerobic membrane bioreactor effluent. *Bioresource Technology* **126**, 247–253.

Santiago, A. F., Calijuri, M. L., Assemany, P. P., Calijuri, M. C. & Reis, A. J. D. 2013 Algal biomass production and wastewater treatment in high rate algal ponds receiving disinfected effluent. *Environmental Technology* **34** (13–14), 1877–1885.

Sialve, B., Bernet, N. & Bernard, O. 2009 Anaerobic digestion of microalgae as a necessary step to make microalgal biodiesel sustainable. *Biotechnology Advances* **27**, 409–416.

USDA 2005 *Agricultural Statistics 2005*. National Agricultural Statistics Service, United States Government Printing Office, Washington, DC.

Uthermöhl, H. 1958 Zur vervollkomnung der quantitativen phytoplankton – methodic mitt (Improving the Quantitative

Phytoplankton Method). *Verhandlungen des Internationalen Verein Limnologie* **9**, 1–38.

Valdés, F. J., Hernández, M. R., Catalá, L. & Marcilla, A. 2012 Estimation of CO_2 stripping/CO_2 microalgae consumption ratios in a bubble column photobioreactor using the analysis of the pH profiles. Application to *Nannochloropsis oculata* microalgae culture. *Bioresource Technology* **119**, 1–6.

Wang, L., Min, M., Li, Y., Chen, P., Chen, Y., Liu, Y., Wang, Y. & Ruan, R. 2010 Cultivation of green algae *Chlorella* sp. in different wastewaters from municipal wastewater treatment plant. *Applied Biochemistry and Biotechnology* **162** (4), 1174–1186.

Wang, M., Kuo-Dahab, W. C., Dolan, S. & Park, C. 2014 Kinetics of nutrient removal and expression of extracellular polymeric substances of the microalgae, Chlorella sp. and Micractinium sp., in wastewater, *Bioresource Technology* **154**, 131–137.

Woertz, I., Feffer, A., Lundquist, T. & Nelson, Y. 2009 Algae grown on dairy and municipal wastewater for simultaneous nutrient removal and lipid production for biofuel feedstock. *Journal of Environmental Engineering-ASCE* **135** (11), 1115–1122.

Xin, L., Hong-Ying, H. & Jia, Y. 2010 Lipid accumulation and nutrient removal properties of a newly isolated freshwater microalga, *Scenedesmus* sp. LX1, growing in secondary effluent. *New Biotechnology* **27** (1), 59–63.

Xu, K., Lv, B., Huo, Y. H. & Li, C. 2018 Toward the lowest energy consumption and emission in biofuel production: combination of ideal reactors and robust hosts. *Current Opinion in Biotechnology* **50**, 19–24.

Zhu, J., Rong, J. & Zong, B. 2013 Factors in mass cultivation of microalgae for biodiesel. *Chinese Journal of Catalysis* **34**, 80–100.

First received 29 October 2017; accepted in revised form 6 April 2018. Available online 17 April 2018

Microalgae cultivation for wastewater treatment and biogas production at Moscow wastewater treatment plant

N. Shchegolkova, K. Shurshin, S. Pogosyan, E. Voronova, D. Matorin and D. Karyakin

ABSTRACT

The process of cultivation of microalgae on purified and clarified wastewater of Kuryanovo wastewater treatment plants (KWWTP) was studied. The studies were conducted on monoculture (*Scenedesmus quadricauda* and *Chlorella sorokiniana*) and on polyculture, the composition of which was formed from microalgae present in the wastewater. The authors created and investigated the columnar photobioreactor (PBR), which acted as a pilot project on the purified and clarified water of KWWTP and allowed the removal of total nitrogen and phosphorus phosphates with an efficiency of up to 90%. The formation of a stable biocenosis from 22 species of algae (with 3–4 dominant species) and 31 species of zooplankton organisms belonging to six systematic subdivisions was recorded. The optimal retention time of the microalgae polyculture for the most effective wastewater treatment has been determined. The conducted studies have shown that the depth of decomposition of ashless matter and the ultimate biogas potential of untreated microalgae biomass is 15% lower than the corresponding values obtained with digestion of activated sludge, which necessitates studies in the field of pretreatment of algal biomass. The paper shows: connections between chlorophyll-a content, algal biomass and fluorescence index F_0 and between biomass increment and Fv/Fm value.

Key words | biogas, biomass, fluorescence, methane, photobioreactor, wastewater treatment

N. Shchegolkova (corresponding author)
D. Karyakin
Institute of Water Problems RAS,
Gubkin Street, 3, 119333 Moscow,
Russia
E-mail: *nshegolkova@gmail.com*

N. Shchegolkova
Faculty of Soil Science,
Lomonosov Moscow State University,
119991, Leninsky Gory, GSP-2, 119234 Moscow,
Russia

K. Shurshin
ERM Eurasia Limited,
Trehprudny Per., 11/13 build. 3, 121001 Moscow,
Russia

S. Pogosyan
E. Voronova
D. Matorin
Faculty of Biology,
Lomonosov Moscow State University,
Leninskiye Gory, GSP-2, 119234 Moscow,
Russia

INTRODUCTION

Numerous studies have shown that the cultivation of microalgae is at the center of solving many environmental problems: the production of renewable biofuels, the production of biomass for livestock feed or fish farms, the assimilation of greenhouse gases, and the wastewater treatment (Benemann 2003; Chen *et al.* 2003; Yen & Brune 2007; Park *et al.* 2013; Marchello *et al.* 2015; Mehrabadi *et al.* 2016). The most important of these problems is the production of biofuels from the biomass of microalgae with simultaneous assimilation of carbon dioxide. Underestimation of the limiting role of nitrogen and phosphorus for obtaining biofuels reduces to 'no' all the grand plans for replacing fossil fuels with biofuels. For the production of microalgae biomass, these elements can be supplied with mineral fertilizers and from various kinds of waste and sewage. Estimated calculation (Shchegolkova 2012) has shown that the energy potential of biomass, which can be obtained by using all nitrogen fertilizers produced in Russia, is about 8% of all energy produced in Russia. This means that the assimilation of CO_2 in the cultivation of microalgae on artificial nutrient media is unlikely to be profitable in the long run. However, nitrogen and phosphorus-containing wastes (primarily domestic wastewater) available in each megapolis are a 'free' source of these elements and, of course, can and should be used to produce biofuel from microalgae. Earlier calculations (Shchegolkova 2012) have allowed estimation of the amount of biogas, which can be obtained from domestic wastewater of such a megacity as Moscow. The total possible production of biogas energy is 1.6–2.2 billion kWh. This makes up 3.2–4.3% of the consumed energy in Moscow. An important point here is that the use of microalgae in the field of wastewater treatment allows the achievement of not only high quality purification, but also an energetic payback of the enterprise, in some

doi: 10.2166/wst.2018.088

cases due to the generation of biogas during anaerobic digestion of the excess algal biomass that is formed. In addition, lately houses with closed cycles of nutrient elements are increasingly spreading. Photobioreactors in these houses are part of the architectural ensemble, a wastewater treatment plant and a biomass source at the same time.

Photobioreactors and biofuel production from algae are being actively studied all over the world. Monoculture of algae can be used in the secondary treatment process and get productivity up to 93 dry weight (DW) g/L/day with 100% reduction of nitrogen and phosphorus (Abdelaziz *et al.* 2014). Strains of algae monoculture can produce biogas – up to 671 mL/L/day (raw biogas) and up to 86 mL/L/day (methane) (Mahdy *et al.* 2016). Several studies have reported the polyculture biomass productivity on pretreated wastewater up to 141 DW g/L/day (Prandini *et al.* 2016) and 12 g/m^2/day (Park *et al.* 2013). Algae strains were carried out removing 68% of nitrites, 54% of nitrates, 45% of ammonium and 94% of phosphates (Marchello *et al.* 2015). Biogas production described as 130 mL/g (Passos *et al.* 2016). Also, studies reported the polyculture productivity on initial (raw) wastewater up to 11 g/m^2/d (Mehrabadi *et al.* 2016). Efficiency of removal by algae achieved 87% for TOC, 85% for nitrogen, 92% for phosphorus and 49% for zinc (Hodges *et al.* 2017). Biogas production achieved from 317 to 1,170 mL/L/day (Yen & Brune 2007) and up to 0.9 g/m^2/day (Mehrabadi *et al.* 2016).

These data were mainly obtained on laboratory devices in short-term experiments. Data on the long-term monitoring of the functioning of the PBR on actual operating water treatment plants is extremely limited.

According to economic analysis conducted by the University of California at Berkeley (Lundquist *et al.* 2010), anaerobic digestion and the generation of biogas are the most beneficial option for utilizing the excess biomass of microalgae. Thus, the efficiency of the entire process and the very possibility of using microalgae directly depend on the completeness of digestion and output of biogas. The depth of the process of anaerobic digestion can be regulated by changing its external conditions (the residence time of the substrate in the reactor) and the properties of the substrate (species composition and pretreatment). The properties of the inoculum (a culture adapted to algomass digestion) are important for initiating the digestion process and the initial stage of the process (Lundquist *et al.* 2010).

Therefore, the study of the processes of microalgae cultivation in domestic wastewater and the processes of digestion of this biomass is of great importance. The main objective of our work was to determine the main technological parameters for the cultivation of microalgae on domestic effluents of a large water treatment plant and to assess the biogas generation potential of the grown biomass of algae. For this purpose, the main characteristics of biomass growth were determined, as well as the biogas composition and the depth of decomposition of the microalgae biomass during digestion. In addition, the biopotential of algomass and the mass of activated sludge as substrates for anaerobic thermophilic digestion were compared.

MATERIALS AND METHODS OF RESEARCH

The research was carried out at KWWTP, which receive sewage from half of the city of Moscow. For the cultivation of microalgae, purified and clarified water with the composition showed in Table 1 was used.

To assess the growth rate of monoculture in the first day, *Chlorella sorokiniana* and *Scenedesmus quadricauda* were used. Cultivation was carried out under artificial illumination in conical flasks with continuous mixing (with a shaker) at different ratios of the initial biomass of algae and wastewater. To an equal volume of biologically purified wastewater (BPW), a suspension of algae *Chlorella sorokiniana* or *Scenedesmus quadricauda* was added in ratios of 7, 13 and 27% of the total volume of liquid in the reactor.

During the 24-h period, the absorption spectra measurements were made. Measurements of absorption spectra of algal suspensions were carried out in the range from 350 nm to 850 nm on a single-beam spectrophotometer with an integrating sphere based on spectrometer USB2000 (Ocean Optics, USA). The absorption spectrum of algal suspension was calculated from the two spectra of light attenuation obtained at the installation of the sample cuvette, placed at different distances from the integrating sphere, that allowed evaluation of light scattering in the sample (Merzlyak & Naqvi 2000). The initial values of the optical density of the algae suspension corresponded to

Table 1 | Chemical composition of biologically purified and clarified wastewater

Indicators	Biologically purified wastewater (BPW)	Clarified wastewater (CW)
N-NH_4, mg/Ll	2.50 ± 0.61 mg/L	24.20 ± 1.31
N-NO_2, mg/L	0.36 ± 0.23 mg/L	0.00 ± 0.00
N-NO_3, mg/L	13.63 ± 1.45 mg/L	0.04 ± 0.01
P-PO_4, mg/L	1.66 ± 0.27 mg/L	5.65 ± 0.56
Chlorophyll *a*, mg/L	0.001 ± 0.001 mg/L	0.001 ± 0.001

Chlorella –2.18, and for *Scenedesmus* – 2.46. The values of the optical density at a wavelength of 678 nm show the relative content of chlorophyll in the suspension of algae, and the optical density at 550 nm demonstrates the relative number of cells in the culture. The values of the optical density normalized at the initial instant of time determine the relative increase in the number and content of chlorophyll 'a' in the suspension.

Laboratory photobioreactor

The pilot plant, which is a photobioreactor with microalgae, was created and put into operation in the Engineering and Technological Center of the MGUP 'Mosvodokanal' on the Kuryanovo Wastewater Treatment Plants. The plant (Figure 1) consists of columns with algal culture; illumination system (3000 lux, determined with the help of a luxometer); an air supply and distribution system (to supply algae with CO_2 and mix biomass within columns); a system for supplying purified wastewater once daily to replace water discharged from the reactor; a system for biomass dehydration by filtering with the use of a vacuum pump (a filter 3 μm pore size). Air is supplied 24 h a day and columns are illuminated 12 h a day; illumination is regulated by a controller. Water temperature in the reactors was 22–25 °C. Columns are completely mixed batch reactors with once daily partial exchange of fluid. In the columns, different regimes were set, differing in the *volumetric exchange ratio* (VER – the volume of the reactor liquid, exchanged for wastewater, in percent from the total volume of the column) – 17, 25, 33, 40 and 50%.

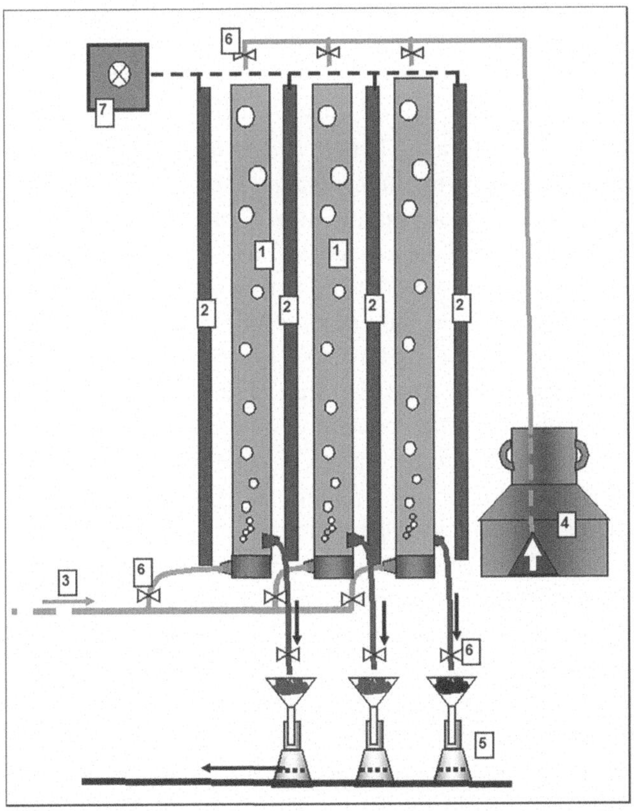

Figure 1 | Diagram of photobioreactor: 1 – reactor columns with algal culture; 2 – illumination system; 3 – the system of air input and distribution for supply of algae with CO_2 and for culture mixing; 4 – the system of treated waste water supply, once daily (reservoir with a pump); 5 – the system for biomass dehydration and culture sampling from cultivators; 6 – valves; 7 – controller.

Determined parameters: the composition of water entering the PBR and outgoing from the PBR (daily: pH, NH_4, NO_3, NO_2, PO_4, suspended matter, chlorophyll; weekly: species composition and the number of microalgae, species composition and the number of zooplankton). The content of chlorophyll 'a' was determined after pigment extraction with organic solvent with subsequent spectrometric measurement. The filtered biomass was weighed.

The amount of biomass was also determined through fluorescence measurements.

Biomass samples were collected into non-sterile, 1-L plastic bottles and analyzed for pigment concentration and fluorescence. Fluorescence was measured within 2 h after sampling. Parameters of chlorophyll fluorescence in algae were measured with a certified pulse fluorometer 'Mega-25' for measurements in highly diluted suspensions of microalgae. The fluorometer has been constructed at the Department of Biophysics, Faculty of Biology, Moscow State University (Pogosyan *et al.* 2006).

Fluorescence was measured using an electronic photomultiplier equipped with light filter CS 18 transmitting light at wavelengths longer than 680 nm. Fluorescence excitation was provided using LEDs having the emission peak at the wavelength of 455 nm. Fluorescence signals from opened (F_0) and closed (F_m) PS II reaction centers were measured at the excitation light intensities with photon densities of 0.8 and 6000 $\mu mol{\cdot}m^{-2}\,s^{-1}$, respectively. The potential efficiency of primary photosynthetic processes was calculated using the formula:

$$\frac{F_m - F_0}{F_m} = \frac{F_v}{F_m},$$

where $F_v = F_m – F_0$ is variable fluorescence.

The efficiency of the primary processes of photosynthesis (Fv/Fm) is a dimensionless characteristic of energy efficiency of photosynthesis, which represents output–input ratio and does not depend on algae species (Rubin 2005). The level of constant fluorescence, Fo, highly correlates with the total content of light-harvesting pigments in the photosynthetic apparatus of phytoplankton, and, correspondingly, it also correlates with the abundance of algal cells (Matorin *et al.* 2004). Hence, it can be used to estimate culture growth. The use of fluorometric method for estimation of chlorophyll content in phytoplanktonic algae does not require any pretreatment of water for measurement. Fluorescence was also measured with a WaterPAM (Walz, Germany) fluorometer.

To study fluorescence, three modes of PBR reactors were chosen: the first reactor was supplied with BPW; culture age was 3 days (one third of reactor volume was replaced daily with biologically purified water), the second and third reactors were supplied with clarified water; culture age in the second reactor – 3 days, in the third reactor – 2 days (VER: half of the reactor volume per day).

Digestion substrate and inoculums for gas generation

A microalgae biomass and activated sludge of KWWTP were used as a digestion substrate. The separation of biomass was carried out by sedimentation. The digested sludge, that was used as a inoculum, was additionally digested in serum bottles under anaerobic (53 °C) conditions for 240 h. Additional digestion was carried out for minimization of an impact of available organic substance in inoculum in general gas generation.

Gas generation experiment conditions

Ultimate biogas potential of algae biomass and activated sludge was determined in 500 cm^3 glass serum bottles. They were filled with inoculum (KWWTP digested sludge) and substrate (suspended algae biomass or KWWTP activated sludge) in a range of ratio (I:S) from 5:1 to 1:1 by ashless substance (AS), and added with liquid nutrient medium for mixed anaerobic cultures (Speece & McCarty 1964; Owen *et al.* 1979). Volume of substrate, inoculum and nutrient solution were calculated to make gas headspace in the bottle of 200 cm^3. 1 g of $NaHCO_3$ was added in each bottle in order to provide medium buffer capacity. The bottles were flushed with nitrogen for 5 min. After that, all bottles were sealed up with rubber plugs and applied with aluminium caps. The bottles were kept in a thermostat at constant temperature 53 °C during 76 days. Biogas volume was calculated by pressure measurement results in the bottles' headspace. Biogas composition was determined in the maximum biogas generation period by gas chromatography (chromatograph Crystal 2000M (Russia), equipped with a conductometer, a column filled with Porapak Q sorbent, carrier gas: He). The measurements were carried out in five replicates. The data were statistically analyzed by Fisher's test.

RESULTS AND DISCUSSION

Investigation of the growth of monoculture on purified wastewater in the first day

Incubation in BPW KWWTP with suspensions of *Chlorella sorokiniana* and *Scenedesmus quadricauda* was carried out at different ratios of algal biomass and wastewater (Figure 2).

During the entire cultivation period, all the reactors retained high fluorescence variable values Fv/Fm (0.62–0.69 for *Chlorella* and 0.53–0.62 for *Scenedesmus*), indicating a good physiological state of the algal culture in the wastewater. During the day, there was a significant decrease in the content of ammonium and nitrate nitrogen in the BPW KWWTP. The greatest increase in the content of chlorophyll 'a' (D678) and the number of cells (D550) in the suspension during the day was observed with a greater dilution of the inoculum in both *Chlorella* and *Scenedesmus* (Figure 2). However, in *Chlorella*, this dependence was expressed much more sharply. If in *Scenedesmus* the difference between the final concentrations in the three variants of the experiments was 0.2–0.3 normalized units of optical density (D678), then for *Chlorella* this difference between the variants at the end of the experiment was 1.0–1.5 normalized units. That is, the optical density at the wavelength of 678 (the content of chlorophyll 'a') doubled.

Within 24 h, the pH values of algal cultures increased significantly in both *Chlorella* and *Scenedesmus* at different amounts of inoculum (Figure 2).

Table 2 presents data on the reduction of ammonium and nitrate nitrogen in the reactors during the first day of cultivation. During 24 h of cultivation, there is a significant decrease in ammonium nitrogen, and the amount of nitrates also decreases. The greatest decrease in the content of ammonium nitrogen was observed for *Chlorella*, which confirms the observed sharp increase in the chlorophyll 'a' content and scattering for this culture. The difference in the rate of growth for these two species of algae is probably due to the reaction of *Scenedesmus* to specific (specifically for this species) growth inhibitors contained in the wastewater.

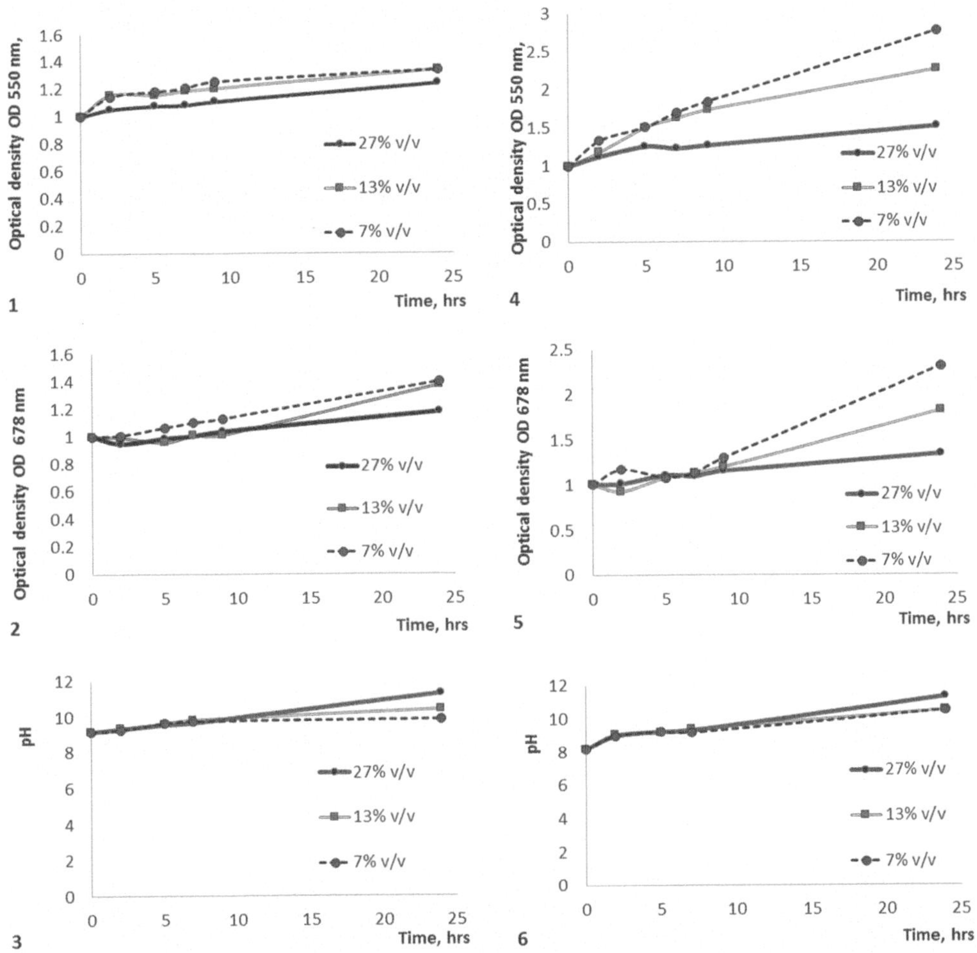

Figure 2 | Changes in the optical density at the wavelength of 678 and 550 nm and pH level of *Scenedesmus* (1)–(3) and *Chlorella* (4)–(6) during 24-hr cultivation (average of five variations).

The incomplete reduction in the amount of nitrates with the growth of culture on BPW KWWTP may depend on the short cultivation time (24 hours) and the small age of the initial culture chosen as an inoculum.

Investigation of the growth of polyculture on purified wastewater in a long-term experiment

Species composition of algae in photobioreactor

The formation of species composition in the reactor was the major problem since the beginning of the work. It is known that reactors with single species of alga are most productive. However, maintaining sterility in a reactor for post-treatment of domestic wastewaters is impossible. Algae, which grow in all wastewater treatment constructions (open channels of supplied water and secondary settling) are constantly coming into reactor.

The experiment lasted about 3 years. Culture of green alga *Scenedesmus* was added to photobioreactor at the very beginning of the experiment. The reactors then functioned as batch complete reactors with one daily wastewater discharge. Routine examination of the species composition of incoming and post-treated water began after the stabilization of the species composition of the PBR, approximately 1 month after the launch of the PBR. This examination has shown that species diversity in the incoming water was significantly higher, than in the reactor. Whereas 89 species were found in incoming water for the total observation time, only 22 species were found in the reactor. In both cases, green and diatomic algae are dominating (Figure 3). Species composition in photobioreactor is rather stable during the whole time of observation.

Dominating species of algae in water, coming into reactors, are, among diatomea: *Nitzschia palea (Kutz.) W. Sm.*, *Stephanodiscus sp*, among green algae: *Scenedesmus quadricauda Chod.*, *Scenedesmus opoliensis P. Richt.*, among

Table 2 | Parameters of ammonium nitrogen proportion and nitrates during 24 h of cultivation – average of five variations (*Scenedesmus* and *Chlorella*)

***Scenedesmus* cultivation time**	**2**	**5**	**7**	**9**	**24**
27 v/v of inoculum					
$N\text{-}NH_4$, mg/L	100	102	82	73	60
$N\text{-}NO_3$, mg/L	100	102	99	99	52
13 v/v of inoculum					
$N\text{-}NH_4$, mg/L	100	96	89	74	39
$N\text{-}NO_3$, mg/L	100	94	88	94	48
7 v/v of inoculum					
$N\text{-}NH_4$, mg/L	100	94	88	84	42
$N\text{-}NO_3$, mg/L	100	104	94	93	50
***Chlorella* cultivation time**	**2**	**5**	**7**	**9**	**24**
27 v/v of inoculum					
$N\text{-}NH_4$, mg/L	100	90	86	63	36
$N\text{-}NO_3$, mg/L	100	101	96	99	73
13 v/v of inoculum					
$N\text{-}NH_4$, mg/L	100	101	82	83	24
$N\text{-}NO_3$, mg/L	100	107	96	94	72
7 v/v of inoculum					
$N\text{-}NH_4$, mg/L	100	96	89	85	14
$N\text{-}NO_3$, mg/L	100	101	96	99	74

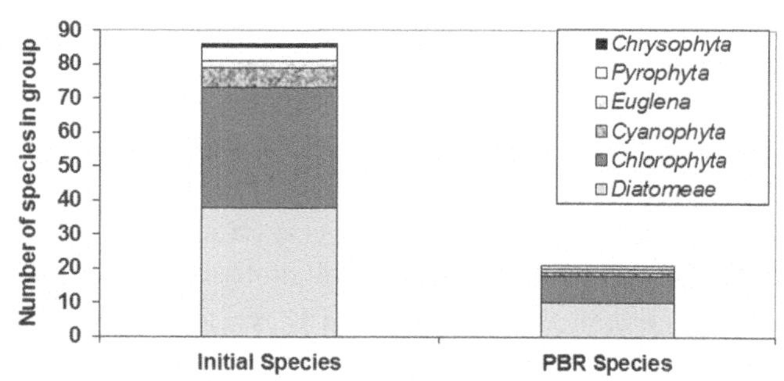

Figure 3 | Algae species characteristic of BPW and CW.

blue-green algae (cyanobacteria): *Oscillatoria amphibia f. tenuis (Anissim.) Elenk., Oscillatoria tenuis Ag. ex Gom, Chroococcus minutus*. Dominating species of algae in the reactors are, among diatomea: *Dicthyosphaerium ehrenbergianum Nag., Navicula viridula Kutz,* among green algae: *Scenedesmus obliquus (Turp.) Kutz., Scenedesmus parvus,* among blue-green algae: *Oscillatoria amphibia f. tenuis (Anissim.) Elenk.*, among pyrrophyta: *Didymocystis lineata Korsch.*, among euglenoids: *Astasia* sp.

As seen from analysis of species composition, structure of algal community in the photobioreactor is altered. Most biomass is composed by the following algae: *Oscillatoria, Scenedesmus* and *Astasia.* Total biomass of these species in discharged water was as high as 500–1,000 g/m^3.

Efficiency of purification from nutrients

Daily measurements were made of water quality and algal biomass growth in each column. Table 3 shows the statistically processed water quality data at the outlet of the reactor for different reactor loading options (with different retention times of the algal culture, which was set by a different fraction of the daily purified water).

The water, which has been purified in the photobioreactor, reaches Russian standards for nitrogen ammonium, nitrates and phosphates. The efficiency of their removal in the experiments was 34–48% for ammonium nitrogen, 33–87% for nitrogen of nitrates, 89–91% for phosphates, and 40–94% for total nitrogen.

The nitrogen content of the nitrite after post-treatment was at the level of 2–11 MPC, increasing with the VERs in the reactors (Table 3). It should be noted that the presence of a relatively high content of nitrites is due to the fact that the nitrification process in the reactor proceeds simultaneously with the assimilation of biogenic elements with microalgae. It was noted that the content of nitrite in the drain water from the reactor reached the normative values when in the storage tank for 1 h. This is due to the presence of a high oxygen content in the drain water (due to photosynthesis), which was spent on the ongoing process of nitrification (second stage).

Table 3 | Chemical composition treated wastewater by photobioreactor after algae biomass separation

Dilution, %	**17**	**25**	**33**	**40**	**50**	**MPC (maximum permissible concentrations, Russia)**
$N\text{-}NH_4$, mg/L	0.20 ± 0.07	0.24 ± 0.05	0.15 ± 0.02	0.04 ± 0.03	0.05 ± 0.04	0.4
$N\text{-}NO_2$, mg/L	0.07 ± 0.06	0.13 ± 0.08	0.15 ± 0.05	0.29 ± 0.08	0.28 ± 0.08	0.024
$N\text{-}NO_3$, mg/L	1.10 ± 0.99	1.55 ± 0.60	4.23 ± 1.35	6.73 ± 2.29	10.63 ± 1.29	9.1
Ntotal, mg/L	1.37 ± 1.01	1.91 ± 0.59	4.54 ± 1.41	7.06 ± 2.29	10.95 ± 1.26	–
$P\text{-}PO_4$, mg/L	0.09 ± 0.02	0.12 ± 0.02	0.13 ± 0.06	0.13 ± 0.04	0.14 ± 0.04	0.20

Removal of nitrogen forms by microalgae is based on the assimilation of this element, whereas purification of water from phosphates occurs both through assimilation and through another process – mediated chemical removal of phosphates. With intensive consumption of CO_2 by microalgae, the pH of the purified water increases (up to 8.0–9.5), which leads to the formation of sparingly soluble phosphate salts. The pH level in the PBR stably kept at values of 9.0–10.5. Thus, both biological and mediated chemical–biological removal of phosphates occurred simultaneously. That is why the efficiency of removal of phosphates in the variants of the experiment (with different biomass increments) did not differ and was 89–91%. The second process 'leveled out' the action of the first. The efficiency of the reactors was estimated by the following technological indicators: the efficiency of total nitrogen removal, the achievement of normative values for nutrient elements, the daily biomass increment per unit volume of the reactor (Table 3, Figure 4).

At VER of 17–33% the highest values of the purification efficiency from total mineral nitrogen (75–95%) are observed. The reduction in the efficiency of nitrogen removal occurred when VER had exceeded a third of the volume of the reactor. If with VER of a third of the volume per day, 75% of nitrogen is removed from the water entering the reactor, then in the case of an exchange rate of a half the volume, only 50% is removed.

VER of a third of the volume of the reactor was characterized by the greatest increase in biomass per unit volume of the reactor: an average of 250 mg/L/day by wet weight of biomass or 50 mg/L/day by dry weight. The maximum increase in raw biomass was 960 g/(m^3 day) or 190 mg/L/day on a dry basis.

As experiments in the pilot photobioreactor have shown, VER of 33% (of the total volume per day) is characterized by the maximum biomass increment with high water purification efficiency.

Disinfection of water and species composition of zooplankton

For three variants of experiments (VER: 17, 25 and 33%), studies were conducted on the efficiency of water disinfection in reactors. The content of total coliform and thermotolerant bacteria was determined in the entering the PBR and purified water (Table 4).

The purified water in three experiments after passing through the reactor has bacterial contamination values corresponding to the standards for wastewater. The disinfecting 'agent' is the organisms of zooplankton inhabiting the reactor. During the observation period, 31 species of zooplankton organisms belonging to six systematic subdivisions have been identified in the reactors: rotifers (16 species), cladoceras (eight species), copepods (one species), crustaceans (one species), nematodes (one species), chironomidae (one view). The dominant species with the maximum biomass (up to 700 mg/m^3) and the number (up to 800,000/m^3) are: crustaceans *Ostracoda*, rotifers *Philodina acuticornis Murray*, *Lecane (Monostyla) bulla Gosse*, *Colurella* sp., *Lepadella ovalis Mull.*, *Lepadella* sp., *Cephalodella gibba Ehrenberg*, *Asplanchna priodonta* Gosse and cladoceras *Oxyurella tenuicaudis Sars.*, *Alona rectangula Sars.* It is

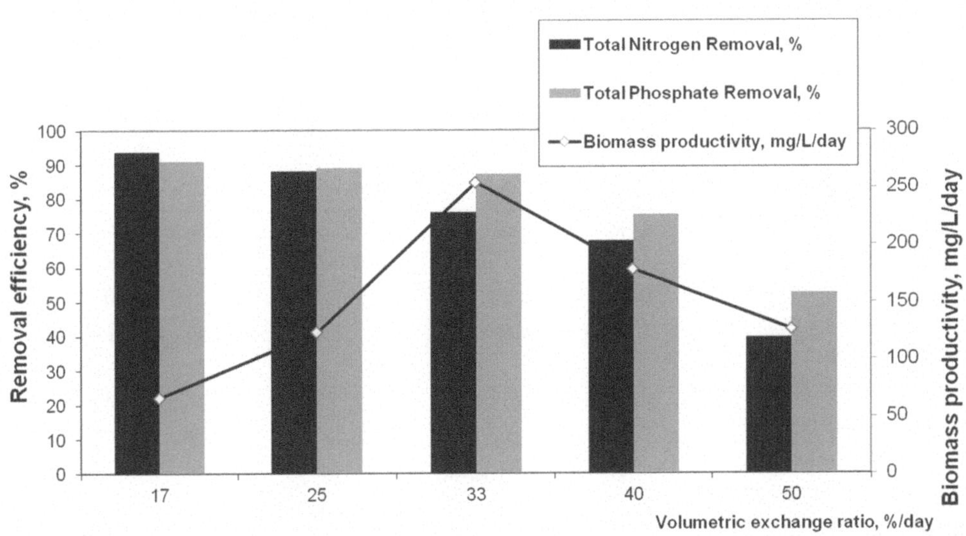

Figure 4 | The main indicators of the photobioreactor for different VER: (1) total nitrogen and phosphate removing efficiency, %; (2) biomass productivity, mg/L/day.

Table 4 | Bacterial characterization of the purified wastewater after photobioreactor

VER. %	Efficiency of disinfection	Total coliforms index, CFU/100 mL	Thermotolerant coliforms index, CFU/100 mL
33	Input value	68,000	44,000
	Output value	**420**	**18**
	Reduction, %	99	100
25	Input value	18,000	13,000
	Output value	**110**	**3**
	Reduction, %	99	100
17	Input value	7,600	5,900
	Output value	**220**	**2**
	Reduction, %	97	100
Russian standard (SanPiN 2.1.5.980-00)		***500***	***100***

these species that primarily disinfect water, consuming bacterial biomass. Their role is also important in the production of algae, since zooplankton transforms organic matter (bacteria, detritus) into nitrogen and phosphorus mineral compounds accessible to algae.

Evaluation of the application of the fluorimetric method for technological control of a photobioreactor

Technological indicators of productivity and fluorescence characteristics

The principal technological parameter of photobioreactor is biomass growth per unit reactor volume per unit time. Productivity depends on several factors: nutrient content, culture age, and the toxicity of incoming water. For 1 month, experiments were conducted in the PBR, where the fluorescence indices were compared for the cultivation of algae on purified and on clarified wastewater. The first and second reactors differed in the quality of incoming water and had the same culture age (3 days). The second and third reactors were supplied with clarified water and had culture age 3 and 2 days, respectively. Purified water (first column) had a lower nutrient content and lack of toxicity. The latter is confirmed by biotesting experiments; toxicity index with infusoria was 0.26 (Kozlov *et al.* 2006). Clarified water contains much nutrients, mostly in the form of ammonia salts, which are first consumed by algae. However, the toxicity level of this water is higher than that of biologically treated water. Toxicity index varied from 0.35 to 0.89 in biotests with infusoria (Kozlov *et al.* 2006). The most toxic water (in accordance with the high content of clarified water) was the water of the third column. All these factors resulted in different average values of productivity of the three reactors: 149, 206, 135 mg/L/day by wet weight with variability of the values not more than 25% of absolute values (Table 1).

Densities of the algal cultures (determined daily in discharge fluid) were also different. Wet weight of algae in the three reactors was 287, 125, and 165 mg/L, respectively.

Fluorescence parameters (Fo) and (Fv/Fm) were measured in samples from reactors, differing in the rate of culture growth and in the extent of growth suppression by toxic substances. It was found that coefficient of correlation between average values of Fo in the three reactors and chlorophyll 'a' concentration was 0.98, and coefficient of correlation between Fo and biomass content was 0.99.

The activity of PS2 reaction centers, related to oxygen evolution in photosynthesis, can be estimated from fluorescence parameter Fv/Fm. This parameter, which reflects the maximum quantum of PS2, can be used to estimate photosynthetic activity of the culture of microalgae in photobioreactor. In natural water bodies with a high nutrient content, Fv/Fm values may be as high as 0.6–0.7. Under the effect of stress conditions and pollutants, this parameter decreases. In dead cells, it is equal to 0.

Fv/Fm values in the three reactors varied within a range of 0.46–0.58. There were no statistically significant differences between the three reactors (Table 1), however, a tendency to direct correlation was found between Fv/Fm and reactor production (Table 5, Figure 5).

Thus, fluorometric method can be used not only for quantitative estimation of algal biomass in photobioreactor (by parameter F_o), but to also assess reactor productivity. An increase in production is associated with a higher Fv/Fm ration. Correspondingly, a decrease in the latter

Table 5 | Chlorophyll concentration, biomass content, fluorescence parameters F_o and photosynthetic activity (Fv/Fm) in PBR

	Column 1	Column 2	Column 3
Kind of wastewater	Biologically purified	Clarified	Clarified
Daily dilution, %	33	33	50
Toxicity of water	no	low-toxic	moderately toxic
Chlorophyll 'a' mg/L	1.48 ± 0.24	0.72 ± 0.39	0.90 ± 0.28
Total biomass, mg/L	287 ± 41	125 ± 28	165 ± 62
Productivity, mg/L/day	149 ± 36	206 ± 48	135 ± 34
F_0	4,475 ± 484	2,029 ± 901	2,918 ± 942
F_v/F_m	0.51 ± 0.07	0.55 ± 0.05	0.50 ± 0.02

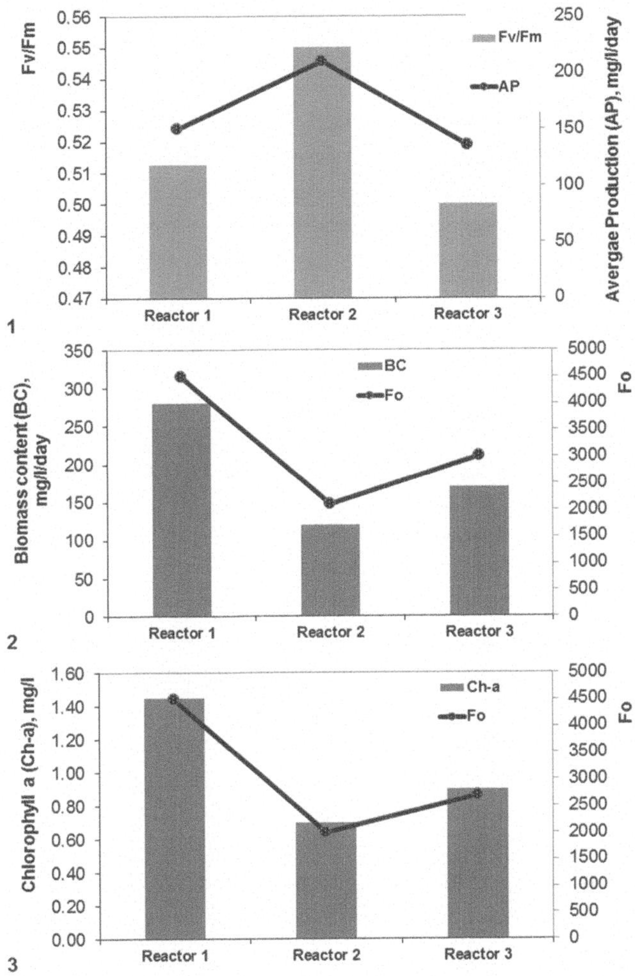

Figure 5 | Value of average production (1), biomass content (2), chlorophyll concentration (3) and fluorescence parameters (Fo and photosynthetic activity Fv/Fm) in photobioreactors.

parameter can be used as an indicator of water toxicity, nutrient deficiency, or other changes in technological mode, which leads to a lower efficiency of photobioreactor operation. Continuous instrumental control of the two parameters will help to timely apply technological measures for stabilization of photobioreactor production.

Evaluation of the energy potential of microalgae biomass

Impact of inoculums: substrate ratio on a value of specific methane yield in anaerobic digestion

The value of specific methane yield (dm^3 per kg of initial AS) depended on I:S ratio and inoculum type. When KWWTP digested sludge was used as inoculum, the increase of I:S from 1:1 to 3:1 resulted in the increase of specific gas yield from 110 to 134 dm^3 CH_4/kg VS, I:S ratio expansion from 3:1 to 5:1 did not result in significant increase of biogas yield (Table 6).

Absolute values of specific methane yield with use of digested sludge varied from 110 to 136 dm^3/kg AS, which corresponds to data (84–217 dm^3/kg AS) of other researchers studying anaerobic digestion of algae (Mussgnug *et al.* 2010; Ras *et al.* 2011; Gonzalez-Fernandez *et al.* 2012; Mehrabadi *et al.* 2016). Comparison of the ultimate biogas yields of algae and active sludge showed that the algae biogas production was 15% lower (at I:S ratio of 3:1).

Thus, methanogenic community of digested sludge is capable of relatively deep digestion of algae biomass. The specific methane yield of algae was found to be slightly less than that of excess activated sludge. An increase in the specific biogas yield and algae biomass degradability can be achieved after biomass pretreatment, such as thermohydrolysis (Gonzalez-Fernandez *et al.* 2012).

Biogas composition. Biogas obtained in the algae digestion process was composed by 57.0–59.7% of methane and by 40.3–43.0% of carbon dioxide. The biogas volume fraction

Table 6 | Parameters of anaerobic digestion of microalgae biomass

			Biogas composition, %	
I:S ratio	Specific methane yield, dm^3/kg AS	AS biodegradability, %	CH_4	CO_2
Algae with digested sludge as inoculum				
1:1	110	14	59.1	40.9
2:1	121	15	58.8	41.2
3:1	134	13	59.7	40.3
5:1	136	22	57.0	43.0
Active sludge with digested sludge as inoculum				
3:1	157	30	67.5	32.5

of methane for algae digestion corresponded to the literature data (Gonzalez-Fernandez *et al.* 2012), but was lower than that in the digestion of activated sludge. Variations in biogas composition may be caused by different chemical composition of digested substrates (microalgae and activated sludge), and also, possibly, by different availability for the methanogenic community of the internal contents of algal cells and activated sludge.

Biodegradability. High cost of waste utilization makes algae biomass decomposition degree one of the most important indicators of the process. Substrate biodegradability (based on AS) varied from 13% to 22% with expansion of I:S ratio. A tendency of increase of microalgae biomass degradability was observed with growth of inoculum ratio in the initial mixture. Active sludge AS degradability was higher, reaching 30%.

Possibility of inhibition with ammonium ion. Destruction of substrate cell walls and further fermentation of cell content in some cases may cause an increase of ammonium concentration to the levels inhibiting methanogenesis. Ammonium nitrogen concentration in centrifuge effluent of digestate (105.0–652.5 mg/dm^3) did not exceed toxic threshold for methanogens (Bruno *et al.* 2009). Thus, possibility of inhibition of anaerobic microalgae digestion process was excluded.

Solving technical problems in the development of an industrial photobioreactor

The technical and economic feasibility of the industrial implementation of the creation of an industrial PBR was assessed according to the following positions:

1. Total energy consumption for the operation of column reactors.
2. Possible volume of capital expenditures of the photobioreactor and the area occupied by the reactor: they are inversely related to the daily water exchange ratio in the columns.
3. Reducing the wastewater discharge fee: this indicator depends on the efficiency of removal of nitrogen and phosphorus.
4. Compensation of a part of the energy costs for water treatment by obtaining fuel from the biomass of algae.

The main energy costs in the pilot project are to illuminate the columns with fluorescent lights. In the pilot project, the specific energy consumption (for lighting and aeration) is about 70 kWh for additional cleaning of 1 m^3 of wastewater. For comparison, the energy costs for the complete treatment of 1,000 m^3 of wastewater in the Moscow treatment plant are 160–180 kWh. Therefore, the industrialization of a large photobioreactor with artificial lighting is unrealizable due to the astronomical high cost of purification. However, there are already ways to replace artificial lighting with natural – solar. In the world, modern technologies such as the 'light well' are actively developing. This technology concentrates the solar stream and distributes it through the system of fiber-optic transmitters to the right spot for lighting. The technology allows the maintenance of sufficient illumination in the room even on cloudy days. Using this technology for lighting, it is possible to reduce the specific energy consumption of a photobioreactor to 1,000 kWh/1,000 m^3 of wastewater with water (VER of 33% of the volume of the reactor and up to 700 kWh/1,000 m^3 of wastewater with VER of 50% of the reactor volume.

Calculate the area occupied by buildings with a system of 2-m column-reactors at the expense of treated waters of about 2 million m^3 per day (similar to a water treatment facility for a city with 10 million inhabitants). Estimated calculations when taking into account the annual dynamics of solar radiation show that at the latitude of Moscow with VER of 33%, the area of buildings will be 3 km^2, with VER of 50% – 2 km^2. For more southern cities of Russia (for example, for the city of Sochi), in order to provide additional treatment of effluents from the city with 0.5 million inhabitants, only 15–20 buildings with an area of about 10–12 thousand m^2 are needed. In warm solar regions, the treatment system does not require the construction of buildings and can be carried out 'in the open.'

Estimated economic calculations show that when using a photobioreactor, wastewater discharge fee on nitrogen and phosphorus compounds are reduced by 100 times according to modern standards. Also, biogas from the biomass of algae produced in the reactor will allow to compensate up to a third of the total energy expended for wastewater treatment.

CONCLUSIONS

The treatment plant in the classical sense can be transformed into an ecological complex for water purification and biogas production from microalgae. The most promising for the development of this technology are warm regions with high solar radiation. The application of this technology in all regions is very promising.

The principal difference between a photobioreactor and other treatment systems is as follows:

- Simultaneously, the water is treated and disinfected to the normative values.
- Purification by this method (using clarified water in PBR) occurs without waste, which must be taken out for deposition, as in the case of activated sludge. The resulting biomass is the raw material for a number of valuable products: biofuel and/or organic fertilizer and/or animal feed.
- The treatment plant has income items in the form of sales (or use for internal needs) of energy from biomass.

The data obtained in the present work demonstrated that the fluorescent method of photosynthesis assessment is applicable for the evaluation of technological processes in growing microalgae biomass. The method allows to significantly reduce the effort required to monitor and implement process control with continuous surveillance.

It was found that the ultimate biogas potential of microalgae is 15% less than that of activated sludge. The composition of the biogas obtained during digestion did not depend on the ratio of I:S and the type of inoculum, while the volume fraction of methane in biogas during digestion of the activated sludge was slightly higher than when digesting the algomass.

According to the obtained data, the optimal ratio of I:S should be recognized as 3:1. Inhibition of methanogenesis by ammonium ions was not observed.

Microalgae have a low potential for digestion, so further studies of increasing the efficiency of digestion should be aimed at finding various methods of preliminary treatment of algomass to increase the availability of organic matter.

ACKNOWLEDGEMENTS

The authors are thankful to the Kuryanovo wastewater treatment plant for their great support and collaboration on this research implementation.

This study was supported by the Russian Science Foundation (project no. 17-17-01204).

REFERENCES

Abdelaziz, A. E., Leite, G. B., Belhaj, M. A. & Hallenbeck, P. C. 2014 Screening microalgae native to Quebec for wastewater treatment and biodiesel production. *Bioresource Technology* **157**, 140–148.

Benemann, J. R. 2003 *Biofixation of CO_2 and Greenhouse gas Abatement with Microalgae – Technology Roadmap. Report No. 7010000926 Prepared for the U.S. Department of Energy.* National Energy Technology Laboratory, Washington, DC, USA.

Bruno, S., Bernet, N. & Bernard, O. 2009 Anaerobic digestion of microalgae as a necessary step to make microalgal biodiesel sustainable. *Biotechnology Advances* **27** (4), 409–416.

Chen, P., Zhou, Q., Paing, J., Le, H. & Picot, B. 2003 Nutrient removal by the integrated use of high rate algal ponds and macrophyte systems in China. *Water Science & Technology* **48** (2), 251–257.

Gonzalez-Fernandez, C., Bruno, S., Bernet, N. & Steyer, P. J. 2012 Thermal pretreatment to improve methane production of *Scenedesmus* biomass. *Biomass and Bioenergy* **40**, 105–111.

Hodges, A., Fica, Z., Wanlass, J., VanDarlin, J. & Sims, R. 2017 Nutrient and suspended solids removal from petrochemical wastewater via microalgal biofilm cultivation. *Chemosphere* **174**, 46–48.

Kozlov, M. N., Danilovich, D. A., Schegolkova, N. M., Filenko, O. F. & Pushkar, V. J. 2006 Investigation of the water quality of Moscow treatment plant by biotesting methods (Original in Russian: Оценка качества очищенной воды Московских очистных сооружений методами биотестирования). *Vodosnabzenie I sanitarnaya tekhnika (Water-Supply and Sanitary Engineering)* **11** (1), 31–39.

Lundquist, T. J., Woertz, I. C., Quinn, N. W. T. & Benemann, J. R. 2010 *A Realistic Technology and Engineering Assessment of Algae Biofuel Production.* Energy Biosciences Institute, University of California, Berkeley, CA, USA.

Mahdy, A., Ballesteros, M. & González-Fernández, C. 2016 Enzymatic pretreatment of Chlorella vulgaris for biogas production: influence of urban wastewater as a sole nutrient source on macromolecular profile and biocatalyst efficiency. *Bioresource Technology* **199**, 319–325.

Marchello, A. E., Lombardi, A. T., Dellamano-Oliveira, M. J. & de Souza, C. W. 2015 Microalgae population dynamics in photobioreactors with secondary sewage effluent as culture medium. *Brazilian Journal of Microbiology* **46** (1), 75–84.

Matorin, D. N., Antal, T. K., Ostrowska, M., Rubin, A. B., Ficek, D. & Majchhrowski, R. 2004 Chlorophyll fluorometry as a method for studying light absorption by photosynthetic pigments in marine algae. *Oceanologia* **46** (4), 519–531.

Mehrabadi, A., Craggs, R. & Farid, M. M. 2016 Biodiesel production potential of wastewater treatment high rate algal pond biomass. *Bioresource Technology* **221**, 222–233.

Merzlyak, M. N. & Naqvi, K. R. 2000 On recording the true absorption and scattering spectrum of a turbid sample: application to cell suspensions of the cyanobacterium *Anabaena variabilis. Journal of Photochemistry and Photobiology B: Biology (B)* **58**, 123–129.

Mussgnug, J. H., Klassen, V., Schlüter, A. & Kruse, O. 2010 Microalgae as substrates for fermentative biogas production in a combined biorefinery concept. *Journal of Biotechnology* **150**, 51–56.

Owen, W. F., Stuckey, D. C., Healy Jr, J. B., Young, L. Y. & McCarty, P. L. 1979 Bioassay for monitoring biochemical methane potential and anaerobic toxicity. *Water Research* **6** (13), 485–492.

Park, J. B. K., Craggs, R. J. & Shilton, A. N. 2013 Enhancing biomass energy yield from pilot-scale high rate algal ponds with recycling. *Water Research* **47** (13), 4422–4432.

Passos, F., Felix, L., Rocha, H., de Oliveira Pereira, J. & de Aquino, S. 2016 Reuse of microalgae grown in full-scale wastewater treatment ponds: thermochemical pretreatment and biogas production. *Bioresource Technology* **209**, 305–312.

Pogosyan, S. I., Kazimirko, Yu. V., Matorin, D. N., Riznichenko, G. Yu. & Rubin, A. B. 2006 Russian Patent Application No. 2354958.

Prandini, J. M., da Silva, M. L. B., Mezzari, M. P., Pirolli, M., Michelon, W. & Soares, H. M. 2016 Enhancement of nutrient removal from swine wastewater digestate coupled to biogas purification by microalgae *Scenedesmus* spp. *Bioresource Technology* **202**, 67–75.

Ras, M., Lardon, L., Bruno, S., Bernet, N. & Steyer, J.-P. 2011 Experimental study on a coupled process of production and anaerobic digestion of *Chlorella vulgaris*. *Bioresource Technology* **102**, 200–206.

Rubin, A. B. 2005 Biophysics of photosynthesis and methods for environment monitoring. *Tekhnologiya zhivykh system (Technology of Living Systems)* **2**, 47–68.

Shchegolkova, N. M. 2012 Utilization of nitrogen and phosphorus-containing waste in the city and the problems of biofuel energy development. *Voda: khimiya i ekologiya (Water: Chemistry and Ecology)* **2**, 38–44.

Speece, R. E. & McCarty, P. L. 1964 Nutrient requirements and biological solids accumulation in anaerobic digestion. *Advances in Water Pollution Research* **2**, 305.

Yen, H. W. & Brune, D. E. 2007 Anaerobic co-digestion of algal sludge and waste paper to produce methane. *Bioresource Technology* **98** (1), 130–134.

First received 30 September 2017; accepted in revised form 14 February 2018. Available online 27 February 2018

Improved methane yield from wastewater grown algal biomass

Mohit Thawani, Nidhi Hans, Saurabh Samuchiwal and Sanjeev Kumar Prajapati

ABSTRACT

Methane production from the algal biomass cultivated in a laboratory scale continuous photobioreactor (PBR) using sewage was evaluated in the present work. During the preliminary experiments, algal biomass reached up to 1.69 ± 0.35 g L^{-1} in 12 days' growth period. Besides, 65 to 100% removal in concentrations of total dissolved phosphorus (TDP), nitrate nitrogen (NO_3–N), total ammoniacal nitrogen (TAN) and soluble chemical oxygen demand (sCOD) was also recorded. The sCOD removal in the reactor was 100%, whereas removal of TDP, NO_3–N and TAN were up to 75, 40 and 92%, respectively. Upon anaerobic digestion, the fresh algal biomass showed methane yield of 180 mL g^{-1} VS_{fed}. Further, algal biomass was stored under natural conditions in open containers (aerobic conditions) in darkness at room temperature (27–30 °C) for 72 h. Interestingly, >48% COD solubilization from algal biomass was observed during storage. Pretreatment through natural storage was further confirmed with qualitative observations including scanning electron and fluorescence microscopic analysis. Moreover, higher methane yield (284.38 mL g^{-1} VS_{fed}) was observed from the samples stored for 60 h. Thus, natural storage for a designated period may be recommended as a prerequisite stage in the process of methane production from wastewater-grown algal biomass.

Key words | algae, anaerobic digestion, native bacteria, pretreatment, sewage

Mohit Thawani
Nidhi Hans
Saurabh Samuchiwal
Sanjeev Kumar Prajapati (corresponding author)
Netaji Subhas Institute of Technology,
Sector 3, Dwarka, New Delhi 110078,
India
E-mail: *sanjeev.kumar@iitp.ac.in*;
sanjukec@gmail.com

Sanjeev Kumar Prajapati
BioResource Engineering Lab,
Chemical and Biochemical Engineering,
Indian Institute of Technology Patna,
06/113, Bihta, Patna, Bihar 801106,
India

INTRODUCTION

Rapid growth of energy demand against the limited availability of conventional fossil fuels has forced human beings to explore alterative renewable energy resources. In the recent past decades, biofuel has emerged as a potential alternate to fossil fuel. Among biofuel options, methane from anaerobic digestion of whole cell algal biomass is considered most feasible (Passos *et al.* 2014). However, the sustainability of the algae-based methane generation process is questionable due to the high cost of nutrients. This limitation is well addressed, and several reports have been published on algae cultivation in wastewaters including industrial effluents and sewage (Rawat *et al.* 2011; Chen *et al.* 2015). Besides, algae cultivation also results in nutrient and pollutant removal, leading to simultaneous wastewater treatment. Hence, through algae cultivation, it is possible to convert wastewater nutrient into useful biomass feedstock.

Management and treatment of sewage produced from human activities is one of the major challenges to avoid environmental pollution. In India, the estimated per capita sewage generation is around 98 L for rural/small town population and 220 L for populations in urban areas including Delhi (Kaur *et al.* 2012). As per the Central Pollution and Control Board (CPCB) report 2016, the estimated sewage generated in India in the year 2015 was more than 61,000 MLD, against the total sewage treatment capacity of around 22,000 MLD. Hence, around 39,000 MLD of untreated sewage is directly discharged to the environment, which is the major cause of environmental deterioration such as eutrophication in fresh water bodies. Interestingly, the nutrient-rich sewage may be utilized as a low cost media for algal cultivation. Several reports are available on utilization of sewage/municipal wastewater for algae cultivation coupled with wastewater treatment (Jiang *et al.* 2011; Cheah *et al.* 2016).

Poor digestibility of the algal cell wall is the major hurdle towards methane generation by anaerobic digestion (Passos *et al.* 2014). Recently, several reports have been published on improving the biodegradability of both micro- and

doi: 10.2166/wst.2018.029

macro-algal biomass under anaerobic conditions through multifarious approaches including physico-chemical as well as biological pretreatment methods (Mendez *et al.* 2014; Passos *et al.* 2014). Moreover, attempts are being made worldwide for development of low cost biological algal biomass pretreatment.

Algae grow in close association with native microorganisms of wastewater, with inter-species transfer of CO_2 and O_2 (Ramanan *et al.* 2016). Moreover, most of the wastewater-grown microorganisms, especially bacteria, are hydrolytic (Gerardi 2006; Krah *et al.* 2016). Hence, it is possible that natural storage of wastewater-grown algae could lead to higher degree of algal cell disintegration due to the presence of hydrolytic bacteria. However, literature reports in this direction are scarce. To the best of our knowledge, a recent paper by Miao *et al.* (2013) reporting 36% improvement in methane yield from Taihu blue algae after 15 d of natural storage, is the only report on this particular research topic. On the other hand, similar biological pretreatment strategies involving activity of microorganisms or their endogenous enzymes during storage (aerobic/micro-aerobic) at temperatures ranging from 50 to 70 °C have been successfully applied for pretreatment of different wastes and biomass such as sewage sludge, food waste and animal waste (Carrere *et al.* 2016). However, for specific case of algae, research is mainly focused on the use of enzymes for biological pretreatment (Passos *et al.* 2014).

Hence, in light of the above discussion, the present study aimed towards the continuous cultivation of algae in sewage for wastewater treatment coupled with algal biomass production. Further, attempts were made to study the effect of natural storage on algal biomass digestibility and methane yield during anaerobic digestion.

MATERIALS AND METHODS

Sewage collection and characterization

Sewage was collected from the continuously running open drainage line near Dwarka Sector 2, New Delhi, India, in the month of March (Supplementary Figure S1, available with the online version of this paper). The drainage line receives sewage from the local community and includes wastewater generated through domestic activities, local markets and some of the small-scale industries such as bakery, milk and dairy processing. The collected sewage was filtered through muslin cloth (pore size 0.5–0.8 mm) in order to separate insoluble particles and solid debris. The filtered sewage was then stored at 4 °C and was used for experiments whenever required. Homogenised samples of filtered sewage (in triplicate) were processed for characterisation through determination of soluble chemical oxygen demand (sCOD), total dissolved phosphorus (TDP), total ammoniacal nitrogen: NH3-N (TAN), nitrate nitrogen (NO3–N), total suspended solids (TSS), total dissolved solids (TDS) and pH.

Algal culture

The native algal culture for the present work was obtained from the sewage itself. The filtered sewage (50 mL) was mixed with equal amounts of nutrient media (BG11) in a conical flask and incubated at 25 °C under continuous illumination (≈6,000 lux) using cool fluorescent lights for 10 d. The culture flasks were manually shaken twice a day to provide mixing. During this incubation, the native algae started growing and culture optical density (measured at 680 nm) reached a stable value of around 1.8. An aliquot (5 mL) was withdrawn from the culture flask for morphological characterisation of algae through a phase contrast microscope (PCM) and scanning electron microscope (SEM). The PCM and SEM analyses were carried out using the methodology reported earlier (Prajapati *et al.* 2015).

Further, there were some bacterial strains associated with the native algal culture. These bacterial strains were isolated from algal cells through streaking on Luria agar plates. The isolated bacterial samples were then subjected to preliminary morphological and biochemical tests including Gram staining, etc., for their identification.

Batch scale studies

For batch studies, 90 mL filtered sewage was inoculated with 10 mL of inoculum culture in a conical flask (in triplicate) and incubated under controlled conditions (see previous section) for 12 d. The flasks containing sewage (100 mL) without inoculum were covered with aluminium foil (to avoid algal growth by blocking light) and incubated under similar controlled conditions (see previous section), used as a negative control. An aliquot (5 mL) of the sample was withdrawn after every 3 d (from both experimental and control flasks) for determination of algal growth, pH and residual nutrient concentrations.

Photobioreactor (PBR) studies

A laboratory scale PBR was fabricated using a 6.00 mm thick transparent acrylic cylinder (internal diameter:

14 cm, height: 30 cm) with a working volume of 3.0 L (Supplementary Figure S2, available with the online version of this paper). The PBR was filled with the filtered sewage, inoculated with a native algae culture (10% v/v) and kept under continuous illumination (≈6,000 lux) using cool fluorescent lights placed on four sides. Air was also bubbled into PBR at 2.0 L min^{-1} using an aquarium pump to provide mixing and to prevent settling of the algae. The PBR was operated in batch mode during the first 12 d and then operated in continuous mode with 12 d hydraulic retention time (HRT) for the next 48 d (four HRT cycles of continuous operation). An aliquot (5.0 mL) from the PBR was withdrawn (after every 24 h for the first 30 d and after 5 days' interval for the next 30 days) for determination of algal growth, pH, residual nutrient concentration and sCOD.

Pretreatment through natural storage

After completion of PBR studies, algal biomass was harvested, in the form of slurry having >80% (w/w) water, through pH assisted auto settling. The algal slurry was then stored in multiple 100 mL open containers (≈50 gm algal slurry in each) in the dark at room temperature (27–30 °C) for 72 h. The algal slurry was manually mixed 2–3 times daily (using a sterile spatula) in order to provide proper aerobic conditions. For qualitative assessment of pretreatment during storage, SEM and fluorescence microscopic analysis of algal samples was done. For fluorescence microscopy, algal samples were stained with Sytox Green® and processed according to the methodology reported earlier (Prajapati *et al.* 2015).

For quantitative analysis, COD solubilisation from algal biomass was studied. Homogenized samples were withdrawn after every 6 h for the first 24 h and after every 12 h for the rest of the storage period. The algal samples were centrifuged at 6,000 g for 10 min and the supernatant was collected for sCOD estimation. The COD solubilization was calculated using the equation adopted from Alzate *et al.* (2012) as given below:

$$COD\ solubilization\ (\%) = \left(\frac{sCOD_t - sCOD_0}{COD_T - sCOD_0}\right) \times 100 \tag{1}$$

where 0 and t denote sCOD of algal slurry initially and sCOD after storage of time t, whereas T denotes the total COD of algal slurry before the storage.

To assess the contribution of cell natural lysis (through natural death and degradation under unfavourable conditions), in pretreatment during storage, a flask containing BF algal slurry (grown aseptically in BG11) was also stored under similar conditions (as mentioned above) and COD solubilization from algal biomass was determined using Equation (1). The bacteria free (BF) algal culture was obtained following the methodology reported earlier by Ferris & Hirsch (1991).

Biomass composition and theoretical methane yield

The sewage grown (SG) algal biomass, collected as a concentrated slurry, was centrifuged at 6,000 g for 10 min at room temperature. The pellet thus obtained was resuspended in distilled water and centrifuged again in order to remove slats or nutrients deposited on the algal biomass. Finally, the pelletized algal biomass was dried at 60 °C until it reached a constant weight and then ground (using a mortar and pestle) to obtain a fine powder of algae. The powdered algal biomass was used for determination of its lipid, protein and carbohydrate content. Lipid was extracted and estimated using the modified Bligh and Dryer's method (Lee *et al.* 2010). Carbohydrate content of the algal biomass was determined through the phenol sulfuric acid method (Dubois *et al.* 1951). Protein was estimated using the elemental nitrogen content and the nitrogen-to-protein conversion factor i.e. 6.25. The biochemical composition based theoretical methane yield was calculated using the equation adopted from Prajapati *et al.* (2013):

$$Theoretical\ methane\ yield\ (m^3\ kg^{-1}\ VS) = \left(\frac{0.851 \times F_P + 1.014 \times F_L + 0.415 \times F_C}{100}\right) \tag{2}$$

where F_P, F_L and F_C are the protein, lipid and carbohydrate content (% total solids, i.e. TS) of algal biomass.

The biochemical methane potential (BMP)

For determination of the methane potential of fresh algal biomass and improvement in methane yield through natural storage of algal biomass (0–72 h), BMP tests (in triplicates) were performed. BMP experiments were carried out in 500 mL amber color reagent bottles with 300 mL working volume. The initial substrate concentration, in terms of algal volatile solids (VS), in the BMP bottles was 5 g L^{-1}, whereas the substrate to inoculum (S/I) ratio was kept at 1:3 on VS basis (Angelidaki *et al.* 2009). The digested slurry collected from a laboratory scale cattle dung-based biogas plant (working volume 18 L; semi-continuous feed with 30 days' HRT) was used as anaerobic inoculum. The VS content and specific methane yield of the inoculum used was 60% (of TS) and 0.178 g COD–CH_4 g^{-1} VS d^{-1}, respectively. The inoculum was aseptically transferred to BMP bottles followed by addition of substrate (biomass), and the volume was made up to 300 mL using tap water whenever required. All the experimental bottles were incubated at 37 ± 1 °C under stationary conditions in a benchtop incubator (Holliger *et al.* 2016). The BMP test was carried out for a 30 d anaerobic digestion period.

An aliquot (10 mL) from each BMP bottle (at the beginning and end of the experiment) was withdrawn and centrifuged at 6,000 g for 40 min. The supernatant thus collected was analyzed for determination of TAN and total volatile fatty acids (TVFA) concentration. TVFA concentration was determined spectrophotometrically (Siedlecka *et al.* 2008). The pH of the digestate was also determined both at the beginning and immediately after the completion of the BMP test.

Analytical methods

Wastewater parameters

The sCOD was determined through the colorimetric method (Hach method 8000). The samples were first digested in Hach Digital Reactor DRB200, and the sCOD concentration was measured through reading the sample in the Hach colorimeter DR890. The TDP and NO_3–N were determined colorimetrically through the molybdovanadate method (Hach method 8114;) and cadmium reduction method (Hach method 8114), respectively. TAN was measured using multi-parameter (HQ40d, Hach) equipped with an ammonium ion selective electrode, and pH was measured with a bench top pH meter (CyberScan PC510, Eutech). TSS and TDS were determined through standard methods for wastewater analysis (Eaton *et al.* 2005).

Biogas volume and composition

The volume of biogas produced from BMP experiments was measured through acidic water (pH <2.0) displacement in every 24 h (Krishania *et al.* 2013). Methane content in the biogas was determined using a gas chromatograph (Agilent 7890A) equipped with a stainless steel column packed with Porapack-Q 80/100 mesh (Supelco) and thermal conductivity detector (TCD). The net methane production (mL g^{-1} VS_{fed}) was determined by subtracting the methane yield of the control bottle (with inoculum only) from the methane yield obtained in experimental bottles.

Algal biomass digestibility during BMP

The biomass digestibility was estimated in terms of % of volatile solids reduction (VSR). VSR was calculated after determination of volatile solid (VS) content of the BMP bottles before and after anaerobic digestion using the following equations:

$$VSR\ (\%) = \left(\frac{VS_f - VS_d}{VS_f}\right) \times 100 \qquad (3)$$

where VS_f and VS_d are the VS concentration (g/L) in the experimental bottles before and anaerobic digestion (30 d), respectively. The VS content of the feed and digestate were estimated using the standard EPA protocols (Telliard 2001) by burning the oven dried samples at and above 550 °C.

RESULTS AND DISCUSSION

Algal culture and remediation studies

The microscopic analysis showed that the algal culture obtained is a consortium of more than one type of algal cell as well as bacteria (Supplementary Figure S3, available with the online version of this paper). The dominating algal species in the consortium were morphologically identified as *Chlorella* and *Chroococcus* spp. Moreover, preliminary biochemical tests indicated the presence of both gram positive and gram negative bacteria including *Azotobacter*, *Bacillus* and *Actinobacteria* in the algal consortium. The native algal consortium thus obtained was used as inoculum for further experimentations.

To assess the native algal biomass production in sewage, batch studies were conducted using sewage as the sole growth medium. The composition of the sewage is shown in Table 1. Under illumination, the growth of algae in the sewage started, and the final biomass concentration after 12 d was 1.69 ± 0.35 g dry weight L^{-1}. Apart from the biomass production, removal of nutrients and sCOD by native algal growth was also studied (Figure 1). A drastic reduction in the sCOD was observed and the sCOD concentration reduced to 0 within 3 d due to the algal growth in sewage under illuminated conditions. Moreover, in the control flask (incubated under similar conditions but covered with aluminum foil to avoid light penetration) the change in sCOD was almost negligible (Figure 1). Further, the concentrations of NO_3–N, TDP

Table 1 | Characterization of collected sewage in terms of concentration of pollutants and nutrients

S. no.	Parameter	Concentration
1.	Total suspended solids (TSS)	1656.40 ± 32.25
2.	Total dissolved solids (TDS)	3509.34 ± 37.52
3.	Total ammoniacal nitrogen (TAN)	58.65 ± 4.20
4.	Soluble chemical oxygen demand (sCOD)	101.55 ± 6.58
5.	Nitrate nitrogen (NO_3-N)	67.02 ± 3.56
6.	Total dissolved phosphorus (TDP)	30.50 ± 2.54
7.	pH	7.15 ± 0.56

Values reported are mean ± Std. Dev. (All values are in mg L^{-1} except pH).

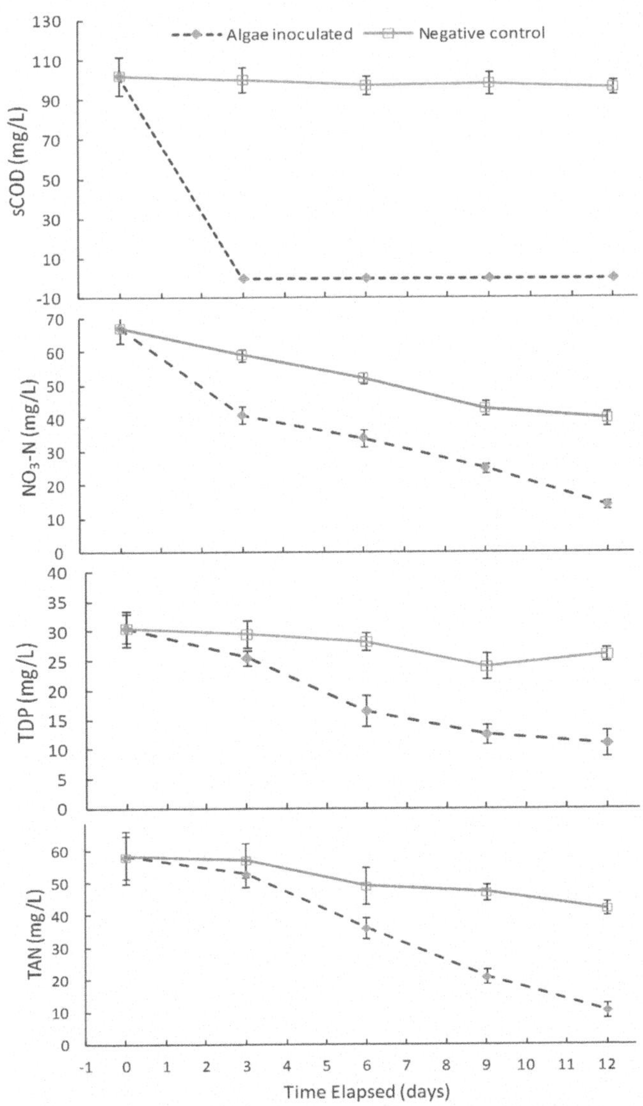

Figure 1 | Variation in residual concentrations of soluble chemical oxygen demand (sCOD), nitrate nitrogen (NO_3-N), total dissolved phosphorus (TDP) and total ammoniacal nitrogen (TAN) with incubation time in control and experimental flask during batch studies.

and TAN in sewage reduced to 14.15, 11.05 and 10.56 mg L^{-1}, respectively, during native algal growth for 12 d. Also, pH of sewage changed from its value near neutral (before algal growth) to alkaline (9.20), which could be due to utilization of carbonates (or dissolved CO_2) from wastewater during native algae growth. Overall, the observed removal efficiency of the native algae were 100, 79, 64 and 82%, for sCOD, NO_3–N, TDP and TAN, respectively. However, the corresponding removal efficiencies observed in the control flask were only 6, 40, 15 and 28%, respectively, which could be due to the growth of native microorganisms other than algae. Similar removal efficiencies for various pollutants from wastewater have been reported for different algal cultures (Chinnasamy *et al.* 2010; Paskuliakova *et al.* 2016).

As in the case of batch flask scale studies, good algal growth and nutrient removal was also observed in PBR (Figure 2). During the batch growth phase, drastic reduction in sCOD was observed. The residual sCOD concentration in the PBR reached zero within 5 d of the experiments. The observed residual concentrations of NO_3–N, TDP and TAN (on 12th day) were 18, 14 and 6.5 mg L^{-1}, respectively. Interestingly, the biomass concentration in the PBR (3.29 ± 0.42 g L^{-1}) during batch growth was higher than the value observed in flask scale studies (1.69 ± 0.35 g L^{-1}). The higher growth rate in the aerated PBR could be due to better light availability, mixing, and the proper CO_2 availability from the bubbled air.

A sudden increase in sCOD along with other nutrients was observed as soon as the PBR was converted to continuous mode. With the addition of fresh sewage (during the first 2 d of continuous operation), the sCOD reached up to 21 mg L^{-1} and again reduced to zero within the next 4 d and remained almost zero thereafter. Similarly, a significant increment in the TAN was also observed with

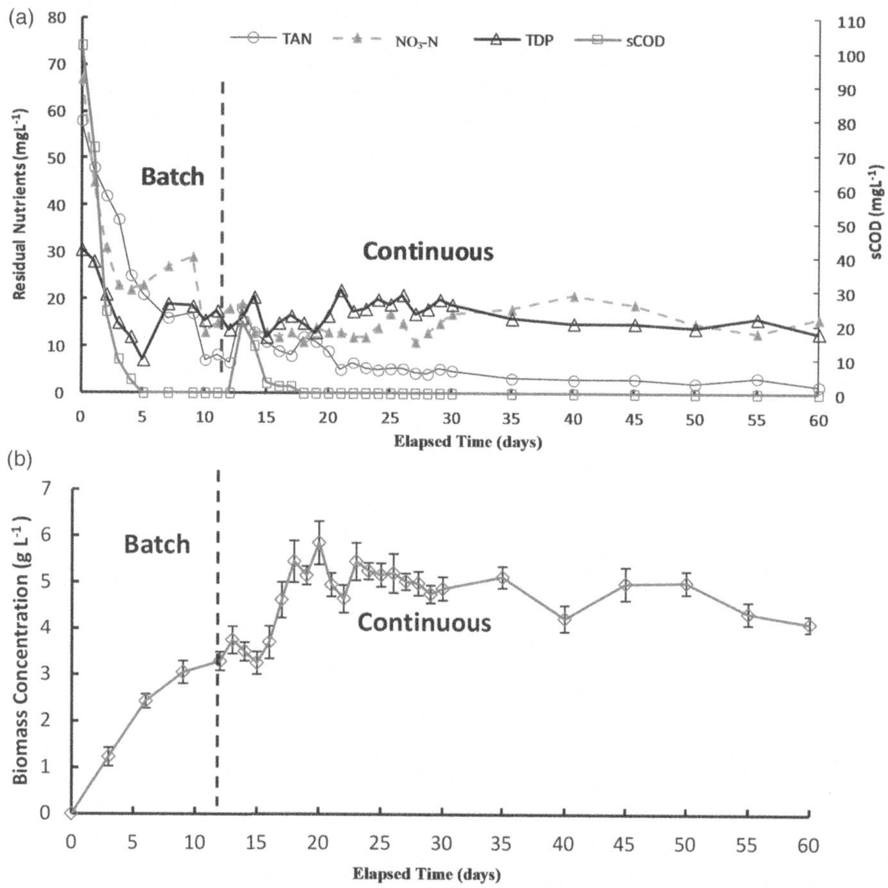

Figure 2 | Variation of (a) residual chemical oxygen demand (sCOD), nitrate nitrogen (NO3-N), total dissolved phosphorus (TDP) and total ammoniacal nitrogen (TAN) and (b) biomass concentration (reported in terms of g dry weight L^{-1}) during treatment of sewage with native algal growth in PBR.

concentrations of up to 18 mg L^{-1} recorded on the 13th day (i.e., the first day of continuous mode). However, there was relatively small change in the concentrations of NO_3–N and TDP, and the respective values observed on the 13th day were 19 and 16 mg L^{-1} (Figure 2(a)). The residual concentration of NO_3–N, TDP and TAN remained in the range of 11–18, 14–20 and 4.8–8.2 mg L^{-1}, respectively, for rest of the period of the continuous mode. The pH in the PBR was raised to 9.2 from an initial value of 7.15, within 3 days of the start of operation, and then remained in the vicinity of 9.0 throughout the experimentation. Moreover, there was a slight decrease in algal biomass concentration (3.15 ± 0.22 g L^{-1}) observed on the 15th day. However, the algae again started growing and reached a maximum biomass concentration of up to 5.45 ± 0.21 g L^{-1} recorded on the 23rd day of the experiment (Figure 2(b)). The average biomass concentration during the continuous mode was 4.64 ± 0.92 g L^{-1}. The observed biomass concentration is in accordance with previously observed values for native algae grown in dairy farm effluent (4.44 g L^{-1}) under outdoor PBR (Prajapati *et al.* 2014).

Methane production from SG algae

The biochemical compositional analyses revealed that the algal biomass harvested after sewage treatment contains up to 22, 16 and 44% carbohydrate, lipid and protein, respectively. Theoretical methane yield calculated using Equation (2) was 628 mL g^{-1} VS. To predict the bioenergy generation potential, anaerobic digestion of harvested algal biomass was carried out. The methane profiles of the algal biomass are shown in Figure 4(b) (shown as a graph for 0 h stored algal biomass). Methane yield from algal biomass was only 180.40 ± 8.92 mL g^{-1} VS_{fed}. Further, the maximum rate of methane production was only 18.92 mL CH_4 d^{-1}, which was observed on the seventh day of the BMP experiment. In contrast, higher cumulative and daily methane yield have been reported in literature (Dębowski *et al.*

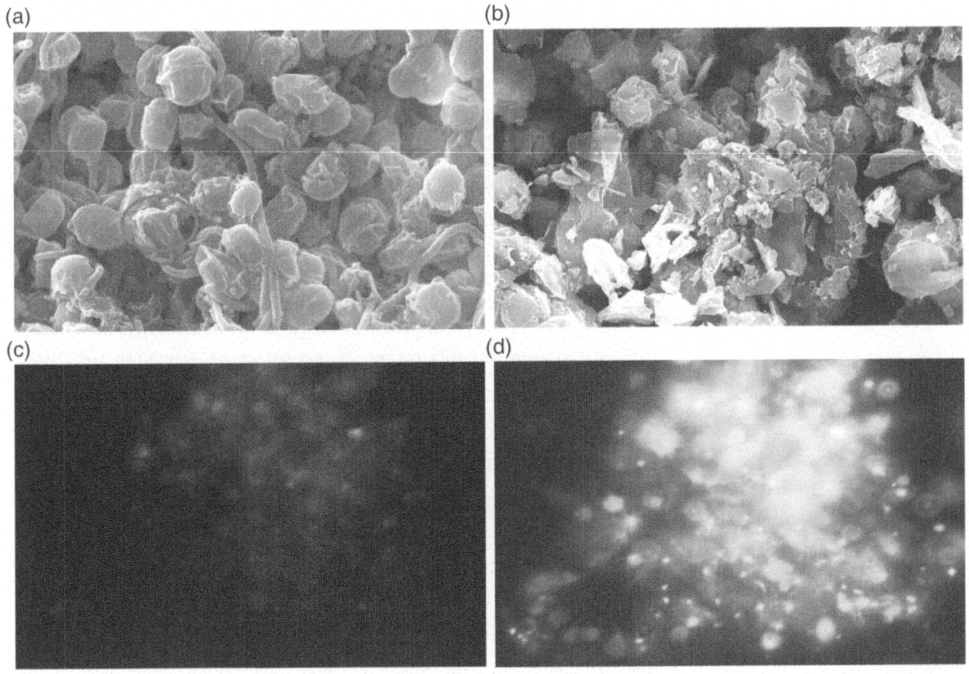

Figure 3 | Scanning electron microscopic (SEM) and fluorescence microscopic images of algal culture at different stages. (a) and (b) SEM (5000x) of the native algal-bacterial consortium grown in the sewage (bacterial cells are highlighted with a circle) and the algal samples stored for 72 h under natural condition (the ruptured cells are highlighted with the arrow) and (c) and (d) fluorescence microscopic images (40x) of Sytox Green® stained algal samples at 0 h (c) and 72 h (d) of storage under natural conditions (live algal cells showed red in colour due to chlorophyll autofluorescence, and cells with ruptured cell wall showed intense green in colour due to fluorescence of DNA bound Sytox Green®). The full colour version of this figure is available in the online version of this paper, at http://dx.doi.org/10.2166/wst.2018.029.

2013). For instance, methane yield in the range of 188–295 mL g^{-1} VS_{fed} was observed for anaerobic digestion of the microalgal mixture (Alzate *et al.* 2012). The low methane yield observed in the present work could be attributed to slow digestibility and degradation of algal biomass. Hence, there are possibilities to improve the methane yield by improving the degradability of the algal biomass.

Algal biomass disruption during storage

Algal biomass solubilization during natural storage was measured in terms of sCOD concentration. The sCOD of stored samples increased dramatically, which indicates release of intracellular constituents from algal biomass, probably due to disruption of the algal cell wall. The SEM image of the culture showed the presence of bacteria with algae (Figure 3(a)), which could have played a major role in pretreated of algal biomass during natural storage. The disruption of the algal cells was clearly observed with SEM and fluorescence microscopy (Figure 3(b)–3(d)). During the first 18 h, the solubilisation of biomass COD was negligible (1–4%). However, after 24 h, the algal biomass solubilization started increasing and the highest algal biomass solubilization with sCOD concentration of up to 7,298 ± 188 mg L^{-1} (≈48% COD solubilization) was observed with a sample incubated for 60 h (Figure 4(a)). Storage beyond 60 h had a negative effect on algal biomass solubilization (sCOD concentration reduced to 6,331 ± 210 mg L^{-1}). This could be due to the utilization of ruptured algal cell content by the native bacteria present in the biomass. In contrast, the BF algal samples, stored under similar natural conditions showed very low change in sCOD (≈3,918 ± 78 mg L^{-1}), with only 10% COD solubilisation even after 72 h.

One of the possible underlying mechanisms for algal cell disruption during storage could be natural cell death under unfavourable conditions for growth, leading to apoptosis (Miao *et al.* 2013). Hence, cell lysis under rugged environment conditions could have led to an increased sCOD concentration due to release of the cytoplasmic constituents of stored algal samples. However, the slow rate of biomass solubilisation in the case of BF samples indicated a minor contribution of natural cell death in algal biomass pretreatment. Another explanation for algal biomass disruption during storage could come from the fact that there was a presence of native bacteria along with the wastewater-grown algal biomass as confirmed with the SEM image (Figure 3(a)). The storage conditions were not favourable for algae growth, as no light and nutrient were supplied,

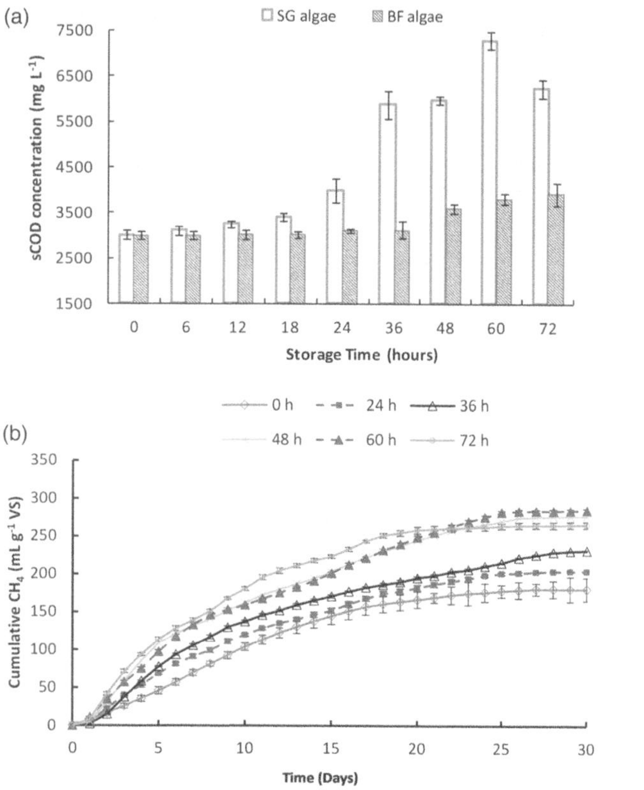

Figure 4 | Effect of natural storage on: (a) algal biomass disruption reported in terms of variation in sCOD concentration with storage time for sewage grown (SG algae) as well as nutrient media grown bacteria free algal biomass (BF algae) and (b) cumulative methane production profile of sewage grown algal biomass stored for 0 h (i.e. freshly harvested algae) to 72 h.

but there was optimal environment (temperature 27–30 °C, moisture content >80% and dead algal biomass as nutrient and carbon source) for native bacteria to grow. Hence, it is possible that native bacteria could have grown on the algal biomass, leading to further algal cell disruption and release of cellular constituents. Indeed, native bacterial cultures isolated from the algal consortium showed good cellulolytic activity on carboxymethylcellulose (CMC) agar plates (results not shown). Hence, the combined effect of algal cell natural death and cell degradation by native bacteria could have led to the high rate of COD solubilization from naturally stored biomass. Similar observations on biomass pretreatment were also made when the algal biomass was stored with the fungi under natural conditions (Prajapati *et al.* 2016). In that study, 50% biomass solubilization was recorded from algal biomass through fungal growth on it. Also, application of bacteria has widely been reported for pretreatment and solubilization of different biomass feedstocks including agro-residue and algae for improving biofuel yield (Keller *et al.* 2003; Muñoz *et al.* 2014; Sharma *et al.* 2017). In line with the above studies, Peces *et al.* (2016) have reported significant improvement in methane yield from primary sludge through semi-aerobic pretreatment at 20 °C for 12–72 h. However, further attempts are crucial to clearly understand the role of bacteria, the underlying mechanisms and the kinetics of algal biomass disruption during natural storage.

It is interesting to note that the observed algal biomass disruption during natural storage is at a par with the previously observed values during enzymatic pretreatment (46–50%) of algal biomass (Prajapati *et al.* 2015). In addition, the observed biomass solubilization in the present work was in line with the previously reported values for different algal cultures pretreated through different thermal, chemical or biological methods (Passos *et al.* 2014). Indeed, biomass solubilization in the present report is higher than that observed during thermal and biological pretreatment of different algal biomass (Alzate *et al.* 2012). Hence, based on these observations it can be concluded that the present method is a simple and efficient approach to replace traditional physico-chemical or biological methods including enzymatic pretreatment of algal biomass.

Natural storage improves the methane yield

The natural storage had a positive impact on methane yield (Figure 4). Significant improvement (as compared to fresh algal biomass i.e., 0 h stored biomass) was observed in the methane production, which reached up to 284.38 ± 11.52 and 276.78 ± 10.24 mL g^{-1} VS_{fed} for samples stored, respectively, for 48 h and 60 h. It was also noticed that storage beyond 60 h has a negative effect on methane yield, and the values observed for samples stored for 72 h were relatively low (265.27 ± 6.20 mL CH_4 g^{-1} VS_{fed}). The BMP of the algal samples stored for less than 24 h was not determined, as the biomass COD solubilization in these samples was very small. The effect of natural storage on algal biomass digestibility was also determined through estimation of VSR. As in the case of methane yield, the highest VSR (78%) was also observed with the samples stored for 60 h. Further, an almost similar VSR (≈76%) was observed for samples stored for 72 h. In contrast, the anaerobic digestion of fresh algae showed only 38% VSR. The VSR for other samples stored for 24–48 h was in the range of 47–65%. Moreover, the digestate obtained after anaerobic digestion was analysed for pH, TAN and TVFA content in order to predict the process stability. The characteristics of the digestate are given in Table 2. The values observed for TAN (209–285 mg L^{-1}) and TVFA (895–1,265 mg L^{-1}) for all the sets

Table 2 | Values recorded for pH, total ammoniacal nitrogen (TAN) and total volatile fatty acid (TVFA) concentrations in the content of BMP bottles at the start (initial) and after completion (final) of anaerobic digestion as well as volatile solid reduction (VSR) during BMP test

Sample \ Parameter	pH		TAN (mg L^{-1})		TVFA (mg L^{-1})		
	Initial	Final	Initial	Final	Initial	Final	VSR (%)
Fresh (0 h)	7.02	7.82	165.32	229.25	280.22	895.55	38.22
Stored (24 h)	7.40	7.21	172.12	284.63	320.12	1025.32	46.58
Stored (36 h)	6.96	7.54	155.14	259.36	312.50	1010.60	52.00
Stored (48 h)	7.12	6.85	162.83	220.06	305.25	1263.40	65.12
Stored (60 h)	6.89	7.44	158.15	209.50	324.35	992.58	77.80
Stored (72 h)	7.14	6.80	156.23	219.62	314.08	1125.56	76.45

Values reported are mean of at least three replicates.

tested were lower than those reported as inhibitory (TVFA > 2,000 mg L^{-1} and TAN > 1,500 mg L^{-1}) for the anaerobic digestion process (Labatut & Gooch 2012). Moreover, that pH was also in the neutral range (6.80–7.82) for all the tested sets.

The observed improvements in the methane yield and the digestibility were in line with algal biomass solubilization measured in terms of sCOD. Further, the observed methane yield and enhancement was at par with the values reported for algal biomass pretreated through various methods. For instance, Alzate *et al.* (2012) reported up to 62% enhancement in the methane yield through thermal hydrolysis of algal biomass. Ben Yahmed *et al.*, (2016) reported methane production up to 260 mL g^{-1} VS through crude enzyme-based pretreatment of *Chaetomorpha linum* biomass. Similarly, more than 50% enhancement in the digestibility and methane yield was observed during novel simultaneous harvesting and pretreatment of *Chroococcus* sp. biomass using fungi as a bioagent (Prajapati *et al.* 2016). Based on these facts, natural storage (for 2–3 d) of wastewater grown algae may be recommended as a prerequisite stage in algae to methane generation by anaerobic digestion.

CONCLUSIONS

It is well documented that in wastewater treatment systems algae and bacteria grow together, with higher growth rates (high biomass production) leading to higher nutrient removal from wastewaters. High removal of nutrients and sCOD (65–100%) along with substantially higher biomass production was observed in both flask and reactor scale during the present studies. Further, up to 48% COD solubilization from SG algal biomass was observed through aerobic treatment when stored for designated periods under natural conditions. The biomass solubilization resulted in >57% enhancement in methane yield from algal biomass stored for 60 h. Overall, the wastewater grown native algal biomass seems to be a good substrate for methane production through natural storage followed by anaerobic digestion. This may further be explored to develop decentralised algae mediated wastewater treatment coupled bioenergy generation systems.

ACKNOWLEDGEMENT

The present work was carried out with financial support from Science and Engineering Research Board, Department of Science and Technology, Government of India (Grant No. ECR/2015/000062). Authors are thankful to Prof. Anushree Malik, AML, IIT Delhi, for providing fluorescence microscope facility for this work.

REFERENCES

Alzate, M. E., Muñoz, R., Rogalla, F., Fdz-Polanco, F. & Pérez-Elvira, S. I., 2012 Biochemical methane potential of microalgae: influence of substrate to inoculum ratio, biomass concentration and pretreatment. *Bioresour. Technol.* **123**, 488–494.

Angelidaki, I., Alves, M., Bolzonella, D., Borzacconi, L., Campos, J. L., Guwy, A. J., Kalyuzhnyi, S., Jenicek, P. & Van Lier, J. B., 2009 Defining the biomethane potential (BMP) of solid organic wastes and energy crops: a proposed protocol for batch assays. *Water Sci. Technol.* **59**, 927–934.

Ben Yahmed, N., Jmel, M. A., Ben Alaya, M., Bouallagui, H., Marzouki, M. N. & Smaali, I., 2016 A biorefinery concept using the green macroalgae *Chaetomorpha linum* for the coproduction of bioethanol and biogas. *Energy Convers. Manag.* **119**, 257–265.

Carrere, H., Antonopoulou, G., Affes, R., Passos, F., Battimelli, A., Lyberatos, G. & Ferrer, I. 2016 Review of feedstock pretreatment strategies for improved anaerobic digestion: from lab-scale research to full-scale application. *Bioresour. Technol.*

Cheah, W. Y., Ling, T. C., Show, P. L., Juan, J. C., Chang, J. S. & Lee, D. J. 2016 Cultivation in wastewaters for energy: a microalgae platform. *Appl. Energy* **179**, 609–625.

Chen, G., Zhao, L. & Qi, Y. 2015 Enhancing the productivity of microalgae cultivated in wastewater toward biofuel production: a critical review. *Appl. Energy* **137**, 282–291.

Chinnasamy, S., Bhatnagar, A., Claxton, R. & Das, K. C. 2010 Biomass and bioenergy production potential of microalgae consortium in open and closed bioreactors using untreated carpet industry effluent as growth medium. *Bioresour. Technol.* **101**, 6751–6760.

Dębowski, M., Zieliński, M., Grala, A. & Dudek, M., 2013 Algae biomass as an alternative substrate in biogas production technologies – review. *Renew. Sustain. Energy Rev.* **27**, 596–604.

Dubois, M., Gilles, K., Hamilton, J. K., Rebers, P. a., Smith, F., Reberes, P. A. & Smith, F. 1951 A colorimetric method for the determination of sugars. *Nature* **168**, 167.

Eaton, A. D., Clesceri, L. S., Greenberg, A. E. & Franson, M. A. H. 2005 *Standard Methods for the Examination of Water and Wastewater.* American Public Health Association, Washington, DC.

Ferris, M. J. & Hirsch, C. F., 1991 Method for isolation and purification of cyanobacteria. *Appl. Environ. Microbiol.* **57**, 1448–1452.

Gerardi, M., 2006 *Wastewater Bacteria,* Vol. 5. John Wiley & Sons, Hoboken, NJ, USA.

Holliger, C., Alves, M., Andrade, D., Angelidaki, I., Astals, S., Baier, U., Bougrier, C., Buffière, P., Carballa, M., De Wilde, V., Ebertseder, F., Fernández, B., Ficara, E., Fotidis, I., Frigon, J.-C., Fruteau De Laclos, H., Ghasimi, D. S. M., Hack, G., Hartel, M., Heerenklage, J., Horvath, I. S., Jenicek, P., Koch, K., Krautwald, J., Lizasoain, J., Liu, J., Mosberger, L., Nistor, M., Oechsner, H., Oliveira, J. V., Paterson, M., Pauss, A., Pommier, S., Porqueddu, I., Raposo, F., Ribeiro, T., Pfund, F. R., Strömberg, S., Torrijos, M., Van Eekert, M., Van Lier, J., Wedwitschka, H. & Wierinck, I., 2016 Towards a standardization of biomethane potential tests. *Water Science & Technology* **74** (11), 2515–2522. doi:10.2166/wst.2016.336.

Jiang, L., Luo, S., Fan, X., Yang, Z. & Guo, R. 2011 Biomass and lipid production of marine microalgae using municipal wastewater and high concentration of CO_2. *Appl. Energy* **88**, 3336–3341.

Kaur, R., Wani, S. P., Singh, A. K. & Lal, K., 2012 Wastewater production, treatment and use in India. In: *2nd Regional Workshop on Safe Use of Wastewater in Agriculture*, 16–18 May, New Delhi, India, pp. 1–13.

Keller, F. A., Hamilton, J. E. & Nguyen, Q. A., 2003 Microbial pretreatment of biomass microbial pretreatment of biomass potential for reducing severity of thermochemical biomass pretreatment. *Appl. Biochem. Biotechnol.* **105**, 105–108.

Krah, D., Ghattas, A.-K., Wick, A., Bröder, K. & Ternes, T. A., 2016 Micropollutant degradation via extracted native enzymes from activated sludge. *Water Res.* **95**, 348–360.

Krishania, M., Vijay, V. K. & Chandra, R., 2013 Methane fermentation and kinetics of wheat straw pretreated substrates co-digested with cattle manure in batch assay. *Energy* **57**, 359–367.

Labatut, R. A. A. & Gooch, C. A. A. 2012 Monitoring of anaerobic digestion process to optimize performance and prevent aystem failure. *Got Manure Tradeshow and Conference*, pp. 209–225.

Lee, J. Y., Yoo, C., Jun, S. Y., Ahn, C. Y. & Oh, H. M. 2010 Comparison of several methods for effective lipid extraction from microalgae. *Bioresource Technology* **101**, S75–S77.

Mendez, L., Mahdy, A., Demuez, M., Ballesteros, M. & González-Fernández, C., 2014 Effect of high pressure thermal pretreatment on *Chlorella vulgaris* biomass: organic matter solubilisation and biochemical methane potential. *Fuel* **117**, 674–679.

Miao, H., Lu, M., Zhao, M., Huang, Z., Ren, H., Yan, Q. & Ruan, W., 2013 Enhancement of Taihu blue algae anaerobic digestion efficiency by natural storage. *Bioresour. Technol.* **149**, 359–366.

Muñoz, C., Hidalgo, C., Zapata, M., Jeison, D., Riquelme, C. & Rivas, M., 2014 Use of cellulolytic marine bacteria for enzymatic pretreatment in microalgal biogas production. *Appl. Environ. Microbiol.* **80**, 4199–4206.

Paskuliakova, A., Tonry, S. & Touzet, N., 2016 Phycoremediation of landfill leachate with chlorophytes: phosphate a limiting factor on ammonia nitrogen removal. *Water Res.* **99**, 180–187.

Passos, F., Uggetti, E., Carrère, H. & Ferrer, I., 2014 Pretreatment of microalgae to improve biogas production: a review. *Bioresour. Technol.* **172**, 403–412.

Peces, M., Astals, S., Clarke, W. P. & Jensen, P. D., 2016 Semi-aerobic fermentation as a novel pre-treatment to obtain VFA and increase methane yield from primary sludge. *Bioresour. Technol.* **200**, 631–638.

Prajapati, S. K., Kaushik, P., Malik, A. & Vijay, V. K., 2013 Phycoremediation coupled production of algal biomass, harvesting and anaerobic digestion: possibilities and challenges. *Biotechnol. Adv.* **31**, 1408–1425.

Prajapati, S. K., Choudhary, P., Malik, A. & Vijay, V. K., 2014 Algae mediated treatment and bioenergy generation process for handling liquid and solid waste from dairy cattle farm. *Bioresour. Technol.* **167**, 260–268.

Prajapati, S. K., Bhattacharya, A., Malik, A. & Vijay, V. K., 2015 Pretreatment of algal biomass using fungal crude enzymes. *Algal Res.* **8**, 8–14.

Prajapati, S. K., Bhattacharya, A., Kumar, P., Malik, A. & Vijay, V. K., 2016 A method for simultaneous bioflocculation and

pretreatment of algal biomass targeting improved methane production. *Green Chem.* **18**, 5230–5238.
Ramanan, R., Kim, B. H., Cho, D. H., Oh, H. M. & Kim, H. S., 2016 Algae-bacteria interactions: evolution, ecology and emerging applications. *Biotechnol. Adv.* **34**, 14–29.
Rawat, I., Ranjith Kumar, R., Mutanda, T. & Bux, F., 2011 Dual role of microalgae: phycoremediation of domestic wastewater and biomass production for sustainable biofuels production. *Appl. Energy* **88**, 3411–3424.
Sharma, H. K., Xu, C. & Qin, W., 2017 Biological pretreatment of lignocellulosic biomass for biofuels and bioproducts: an overview. *Waste and Biomass Valorization* 1–17.
Siedlecka, E. M., Kumirska, J., Ossowski, T., Glamowski, P., Gołębiowski, M., Gajdus, J., Kaczyński, Z. & Stepnowski, P., 2008 Determination of volatile fatty acids in environmental aqueous samples. *Orig. Res. Polish J. Environ. Stud* **17**, 351–356.
Telliard, W. A. 2001 Method 1684. Total, Fixed, and Volatile Solids in Water, Solids, and Biosolids.

First received 1 November 2017; accepted in revised form 8 January 2018. Available online 19 January 2018

Enhanced biogas production potential of microalgae and swine wastewater using co-digestion and alkaline pretreatment

K. Panyaping, R. Khiewwijit and P. Wongpankamol

ABSTRACT

Biogas yield obtained from anaerobic digestion of swine wastewater (SWW) needs to be increased to produce electrical energy. To enhance biogas and prevent pollution, use of mixed culture microalgae grown in wastewater (MWW) with SWW has attracted a lot of interest. This research was focused on the possibility of utilizing MWW. Six experiments using raw SWW and MWW, and their co-digestion were conducted on a laboratory scale in one-litre reactors with the ratio of inoculum and substrate of 70:30 under without and with alkaline pretreatment (using 3% NaOH for pH adjustment every 15 min at pH 11 for 3 h). The results showed that co-digestion had the major effect on increasing biogas and methane yields (0.735 and 0.326 m^3/kg of volatile solids (VS) removed), and the highest chemical oxygen demand and VS removal (60.29% and 63.17%). For pretreatment, the effect of ammonia inhibition at a high pH of 11 had more influence on biodegradation than the effect of destruction of MWW's cell walls, resulting in a low biogas production of pretreated MWW and pretreated co-digestion. These findings affirm the potential of co-digestion, and the possibility of using both single and co-substrate MWW. Pretreatment could be improved at a lower alkaline pH condition. A pilot scale of co-digestion should be performed.

Key words | alkaline pretreatment, biogas, co-digestion, mixed culture microalgae, swine wastewater

K. Panyaping (corresponding author)
R. Khiewwijit
P. Wongpankamol
Department of Environmental Engineering, Faculty of Engineering, Rajamangala University of Technology Lanna, 128 Huay Kaew Rd., Chiang Mai 50200, Thailand
E-mail: *klin41@hotmail.com*; *klinpratoomp@yahoo.com*

INTRODUCTION

Globally, swine farms are a very large industry with many related products and waste businesses such as animal feed, medical supplies, and processed animal products, for example in China, Europe, US, and other Asian countries (McOrist *et al.* 2011; UNITC 2014).

In Thailand, the major problem for swine farms is the proper management of wastewater according to the environmental standard of the Pollution Control Department (PCD 1999).

Swine wastewater (SWW) is a waste mixture of feces (45%), urine (55%), and washing water, which is characterized by a high content of organic matter and nutrients such as biochemical oxygen demand (BOD) 14.2–26.3 g/l, nitrogen (N) 2.4–6.0 g/l, phosphorus (P) 0.6–1.4 g/l, ammonium (NH_4^+-N) 2.3–5.1 g/l, nitrate (NO_3^--N) 0.03–0.07 g/l, and other compounds (Moral *et al.* 2008). The use of swine manure and SWW for biogas production using anaerobic digestion (AD) technology can produce a total of about 320.94 million m^3/y of biogas that can be used as a fuel and for electricity generation (DEDE 2011). Thus, there is a high potential for clean and renewable energy production from SWW as electricity for use in both farms and for sale according to the Thai energy policy on renewable energy production in 10 years.

Microalgae grown in wastewater (MWW) are well-known as fast-growing photosynthesis organisms that can accumulate a large amount of carbohydrate in the form of starch, cellulose, polysaccharides, etc. (Ward *et al.* 2014; Gonzalez-Fernandez *et al.* 2015). These are microscopic organisms such as the microalga *Chlorella* sp. which has high efficiency to decrease nutrients such as N and P in wastewater. The carbohydrates appearing in microalgae are considered to be an appropriate feedstock for various kinds of energy sources such as bioethanol and bio-methane (Ward *et al.* 2014; Gonzalez-Fernandez *et al.* 2015). Thus, microalgal biomass has a high potential for use as a

doi: 10.2166/wst.2018.077

substrate or co-substrate for biogas generation, including reducing CO_2 from organic digestion and fuel combustion (Gonzalez-Fernandez *et al.* 2015; Hwang *et al.* 2016; Molinuevo-Salces *et al.* 2016). In fact, several aspects of microalgae were studied such as species, renewable energy, type of pretreatment method, and the related effects (Kim *et al.* 2003; de Schamphelaire & Verstraete 2009; Luna-delRisco *et al.* 2011; Esposito *et al.* 2012; Milledge & Heaven 2014; Stina 2015; Kumar *et al.* 2016; Passos *et al.* 2016; Zuliani *et al.* 2016).

As microalgae have cell wall resistance due to some compounds such as cellulose and phospholipids, it is necessary to use a pretreatment method that can facilitate the degradability of MWW and co-digestion for improving the availability of the nutrients and various trace elements (Khiewwijit *et al.* 2015; Calicioglu & Demirer 2016).

Therefore, this research investigated biogas and methane productivity from a single substrate of raw MWW and a co-substrate by co-digestion of MWW and SWW without pretreatment as compared to single substrate of raw SWW. In addition, pretreatment of co-digestion of MWW and SWW was conducted for comparison with non-pretreated co-digestion and pretreated MWW using 3% NaOH (w/v) for pH adjustment every 15 min at pH 11 for 3 h.

It is expected that extensive information will be obtained to develop an alternative bioenergy production and a wastewater management option for enhancing biogas yield in swine wastewater treatment (WWT).

In this research, one example of a large-sized swine farm of type A, with a total number of more than 5,000 sows or 600 units of livestock (1 livestock unit = 500 kg of live weight) (DEDE 2011), using 'covered lagoon' technology for treating SWW, was used as a case study. There was a raw wastewater problem because the quality of some wastewater, for example in terms of total Kjeldahl nitrogen (TKN) and BOD, was still over the standard for treated wastewater resulting in the growth of MWW in stabilization ponds (Four T Farm 2016).

The results of this study will be used to improve the biogas yield and treat SWW in this swine farm. Besides this, the results could be applied to any livestock or domestic WWT system world-wide.

METHODS

Microalgae grown in wastewater and SWW

The mixed culture MWW were collected by grab sampling from the mixed liquid and MWW in a stabilization pond in Doi Lor swine farm, Chiang Mai, Thailand. These microorganisms contained the total species of microalgae that had grown in wastewater in a pond. The dominant group of MWW was green algae, which mostly contained *Chlorella* sp., *Scenedesmus* sp., and pennate diatoms based on their robustness in wastewater (González-Fernández *et al.* 2011; Emeka *et al.* 2016).

As for SWW, it was collected from an effluent pond before passing through the WWT by grab sampling.

The characteristics of MWW and SWW are shown in Table 1.

Microorganisms used as inoculum

Seeding sludge was also collected by grab sampling from a drainage channel of an anaerobic digester at a SWW treatment in Doi Lor swine farm, Chiang Mai, Thailand.

This AD system called 'covered lagoon', which generated biogas yields at a mesophilic temperature (25–45 °C), created a type of biogas plant that was the same as a WWT plant. These biogas yields were used as the fuel for electricity generation in the swine farm with the capacity of 1.5–1.7 kW/h for 1 m^3 of biogas. The biogas production efficiency from this plant was 70%. There was a methane content of 50–65%. The removal efficiency of this AD system was in the range of 85–90% for both chemical

Table 1 | Characterization of inoculum and substrates

	Physico-chemical characteristic					
Substrate	**pH**	**TS (mg/l)**	**VS (mg/l)**	**COD (mg/l)**	**Density (g/l)**	**Ratio of COD of substrate to VS of inoculum**
Inoculum	6.95 (± 0.04)	52.465.00 (± 2.708.22)	25.470.00 (± 1.230.37)	41.760.00 (± 2.009.18)	–	–
SWW	7.44 (± 0.05)	3.723.33 (± 360.05)	2.673.33 (± 164.42)	10.560.00 (± 2.009.18)	0.833	0.49
MWW	7.10 (± 0.01)	9.400.00 (± 887.07)	5.633.33 (± 365.56)	12.480.00 (± 0.00)	0.714	0.41

Note: MWW, mixed culture microalgae grown in wastewater; SWW, swine wastewater; number of samples = 3.

oxygen demand (COD) and BOD, and in the range of 25–60% for TKN. The highest removal efficiency in terms of total suspended solids (TSS) was in the range of 90–99%. The sludge was used for composting after drying in sunlight (Four T Farm 2016).

The seeding sludge sample was immediately used and analyzed for total solids (TS) and total volatile solids (TVS) contents, total COD (COD_t), density, and the ratio of COD of substrate to volatile solids (VS) of inoculum as shown in Table 1.

Reactor component and experimental procedure

The batch reactors with gas collectors were operated on a laboratory scale at room temperature (25–32 °C).

Each reactor was designed using acrylic plastic for a container with a width of 28 × 40 cm, and a length of 45 × 40 cm, with two outlets used for sampling (Figure 1). Each gas collector contained an L-shape pipe to collect the gas in 500 ml cylinders above a pipe inside a container. Each rubber stopper and stainless steel pipe were completely sealed by Whatman laboratory sealing film before starting digestion. The biogas produced was measured by volume displacement system.

A series of six experiments were conducted in six reactors using the ratio of inoculum and substrate of 70:30. Each reactor was operated in triplicate with an effective volume of 800 ml in one-litre test bottles. Another 200 ml of sample from each experiment was siphoned off in order to determine all the related parameters before starting digestion.

The details of the six experiments are as below.

- Reactor 1 consisted of no substrate, and only 70% of inoculum and 30% of distilled water was added in order to determine the biogas and methane production from inoculum as a control.

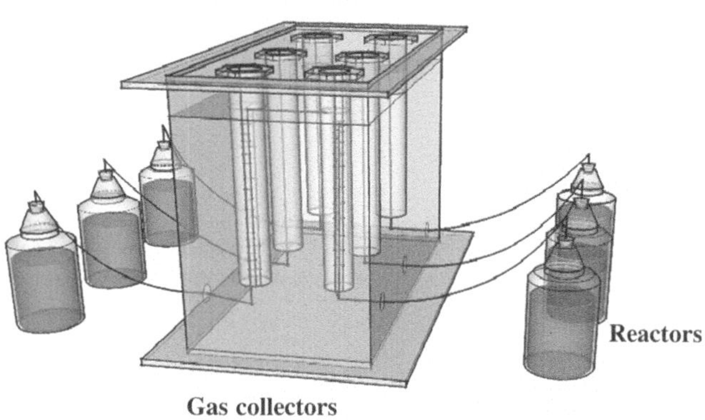

Figure 1 | A container of six batch reactors and gas collectors.

- Reactor 2 consisted of a single substrate or non-pretreated SWW. There was 70% of inoculum and 30% of SWW.
- Reactor 3 consisted of the non-pretreated mixed culture MWW. There was 70% of inoculum and 30% of MWW.
- Reactor 4 consisted of the pretreated MWW. There was 70% of inoculum, and 30% of pretreated MWW.
- Reactor 5 consisted of the co-digestion of non-pretreated MWW and SWW. There was 70% of inoculum, 15% of MWW, and 15% of SWW.
- Reactor 6 consisted of the pretreated co-digestion of pretreated MWW and SWW. There was 70% of inoculum, 15% of pretreated MWW, and 15% of pretreated MWW.

All reactors were mixed before starting digestion, and mixing was performed once a day for 10 min. Biogas production and its composition were determined based on the stability of gas volume until the end of the experiment.

The details of all experiments are shown in Table 2.

Pretreatment procedure

Reactor 4, raw substrate (MWW), and Reactor 6, co-substrate (MWW) with SWW, were added to a container for pretreatment conditions, respectively. No additional nutrient or chemical compound was added to the reactors. Only 3% NaOH (w/v) was added every 15 min at pH 11 for 3 h during preparation of the sample. After that, all pretreated substrates were mixed with inoculum and added to each reactor for three replicates for operating the AD.

Operational performance

Before starting an operation process in the reactors, the ratio of food to microorganisms, in the form of COD of each substrate and VS of inoculum, should be in the range of 0.4–0.6 for efficient degradability (Angelidaki *et al.* 2007). This initial ratio can be calculated according to the following equations:

For SWW: the ratio of food to microorganisms

$$= \frac{COD \text{ of SWW}}{VS \text{ of inoculum}} \tag{1}$$

where COD and VS are in mg/l.

For MWW: the ratio of food to microorganisms

$$= \frac{COD \text{ of MWW}}{VS \text{ of inoculum}} \tag{2}$$

where *COD* and *VS* are in mg/l.

Table 2 | Details of all experiments

Experiment	Detail of each experiment
Reactor 1; Control	Inoculum (560 ml) + distilled water (240 ml)
Reactor 2; SWW	Inoculum (560 ml) + swine wastewater (240 ml)
Reactor 3; MWW	Inoculum (560 ml) + mixed culture microalgae grown in wastewater (240 ml)
Reactor 4; pretreated MWW	Inoculum (560 ml) + [pretreated mixed culture microalgae grown in wastewater at pH =11 (240 ml)]
Reactor 5; MWW and SWW	Inoculum (560 ml) + swine wastewater (120 ml) + mixed culture microalgae grown in wastewater (120 ml)
Reactor 6; Pretreated MWW and SWW	Inoculum (560 ml) + [pretreated swine wastewater + pretreated mixed culture microalgae grown in wastewater at pH 11 (120 ml)]

Note [ ]: alkaline pretreatment.

If this ratio is not in the efficient range of biodegradation, the substrate and inoculum are either diluted or concentrated before analysis, and the above equations are used to calculate the ratio above. In this research, this value was 0.41 and 0.49 for MWW and SWW, respectively, as shown in Table 1. When the proper ratio was obtained, the related parameters of all substrates were determined. Inoculum and substrate were added to reactors as shown in Figure 1 and Table 2.

The calculation of VS of feed stock and inoculum in each reactor can be calculated as below:

$$\text{Weight of VS of feedstock(kg)} = (VS \text{ of \% feedstock, mg}) \times Density \times 10^{-6}\text{kg} \quad (3)$$

where *VS* of % feedstock is volatile solids of the single substrates SWW and MWW (30%), and volatile Solids of the substrates in co-digestion of MWW (15%) and SWW (15%). *Density* is 0.714 for MWW, and 0.833 for SWW, g/l.

$$\text{Volume of VS of inoculum (ml)} = VS \text{ of inoculum 70\%, or 700 ml in 1 l} \quad (4)$$

where *VS* is in ml.

$$\text{Weight of } VS_{removed} \text{ of feedstock (kg)} = \frac{(VS_{influent} - VS_{effluent}, \text{mg})}{HRT} \times Density \times 10^{-6}\text{ kg} \quad (5)$$

where $VS_{influent}$ is the VS of feedstock in influent, excluding VS of inoculum, mg/l
$VS_{effluent}$ is the VS of feedstock in effluent, excluding VS of inoculum, mg/l
HRT is the hydraulic retention time, days
Density is 0.714 for MWW, and 0.833 for SWW, g/l

In addition, the effect of biogas and methane yields generated from inoculum was used as a control set in Reactor 1. These biogas and methane yields were used as a blank by subtracting them from the actual measurement of biogas yields in Reactor 2 to Reactor 6.

Gas volume and temperature were measured and recorded every day. However, biogas composition was measured when gas volume was stable. The volume of biogas and methane yields were expressed in m^3/kg $VS_{removed}$.

The digestion performance of the reactors was ended if there was constant gas volume, and no biogas yield increased. After the end of operation, the sample in each set of experiments was analyzed for related parameters.

All data were statistically analyzed.

Analytical methods

The pH, TS, TSS, VS, BOD, COD_t and TKN were examined in triplicate according to the standard methods (APHA *et al.* 2005). Temperature and pH were measured on a daily basis (24 h) using a pH meter and thermometer. CH_4 content was quantified by a gas chromatograph (Model 7890, Agilent, USA) equipped with a thermal conductivity detector (TCD). Helium gas was used as a carrier gas at flow rate 10 l/min. The temperatures of injector, oven, and detector were 120 °C, 150 °C, and 250 °C, respectively.

Scanning electron microscopy analysis

Scanning electron microscopy (SEM) (LV 5910 model, JEOL) was used to analyze the samples of raw MWW and pretreated MWW in order to observe the surface change of cell walls.

Statistical analysis

T-test was used to determine the statistical significance ($p < 0.05$) of differences between without and with pretreatment groups by SPSS 22.0.

RESULTS AND DISCUSSION

Characteristics of MWW, SWW, and inoculum

As shown in Table 1, TS and VS of MWW were higher than those of SWW and inoculum. The density of MWW (0.714 g/l) was lower than that of SWW (0.833 g/l) due to the light component weight of MWW as compared to SWW. These organic matters in all substrates were associated with the different biogas and methane yields as well as the digestion time, as was found in some previous studies (Pavlostathis & Giraldogomez 1991; Calicioglu & Demirer 2016).

The pH values of all substrates were in the range of 7.28–8.00. These values were in the optimum range for degradation (pH 6–8) (DEDE 2011). As for temperature, there was an extreme change of the weather in winter for a few days, resulting in biogas yield, but it was in the range of 25–32 °C which was still in the optimum range (25–40 °C) for microbial degradation by mesophilic bacteria (Ward *et al.* 2014; Gonzalez-Fernandez *et al.* 2015).

In addition, there was quite a high TKN (299.17 mg/l) in SWW after treatment in the stabilization pond of MWW (data not shown). This can cause the ammonia inhibition during degradation at high pH conditions. For example, the results from previous studies by Guerrero *et al.* (1997) and Chen *et al.* (2008) showed that when SWW had high nitrogen (N) concentrations (>200 mg/l) with a high pH >7.5, it can cause the presence of free ammonia resulting in the inhibition of some bacterial methanogens such as aceticlastic and hydrogenotrophic methanogens.

Digestion performance

As shown in Table 3, TS, VS, COD, and VS/TS of all substrates mixed with inoculum in the influent before starting the digestion were higher than those of the effluent. This implied that substrates were degraded by the hydrolysis, acidogenesis, and methanogenesis stages of the AD process with varying results.

High TKN in both raw SWW (667.49 mg/l) and the stabilization pond of MWW (299.27 mg/l) could be the cause of methanogen inhibition at a high pH. TKN was the sum of nitrogen in organic substances in the form of ammonia (NH_3), and ammonium (NH_4^+): thus, NH_3, and NH_4^+ were released during nitrogen degradation resulting in pH rise and toxicity on the methanogens (Guerrero *et al.* 1997; Chen *et al.* 2008).

TS and COD of co-digestion of MWW and SWW were the highest values. In addition, TS, VS, and COD of MWW were higher than those of SWW. This was associated with the different composition of MWW, which consisted of starch, cellulose, protein, lipids, and trace elements.

Table 3 | Characteristics of influent and effluent during digestion in the reactors

Experiment	Inf. and Eff.	pH	TS (mg/l)	VS (mg/l)	VS/TS	COD (mg/l)
Reactor 1; Control	Inf.	7.45	44.566	26.998	0.61	27.970
	Eff.	7.21	34.617	16.990	0.49	18.133
Reactor 2; SWW	Inf.	7.36	39.001	26.477	0.68	31.000
	Eff.	7.32	28.490	11.267	0.40	13.591
Reactor 3; MWW	Inf.	7.28	42.300	30.373	0.72	31.600
	Eff.	7.26	31.127	16.090	0.52	17.273
Reactor 4; PretreatedMWW	Inf.	7.99	43.980	29.990	0.68	31.995
	Eff.	7.25	35.000	20.395	0.58	22.201
Reactor 5; MWW and SWW	Inf.	7.30	45.500	30.000	0.66	34.000
	Eff.	7.28	33.550	11.050	0.33	13.500
Reactor 6; Pretreated MWW and SWW	Inf.	7.96	44.500	29.900	0.67	34.690
	Eff.	7.38	35.900	15.001	0.42	18.003

Note: Inf., influent before digestion; Eff., effluent after digestion.

Effect of the single substrates of MWW and SWW without pretreatment and MWW with pretreatment

Biogas and methane production

As shown in Figure 2, in Reactor 2, raw SWW provided the highest cumulative biogas and methane yields (0.670 m^3/kg $VS_{removed}$, and 0.314 m^3/kg $VS_{removed}$ with the same hydraulic retention time of 60 days, respectively). This implied that there was appropriate AD in reactors for all non-treated substrates as pH was close to neutrality. Thus, there was no ammonia inhibition from free ammonia (NH_3) on methanogens due to a low pH (<8). Even SWW had quite a high TKN that consisted of several forms of nitrogen compound.

Meanwhile, for Reactor 3, non-pretreated MWW provided lower cumulative biogas and methane yields (0.580 m^3/kg $VS_{removed}$, and 0.274 m^3/kg $VS_{removed}$ with the same hydraulic retention time of 60 days, respectively). This might be the effect of the structural differences of mixed culture microalgal cell walls (Allard *et al.* 2002; Versteegh & Blokker 2004; Ward *et al.* 2014) in that some of the microalgal culture had multiple layers resulting in difficult accessibility of enzyme for degradation of microalgal cell walls.

MWW had TS, VS and COD higher than those of SWW because most microalgal cell walls contained cellulose and lipids. Besides this, some species of MWW had an additional tri-laminar sheath (TLS) containing algaenan, which is a substance known for its resistance to degradation, resulting in difficult degradation (Allard *et al.* 2002; Versteegh & Blokker 2004; Ward *et al.* 2014).

The results from SEM of the raw MWW and pretreated MWW emphasized the effect of pretreatment condition on the surface of cell walls as shown in Figure 3. It was observed that the whole cell in raw MWW had smooth surface cell walls at a higher magnification (2,000×). When using alkaline pretreatment, the cell walls of MWW were destroyed, and the surface of cell walls at a lower magnification (1,000×) were changed as compared to that of raw MWW. This confirmed that the cell walls of MWW, which

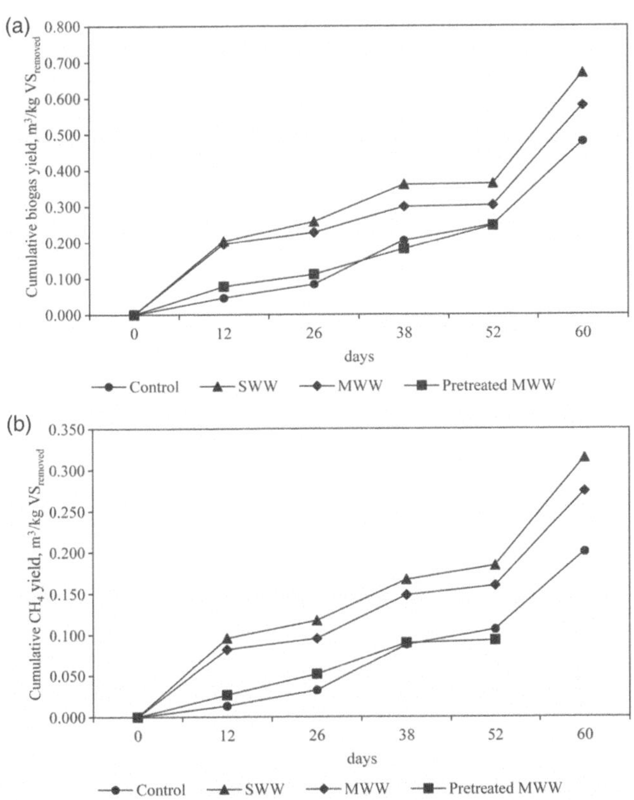

Figure 2 | Cumulative biogas and methane yield from without and with pretreatment of the single substrates. (a) Cumulative biogas yield of the single substrates. (b) Cumulative methane yield of the single substrates. *Note:* Significant at $p < 0.05$.

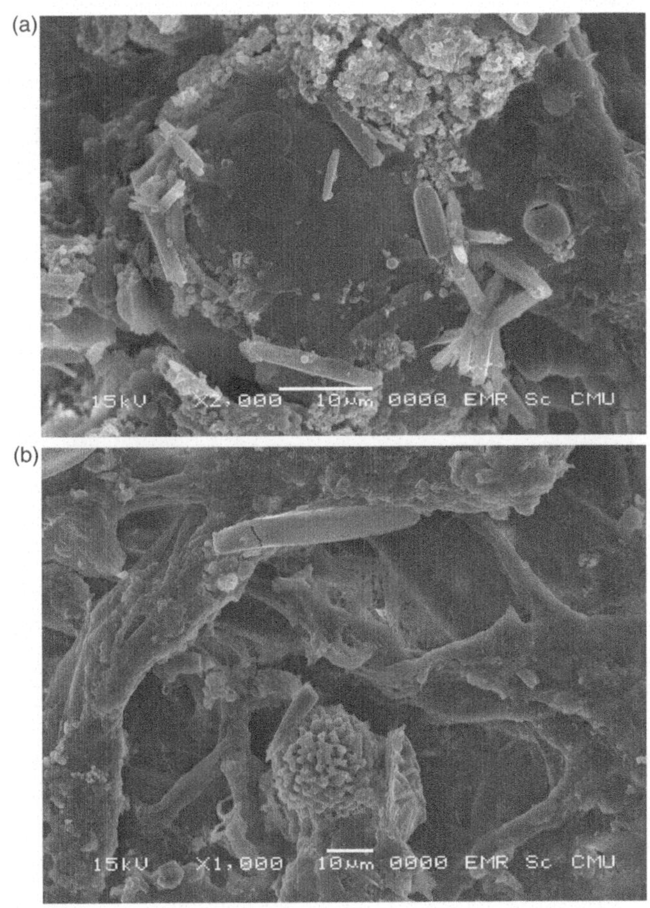

Figure 3 | Surface of the whole cell in raw MWW (a) and pretreated MWW (b).

are composed of starch, cellulose, polysaccharides, lipids, etc. called 'extracellular polymeric substances', were affected by alkaline pretreatment. This meant that microbial enzymes could degrade cell walls more easily, and biogas production could be increased.

As a matter of fact, raw MWW was contaminated with a slightly high nitrogen content in the stabilization pond (261–331 mg/l). This contamination could have much more adverse effects on microbial enzymes. Thus, it was the main cause of ammonia inhibition for pretreated MWW digestion in Reactor 4, resulting in the low biogas and methane yields due to the high pH adjustment at pH 11 (Chen *et al.* 2008; Montingelli *et al.* 2015). Biogas and methane yields from Reactor 4 with pretreated MWW was, therefore, lower (0.247 m^3/kg $VS_{removed}$ and 0.093 m^3/kg $VS_{removed}$ with the same hydraulic retention time of 52 days, respectively) than those of control Reactor 1 (0.479 m^3/kg $VS_{removed}$ and 0.257 m^3/kg $VS_{removed}$ with the same hydraulic retention time of 60 days, respectively).

It is important to note that the effect of ammonia inhibition during pH adjustment had more influence on microbial degradation than the effect of destruction of MWW structural cell walls, resulting in a low biogas production. In addition, non-pretreated single substrates of SWW and MWW, including control reactor, and pretreated MWW provided different biogas and methane yields at the statistical significance level of 0.05. This implied that the single substrate digestion of SWW had an influence on biogas production.

Removal efficiency

It was found that Reactor 3 with raw MWW had higher COD and VS removal efficiency (45.34% and 47.03%, respectively) than those of Reactor 4 with pretreated MWW (30.61% and 31.99%, respectively). In particular, raw SWW provided the highest removal efficiency of COD and VS (56.16% and 57.45%, respectively) for the single substrate digestion (Table 5). These results were consistent with the results of biogas and methane yields obtained from the single substrates of SWW and MWW digestion. According to the standard method, VSS and volatile dissolved solids are included in TVS, and VSS and fixed suspended solids are included in TSS. Thus, the low VS removal efficiency of Reactor 4 implied that there was a high possibility of sludge occurring from the single substrate of pretreated MWW digestion in the WWT system, which may affect operation performance of WWT (Kim *et al.* 2003; Dumas *et al.* 2010).

Effect of co-digestion of MWW and SWW without and with pretreatment

Biogas and methane production

As shown in Figure 4 for Reactor 5, non-pretreated co-digestion of MWW and SWW provided higher biogas and methane yield (0.735 m^3/kg $VS_{removed}$ and 0.326 m^3/kg $VS_{removed}$ with the same hydraulic retention time of 60 days, respectively) than those of Reactor 6 with pretreated co-digestion of SWW and MWW (0.620 m^3/kg $VS_{removed}$ and 0.292 m^3/kg $VS_{removed}$ with the same hydraulic retention time of 60 days, respectively). The lower amounts obtained from pretreated co-digestion may be more associated with the effect of ammonia inhibition during pH adjustment than the effect of MWW cell wall destruction, which was the same as those amounts obtained from pretreated MWW. Therefore, there was no difference between non-pretreated co-digestion and pretreated co-digestion of MWW and SWW on biogas and methane yields at the statistical significance level of 0.05.

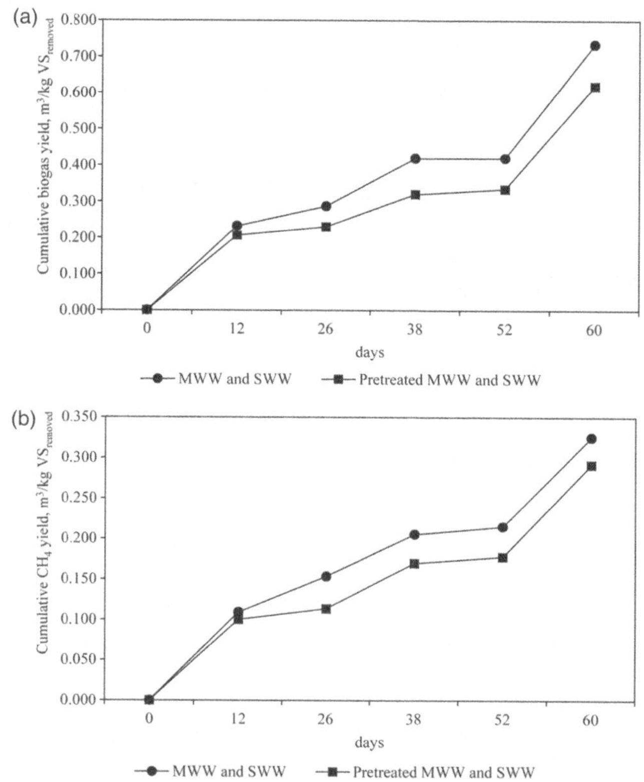

Figure 4 | Cumulative biogas and methane yield from without and with pretreatment of the co-digestion of MWW and SWW. (a) Cumulative biogas yield of the co-digestion. (b) Cumulative methane yield of the co-digestion. *Note:* Non-significant at $p < 0.05$.

The high nitrogen content in SWW (579–828 mg/l) and MWW (261–331 mg/l) of the swine farm (data not shown) seemed to hamper an efficient biodegradation to methane. In this circumstance, the concentrations of free ammonia (NH_3) and ammonium (NH_4^+) as functions of pH at equilibrium were changed when using alkaline pretreatment at a high pH 11 (Nielsen & Angedaki 2008; Olsson *et al.* 2014).

Free NH_3 concentrations from this study were calculated from the total nitrogen (NH_3 and NH_4^+) in mg/l of SWW and MWW for pH at 11 and temperature 25 °C to 32 °C by the Alleman calculator method (Alleman 1998). The high amounts of free NH_3 were obtained in the range of 655.96–666.33 mg/l from SWW and 225.21–226.71 mg/l from MWW. Besides this, the percentage of free NH_3 could be estimated from the pKa value at various pH and temperatures. For example, for pH at 10 and temperature 24 °C to 32 °C, there was free NH_3 of about 84–90.6% (Belmonte *et al.* 2011).

When using pH at 11, free NH_3 was 95%, which is completely toxic to methanogens.

This emphasized the effect of free NH_3 inhibition on degradation of pretreated co-digestion. In order to avoid a high percentage of free NH_3, the pH used for this pretreatment should be 9 with the same temperature, because free NH_3 of not more than 50% (49%) can occur, and there is no toxicity to methanogens (Belmonte *et al.* 2011; CCR 2013).

Thus, free NH_3 can occur in Reactor 6. This ammonia is the most toxic to methanogens during methanogenesis, resulting in the low cumulative biogas and methane yields (Nielsen & Angedaki 2008; Belmonte *et al.* 2011).

As compared to a previous study by Guerrero *et al.* (1997), the results showed that methanogenic toxicity was generated by the concentration of free NH_3 between 35 mg/l and 140 mg/l. Moreover, the results of a study by Nielsen & Angedaki (2008) indicated that ammonia inhibition was caused by free NH_3 between 80 mg/l and 100 mg/l. Furthermore, Belmonte *et al.* (2011) found that a reduction of methanogenic activity was related to the presence of free NH_3 >40 mg/l in the system. It was seen that free NH_3 concentration occurring from pH adjustment of the pretreated experiments was at a high value compared to the previous studies (Nielsen & Angedaki 2008; Belmonte *et al.* 2011). To overcome ammonia inhibition, a very high alkaline pH adjustment should be avoided (Nielsen & Angedaki 2008).

Pretreatment of substrates using 3% NaOH (w/v) for pH adjustment every 15 min at pH 11 for 3 h should be improved by considering a lower alkaline pH such as pH 9 for the temperature in the range of 25–32 °C in order to avoid ammonia inhibition of methanogens. In addition, substrates and inoculum should be mixed during pH adjustment, as substrates could be degraded by microbial enzymes.

Removal efficiency

Co-digestion of MWW and SWW provided the highest COD and VS removal efficiency (60.29% and 63.17%, respectively). The results were consistent with the results of biogas and methane yields obtained from this co-digestion. This indicates that although MWW has high polysaccharides and phospholipids, etc., some nutrients in MWW can help to increase biodegradation, resulting in a higher removal efficiency and biogas production (Bux & Chisti 2016).

In addition, a high VS removal efficiency, which included VSS, implied that the accumulation of sludge from co-digestion will not rapidly increase (Kim *et al.* 2003; Dumas *et al.* 2010). This results in a stable process and saves operating costs of the WWT.

Enhanced biogas and methane production from co-digestion of SWW and MWW without pretreatment

Biogas and methane production

As shown in Figure 5 and Table 4 for Reactor 5, non-pretreated co-digestion of MWW and SWW provided the highest cumulative biogas and methane yields (0.735 $m^3/kg\ VS_{removed}$, and 0.326 $m^3/kg\ VS_{removed}$ with hydraulic retention time of 60 days, respectively). This result indicated that the organic matters in the form of carbon content, nutrient availability, and trace elements were optimized for process degradability during digestion performance (Wang *et al.* 2013; Ward *et al.* 2014).

At the same time, mixing a more easily degradable substrate such as SWW with a more difficult to biodegrade substrate such as MWW can help to increase process stability.

In addition, the study of Wang *et al.* (2013) showed that the co-digestion of different species of microalgae with waste activated sludge had a variety of results.

Moreover, the biogas and methane yields obtained from non-pretreated co-digestion of MWW and SWW (8–25%) were higher than those of the single substrates of SWW and MWW. Besides this, these results were also high as compared to the previous studies (8.6–19%) by Wang *et al.* (2013) and Olsson *et al.* (2014).

Furthermore, the COD and VS removal efficiency of co-digestion of MWW and SWW was the highest (Table 5). It is

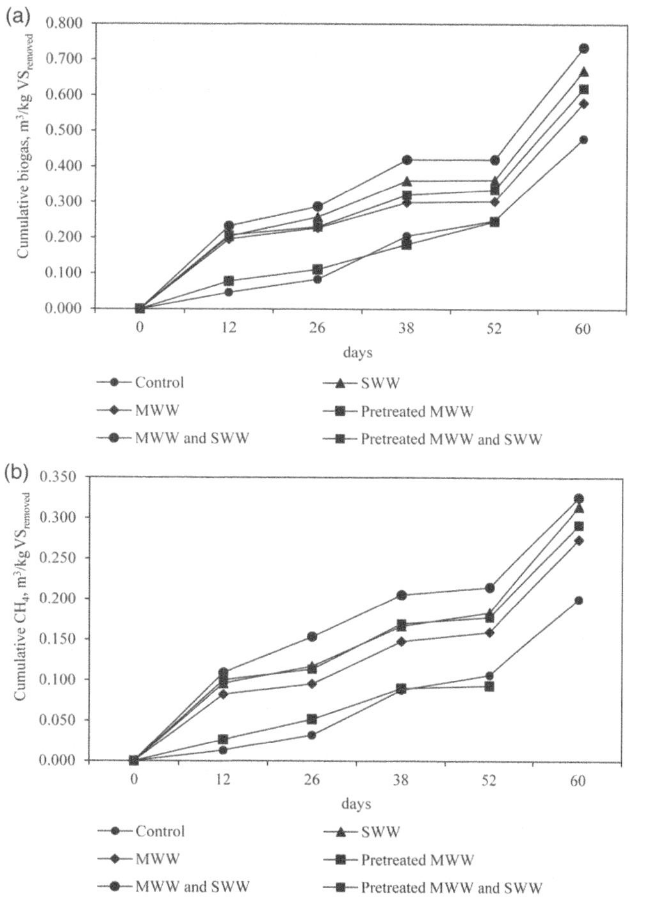

Figure 5 | Cumulative biogas and methane yield from without and with pretreatment of all experiments. (a) Cumulative biogas yield of all experiments. (b) Cumulative methane yield of all experiments. *Note:* Significance at $p < 0.05$.

beneficial to operate the WWT system in respect to efficient sludge treatment.

Therefore, Reactor 4, non-pretreated co-digestion of MWW and SWW, provided the best cumulative biogas and methane yields. This reactor could be used to enhance biogas production of SWW for further application.

Comparison of results of single substrates and the co-digestion without and with pretreatment

As shown in Tables 4 and 5, cumulative biogas and methane yields from the co-digestion of SWW and MWW were higher than those of single substrates of SWW (8% and 3%), MWW (25% and 16%), and pretreated MWW (60% and 71%), respectively. In addition, biogas and methane yields from pretreated co-digestion of SWW and MWW were higher than those of pretreated MWW alone (60%, and 68%, respectively). This was due to the pretreatment at a high pH as mentioned above. The results indicated that co-digestion was a good alternative solution for improving biogas production in WWT of a swine farm.

As compared to biogas and methane yields of pretreated co-digestion and co-digestions from the previous studies by Wang *et al.* (2013), Olsson *et al.* (2014) and Ward *et al.* (2014), these results obtained from this study were quite high. However, the results from a previous study by Montusiewicz & Lebiocka (2010) showed that the biogas yields obtained from the co-digestion of a sewage sludge and intermediate leachate at the ratio of 20:1 and 10:1 were 1.30 and 1.24 $m^3/kg\ VS_{removed}$, respectively, higher than those obtained from this study due to the higher organic contents of those substrates compared with those of MWW and SWW.

Pretreated co-digestion could not increase biogas and methane yields as compared to non-pretreated co-digestion due to the same reasons as mentioned for the effect of the co-digestion without pretreatment. Besides this, pretreated co-digestion of MWW and SWW had more influence on biogas production as compared to a pretreated single substrate of MWW. Moreover, biogas and methane yields obtained from the single substrates and co-digestion both without and with pretreatment provided different biogas and methane yields at the statistical significance level of 0.05.

Table 4 | Summary of biogas and methane efficiency from all experiments

Substrate	Biogas yield ($m^3\ kg/VS_{removed}$)	Methane ($m^3\ kg/VS_{removed}$)	%COD removal efficiency	%VS removal efficiency	Temperature (°C)	Duration time (days) for	
						Biogas	Methane
Reactor 1; Control	0.479	0.200	35.17	37.07	25–30	60	60
Reactor 2; SWW	0.670	0.314	56.16	57.45	25–31	60	60
Reactor 3; MWW	0.580	0.274	45.34	47.03	25–31	60	60
Reactor 4; Pretreated MWW	0.247	0.093	30.61	31.99	25–32	52	52
Reactor 5; MWW and SWW	0.735	0.326	60.29	63.17	25–31	60	60
Reactor 6; Pretreated MWW and SWW	0.620	0.292	48.10	28.05	25–31	60	60

Table 5 | Comparison of the major results from all experiments

Efficiency and others	Co-digestion of MWW and SWW without alkaline pretreatment	Co-digestion of MWW and SWW with alkaline pretreatment
Biogas production	Increase by 8% and 25% of the single substrate of SWW and MWW, respectively, and 60% of pretreated MWW	Increase by 60% of pretreated MWW
Methane production	Increase by 3% and 16% of the single substrate of SWW and MWW, respectively, and 71% of pretreated MWW	Increase by 68% of pretreated MWW
%COD removal	Increase by 7% of SWW	Increase by 36% of pretreated MWW
%VS removal	Increase by 9% of SWW	Increase by 36% of pretreated MWW
Operation performance	VS/TS reduction 0.5 times	VS/TS reduction 0.37 times
Inhibition	–	Free ammonia inhibition
Chemical cost	–	Alkaline chemical cost

The values of %COD and %VS removal obtained from non-pretreated co-digestion were higher than those of SWW (7% and 9%, respectively). Meanwhile, both %COD and % VS removal of pretreated co-digestion were higher than those of pretreated MWW (36%). However, the co-digestion of MWW and SWW provided the highest %COD and %VS removal (60.29%, and 63.17%, respectively). In addition, the operation performance showed that biodegradation of co-digestion could be operated at a higher VS/TS ratio (0.5 times) than that of pretreated co-digestion (0.37 times). Besides this, pretreated co-digestion had no alkaline chemical cost for pretreatment. The results implied that co-digestion was a good option for management of waste and sludge in a small to medium WWT system of a swine farm.

This is a good wastewater management practice for enhancing biogas production of SWW.

Therefore, co-digestion of MWW and SWW without pretreatment is a high possibility for utilization in a swine farm. It is recommended that a pilot plant should be operated before implementation in a WWT system of a swine farm.

CONCLUSIONS

The results showed the possible potential of enhancing biogas production from co-digestion of MWW and SWW. Co-digestion had a significant effect on biogas production and removal efficiency. MWW could be used as a co-substrate with SWW. For pretreatment, the effect of ammonia inhibition at a high pH of 11 had more influence on biodegradation than the effect of destruction of MWW's cell walls, resulting in a low biogas production of pretreated MWW and pretreated co-digestion. The results suggest that alkaline pretreatment should be improved by considering a lower pH condition. A pilot scale study of co-digestion should be performed before implementation.

ACKNOWLEDGEMENTS

This work was the research project to entrepreneur supported by the Rajamangala University of Technology Lanna and Four T Farm (59R2EENCM202). The authors would like to thank all related staffs for their help in collecting samples and their discussions.

REFERENCES

Allard, B., Rager, M. N. & Templier, J. 2002 Occurrence of high molecular weight lipids (c80^{+}) in the trilaminar outer cell walls of some freshwater microalgae: a reappraisal of algaenan structure. *Org. Geochem.* **33**, 789–801.

Alleman, J. E. 1998 *Free Ammonia-Nitrogen Calculator & Information* Iowa University, http://home.eng.iastate.edu/~jea/w3-research/free-ammonia/nh3.html/ (accessed 13 June 2017).

Angelidaki, I., Alves, M., Bolzonella, D., Borzacconi, L., Campos, L., Guwy, A., Jenicek, P., Kalyuzhnui, S. & Van Lier, J. 2007 Anaerobic biodegradation, activity and inhibition (ABAI). In: *Manual by the Task Group Meeting, 9–10 October, 2006, Prague, Republic of Czechoslovakia*.

APHA/AWWA/WEF 2005 *Standard Methods for the Examination of Water and Wastewater,* 21st edn. American Public Health Association/American Water Works Association/Water Environment Federation, Washington, DC, USA.

Belmonte, M., Hsieh, C.-F., Figueroa, C., Campos, J. L. & Vidal, G. 2011 Effect of free ammonia nitrogen on the methanogenic activity of swine wastewater. *Elec. J. Biotechnol.* **14** (3). http://dx.doi.org/10.2225/vol14-issue3-fulltext-11/ (accessed 12 June 2017).

Bux, F. & Chisti, Y. 2016 *Algae Technology: Products and Processes*. Springer, Cham, Switzerland.

Calicioglu, O. & Demirer, N. G. 2016 Biogas production from waste microalgal biomass obtained from nutrient removal of domestic wastewater. *Waste Biomass Valor.* **7**, 1397–1408.

Chen, Y., Cheng, J. J. & Creamer, K. S. 2008 Inhibition of anaerobic digestion process: a review. *Bioresour. Technol.* **99** (10), 4044–4064.

CCR 2013 *The Relationship between NH_4^+, NH_3, pH, Temperature.* http://www.nico2000.net/analytical/ammonium/NH4lib.html/ (accessed 12 June 2017).

DEDE 2011 *Handbook of Development and Investment for Production of Biogas Energy.* Department of Alternative Energy Development and Efficiency, Bangkok, Thailand.

de Schamphelaire, L. & Verstraete, W. 2009 Revival of the biological sunlight-to-biogas energy conversion system. *Biotech. Bioeng.* **103** (2), 296–304.

Dumas, C., Perez, S., Paul, E. & Lefebvre, X. 2010 Combined thermophilic aerobic process and conventional anaerobic digestion: effect on sludge biodegradation and methane production. *Bioresour. Technol.* **101**, 2629–2636.

Emeka, G. N., Jeremy, M. A., Navid, R. M., Benjamin, E. U. & James, C. O. 2016 Growth comparison of microalgae in tubular photobioreactor and open pond for treating anaerobic digestion piggery effluent. *Algal Res.* **17**, 268–276.

Esposito, G., Frunzo, L., Liotta, F., Panico, A. & Pirozzi, F. 2012 Bio-methane potential tests to measure the biogas production from the digestion and co-digestion of complex organic substrates. *TOEE* **5**, 1–8.

Four T Farm 2016 *Data of Swine Farm, Wastewater Treatment System, and Biogas Production.* Four T Farm, Chiang Mai, Thailand. http://www.fourt.co.th/home (accessed 24 May 2016).

González-Fernández, C., Riaño-Irazábal, B., Molinuevo-Salces, B., Blanco, S. & García- González, M. C. 2011 Effect of operational conditions on the degradation of organic matter and development of microalgae-bacteria consortia when treating swine slurry. *Appl. Microbiol. Biotechnol.* **90**, 1147–1153. doi: 10.1007/s00253-011-3111-z.

Gonzalez-Fernandez, C., Sialve, B. & Molinuevo-Salces, B. 2015 Anaerobic digestion of microalgal biomass: challenges, opportunities and research needs. *Bioresour. Technol.* **198**, 896–906.

Guerrero, L., Omil, F., Mendez, R. & Lema, J. M. 1997 Treatment of saline wastewaters from fish meal factories in an anaerobic filter under extreme ammonia concentrations. *Bioresour. Technol.* **61** (1), 69–78.

Hwang, J. H., Church, J., Lee, S. J., Park, J. & Lee, W. H. 2016 Use of microalgae for advanced wastewater treatment and sustainable bioenergy generation. *Environ. Eng. Sci.* **33** (11), 882–897.

Khiewwijit, R., Temmink, H., Labanda, A., Rijnaarts, H. & Keesman, K. J. 2015 Production of volatile fatty acids from sewage organic matter by combined bioflocculation and alkaline fermentation. *Bioresour. Technol.* **197**, 295–301.

Kim, J., Park, C., Kim, T-H., Lee, M., Kim, S., Kim, S-W. & Leej, J. 2003 Effects of various pretreatments for enhanced anaerobic digestion with waste activated sludge. *Biosci. Bioeng.* **95**, 271–275.

Kumar, G., Sivagurunathan, P., Thi, N. B. D., Zhen, G., Kobayashi, T., Kim, S. H. & Xu, K. 2016 Evaluation of different pretreatments on organic matter solubilization and hydrogen fermentation of mixed microalgae consortia. *Int. J. Hydrogen Energy* **41** (46), 21628–21640. http://dx.doi.org/10.1016/j.ijhydene.2016.05.195.

Luna-delRisco, M., Normak, A. & Orupold, K. 2011 Biochemical methane potential of different organic wastes and energy crops from Estonia. *Agron. Res.* **9** (1–2), 331–342.

McOrist, S., Khampee, K. & Guo, A. 2011 Modern pig farming in the People's Republic of China: growth and veterinary challenges. *Rev. Sci. Tech. Off. Int. Epiz.* **30** (3), 961–968.

Milledge, J. J. & Heaven, S. 2014 Methods of energy extraction from microalgal biomass: A review. *Rev. Environ. Sci. Bio.* **13** (3), 301–320.

Molinuevo-Salces, B., Mahdy, A. & Ballesteros, M. 2016 From piggery wastewater nutrients to biogas: microalgae biomass revalorization through anaerobic digestion. *Renew. Energy* **96**, 1103–1110.

Montingelli, M. E., Tedesco, S. & Olabi, A. G. 2015 Biogas production from algal biomass: a review. *Renew. Sustainable Energy Rev.* **43**, 961–972.

Montusiewicz, A. & Lebiocka, M. 2010 Co-digestion of intermediate landfill leachate and sewage sludge as a method of leachate utilization. *Bioresour. Technol.* **102** (3), 2563–2571.

Moral, R., Perez-Murcia, M. D., Perez-Espinosa, A., Moreno-Caselles, J., Paredes, C. & Rufete, B. 2008 Salinity, organic content, micronutrients and heavy metals in pig slurries from South-eastern Spain. *Waste Manage.* **28**, 367–371.

Nielsen, H. B. & Angedaki, I. 2008 Strategies for optimizing recovery of the biogas process following ammonia inhibition. *Bioresour. Technol.* **99** (17), 7995–8001.

Olsson, J., Feng, X. M., Ascue, J., Gentili, F. G., Shabiimam, M. A., Nehrenheim, E. & Thorin, E. 2014 Co-digestion of cultivated microalgae and sewage sludge from municipal wastewater treatment. *Bioresour. Technol.* **171**, 203–210.

Passos, F., Felix, L., Rocha, H., Pereira, J. O. & Aquino, S. 2016 Reuse of microalgae grown in full-scale wastewater treatment ponds: thermochemical pretreatment and biogas production. *Bioresour. Technol.* **209**, 305–312.

Pavlostathis, S. G. & Giraldogomez, E. 1991 Kinetics of anaerobic treatment: a critical review. *Water Sci. Technol.* **24**, 35–59.

PCD 1999 *Thai Standard of Wastewater Quality From Swine Farm.* Pollution Control Department, Bangkok, Thailand.

Stina, L. 2015 *Renewable Energy from Wastewater Grown Microalgae.* MS thesis, Lund University, Lund, Sweden.

UNITC 2014 *Pork and Swine Industry & Trade.* Report UNITC. Publication ITS-11, Washington, DC, USA.

Versteegh, G. J. M. & Blokker, P. 2004 Resistant macromolecules of extant and fossil microalgae. *Phycol. Res.* **52**, 325–339.

Wang, M., Sahu, A. K., Rusten, B. & Park, C. 2013 Anaerobic co-digestion of microalgae *Chlorella* sp. and waste activated sludge. *Bioresour. Technol.* **142**, 585–590.

Ward, A. J., Lewis, D. M. & Green, F. B. 2014 Anaerobic digestion of algae biomass: a review. *Algal Res.* **5**, 204–214.

Zuliani, L., Frison, N., Jelic, A., Fatone, F., Bolzonella, D. & Ballottari, M. 2016 Microalgae cultivation on anaerobic digestate of municipal wastewater, sewage sludge and agro-waste. *Int. J. Mol. Sci.* **17** (10), 1692–1709. doi:10.3390/ijms17101692.

First received 1 November 2017; accepted in revised form 12 February 2018. Available online 21 February 2018

Biogas from mono- and co-digestion of microalgal biomass grown on piggery wastewater

P. Carminati, D. Gusmini, A. Pizzera, A. Catenacci, K. Parati and E. Ficara

ABSTRACT

Biogas production has been suggested as a valid valorization solution for microalgal/bacteria biomass (MAB) grown on wastewater. This research is aimed at assessing the feasibility to use MAB grown in an outdoor raceway fed on piggery wastewater for biogas production. Batch and continuous anaerobic tests were conducted on the sole MAB and on a blend of MAB and carbonaceous substrates (deproteinated cheese whey and cellulose) to improve the carbon/nitrogen ratio. Results of batch biochemical methane potential tests confirmed that the sole microalgal/bacteria biomass was poorly degradable (119 $NmL_{CH4}·g_{COD}^{-1}$), while blending it with deproteinated cheese whey or cellulose (80% of carbonaceous material and 20% of MAB, as chemical oxygen demand (COD)) had no synergistic effects on the methane yield, although slight improvements in the degradation kinetics were observed. Continuous anaerobic degradation tests (at an organic loading rate of 1.5 $g_{COD}·L^{-1}·d^{-1}$, 35 °C and 30 days of hydraulic retention time) increased the overall methane yield from 81 $NmL_{CH4}·g_{COD}^{-1}$ (sole MAB) to 216 $NmL_{CH4}·g_{COD}^{-1}$ (MAB and deproteinated cheese whey) and 122 $NmL_{CH4}·g_{COD}^{-1}$ (MAB and cheese whey). However, data confirm that no evident synergistic effects were obtained.

Key words | biogas, cellulose, cheese whey, co-digestion, microalgae, piggery wastewater

P. Carminati
D. Gusmini
A. Pizzera
A. Catenacci (corresponding author)
E. Ficara
Department of Civil and Environmental Engineering (DICA),
Politecnico di Milano,
P.zza L. da Vinci 32, 20133 Milano,
Italy
E-mail: *arianna.catenacci@polimi.it*

K. Parati
Istituto Sperimentale Italiano Lazzaro Spallanzani,
Localita' La Quercia, Rivolta d'Adda, Cremona,
Italy

INTRODUCTION

In response to the need for more effective alternatives to wastewater treatment enabling the implementation of the circular economy concept, microalgae-based processes are being tested. Indeed, those processes are less energy intensive and produce valuable biomass while removing nutrients and other contaminants from waste streams (Olguín (2012) among others).

In Northern Italy, and in other similar European areas such as Flanders and Brittany, intensive animal breeding has become the reference farming system. More than half of the overall piggery production is concentrated in large farms, breeding 1,800 pigs on average. However, increasing the size of a pig farm, while improving productivity, increases wastewater and greenhouse gas emissions to be managed within a localized area (Girard *et al.* 2009). Indeed, piggery wastewaters (PWWs) carry important loads of ammoniacal nitrogen and must be carefully handled in order to prevent environmental damage, and to comply with limitations set out by the European Directive 91/676/EEC (EC 1991) and its national/regional transposition.

Microalgae-based treatment processes have been tested on different kinds of wastewaters, including domestic, industrial and agricultural. Since many species are able to utilize organic molecules as substrate or as growth factors, microalgal mixotrophic and heterotrophic growth can also be exploited to remove organics (Skorupskaite *et al.* 2015). Moreover, oxygen released during photosynthesis normally fosters bacterial activity. Indeed, microalgae-based systems have been suggested as capable of significantly reducing both organic matter and nutrients in municipal and piggery wastewater at minimal energy cost (González *et al.* 2008).

What makes microalgal-based treatment interesting are the reduced energy costs and the production of a biomass that can be further converted into valuable products. To make these treatments more convenient, both aspects are to be carefully studied. When microalgal biomass coming from wastewater treatment is considered, valorization routes include biofuel production or fertilizer/soil amendment production, and those two alternatives may be also combined (Park *et al.* 2011).

doi: 10.2166/wst.2018.134

A promising biofuel that can be produced from microalgal biomass is biogas, allowing for the integration of algal biomass into existing agro-industrial digesters. Such installations have spread remarkably in recent years, most of them fed on energy crops leading to the well-known food versus fuel debate. Therefore, in 2015 algal biomass was suggested as a third-generation feedstock by the European Commission.

Studies on anaerobic digestion of algal biomass peaked in the late 1970s as a consequence of the first oil crises, and species under investigation included mainly macroalgae (Mussgnug *et al.* 2010). Recently, the identification of microalgae with promising characteristics and the technological advances in cultivation and harvesting techniques have raised the interest to use them for bioenergy generation. Although research related to microalgae as substrates for biogas production is limited, recent theoretical calculations indicated their potential (Sialve *et al.* 2009).

Nonetheless, biogas production from microalgal biomass needs to overcome some other obstacles such as the cost of biomass harvesting, the low concentration of algal suspension limiting the applicable organic loading rate (OLR), and the algal biomass biogas yield (Ras *et al.* 2011). Indeed, the scientific literature reports highly strain-specific methane yields and kinetics; however, a high variability (153–600 $NL_{CH4}{\cdot}kg_{VS}^{-1}$), even within the same strain, is also reported. The macromolecular composition and the cell wall characteristics of each microalgae species greatly affect the biogas production. Since the hydrolysis is considered the main limiting step, many pretreatment methods (thermal, mechanical, chemical, biological) for microalgae cell wall disruption have been investigated, aimed at being both efficient and cost-effective in terms of energy balance of the overall process (Bohutskyi *et al.* 2014).

Another crucial parameter is the wide variation and unbalanced nutrients content in microalgal biomass, which is related to several environmental factors (season, growing substrate and ecological conditions) and that affects the anaerobic digestion process. The optimal carbon to nitrogen (C/N) ratio was observed between 20 and 30, but in algal biomass it ranges 6–10, which seems too low for an effective digestion (Sialve *et al.* 2009): in such conditions, inhibition from free-ammonia and, consequently, volatile fatty acids (VFAs) and sulfide (H_2S) accumulation may occur. Therefore, co-digestion with carbon-rich materials can help improve the digestion performance. Synergistic effects in co-digestion of microalgae are usually attributed to a balanced C/N ratio; in addition, most of the trace elements with positive effects on the anaerobic biodegradability (Fe, Co, Zn) are naturally available in the microalgal biomass. Enhanced process stability may result in higher OLRs and lower digester volumes (Herrmann *et al.* 2016). Furthermore, Wang *et al.* (2013) found that anaerobic co-digestion of algae and sludge could also improve the digestate dewaterability. Yet, results in literature are partly contradictory, and co-digestion of microalgae with no significant synergistic or antagonistic effect on the methane yield has also been reported. Literature studies about a wide variety of co-substrate are available, such as piggery and bovine sewage (González-Fernández *et al.* 2011), municipal sludge (Wang *et al.* 2013), agricultural and livestock residues (Herrmann *et al.* 2016), and waste paper (Yen & Brune 2007).

Within this framework, the general objective of this research was to assess the feasibility of feeding the algal/bacteria suspension, grown in an outdoor raceway fed on a PWW, to agro-industrial digesters for biogas production. This objective was targeted by:

- exploring the anaerobic degradability of the microalgal/bacteria consortium;
- comparing algal biomass mono-digestion to co-digestion with carbonaceous materials.

METHODS

MAB origin: pilot plant and the PWW

The microalgal biomass (MAB, in the following) was collected from a pilot-scale raceway pond (RW, from Aqualgae S.L. Spain), installed outdoors at a piggery farm breading 20,000 pigs in Northern Italy (Cremona). The RW had a surface of 3.8 m^2 and a water depth of 20 cm, and it was featured with an integrated sump for a pH-regulated CO_2 injection; a speed-adjustable paddlewheel rotating at 20 rpm and moving the flow at a rate of approximately 0.15 $m{\cdot}s^{-1}$; temperature, dissolved oxygen and pH probes equipped with data logging; an adjustable flow rate (40–90 $L{\cdot}d^{-1}$) peristaltic pump for continuous feeding; and a gravity-driven overflow to divert the RW effluent to a storage tank. The hydraulic retention time (HRT) was maintained at around 8 days although it varied between 7 and 14 days.

The RW was continuously fed with a PWW collected downstream to a primary solid/liquid separation by dissolved air flotation. A mixed storage tank of 1 m^3 was available to ensure continuous feeding upon weekly wastewater supply.

MAB harvesting was achieved by using a centrifuge (10,000 G from Filtermaxx, Jacksonville USA) operated at 3,000–3,100 rpm, in semi-continuous mode with a feeding rate of 3.35 $L \cdot min^{-1}$.

Biochemical methane potential tests

Biochemical methane potential (BMP) tests in mesophilic conditions ($35 \pm 0.5\,°C$) were performed in batch mode, under constant mixing, according to a manometric method. Glass bottles (570 mL) closed with rubber septa and maintained in place by an aluminum cap were used as anaerobic reactors. As inoculum, pre-incubated digestate from a mesophilic digester fed on municipal waste sludge was used. The inoculum to substrate ratio was 0.3 as volatile solids (VS). A mineral medium containing macro- and micro-nutrients was dosed as suggested by OECD (2006), Test No. 311, and tap water was finally added to reach the desired working volume (165 mL). N_2 was bubbled to ensure initial anaerobic conditions. Each substrate was tested in duplicate. Additionally, two bottles serving as a blank were prepared (inoculum, mineral medium and tap water). The manometric measures were performed using a digital manometer (Keller LEO 2) by puncturing the rubber septum. The volume of gas produced by the anaerobic degradation was computed from pressure data, according to the ideal gas law. The methane content in the biogas was measured by gas chromatography. The methane yield (*BMP*, expressed in $NmL_{CH4} \cdot g_{VS}^{-1}$) was calculated as follows (Equation (1)):

$$BMP = \frac{(V_{CH4,S} - V_{CH4,B})}{M_S} \qquad (1)$$

where $V_{CH4,S}$ and $V_{CH4,B}$ (NmL_{CH4}) are the accumulated volume of methane at standard conditions (273 K, 1 atm) coming from the substrate and the blank samples, respectively; M_S (g_{VS}) is the mass of substrate in terms of VS.

BMP-1 tests lasted 67 days and were performed on MAB, deproteinated cheese whey (DP-CW) produced by ultrafiltration and a mixture of MAB and DP-CW (17% and 83% as VS), in order to raise the C/N ratio to the optimal range of 20–30. BMP-2 tests lasted 29 days and were performed on MAB, cellulose (CEL) and a mixture of MAB and CEL (16% and 84% as VS). Both mixtures correspond to 20% algal biomass and 80% carbonaceous co-substrate in terms of chemical oxygen demand (COD).

Continuous anaerobic digestion (CD) tests

Tests were performed by using two reactors (named 'A' and 'B'), both simulating a continuous anaerobic digester, with an active volume of 1.5 L (0.5 L of tap water; 1 L of inoculum, same mesophilic anaerobic sludge used for BMP tests). To limit potential photosynthetic activity, bottles were wrapped in aluminum foil. Feeding and discharging glass pipes, both provided with valves, were inserted into two openings on the top of each reactor. Each reactor head was closed by means of a rubber stopper equipped with a stirring rod and a tygon tube connected to the gas measuring unit. A CO_2 trap, filled with a 3 M NaOH solution, was located before the volumetric device for methane measurement (AMPTS II, Bioprocess Control®). The reactors were kept at $34 \pm 1\,°C$ by a thermostatic bath.

Reactors were fed twice a week. In correspondence with each feeding, digestate was also discharged and analyzed for pH and FOS/TAC. Once a week, total solids (TS), VS, total and soluble COD, total Kjeldahl nitrogen (TKN), ammoniacal nitrogen, total phosphorous (TP) and speciation of VFAs were assessed. The biogas composition was analyzed weekly.

Two types of experiments were performed:

(i) Experiment CD-1: co-digestion of MAB and DP-CW (17% and 83% as VS) was tested on both reactors, in order to verify repeatability; this test lasted 44 days.
(ii) Experiment CD-2: mono-digestion of MAB in the reactor A and co-digestion of MAB and CEL (16% and 84% as VS) in the reactor B were tested; this test lasted 60 days.

For both experiments, a startup phase of about 3 weeks was first applied by increasing step by step the OLR from 0.75 to 1.5 $g_{COD} \cdot L^{-1} \cdot d^{-1}$. The HRT of the reactors was kept at 30 days for the entire duration of the tests.

Analytical methods

Solid content (TS and VS) and suspended solid content (total (TSS) and volatile (VSS)) were determined in duplicate according to Standard Methods 2540 (APHA 2005). The pH and the electrical conductivity (EC) were directly measured in liquid samples by means of portable multi-probe meter (Hach-Lange, HQ40D). The VFAs (acetic, propionic, isobutyric, butyric, isovaleric and valeric) concentrations were determined according to Standard Methods 5560 (APHA 2005), using a gas chromatograph (DANI Master GC) coupled with a flame ionization detector. The Hach-Lange Biogas Titrator was used for

VFA/alkalinity (so-called FOS/TAC) analyses according to the Nordmann method. Ammoniacal nitrogen ($N\text{-}NH4^+$), nitrite ($N\text{-}NO_2^-$), nitrate ($N\text{-}NO_3^-$), TP and soluble COD (sCOD) were measured using spectrophotometric test kits (Hach-Lange) on 0.45 μm filtered samples. Total COD was determined according to Standard Methods 5220 (APHA 2005). TKN was measured according to ISO 5663:1984 (International Organization for Standardization 1984). Biogas composition (CO_2, CH_4, H_2, O_2, N_2) was determined by using a gas chromatograph (DANI Master GC) equipped with an Hayesep Q column and a Molesieve 5A column. An optical microscope (BM60 Optech, Exacta+Optech Labcenter S.p.A., Italy) was used to view the algal biomass and images were taken in contrast phase with a 40× objective. A haemacytometer counting chamber (Bürker) was used to visually determine under a microscope the total algal count. The optical density at 680 nm (OD@680) was measured with the Hach-Lange DR6000TM UV–visible spectrophotometer.

RESULTS AND DISCUSSION

Microalgal/bacteria biomass characteristics

The microalgal/bacteria biomass was sampled from the RW which was continuously operated from May to October 2017. Although a detailed discussion on the efficiency of the system is beyond the scope of this paper, relevant characteristics of the PWW compared to MAB are summarized in Table 1. A stable algal community has developed in the system, as demonstrated by both the total algal count and the OD at 680 nm. However, microscope observations as well as the high VSS/OD ratio suggest the presence of flocs composed of organic debris and bacteria. Moreover, some filamentous algae (genus *Tribonema*) were also found both colonizing the raceway and floating in the suspension. The microalgal community included algae from the families Scenedesmaceae (mainly *Scenedesmus acuminatus* and *Scenedesmus quadricauda*) and Chlorellaceae; they both belong to the Chlorophyta phylum which is known to include very tolerant organisms, typically observed in microalgal systems treating wastewaters. *Chlamydomonas* spp. and *Navicula* spp. were also sporadically spotted but did not significantly contribute to the overall algal count.

The average areal productivity of the RW was 12 $g_{VSS} \cdot m^{-2} \cdot d^{-1}$ (including algae and bacteria biomass), with average values for irradiance and temperature of 251 $W \cdot m^{-2}$ and 24 °C, respectively.

As shown by data in Table 1, the raceway effectively reduced the ammoniacal load of the PWW, which was both assimilated by MAB and nitrified. In addition, though to a lower extent, sCOD and P concentrations were decreased.

The average COD/VSS and N/VSS ratios of the MAB collected by centrifugation were 1.51 $g \cdot g^{-1}$ and 0.10 $g \cdot g^{-1}$, respectively. Nitrogen and COD contents match typical compositions reported in literature. According to Heaven & Zhang (2011), the empirical elemental formula of *Chlorella*

Table 1 | Chemical characteristics of the PWW and the MAB

Parameter	Unit	PWW			MAB		
		Min	Average	Max	Min	Average	Max
TSS	$g \cdot L^{-1}$	0.09	0.20	0.32	N.A.	N.A.	N.A.
VSS	$g \cdot L^{-1}$	N.A.	N.A.	N.A.	0.28	0.77	1.4
$N\text{-}NH_4^+$	$mg \cdot L^{-1}$	108	195	313	1.10	23.5	73.2
$N\text{-}NO_3^-$	$mg \cdot L^{-1}$	N.A.[a]	N.A.	N.A.	3.42	139	244
$N\text{-}NO_2^-$	$mg \cdot L^{-1}$	N.A.	N.A.	N.A.	0.06	24.4	118
$P\text{-}PO_4^{3-}$	$mg \cdot L^{-1}$	5.55	20.7	51.6	1.89	12.7	29.7
COD	$mg \cdot L^{-1}$	323	924	1,610	N.A.	N.A.	N.A.
sCOD	$mg \cdot L^{-1}$	176	658	1,090	140	251	404
EC	$mS \cdot cm^{-1}$	3.1	3.5	3.9	N.A.	N.A.	N.A.
OD@680	[–]	N.A.	N.A.	N.A.	0.11	0.38	0.84
Total algal count	$cell \cdot mL^{-1}$	N.A.	N.A.	N.A.	$6.0 \cdot 10^5$	$4.0 \cdot 10^6$	$1.3 \cdot 10^7$

Data refer to 150 days of experimentation with samples collected every 2 weeks on average for the PWW, and once or twice a week, depending on the parameter, for the MAB.
N.A.: not available.
[a]The contribution of nitrate was negligible (always lower than 2% of the ammoniacal nitrogen).

vulgaris ranges from $C_{4.3}H_8N_{0.6}O_{1.8}$ to $C_{3.3}H_{6.3}N_{0.5}O_{1.5}$, while for *Scenedesmus obliquus* it varies between $C_{3.1}H_{5.8}N_{0.5}O_{1.4}$ and $C_{3.7}H_7N_{0.6}O_{1.7}$. The average COD/VSS ratio is also comparable to experimental data from Ras *et al.* (2011) and Ramos-Suárez & Carreras (2014), who reported about 1.33–1.43 $g{\cdot}g^{-1}$ for *Chlorella vulgaris* and 1.65 $g{\cdot}g^{-1}$ for *Scenedesmus*, respectively.

Batch BMP tests

In Table 2, results of mono- and co-digestion BMP tests are summarized, while Figure S1 (Supplementary Material, available with the online version of this paper) shows the time trends of the net specific biogas production (*SBP*). Experimental data of the cumulative biogas volumes (V_t) versus t were fitted with an exponential model (Equation (2)) to assess the hydrolysis kinetic constant (k_h):

$$V_t = V_\infty \times [1 - \exp(-k_h \times t)] \qquad (2)$$

where V_∞ is the asymptotic biogas production. The minimum square error criterion was used to assess the optimal values of k_h and V_∞.

The overall repeatability of duplicate bottles was fully satisfactory, with standard deviation always lower than 10%. The different biogas production for MAB in BMP-1 and BMP-2 trials is primarily due to the different test duration: 67 days in the first case and 1 month in the second case. However, by comparing the two sets of data after 29 days, similar BMP values were obtained (124 and 108 $NmL_{CH4}{\cdot}g_{VS}^{-1}$, respectively).

These results are comparable to data found in literature for the genera *Scenedesmus* and *Chlorella*. Mussgnug *et al.* (2010) and Ramos-Suárez & Carreras (2014) obtained 178 and 140 $NmL_{CH4}{\cdot}g_{VS}^{-1}$ for *Scenedesmus obliquus*, respectively, while similar or higher values are reported for *Chlorella vulgaris*: 191 $NmL_{CH4}{\cdot}g_{VS}^{-1}$ by Mahdy *et al.* (2014) and up to 337 $NmL_{CH4}{\cdot}g_{VS}^{-1}$ by Zhao *et al.* (2014). As for the methane content in biogas, values shown in Table 2 fall within the range already found by Mussgnug *et al.* (2010) and Sialve *et al.* (2009): 61–67% and 69–75%, respectively.

Moving to tests on carbonaceous substrates, the degradation of the DP-CW was almost complete, producing 294 of the maximum theoretical value of 350 $NmL_{CH4}{\cdot}g_{COD}^{-1}$. The remaining 16% was likely to be uptaken for bacteria growth. As for the cellulose degradability, comparable BMP values are reported in literature by Pham *et al.* (2013) (254–378 $NmL_{CH4}{\cdot}g_{VS}^{-1}$); the methane content too is close to the 50%, which is the typical value obtained for carbohydrates or sugars.

The time required to achieve 90% of the final methane yield is quite different between the three substrates (55 days for MAB, 7 days for DP-CW and 11 days for CEL), suggesting that the algal biomass is the slowest to be degraded, as confirmed by the first order degradation kinetic constant, which is one order of magnitude higher for the DP-CW than for the MAB.

Co-digestion tests were performed in order to assess possible synergistic effects ascribable to more suitable or balanced nutrient ratios. However, hypothesis were not confirmed by results. As for the co-digestion of MAB and DP-CW, the experimental value of 237 ± 6 $NmL_{CH4}{\cdot}g_{COD}^{-1}$ is lower than the expected one, computed by the linear combination of methane yields obtained in mono-digestion tests ($119 \times 0.2 + 294 \times 0.8 = 259$ $NmL_{CH4}{\cdot}g_{COD}^{-1}$). A similar conclusion can be drawn for co-digestion tests using cellulose: the experimental value of 220 ± 10 $NmL_{CH4}{\cdot}g_{COD}^{-1}$ is slightly lower than the expected one, evaluated as previously explained ($72 \times 0.2 + 270 \times 0.8 = 230$ $NmL_{CH4}{\cdot}g_{COD}^{-1}$). Nonetheless, a faster kinetics is observed in co-digestion tests, possibly

Table 2 | Final results of biogas (*SBP*) and methane (*BMP*) production in mono- and co-digestion BMP tests; the hydrolysis kinetic constant (k_h) is evaluated by fitting data with an exponential model

			BMP			
Trial	**Tested substrate**	***SBP*** **($NmL{\cdot}g_{VS}^{-1}$)**	**($NmL_{CH4}{\cdot}g_{VS}^{-1}$)**	**($NmL_{CH4}{\cdot}g_{COD}^{-1}$)**	**CH_4/Biogas (%)**	**k_h (d^{-1})**
BMP-1	MAB	285 ± 4	174 ± 2	119 ± 1	61.0	0.04[a]
BMP-1	DP-CW	683 ± 10	365 ± 3	294 ± 3	53.4	0.41
BMP-1	MAB (20%) + DP-CW (80%)	566 ± 12	302 ± 7	237 ± 6	53.4	0.48
BMP-2	MAB	151 ± 14	108 ± 8	72 ± 5	71.3	0.04[a]
BMP-2	CEL	579 ± 11	320 ± 6	270 ± 5	55.2	0.15
BMP-2	MAB (20%) + CEL (80%)	505 ± 24	272 ± 12	220 ± 10	53.9	0.16

[a]Parameters were obtained from aggregated data of both BMP tests.

leading to higher methane productions during continuous tests. This evidence is in agreement with that observed by Ramos-Suárez & Carreras (2014).

However, because of the batch nature of the BMP tests, their results may fail in fully evidencing co-digestion synergies as suggested by other researchers (Ramos-Suárez & Carreras 2014; Herrmann *et al.* 2016). Continuous co-digestion tests were therefore planned and results are reported in the following section.

Continuous anaerobic digestion

Chemical characteristics of the effluents from reactors A and B in CD-1 experiments (co-digestion of MAB and DP-CW) are shown in Figure 1, as well as cumulated methane production over time (note that continuous data acquisition started from day 9); VFAs composition is reported in Table S1 (Supplementary Material, available online).

Reproducible results and time trends for methane production, VS degradation and nutrients solubilization are shown for the two reactors, thus suggesting that the equipment used is appropriate for anaerobic co-digestion tests. More significant differences are observed in the sCOD and VFAs data, indicating that intermediate compounds in the anaerobic degradation chain may be more affected by slight differences, between parallel reactors, in the environmental conditions such as the pH.

According to biogas production and VS degradation, anaerobic co-digestion of MAB and DP-CW in semi-continuous mode achieved a pseudo-steady-state condition in about 23 days, corresponding to an OLR of $1.5\,g_{COD}{\cdot}L^{-1}{\cdot}d^{-1}$ (HRT of 30 days) for both reactors. Indeed, when reactions take place in a continuous stirred tank reactor (CSTR), the steady-state condition can be achieved in much less than $3 \times$ HRT, which is the time required to achieve the technical steady-state condition for a non-reactive CSTR.

The average methane yield in stationary conditions (days from 27 to 41) was evaluated as 222 and 210 $NmL_{CH4}{\cdot}g_{COD}{}^{-1}$ for the reactors A and B respectively, corresponding to 94% and 89% of the BMP value obtained from BMP-1 tests with the same substrates. The methane content in the biogas ranged 59–70%, with an average of 66%.

The stability of the process was controlled by measuring FOS/TAC and pH in order to verify the possible VFAs accumulation and the consequent inhibitory effect on methanogens due to the rapid biodegradability of the DP-CW. It is worth noting that alkalinity was added to the feed (5.7 $mmol{\cdot}L^{-1}{\cdot}d^{-1}$ as $NaHCO_3$) to improve the FOS/TAC ratio. Following a first stage of acclimation, the pH stabilized at around 7.5–7.6, i.e. within the optimal range for the biomass activity; at the same time, the FOS/TAC ratio returned to values lower than $0.3\,g_{CH3COOH}{\cdot}g_{CaCO3}^{-1}$, accounted in literature as an indication of stability (Drosg 2013).

As for nutrients, the protein-associated organic nitrogen in the MAB was only partially released as NH_4^+ because of the scarce biodegradability of microalgae; indeed, ammoniacal nitrogen decreased from its initial value to about $100\,mg{\cdot}L^{-1}$, largely below inhibition thresholds already indicated in literature ($>1.7\,g{\cdot}L^{-1}$, Chen *et al.* 2008). Phosphorus was mainly introduced in the reactors with the DP-CW ($184\,mg{\cdot}L^{-1}$), and stable values of P-PO_4^{3-} concentrations in the range 100–120 $mg{\cdot}L^{-1}$ were achieved.

The accumulation of COD observed in both reactors was likely due to the undegraded algal biomass. In addition, the COD mass balance, evaluated between days 27 and 41, confirmed that part of the COD fed to reactors was not fully retrieved in the biogas as methane or in the digestate (16% and 20% for reactors A and B, respectively), in good accordance with the increase of COD in the reactors.

Compared to reactor B, higher concentrations of sCOD were measured in reactor A, probably related to VFAs. Indeed VFAs maximum concentations in reactors A and B after 41 days were 751 and $186\,mg_{COD}{\cdot}L^{-1}$, respectively. Such a difference ($560\,mg_{COD}{\cdot}L^{-1}$) is clearly related to the different concentrations of sCOD measured at the same time (day 41), and resulting in a gap of about $536\,mg_{COD}{\cdot}L^{-1}$ between the two reactors. As for the VFAs composition, acetic acid accounts for more than 92% (A) and 89% (B) of the total VFAs, while propionic and isobutyric acids make up the remaining 8–11%.

Chemical characteristics of the effluents from reactors A and B in CD-2 experiments, together with cumulated methane production, are shown in Figure 2; VFAs composition is reported in Table S2 (Supplementary Material, available online).

In this test, reactors A and B were used for the mono-digestion of MAB and the co-digestion of MAB and CEL, respectively. Both reactors were initially fed with microalgae and cellulose to acclimate the biomass and verify once again the reproducibility of the digestion process. Starting from day 14, reactor A was fed with the MAB only, while reactor B was carried on in co-digestion mode. The applied OLR was increased step by step from 0.75 to $1.5\,g_{COD}{\cdot}L^{-1}{\cdot}d^{-1}$ during the initial 22 days and then stabilized, corresponding to an HRT of 30 days.

The methane yield was evaluated for two time periods: days 28–42 and days 46–60. In relation to reactor A (mono-digestion), the yield was 36 and 81 $NmL_{CH4}{\cdot}g_{COD}{}^{-1}$, respectively, equivalent to 30% and 68% of the BMP value

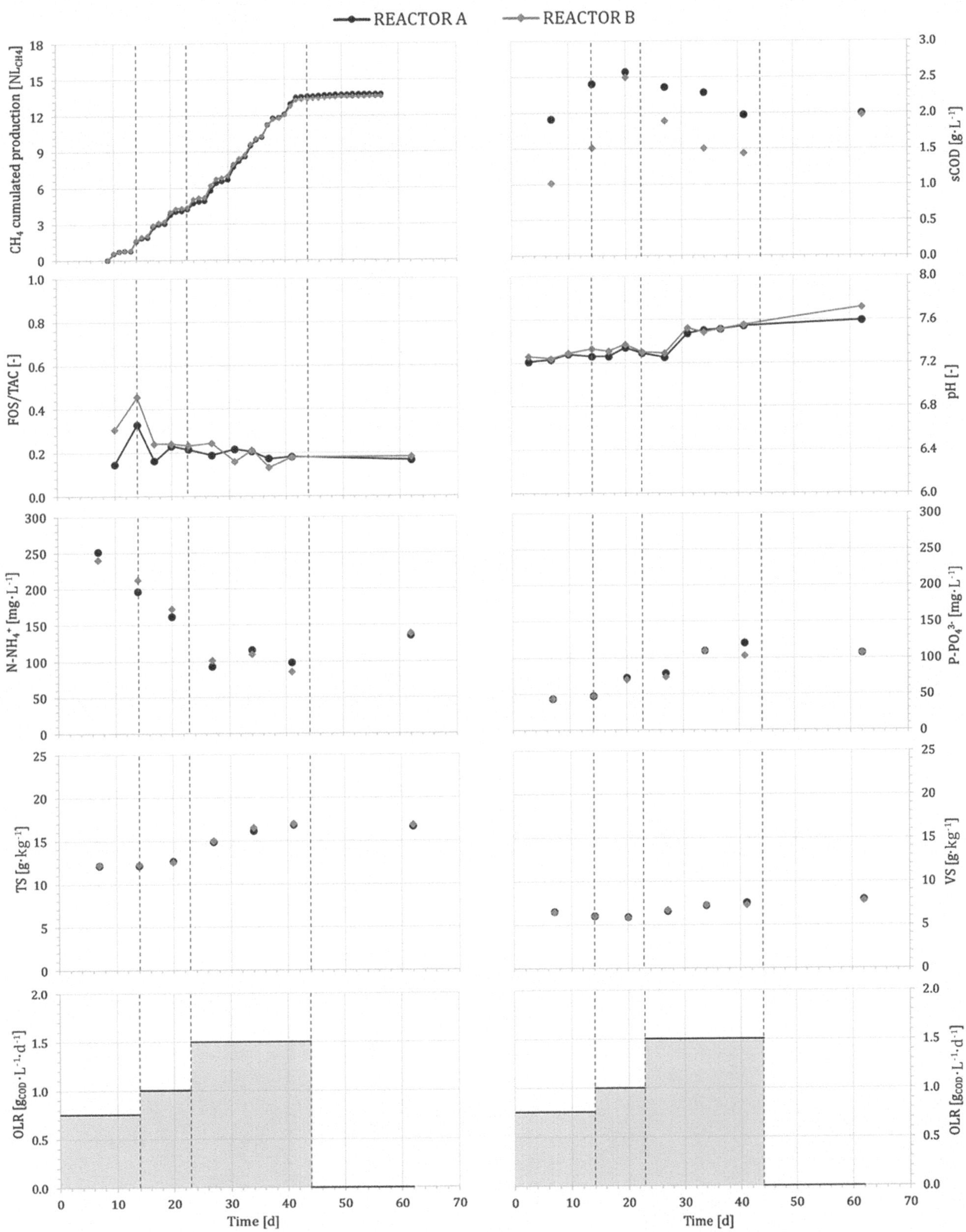

Figure 1 | Time trends of effluent concentrations and cumulated methane production for reactors A (black circles) and B (grey diamonds) during CD-1 experiments (co-digestion of MAB and DP-CW), at the applied OLR.

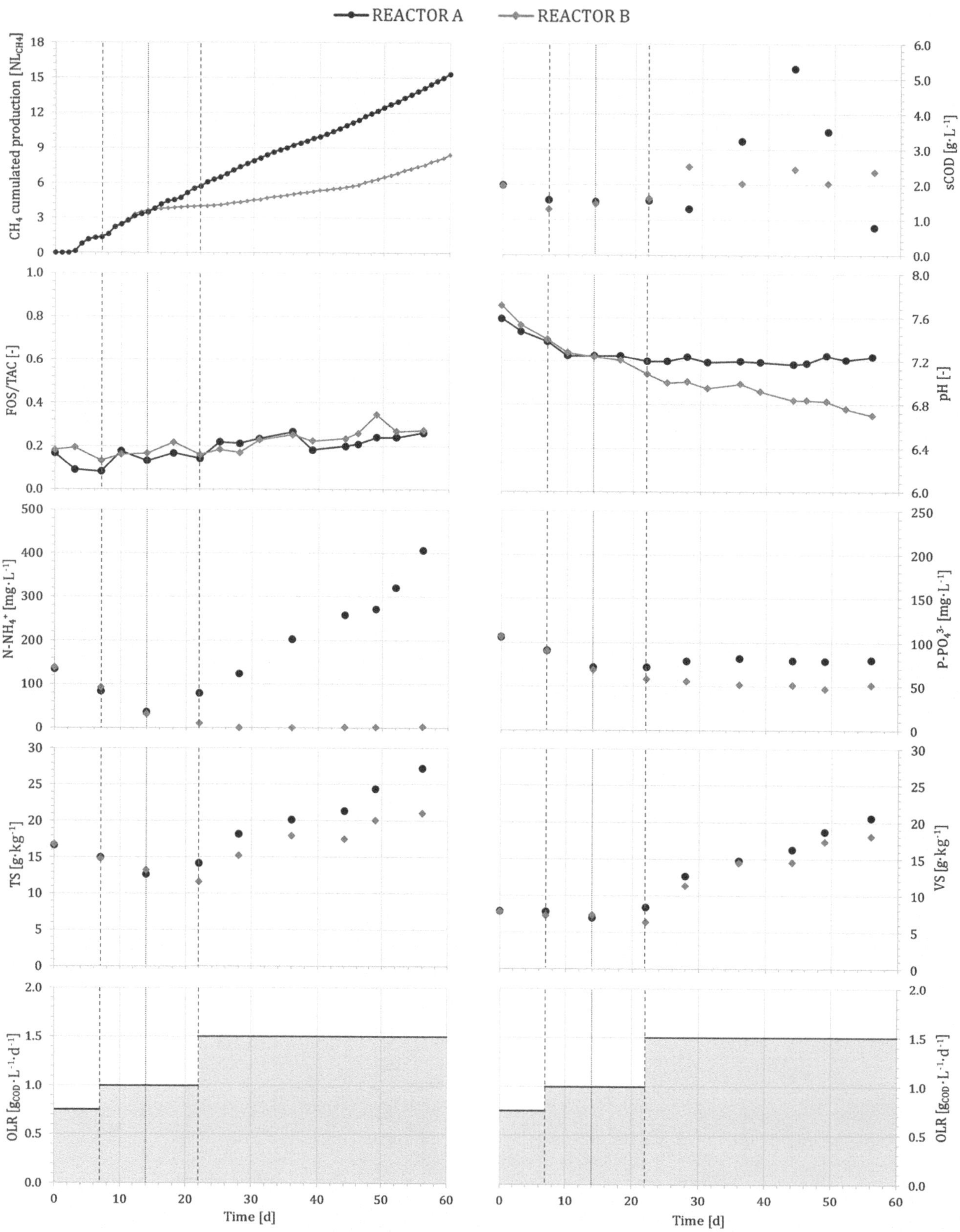

Figure 2 | Time trends of effluent concentrations and cumulated methane production for reactors A (black circles, MAB mono-digestion) and B (grey diamonds, MAB + CELL co-digestion) during CD-2 experiments, at the applied OLR.

obtained from BMP-1 tests with the sole MAB. As for the reactor B, the methane productions in the same periods were 97 and 122 $NmL_{CH4}{\cdot}g_{COD}^{-1}$, corresponding to 44% and 55% of the methane yield measured in BMP co-digestion tests. The average methane content in the biogas was 67% and 51% for the mono- and the co-digestion, respectively. The increased biogas yield in the final stage of the experiment (day 46–60) can be explained by the acclimation of the anaerobic microbial consortium to the substrates or to a more degradable harvested algal biomass.

Ras *et al.* (2011) studied the continuous anaerobic digestion of *Chlorella* under quite similar conditions to those applied to the reactor A ($OLR = 1.0\ g_{COD}{\cdot}L^{-1}{\cdot}d^{-1}$ and HRT = 28 d). They found a specific methane yield of 180 $NmL_{CH4}{\cdot}g_{COD}^{-1}$, significantly higher than that obtained in the present study. This may be partially due to differences in the microalgal community. Indeed, in the present experimentation, the presence of *Scenedesmus* can explain a poorer degradation efficiency, since the microalgal biomass is known for being refractory to the anaerobic degradation owing to the cell wall resistance to hydrolysis (Mussgnug *et al.* 2010; Ramos-Suárez & Carreras 2014). Optical microscopic analysis revealed the presence of intact *Scenedesmus* cells in the digestate exiting the reactors (see Figure S2, Supplementary Material, available online).

The study of Yen & Brune (2007), in relation to mono-digestion experiments of algal biomass composed of *Scenedesmus* and *Chlorella* ($OLR = 2.0\ g_{VS}{\cdot}L^{-1}{\cdot}d^{-1}$ and HRT = 10 d) and co-digestion with waste paper ($OLR = 4.0\ g_{VS}{\cdot}L^{-1}{\cdot}d^{-1}$ and HRT = 10 d) at different ratios, reports a specific methane yield of 90 $NmL_{CH4}{\cdot}g_{VS}^{-1}$ in mono-digestion, similar to the production of reactor A during the last experimental period of 46–60 days (123 $NmL_{CH4}{\cdot}g_{VS}^{-1}$). As for the methane yield in co-digestion (25% algae and 75% waste paper, as VS), they found a production of 79 $NmL_{CH4}{\cdot}g_{VS}^{-1}$ compared to 120 $NmL_{CH4}{\cdot}g_{VS}^{-1}$ observed in reactor B (days 28–42); this higher yield is likely to be attributable to the longer HRT applied in CD-2 experiments, compared to the conditions used in the mentioned study found in literature.

As shown in Figure 2, the degradation process in reactor A was stable, as suggested by the FOS/TAC trend which remained within the range 0.2–0.3. Alkalinity stabilized to values of 3,300–3,500 $mg_{CaCO3}{\cdot}L^{-1}$, and the pH remained around 7.2. Differently, in the co-digestion process (reactor B), alkalinity dropped to 1,400 $mg_{CaCO3}{\cdot}L^{-1}$, not enough to prevent the pH from decreasing below the value of 6.7. However, the low concentration of VFAs measured in the reactor B (see Table S2, Supplementary Material), led to a limited increase in the FOS/TAC ratio, thus preventing the process from severe inhibition.

In reactor B, the ammonium concentration lowered to values close to zero, probably because of the high C/N ratio (about 30) of the feeding substrate, and the scarce degradability of MAB. Low NH_4^+ concentrations negatively affected the alkalinity content, as already stated. It is well known that ammonium is required to sustain microbial growth (50–200 $mg{\cdot}L^{-1}$ as $N\text{-}NH_4^+$ according to McCarty (1964); therefore ammonium scarcity is expected to play a major role in the co-digestion of MAB and CEL. Indeed, ammonium limitation was already identified by Yen & Brune (2007), while testing the co-digestion of algae and waste paper. Similarly, in reactor B the methane production was almost double (187 $NmL_{CH4}{\cdot}g_{COD}^{-1}$), when the OLR was $1.0\ g_{COD}{\cdot}L^{-1}{\cdot}d^{-1}$ (days 7–21) and the ammonia content was still non-limiting (10–100 $mg{\cdot}L^{-1}$ as nitrogen).

Conversely, in the MAB mono-digestion reactor, a significant increase in the released ammonia was observed. In this case, the low methane yield is mainly related to the low degradability of the substrate, since the ammonium nitrogen, originating from the hydrolysis of proteins, was within the optimal range (non-limiting and non-inhibiting). Phosphate concentrations stabilized at 50 and 80 $mg{\cdot}L^{-1}$ as $P\text{-}PO_4^{3-}$ for reactor A and B, respectively.

As already observed for CD-1 experiments, TS and VS accumulated in both reactors and led to an increase in COD concentrations, particularly in the mono-digestion reactor. This is in accordance with that assessed by Golueke & Oswald (1959), who found that solids accumulation in the digester was due to the slow hydrolysis rate of the algal biomass.

VFAs concentrations remained low (<250 $mg_{COD}{\cdot}L^{-1}$) in both reactors and acetic acid was the main component, with percentages always above 81% and 76% for the two reactors.

In Table 3, relevant aspects of algal mono-digestion and co-digestion with DP-CW and cellulose are summarized for comparison. The biogas production efficiency (*B*), is calculated by dividing the biogas yield measured in continuous test by the *SBP* evaluated through BMP tests. Similarly, the methane production efficiency (*M*) is the ratio between the methane yield in continuous experiments and the *BMP* value. Such parameters can be compared to the value expected from the k_h data obtained from BMP (and reported in Table 2). By considering the lab-reactors as ideal CSTRs in which a first order degradation process takes place, the theoretical biogas production efficiency

Table 3 | Comparison of main parameters evaluated for MAB mono-digestion and co-digestion tests operated in continuous mode

Parameter	MAB mono-digestion (days 46–60)	Co-digestion with DP-CW (days 27–41)[a]	Co-digestion with CEL (days 46–60)
Methane yield ($NmL_{CH4}{\cdot}g_{COD}^{-1}$)	81	216	122
Methane production efficiency (M)	68%	91%	55%
Biogas production efficiency (B)	63%	77%	59%
Theoretical biogas production efficiency (B_{th})	54%	94%	82%
Average VFA concentration ($mg_{COD}{\cdot}L^{-1}$)	235	302	213
Stability	Stable process	Stable with external alkalinity addition	Continuous decrease of TAC and pH
Nutrients availability	Continuous increase of $N\text{-}NH_4^+$ concentration	Sufficient availability without ammonia inhibition	Insufficient availability of $N\text{-}NH_4^+$

[a]Average data of reactors A and B.

(B_{th}) can be computed as in Equation (3):

$$B_{th} = 1 - 1/(1 + k_h \times HRT) \quad (3)$$

Algal mono-digestion was more effective under continuous tests ($B > B_{th}$), thus suggesting that a more adapted and efficient microbial consortium was grown during the continuous test. In contrast, co-digestion with DP-CW was slightly less effective ($B < B_{th}$) and this may be due to sub-optimal degradation conditions compared to those obtained in BMP tests. A much lower experimental efficiency was finally obtained in continuous co-digestion tests with cellulose, and this is likely due to the previously mentioned nitrogen limitation. Overall, none of the co-digestion conditions seem to be optimal, since no synergistic effects could be evidenced.

However, it is worth noting that real biodegradability of the microalgal biomass could have been underestimated since the inoculum used, although being fed on a complex mixture of agro-wastes, was not previously acclimated to such biomass (Passos *et al.* 2015; Koch *et al.* 2017). The limited length of the trials has to be taken into consideration when interpreting the experimental results and long-term adaptation to the algal biomass could result in better performances.

CONCLUSIONS

In this work, the anaerobic degradability of an algal suspension grown on an outdoor RW fed on PWW was studied. Results of batch mesophilic BMP tests confirmed that this algal suspension was poorly degradable (119 ± 1 $NmL_{CH4}{\cdot}g_{COD}^{-1}$ measured with a 67-day BMP test) under the tested conditions, in agreement with previous literature data. A maximum methane yield of 81 $NmL_{CH4}{\cdot}g_{COD}^{-1}$ was then obtained in a continuous test at an OLR of 1.5 $g_{COD}{\cdot}L^{-1}{\cdot}d^{-1}$ and an HRT of 30 days. Stability issues were not observed in this trial.

Then the algal biomass was tested in co-digestion with nitrogen-poor substrates, one containing soluble sugars (deproteinated cheese whey) and one containing complex carbohydrates (cellulose), to evaluate the feasibility of using microalgae to correct the C/N ratio. No synergistic effects were proven either in batch or in continuous trials although a slight improvement in the degradation kinetics was observed in batch tests but not confirmed in continuous mode. The limited bioavailability of microalgal nitrogen may have limited bacterial activity when co-digesting algal biomass and cellulose in continuous mode.

According to these results, methane production does not seem to be an interesting alternative for the valorization of algal consortia grown in agro-wastewaters, and other alternatives, such as the reuse in agriculture as soil amendment, could be explored. However, long-term experiments allowing for inoculum adaptation are needed to validate these short-term results.

ACKNOWLEDGEMENTS

The authors wish to thank Fondazione Cariplo for funding the project (Polo delle Microalghe) and Eng. Giorgio Tornotti and Mr Mario Drago from Società Agricola Cortegrande for their valuable support.

REFERENCES

APHA 2005 *Standard Methods for the Examination of Water and Wastewater*, 21st edn. American Public Health Association/ American Water Works Association/Water Environment Federation, Washington, DC, USA.

Bohutskyi, P., Betenbaugh, M. J. & Bouwer, E. J. 2014 The effects of alternative pretreatment strategies on anaerobic digestion and methane production from different algal strains. *Bioresource Technology* **155**, 366–372.

Chen, Y., Cheng, J. J. & Creamer, K. S. 2008 Inhibition of anaerobic digestion process: a review. *Bioresource Technology* **99** (10), 4044–4064.

Drosg, B. 2013 *Process Monitoring in Biogas Plants*. IEA Bioenergy, http://www.rohkraft.net/wp-content/uploads/2014/03/20140319_DROSG_process_montoring_RZ_web1.pdf.

EC (European Community) 1991 Council Directive 91/676/EEC of 12 December 1991 concerning the protection of waters against pollution caused by nitrates from agricultural sources. http://eur-lex.europa.eu/legal-content/en/ALL/?uri=CELEX:31991L0676.

Girard, M., Nikiema, J., Brzezinski, R., Buelna, G. & Heitz, M. 2009 A review of the environmental pollution originating from the piggery industry and of the available mitigation technologies: towards the simultaneous biofiltration of swine slurry and methane. *Canadian Journal of Civil Engineering* **36** (12), 1946–1957.

Golueke, C. G. & Oswald, W. J. 1959 Biological conversion of light energy to the chemical energy of methane. *Applied Microbiology* **7** (4), 219–227.

González, C., Marciniak, J., Villaverde, S., García-Encina, P. A. & Muñoz, R. 2008 Microalgae-based processes for the biodegradation of pretreated piggery wastewaters. *Applied Microbiology and Biotechnology* **80** (5), 891–898.

González-Fernández, C., Molinuevo-Salces, B. & García-González, M. C. 2011 Evaluation of anaerobic codigestion of microalgal biomass and swine manure via response surface methodology. *Applied Energy* **88** (10), 3448–3453.

Heaven, S. & Zhang, Y. 2011 Comments on anaerobic digestion of microalgae as a necessary step to make microalgal biodiesel sustainable. *Biotechnology Advances* **29** (1), 164–167.

Herrmann, C., Kalita, N., Wall, D., Xia, A. & Murphy, J. D. 2016 Optimised biogas production from microalgae through co-digestion with carbon-rich co-substrates. *Bioresource Technology* **214**, 328–337.

International Organization for Standardization 1984 ISO 5663:1984 Water quality – determination of Kjeldahl nitrogen – method after mineralization with selenium. International Organization for Standardization, Geneva, Switzerland.

Koch, K., Lippert, T. & Drewes, J. E. 2017 The role of inoculum's origin on the methane yield of different substrates in biochemical methane potential (BMP) tests. *Bioresource Technology* **243**, 457–463.

Mahdy, A., Mendez, L., Ballesteros, M. & González-Fernández, C. 2014 Autohydrolysis and alkaline pretreatment effect on *Chlorella vulgaris* and *Scenedesmus* sp. methane production. *Energy* **78**, 48–52.

McCarty, P. L. 1964 Anaerobic waste treatment fundamentals. *Chemistry and Microbiology* **95** (9), 107–112.

Mussgnug, J. H., Klassen, V., Schlüter, A. & Kruse, O. 2010 Microalgae as substrates for fermentative biogas production in a combined biorefinery concept. *Journal of Biotechnology* **150** (1), 51–56.

OECD 2006 Test No. 311: Anaerobic biodegradability of organic compounds in digested sludge: by measurement of gas production. *OECD Guidelines for the Testing of Chemicals*, Section 3, No. 311. OECD Publishing, Paris, France.

Olguín, E. J. 2012 Dual purpose microalgae-bacteria-based systems that treat wastewater and produce biodiesel and chemical products within a biorefinery. *Biotechnology Advances* **30** (5), 1031–1046.

Park, J. B. K., Craggs, R. J. & Shilton, A. N. 2011 Wastewater treatment high rate algal ponds for biofuel production. *Bioresource Technology* **102** (1), 35–42.

Passos, F., Gutiérrez, R., Brockmann, D., Steyer, J. P., García, J. & Ferrer, I. 2015 Microalgae production in wastewater treatment systems, anaerobic digestion and modelling using ADM1. *Algal Research* **10**, 55–63.

Pham, C. H., Triolo, J. M., Cu, T. T. T., Pedersen, L. & Sommer, S. G. 2013 Validation and recommendation of methods to measure biogas production potential of animal manure. *Asian-Australasian Journal of Animal Sciences* **26** (6), 864–873.

Ramos-Suárez, J. L. & Carreras, N. 2014 Use of microalgae residues for biogas production. *Chemical Engineering Journal* **242**, 86–95.

Ras, M., Lardon, L., Sialve, B., Bernet, N. & Steyer, J. P. 2011 Experimental study on a coupled process of production and anaerobic digestion of *Chlorella vulgaris*. *Bioresource Technology* **102** (1), 200–206.

Sialve, B., Bernet, N. & Bernard, O. 2009 Anaerobic digestion of microalgae as a necessary step to make microalgal biodiesel sustainable. *Biotechnology Advances* **27** (1), 409–416.

Skorupskaite, V., Makareviciene, V. & Levisauskas, D. 2015 Optimization of mixotrophic cultivation of microalgae *Chlorella* sp. for biofuel production using response surface methodology. *Algal Research* **7**, 45–50.

Wang, M., Sahu, A. K., Rusten, B. & Park, C. 2013 Anaerobic co-digestion of microalgae *Chlorella* sp. and waste activated sludge. *Bioresource Technology* **142**, 585–590.

Yen, H. W. & Brune, D. E. 2007 Anaerobic co-digestion of algal sludge and waste paper to produce methane. *Bioresource Technology* **98** (1), 130–134.

Zhao, B., Ma, J., Zhao, Q., Laurens, L., Jarvis, E., Chen, S. & Frear, C. 2014 Efficient anaerobic digestion of whole microalgae and lipid-extracted microalgae residues for methane energy production. *Bioresource Technology* **161**, 423–430.

First received 1 November 2017; accepted in revised form 13 March 2018. Available online 27 March 2018

Start-up of a microalgae-based treatment system within the biorefinery concept: from wastewater to bioproducts

Enrica Uggetti, Joan García, Juan Antonio Álvarez and María Jesús García-Galán

ABSTRACT

Within the European project INCOVER, an experimental microalgae-based treatment system has been built for wastewater reuse and added-value products generation. This article describes this new experimental plant and the start-up stage, starting from the new design of three semi-closed horizontal photobioreactors with low energy requirements for microalgae cultivation (30 m^3 total), using agricultural runoff and urban wastewater as feedstock. The inflow nutrients concentration is adjusted to select cyanobacteria, microalgae able to accumulate polyhydroxybutyrates, which can be used for bioplastics production. Part of the harvested biomass is used as substrate for anaerobic co-digestion (AcoD) with secondary sludge to obtain biogas. This biogas is then cleaned in an absorption column to reach methane concentration up to 99%. The digestate from the AcoD is further processed in sludge wetlands for stabilization and biofertilizer production. On the other hand, treated water undergoes ultrafiltration and disinfection through a solar-driven process, then it is pumped through absorption materials to recover nutrients, and eventually applied in an agricultural field to grow energy crops by means of a smart irrigation system. This plant presents a sustainable approach for wastewater management, which can be seen as a resource recovery process more than a waste treatment.

Key words | bioplastics, circular economy, closed systems, microalgal biomass, wastewater

Enrica Uggetti (corresponding author)
Joan García
María Jesús García-Galán
GEMMA – Environmental Engineering and Microbiology Research Group, Department of Civil and Environmental Engineering,
Universitat Politècnica de Catalunya-BarcelonaTech,
c/ Jordi Girona 1-3, Building D1, Barcelona E-08034, Spain
E-mail: *enrica.uggetti@upc.edu*

Juan Antonio Álvarez
AIMEN Technology Centre,
c/ Relva, 27 A – Torneiros, Porriño, Pontevedra 36410,
Spain

INTRODUCTION

Basically, all kind of human activities are putting water resources under a constant pressure: global warming, over-exploitation and pollution of freshwater resources, urbanization and also an increasing competition between various user groups. The 2007 European Communication on Water Scarcity and Droughts (EC 2007) highlighted that climate change and increasing population would raise frequency and severity of water scarcity and drought events. Over the past 30 years, droughts have dramatically increased in number and intensity in the EU and at least 11% of the European population and 17% of its territory have been affected by water scarcity to date (EC 2007). Moreover, only half of the European countries meet the Water Framework Directive (EC 2000), that foresaw a good ecological status of water bodies by 2015 through an adequate wastewater treatment, making major additional wastewater treatment solutions necessary. It is therefore mandatory to put effort in the wastewater treatment sector to look for new technological solutions environmentally and economically feasible in order to overcome this gap.

Nowadays, treated water is mostly discharged into surface waters or into the ground instead of being reused. Despite that during the last years reclaimed water reuse has been encouraged for agricultural irrigation as well as other uses, only a small amount of treated wastewater is currently recycled. At present, about 1 billion m^3 of treated urban wastewater is reused annually, which accounts for approximately 2.4% of the treated urban wastewater effluents and less than 0.5% of annual EU freshwater withdrawals. However, the EU potential is estimated in the order of 6 billion m^3 – six times the current volume. Cyprus and Malta already reuse more than 90% and 60% of their wastewater respectively, while Greece, Italy

doi: 10.2166/wst.2018.195

and Spain reuse between 5 and 12% of their effluents, clearly indicating a huge potential for further uptake (EC 2017).

In this context, the solution to the increasing water scarcity lies partially in the search for and the implementation of new alternative wastewater technologies with low cost and energy consumption, able to generate reusable water and new products and resources instead of residues. The need for a radical change in the wastewater sector has driven the European Commission to promote innovative ideas in this field. A recent example is the project INCOVER: 'Innovative Eco-Technologies for Resource Recovery from Wastewater' (http://incover-project.eu/ GA 689242), that aims to resolve the need for new wastewater technologies by promoting water reuse, changing the paradigm of considering wastewater as only a waste and regarding it as a valuable resource, from which new added-value products can be obtained. The project is coordinated by the AIMEN Technology Center (Spain) and is being carried out through the collaboration of 18 partners, including companies, universities, and research and technological centers from seven European countries, with a proven expertise in wastewater and organic waste treatment and bioproducts generation. The project started in June 2016 and research and demonstration activities will be developed during the following 3 years in three different experimental sites located in Spain (Barcelona, Almería-Cádiz) and Germany (Leipzig). The project 'technology readiness level' to be achieved is 7, implying a complete demonstration in a real environment.

The aim of this article is to describe the implementation and start-up stage of the experimental plant designed and built by the Group of Environmental Engineering and Microbiology of the Universitat Politècnica de Catalunya-BarcelonaTech (GEMMA-UPC), and located at the UPC site Agrópolis (Viladecans, Barcelona). Special focus will be given to the design and operation of the microalgae production system. The results obtained during the first months of operation of the different technologies installed are shown and discussed.

The novelty of this plant resides in changing the paradigm of considering wastewater as a disposable waste, to start regarding it as a valuable resource, from which new added-value products can be obtained and proficiently used. To our knowledge this is the first plant built at this scale coupling microalgae-based wastewater treatment with generation of bioplastics, biomethane, biofertilizars and clean water coupled with smart irrigation.

MATERIALS AND METHODS

Agròpolis experimental plant

In the Agròpolis plant (Figure 1), a combination of agricultural runoff and urban wastewater is treated by means of a mixed culture containing mostly microalgae and cyanobacteria growing in three horizontal hybrid (semi-closed) tubular photobioreactors (PBRs). In the culture, heterotrophic bacteria are also growing but at lower rates than the other microbial groups due to the low organic matter concentration of the influent. After biomass separation, treated water is submitted to solar energy-powered ultrafiltration (UF) and disinfection, then to nutrient recovery by means of

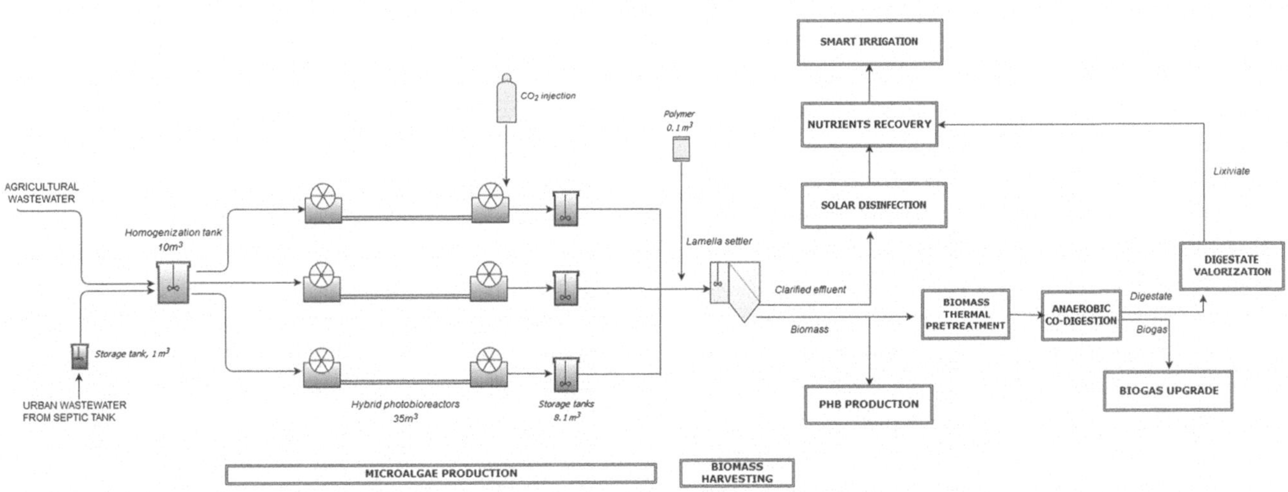

Figure 1 | Scheme of the INCOVER plant located at the Agròpolis campus, Viladecans, Barcelona.

three adsorption columns and finally reused in a smart irrigation system to grow energy crops (i.e. rapeseed). Concomitantly, the biomass obtained is submitted to anaerobic co-digestion (AcoD) with secondary sludge to produce biogas, which will be enriched in an absorption column filled with the mixed liquor from one of the PBRs. The digestate produced in the anaerobic digester is also treated in sludge treatment wetlands in order to produce biofertilizers.

Part of the biomass obtained from the PBRs is analyzed in the laboratory to determine the abundance of cyanobacteria and the content of bioplastics (in the form of polyhydroxybutyrate (PHB)) accumulated. Experiments regarding the best conditions to obtain cyanobacteria-dominated cultures and to achieve the highest accumulation of PHBs are currently being carried out. The operation of the PBRs will be adjusted after the results are obtained.

Photobioreactor design and operation

The PBRs were conceived, designed and constructed by the GEMMA Research Group (Universitat Politècnica de Catalunya-BarcelonaTech) in collaboration with the company Disoltech S.L. Each reactor (Figure 2) consists of two open tanks made from polypropylene (5 m length × 1 m width × 0.6 m height) connected by 16 low density polyethylene tubes (125 mm diameter and 47 m length). The useful volume is approximately 11.7 m^3. The tubes lie on a waterproof covering sheet in order to ensure separation from the ground. Both open tanks ensure and favor the homogeneous distribution and mixing of the liquor and also the release of the excess dissolved oxygen accumulated along the closed tubes. In each open tank, a paddle wheel with six blades (1 m width × 0.35 m length) is installed 1.8 m away from the external edge and at 3 cm height from the bottom. An engine (0.35 kW) connected to each paddle wheel provides a turning speed which can be changed from 0 to 12 rpm. Usually PBRs are operated with turning speeds ranging from 9 to 12 rpm to ensure turbulent flow inside the tubes. The water level is fixed at approximately 0.15 m before the paddle wheel and 0.35 m after it. The total working volume in each open tank is 1.25 m^3 (approximately 20% of the volume of the PBR is due to the tanks). Each tank has a dam, which assists in maintaining two different surface water levels within the tank. In each tank, the mixed liquor is moved by the paddle wheels from the shallow water lever sector to the deep one. Afterwards, thanks to the different water levels (approximately 0.20 m of variation), the mixed liquor flows by gravity through eight tubes from the deep side of one tank to the shallow side of the opposite one. Here the flow is moved by the paddle wheels to the deeper part of the tank and then it returns to the shallow side of the first tank through the other eight tubes, and so on. Technical characteristics of PBRs are summarized in Table 1.

In order to control PBR operation, online sensors of pH, dissolved oxygen and temperature (Hach Lange Spain S.L.) were installed in one of the two open tanks of each PBR. In the mixed liquor, the pH value is approximately maintained below 8.5 through controlled CO_2 addition. This pH boundary was selected based on the results of previous works that reported a pH preference of cyanobacteria ranging from 8 to 9 (Yamamoto & Nakahara 2005; Unrein *et al.* 2010). A recent study by Ji *et al.* (2017) demonstrated that green algae such as *Scenedesmus obliquus* or *Chlorella vulgaris* outcompeted cyanobacteria (*Microcystis aeruginosa*) under low CO_2 conditions and pH levels above 10. The injection is made through an air sparger placed in one of the open tanks after the paddle wheel. The pH of the mixed liquor is measured every 5 seconds with a pH probe, and when the measurement exceeds the pH 8.5 set point, a valve opens and CO_2 is bubbled into the PBR tubes. CO_2 injection stops as soon as pH 8.5 is reached again.

The three PBRs were installed in winter 2017. They were inoculated at the end of April with a mixed culture of microalgae and bacteria grown in urban wastewater. Approximately, 10 L was added to each PBR, with a volatile suspended solids (VSS) concentration of 223 mg/L. From that moment on, PBRs have been operating in parallel, fed with 2.3 m^3/d each (6.9 m^3/d in total) of a mixture of agricultural and urban wastewater at a ratio of approximately 5:1.

Every day, from 2 a.m. to 5 a.m., 6.4 m^3 of agricultural wastewater and 0.5 m^3 of urban wastewater (the latter has been previously treated in an aerated septic tank) is pumped into the homogenization tank, where they are mixed in order to reach an influent with the suitable nutrients concentration for cyanobacteria growth. Then, at 5 a.m. the feeding operation starts: firstly, 2.3 m^3/d of mixed liquor are pumped out from each PBR to one storage tank. After that (from 7 a.m. on), the same volume of influent from the homogenization tank is pumped into each PBR. This operation takes place in the early morning in order to have nutrients available for biomass growth during the day.

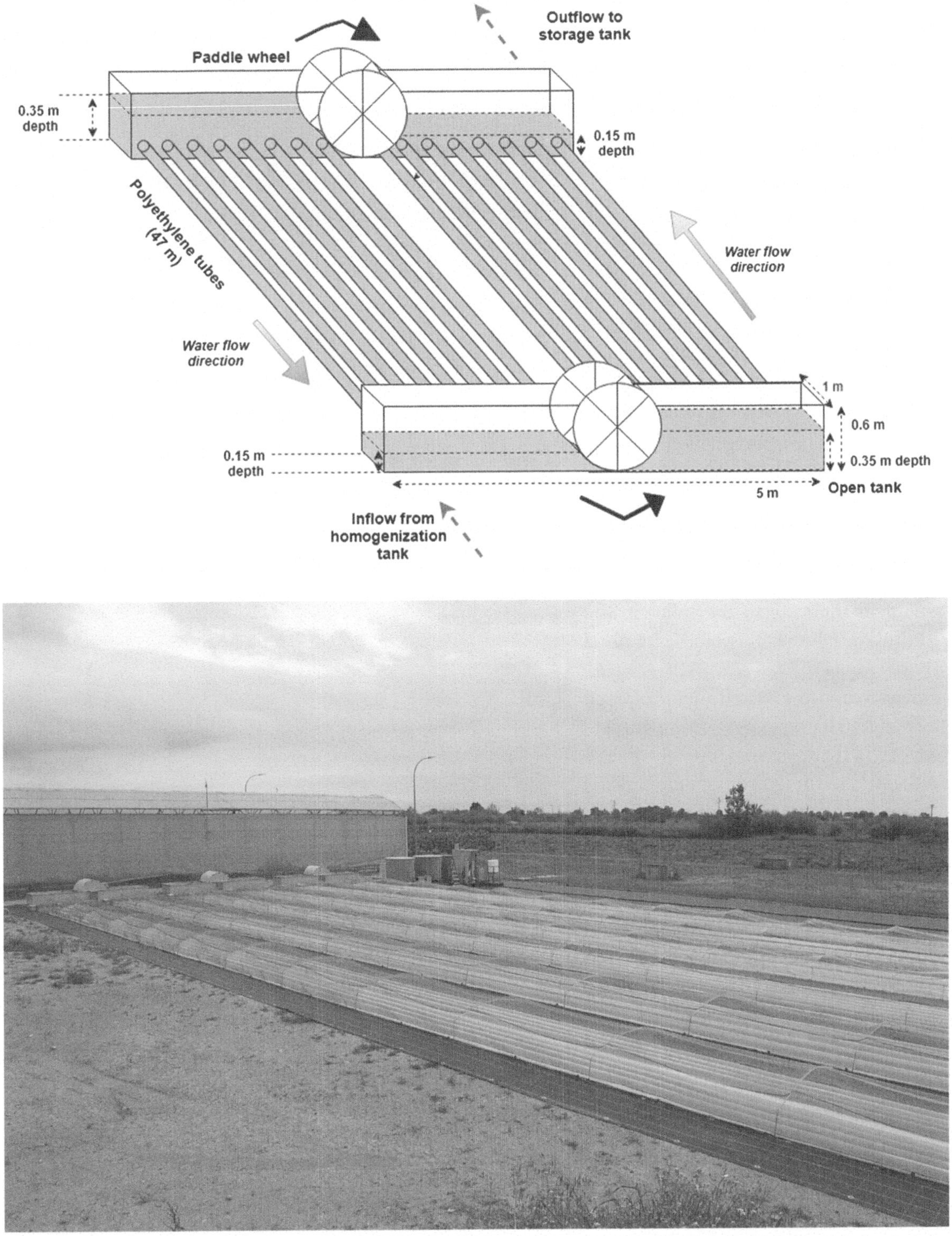

Figure 2 | Scheme of one photobioreactor and picture of the experimental plant with the three photobioreactors installed and operating.

During the microalgae-based wastewater treatment, microalgae biomass grows thanks to the solar radiation and the nutrients (mostly nitrogen (N) and phosphorus (P)) present in the influent wastewater. At the same time and through photosynthesis, microalgae generate the oxygen needed by the bacteria to aerobically degrade the organic contaminants present in that medium. Thus, PBRs have the dual advantage of simultaneously producing microalgae biomass and treating wastewater without requiring external aeration.

Analytical methodology and PBRs follow-up

Samples from the influent and from the three PBRs are collected weekly and immediately taken to the laboratory, where they are analyzed to determine: (1) biomass concentration and production as VSS and (2) water quality in terms of ammonium and phosphorus concentrations. Sampling was paused during the month of August.

All parameters, with the exception of chlorophyll *a*, were analyzed in both influent wastewater and effluent (mixed liquor of each PBR). Orthophosphate (dissolved reactive phosphorus) ($P\text{-}PO_4^{3-}$), nitrite ($N\text{-}NO_2^-$) and nitrate ($N\text{-}NO_3^-$) were measured using an ion chromatograph, Dionex ICS1000 (Thermo Scientific, USA). Alkalinity was determined using the titration method 2320 B of *Standard Methods* (APHA *et al.* 2001). Total suspended solids (TSS) and VSS were measured in triplicate in the influent and effluent of each PBR following the gravimetric method 2540 C and 2540 D in *Standard Methods* (APHA *et al.* 2001). Turbidity was measured in the mixed liquor of each PBR with a turbidity-meter (Hanna, USA). Temperature, pH and dissolved oxygen were measured directly in the open tanks of each PBR, where the corresponding sensors are introduced in the mixed liquor. Chlorophyll *a* (procedure 10200 H) and total and soluble chemical oxygen demand (CODs) were determined also following *Standard Methods* (APHA *et al.* 2001).

Mixed liquor samples were regularly examined under an optic microscope (Motic, China) for qualitative evaluation of microalgae populations and to determine the cyanobacteria and microalgae abundance. Taxonomic books were employed for identification of both microalgae and cyanobacteria species (Palmer 1962; Bourrelly 1985).

Table 1 | Technical characteristics of each of the photobioreactors

Parameter	Value
Number of tubes	16
Number of tanks	2
Total volume of tanks (m^3)	2.5
Tank surface (m)	5 × 1
Total volume of tubes (m^3)	9.2
Tube diameter (m)	0.125
Tube length (m)	47
Total PBR volume (m^3)	11.7
Design velocity inside tube (m/s)	0.25
Number of engines	2
Engine power (kW)	0.35
Retention time within the tubes at design velocity (min)	3.13
Retention time within the open tanks at design velocity (min)	0.87

Table 2 | Physico-chemical characterization of the influent wastewater (from April until October 2017, $n = 19$)

Parameter	Value
pH	8.3 ± 0.3
Electrical conductivity (mS/cm)	2.5 ± 0.4
T (°C)	24.2 ± 2.0
NH_4^+-N (mg/L)	3.8 ± 3.4
NO_2^--N (mg/L)	0.9 ± 1.4
NO_3^--N (mg/L)	8.4 ± 2.1
PO_4^{3-}-N (mg/L)	0.8 ± 1.1
TSS (mg/L)	78.5 ± 48.9
VSS (mg/L)	27.3 ± 12.0

PRELIMINARY RESULTS

The physico-chemical characterization of the influent is shown in Table 2.

Figure 3 shows the irradiance, temperature and precipitation registered on site during this first period. The biomass concentration during the first months of operation (April to October) is shown in Figure 4 and expressed as VSS (mg/L). After inoculation, around 20 days were needed to detect an increase in the volatile solids concentration; from that day onwards, the concentration increased constantly until reaching concentrations up to 320 mg VSS/L (Figure 4). Considering the operation of the three PBRs, this would correspond to a biomass production of almost 2.2 kg VSS/d.

Concerning wastewater treatment, the concentration of nutrients in the influent was always <2 mg/L for phosphates, <10 mg/L for ammonia and <15 mg/L for nitrites and nitrates. Both N and P present in the influent were completely removed in the PBRs (Figure 5).

As observed in Table 3, the CODs concentrations measured in the mixed liquor were usually higher than

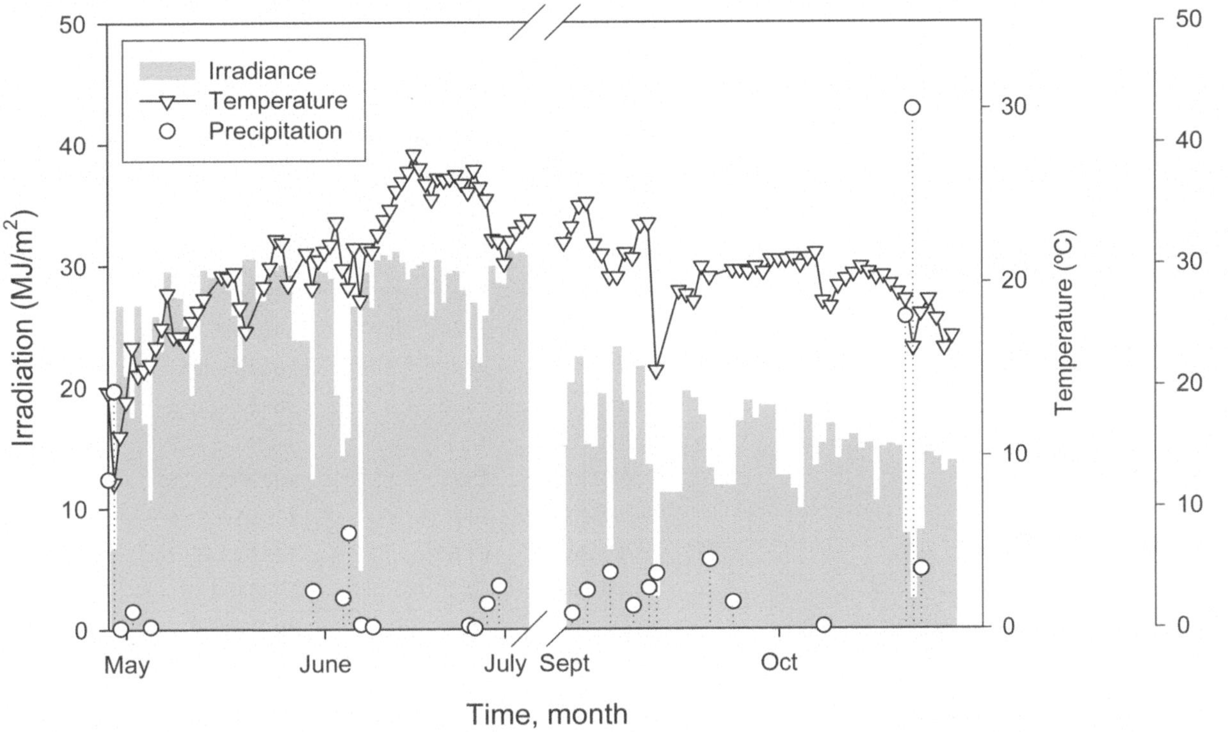

Figure 3 | Daily irradiance, average air temperature and precipitation registered during the first 6 months of operation of the plant (*Source:* www.meteo.cat/observacions/xema/dades).

the total COD concentrations at the influent feedstock, and so elimination rates were negative. It has been demonstrated that a fraction of photosynthetically fixed carbon is released during microalgae growth as dissolved carbon, and it usually corresponds to 5–30% of the carbon fixed by photosynthesis. The higher COD values in the mixed liquor of the PBR could be attributed to this dissolved organic matter exudation, but also to the low organic matter biodegradability of the wastewater influent. Similar results were observed in recent studies (Zouhayr *et al.* 2013; García-Galán *et al.* 2018).

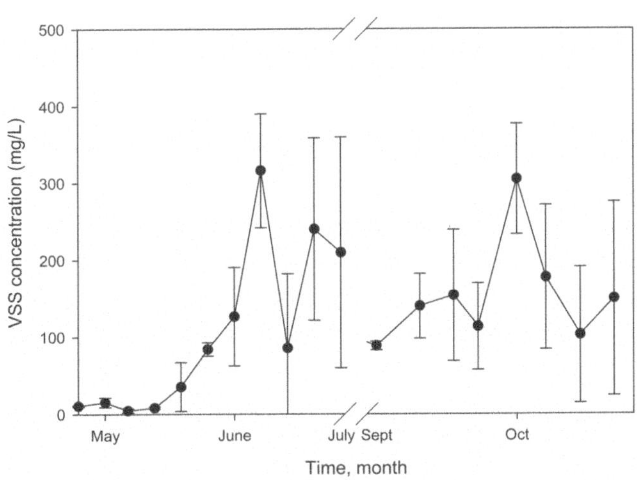

Figure 4 | Biomass concentration measured as VSS (mean and standard deviation) achieved in the three photobioreactors during the first 6 months of operation.

Regarding the species observed in the mixed liquor of the PBRs by optical microscopy, qualitative identification showed that the most frequent species were the diatom *Cyclotella* sp., the green algae *Oocystis* sp. and the cyanobacteria *Synechocystis* sp.

In order to separate the biomass produced and to clarify the water treated, a lamella settler was installed after the three PBRs. The water stored in the three storage tanks is then pumped through the lamella settler, which is divided into two chambers: the first one is a mixing chamber, where the water gets in contact with a liquid coagulant (polyaluminum chloride liquid, with 9% of aluminum (PAX-18), provided by Kemira Water Solutions, Spain). The optimum dose is estimated weekly by performing jar tests with the corresponding mixed liquor, and it is usually between 2 and 5 mg/L. After the mixing chamber, the water passes to a second chamber where the lamella favors the solids settling at the bottom whereas water is discharged in the upper part of the settler (Figure 6). The settler currently works at a rate of 400 L per hour.

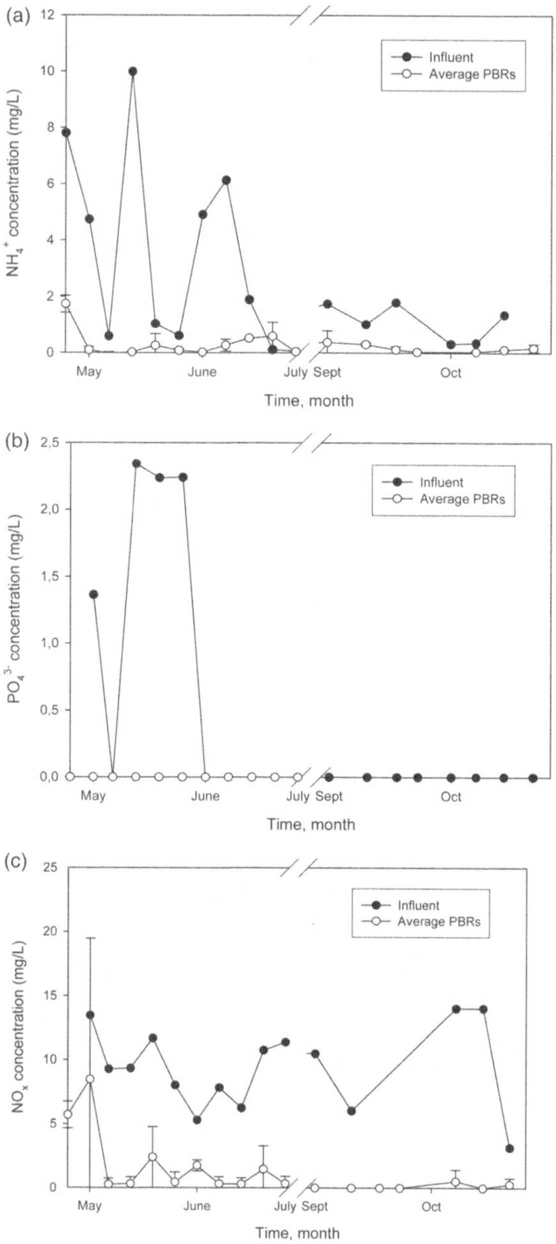

Figure 5 | Average concentrations of (a) ammonium, (b) phosphorus and (c) inorganic oxidized nitrogen (NO_X) recorded in the influent and in the three photobioreactors during the first months of operation. *Note:* the standard deviation is not shown for the influent because only one sample was analyzed.

Table 3 | Total and soluble COD values observed in the mixed influent wastewater and the mixed liquor of the PBRs (given as average value) (from April until October 2017, $n = 16$)

Date	Influent total COD (mg/L)	Effluent soluble COD (average value) (mg/L)
03/05/2017	62.1	33.37 ± 8.1
09/05/2017	154	181.10 ± 105.2
16/05/2017	–	85.13 ± 29.8
23/05/2017	110.5	137.60 ± 38.2
30/05/2017	157.7	157.37 ± 18.9
06/06/2017	98.4	162.43 ± 1.3
13/06/2017	139.3	260.67 ± 10.1
20/06/2017	58.5	142.47 ± 34.2
27/06/2017	159.3	107.40 ± 22.8
04/07/2017	144	122.93 ± 26.6
05/09/2017	64.6	148.93 ± 39.2
14/09/2017	50.3	59.47 ± 12.5
26/09/2017	25.6	40.77 ± 8.6
04/10/2017	–	44.50 ± 12.5
10/10/2017	69	75.90 ± 5.5
24/10/2017	68.5	54,00 ± 21.4

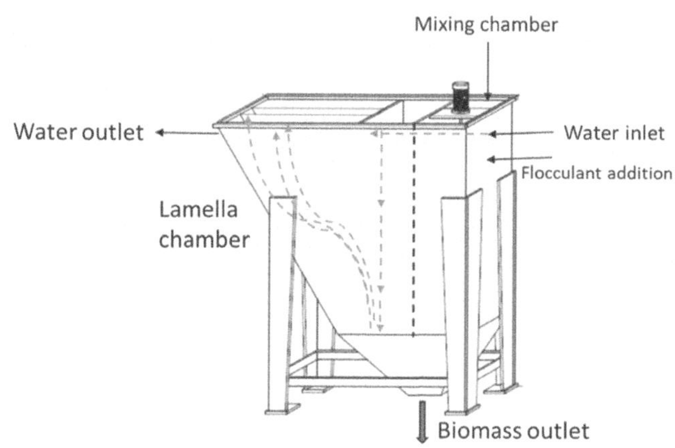

Figure 6 | Scheme of the lamella settler.

BIOPRODUCTS GENERATION

Bioplastic accumulation in wastewater-borne cyanobacteria

As mentioned previously, the three PBRs in the Agròpolis site are fed with a mixture of urban and agricultural wastewater in order to obtain a concentration of nutrients appropriate to select cyanobacteria. The interest in growing cyanobacteria is because these prokaryotic photosynthetic microorganisms have the potential to assimilate and store glycogen, cyanophycin (amino acid polymer), polyphosphates and polyhydroxyalkanoates (PHAs) (Stal 1992). PHAs are an interesting alternative to ordinary plastics from the petrochemical industry, and are nowadays used for packaging, and their use in biomedicine for prostheses is now being investigated. The most commonly occurring polymer within the PHAs family is polyhydroxybutyrate (PHB) and its accumulation has been demonstrated for several cyanobacteria such as *Spirulina* sp., *Aphanothece* sp., *Gloeothece* sp., and *Synechococcus* spp. (Balaji *et al.* 2013). The novelty in the INCOVER plant resides in the use of wastewater to select cyanobacteria from a mixed microalgae culture, instead of using pure or genetically modified cultures in usually expensive processes. Indeed, most of the studies dealing with the production of PHBs from cyanobacteria are based on pure or genetically modified cultures (Drosg 2015; Koller & Marsalek 2015), using sterile medium substrates in expensive and highly controlled processes. Once the cyanobacteria are selected in the PBRs, they will be submitted to nutrients stress in lab experiments in order to discern the best conditions to promote PHB accumulation, following the recent study by (Arias *et al.* 2017a).

Biomethane

The biomass obtained in the PBRs is submitted to a thermal pretreatment and subsequently used as substrate in an AcoD process. A thickener is used after the lamella settler to increase solids concentration of the separated biomass; this harvesting unit (lamella settler and thickener) has been designed in order to increase TSS concentration from 3–8 to 30 g/L. After this step, the biomass is directed to the AcoD unit, which consists of a thermal pretreatment tank (19 L) and an anaerobic digester (1,000 L), together with the corresponding storage tanks for the biomass and co-substrate. Due to the proximity of the wastewater treatment plant of Gavá, secondary sludge was selected as co-substrate. The inoculum was sewage activated sludge from the same plant (100 L). Considering the biomass volume obtained from the PBR, the digester is now operated at a working volume of 400 L. The microalgae biomass undergoes thermal pretreatment at 75 °C during 20 hours before being introduced in the digester, currently at a loading rate of 7.5 L/d together with 7.5 L/d of co-substrate (1:1, v/v), resulting in a hydraulic retention time of 20 days in the digester. Such a proportion was selected during the start-up in accordance with the results obtained by Arias *et al.* (2017b), who demonstrated good results in term of methane content and biogas production rates working at 50% of microalgae and 50% of co-substrate (activated sludge). It has also been demonstrated in previous studies that thermal pretreatment improves the solubility of the biomass, leading to an increase of 90% in the biogas production during the AcoD process (Passos *et al.* 2013, 2015).

During the start-up stage, the organic loading rate introduced to the digester was low (0.15–0.30 g VS/L·d). This amount could be increased using settled secondary sludge as co-substrate, with a higher concentration of organic matter and also improving the operation of the PBRs and the harvesting of the biomass achieved in the settler. Characterization of the feedstock and products of the AcoD unit is given in Table 4.

It has recently been shown that microalgae biomass can be a potential co-substrate for biogas production together with municipal wastewater sludge, significantly enhancing CH_4 yields (Mahdy *et al.* 2015; Arias *et al.* 2018). The biogas produced is stored in a gasometer (1,000 L), before the upgrading process, aimed at increasing the CH_4 content of the biogas from 60–70% up to 99%, so that it can be used directly as biofuel. In the upgrading process, mixed liquor of the PBR is used as absorption solution to simultaneously remove CO_2, H_2S, NH_3 and volatile organic chemicals from biogas at a low energy cost. The biogas produced at the INCOVER plant is therefore sparged through a fine bubble plate diffuser in an absorption bubble column (PVC, 4 m high, 20 cm diameter) interconnected to one of the open tanks of one PBR that provides the mixed liquor. The biogas is injected at the bottom of the column, while mixed liquor from the PBR is circulating countercurrent from the top to the bottom. Within the column, photosynthetic CO_2 assimilation concomitant with the aerobic

Table 4 | Characterization of the different feedstocks and products used in the AcoD unit

	Total solids (g/L)	VS (g/L)	COD (g/L)	N-NH_4^+ (mg/L)	Total P (mg/L)	TOC (mg/L)
Algae	2.85–11.6	1.23–40.8	0.38–7.01	0.5–84.1	1.5–80.2	93.7–170
Pretreated algae	2.75–14.7	1.14–5.74	1.29–6.31	4–29.1	41.5–110.1	67.8–553.4
Secondary sludge	3.06–7.44	1.29–6.14	0.25–24.9	3.3–135.9	11.8–307.9	311.4–476.4
Digestate	3.25–18.2	1.28–11.9	7.05–15.2	35.9–1115.32	77.8–408.7	343.2–827

Analyses were carried out from 20 June to 7 November.

oxidation of the other trace biogas pollutants aforementioned will take place (Toledo-Cervantes *et al.* 2017). The absorption column has been designed by a research group of the Universidad de Valladolid (partners of the project).

Biofertilizers

The sludge produced from the anaerobic digester (digestate) is a product rich in nutrients. The recent study from Solé-Bundó *et al.* (2017) evaluated the quality of microalgae digestate for agricultural reuse highlighting that, even if the digestate seems to be apt for agricultural reuse, a higher stabilization degree would be suitable before its application. For this reason, the digestate produced in the INCOVER plant is treated in a sludge treatment wetland (6 m^2), which has proved to be highly efficient for sludge dewatering and stabilization in previous studies (Uggetti *et al.* 2010). The system, designed by Centre for Recirkulering (partner of the project), consists of a granular medium (0.45 m depth) planted with *Phragmites australis* (5 plants/m^2). The drainage system for the discharge effluent is installed embedded in the coarse gravel layer at the bottom of the bed. In addition, vertical pipes for the aeration of the gravel bed are connected to the drainage system to allow the flow of air to improve aeration in the bed. Currently, the system has been loaded with water to ensure the establishment of the plants; once this stage is over, the planned solids load rate will be low (total solids of around 35 kg/m^2 year) in order to ensure the proper plant acclimation, but it will be gradually increased.

Reclaimed water

Getting back to the water treatment line, the clarified water from the lamella settler and thickener is pumped to a solar-driven disinfection system designed and developed by Solar-Spring Gmbh (partners of the project) to provide drinking water. The process is based on low pressure UF, which requires low energy. The system consists of different capillary membranes made of polyethersulfone with a capillary diameter of 0, 9 and 1.5 mm and a pore diameter around 0.02 μm (total membrane surface up to 6 m^2). For both membranes, a pretreatment is set up based on a pre-filter (100 μm) and a media filter (5 μm). After the UF, water passes through a post-treatment consisting of activated carbon filters and an integrated UV-C reactor for disinfection, finally reaching a storage tank. There is also a solar generator for independent power supply and various sensors for online monitoring of pressure, flow, water level, turbidity and UV-radiation. The working flux is 500 L/h divided into two parallel lines (dual system), one used as a reference system and monitored, and the other used for potential implementations depending on the results. This way, efficiencies under different settings can be compared.

Nutrients recovery

After the solar-driven UF and disinfection, the water treated in the PBRs is fed to three adsorption columns filled with sol–gel coating developed by researchers from Aarhus University and the Danish Technological Institute (partners of the project) in order to recover N and P. Adsorption processes are attractive for nutrient removal due to the simplicity of operation, their low cost and the possibility of resource recovery from wastewaters (Bhatnagar & Sillanpää 2010), and they have already been proved as a cost-effective solution for nutrients removal from wastewaters using industrial by-products (Garfí & Puigagut 2016). Calcite and crushed autoclaved aerated concrete were tested as base materials in the INCOVER plant, due mainly to their high affinity for P and their low cost. Two different coating compositions were developed and tested, an inorganic- and an organic-based one, both having the property of stabilizing the adsorptive base material, avoiding its disintegration in contact with water. They also allow the penetration and absorption of P, and even enhance the capacity of the base material, and prevent the biofouling of the adsorptive material.

Finally, the water produced is used to irrigate a small area of agricultural land within the Agròpolis facilities (around 250 m^2). A smart irrigation system has been deployed by FINT (partners of the project) that allows minimization of the water use and the energy requirements. The system allows real time water needs to be known, and cultivation key performance indicators such as pH, soil and air temperature, soil characteristics and plant response to be controlled by means of a wireless sensor network (based on system-on-a-chip devices) capable of reading and transmitting the variables' values. The crop selected is rapeseed (*Brassica napus*), an autumn–winter cycle crop already known for producing food (mustard) and feed oil from its oilseeds, and currently considered a new crop for bioenergy production. Sowing took place at the end of September 2017 and the rapeseed population is initially aimed to be 40–45 plants/m^2. A recommended amount of 300–400 g/ 1,000 m^2 was planted at a depth of 1.5 cm, in sowing lines spaced 15 cm apart.

MONITORING AND DECISION SUPPORT SYSTEM DECISIONS

The operation and maintenance of the plant in Agrópolis is under constant optimization for biomass, bioplastics and biomethane production. Novel sampling methodologies, based on optical sensing, are currently being implemented as *in-situ* sensors which will allow for a constant monitoring of physico-chemical parameters, aiming for a better system control and data management. This also implies a significant reduction of the energy costs of each system.

Finally, the data generated throughout the length of the project will be used to carry out life cycle assessments of the technological solutions proposed, in order to assess the impacts associated with the functioning, economical feasibility and productivity of each study (bioplastics production, treated wastewater, etc.). Previous studies focusing on the life cycle assessment of microalgal-based wastewater systems have already shown encouraging results (Garfí *et al.* 2017; Terumi Arashiro *et al.* 2018). For instance, Terumi Arashiro *et al.* (2018) highlighted that microalgae systems were more environmentally friendly when coupled with biogas production than when coupled with biofertilizer production. From a wastewater point of view, when comparing microalgae systems with activated sludge and constructed wetlands, microalgae high rate algal ponds appeared as the less expensive alternative, being the most suitable solution from an economic point of view (Garfí *et al.* 2017). The data gathered from the INCOVER plant will be useful to further improve information about the impact of the whole technology.

Eventually, and following the life cycle approaches, a decision support system will be developed within the framework of the INCOVER project, to favor the selection of a low cost, feasible wastewater treatment, always considering a holistic perspective for water management.

CONCLUSIONS

This paper presents the experimental microalgae-based treatment plant built under the European project INCOVER. The preliminary results are encouraging, showing biomass production of almost 2.2 kg VSS/d and satisfactory wastewater treatment performances (<2 mg/L for phosphates, <10 mg/L for ammonia and <15 mg/L for nitrites and nitrates). Further good results are expected in terms of bioplastics, biomethane and biofertilizer production. This plant presents a sustainable approach for wastewater management, by changing the paradigm of considering wastewater as a disposable waste, to start regarding it as a valuable resource, from which new added-value products can be obtained and proficiently used.

ACKNOWLEDGEMENTS

The INCOVER project has received funding from the European Union's Horizon 2020 research and innovation programme under grant agreement No. 689242. The dissemination of results herein reflects only the authors' view and the Commission is not responsible for any use that may be made of the information it contains. M. J. García and E. Uggetti would like to thank the Spanish Ministry of Industry and Economy for their research grants (IJCF-2014-22767 and IJCI-2014-21594, respectively).

REFERENCES

APHA, AWWA & WPCF 2001 *Standard Methods for the Examination of Water and Wastewater*, 20th edn. Washington.

Arias, D. M., Uggetti, E., García-Galán, M. J. & García, J. 2017a Production of polyhydroxybutyrates and carbohydrates in wastewater-borne cyanobacteria: effect of nutrients limitation and photoperiods. *New Biotechnology* **42**, 1–11.

Arias, D. M., Uggetti, E., García-Galán, M. J. & García, J. 2017b Cultivation and selection of cyanobacteria in a closed photobioreactor used for secondary effluent and digestate treatment. *Science of the Total Environment* **587–588**, 157–167.

Arias, D., Solé-Bundó, M., Garfí, M., Ferrer, I., García, J. & Uggetti, E. 2018 Integrating microalgae tertiary treatment into activated sludge systems for energy and nutrients recovery from wastewater. *Bioresource Technology* **247**, 513–519.

Balaji, S., Gopi, K. & Muthuvelan, B. 2013 A review on production of poly β hydroxybutyrates from cyanobacteria for the production of bio plastics. *Algal Research* **2**, 278–285.

Bhatnagar, A. & Sillanpää, M. 2010 Utilization of agro-industrial and municipal waste materials as potential adsorbents for water treatment – a review. *Chemical Engineering* **157**, 277–296.

Bourrelly, P. 1985 Les algues d'eau douce. Les algues vertes. *Reviue Hydrobiologie Tropical* **17** (2), 101–115.

Drosg, B. 2015 Photo-autotrophic production of poly (hydroxyalkanoates) in cyanobacteria. *Chemical Biochemistry Engineering Quarterly* **29**, 145–156.

EC 2000 *Water Framework Directive 2000/60/EC of the European Parliament and of the Council Establishing a Framework for the Community Action in the Field of Water Policy*. http://ec.europa.eu/environment/water/water-framework/index_en.html.

EC 2007 *Communication from the Commission to the European Parliament and the Council – Addressing the Challenge of*

Water Scarcity and Droughts in the European Union (COM/2007/0414 Final). http://ec.europa.eu/environment/water/quantity/eu_action.htm#2007_com.

EC 2017 *EC Water Reuse: Background and Policy Context, 2017.* http://ec.europa.eu/environment/water/reuse.htmc.

García-Galán, M. J., Gutiérrez, R., Uggetti, E., Maramoros, V., García, J. & Ferrer, I. 2018 Use of full-scale hybrid horizontal tubular photobioreactors to process agricultural runoff. *Biosystem Engineering* **116**, 138–149.

Garfí, M. & Puigagut, J. 2016 Reusing industrial by-products to enhance phosphorus removal in waste stabilization ponds: laboratory approach. *Desalination and Water Treatment* **57**, 1857–1864.

Garfí, M., Flores, L. & Ferrer, I. 2017 Life cycle assessment of wastewater treatment systems for small communities: activated sludge, constructed wetlands and high rate algal ponds. *Journal of Cleaner Production* **161**, 211–219.

Ji, X., Verspagen, J., Stomp, M. & Huisman, J. 2017 Competition between cyanobacteria and green algae at low versus elevated CO_2: who will win, and why? *Journal of Experimental Botany* **68** (14), 3815–3828.

Koller, M. & Marsalek, L. 2015 Cyanobacterial polyhydroxyalkanoate production: status quo and quo vadis? *Current Biotechnology* **4**, 1–1.

Mahdy, A., Mendez, L., Ballesterosa, M. & González-Fernández, C. 2015 Algaculture integration in conventional wastewater treatment plants: anaerobic digestion comparison of primary and secondary sludge with microalgae biomass. *Bioresource Technology* **184**, 236–244.

Palmer, C. M. 1962 Algas en los abastecimientos de agua. Manual ilustrado acerca de la identificación, importancia y control de las algas en los abastecimientos de agua. Editorial Interamericana, Mexico.

Passos, F., García, J. & Ferrer, I. 2013 Impact of low temperature pretreatment on the anaerobic digestion of microalgal biomass. *Bioresource Technology* **138**, 79–86.

Passos, F., Carretero, J. & Ferrer, I. 2015 Comparing pretreatment methods for improving microalgae anaerobic digestion: thermal, hydrothermal, microwave and ultrasound. *Chemical Engineering Journal* **279**, 667–672.

Solé-Bundó, M., Cucina, M., Folch, M., Tàpias, J., Gigliotti, G., Garfí, M. & Ferrer, I. 2017 Assessing the agricultural reuse of the digestate from microalgae anaerobic digestion and co-digestion with sewage sludge. *Science of the Total Environment* **586**, 1–9.

Stal, L. 1992 Poly(hydroxyalkanoate) in cyanobacteria: an overview. *FEMS Microbiology Letters* **103**, 169–118.

Terumi Arashiro, L., Montero, N., Ferrer, I., Acién, F. G., Gómez, C. & Garfí, M. 2018 Life cycle assessment of high rate algal ponds for wastewater treatment and resource recovery. *Science of the Total Environment* **622–623**, 1118–1130.

Toledo-Cervantes, A., Lebrero, R., Cavinato, C. & Muñoz, R. 2017 Biogas upgrading using algal-bacterial processes. In: *Microalgae-Based Biofuels and Bioproducts: From Feedstock Cultivation to End Products*, 1st edn (R. Muñoz & C. Gonzalez-Fernandez, eds). Woodhead Publishing, Cambridge, UK, pp. 283–304.

Uggetti, E., Ferrer, I., Llorens, E. & García, J. 2010 Sludge treatment wetlands: a review on the state of the art. *Bioresource Technology* **101** (9), 2905–2912.

Unrein, F., O'Farrell, I., Izaguirre, I., Sinistro, R., dos Santos Afonso, M. & Tell, G. 2010 Phytoplankton response to pH rise in a N-limited floodplain lake: relevance of N2-fixing heterocystous cyanobacteria. *Aquatic Sciences* **72** (2), 179–190.

Yamamoto, Y. & Nakahara, H. 2005 The formation and degradation of cyanobacterium *Aphanizomenon flos-aquae* blooms: the importance of pH, water temperature, and day length. *Limnology* **6** (1), 1–6.

Zouhayr, A., Ruiz, J., Álvarez-Díaz, P., Garrido-Pérez, C., Barragan, J. & Perales, J. A. 2013 Effect of pH control by means of flue gas addition on three different photo-bioreactors treating urban wastewater in long-term operation. *Ecological Engineering* **57**, 226–235.

First received 31 October 2017; accepted in revised form 16 April 2018. Available online 27 April 2018

Co-digestion of microalga-bacteria biomass with papaya waste for methane production

Glenda Cea-Barcia, Jaime Pérez and Germán Buitrón

ABSTRACT

The anaerobic co-digestion of microalga-bacteria biomass and papaya waste (MAB/PW) was evaluated under semi-continuous conditions. Microalgae-bacteria biomass was obtained from a high rate algal pond fed with municipal wastewater and artificially illuminated. The co-digestion of MAB/PW was evaluated using a 1:1 (w/w) ratio and an organic loading rate of 1.1 ± 0.1 g COD/L/d. Enzymatic activity assays of papain were performed in the feeding to determine the activity of this enzyme in the substrate mixture. A methane yield of 0.55 L CH_4/gVS and 68% of total volatile solid removal were observed. The volumetric productivity was 0.30 ± 0.03 L CH_4/L/d with a methane content of 71%. It was observed that papaya waste was a suitable co-substrate because it maintained a low ammonium concentration, decreasing the risk of inhibition due to ammonia and then increasing the methane yield of the microalgae-bacteria biomass compared to the biomass alone. The pretreatment effect by the addition of papaya waste on the microalgae-bacteria biomass was supported by the papain activity remaining in the substrate.

Key words | co-digestion, methane, microalgae, papain, pretreatment

Glenda Cea-Barcia
Jaime Pérez
Germán Buitrón (corresponding author)
Laboratory for Research on Advanced Processes for Water Treatment, Academic Unit Juriquilla, Instituto de Ingeniería, Universidad Nacional Autónoma de México,
Blvd. Juriquilla 3001, Querétaro 76230, México
E-mail: *gbuitronm@iingen.unam.mx*

INTRODUCTION

Microalgae feedstock appears as a promising alternative to CH_4 production because of the high lipids and carbohydrates accumulation during starvation as well as the high biomass productivity (Bahadar & Bilal Khan 2013). Currently, the use of microalgae-bacteria systems for wastewater treatment is particularly attractive because of their ability to produce inexpensive O_2, to remove nutrients, pathogens, and heavy metals, and to fix CO_2 during a photoautotrophic process (Arcila & Buitrón 2016). That represents a considerable gain in the carbon available for CH_4 production compared to classic aerobic processes and an inexpensive alternative to biomass production (Zamalloa *et al.* 2011; Chisti 2013).

The theoretical methane yield for different microalgae species varies between 0.47–0.80 L CH_4/g VS depending on the composition of the microalgal biomass (Sialve *et al.* 2009). However, experimental studies have shown lower methane yields (MY) ranging from 0.09 to 0.45 L CH_4/g VS, depending on the species and culture conditions (Golueke & Oswald 1959; Sialve *et al.* 2009; Alzate *et al.* 2012). Because of the cellulose cell wall of microalgae, the hydrolysis of the macromolecules present in the microalgae structure can be difficult, lowering the MY. The experimental MY can be increased by a pretreatment step to disrupt the microalgae cell wall before anaerobic digestion. However, the excessive energy input to maximize the methane conversion might negatively impact the feasibility of this technology.

Co-digestion of organic substrates has been described as an attractive strategy to optimize methane production (Mata-Alvarez *et al.* 2014). The addition of highly biodegradable substrates increases the organic load and supplies additional nutrients, which can attenuate the inhibition that would occur during digestion of the individual substrate. For example, Yen & Brune (2007) reported that the addition of waste paper to a mixture of *Scenedesmus* and *Chlorella* resulted in an improved MY due to an increased cellulase activity. Park & Li (2012) achieved an increase in the organic loading rate (OLR) up to 3 g VS/L d with a specific MY of 0.54 L CH_4/g VS in the anaerobic co-digestion of algae biomass with lipid waste. In contrast, combining swine manure and microalgal biomass, both with high nitrogen contents, did not result in significant MY improvement (González-Fernández *et al.* 2011). The best co-substrate for microalgae anaerobic co-digestion should have a high carbon-to-nitrogen ratio to minimize the inhibitory effects

doi: 10.2166/wst.2018.320

of the ammonia released. Reducing ammonia inhibition is a key factor in the anaerobic digestion of microalgae (Vargas *et al.* 2016). Fruit wastes can be an appropriate co-substrate for microalgae because of the high biodegradability and carbon-to-nitrogen ratio (Ge *et al.* 2014). Among different fruit wastes, papaya waste is an interesting alternative due to the presence of the papain enzyme. Previously, papaya waste was proven to have high biodegradability and to be suitable to be used even in the composting process (Lim *et al.* 2011). Raw papain consists of a mixture of carbohydrates and enzymes, of which the major enzyme is the cysteine protease papain. Papain digests most protein substrates more extensively than the pancreatic proteases. Papain exhibits broad specificity, cleaving peptide bonds of basic amino acids, leucine, or glycine. It also hydrolyzes esters and amides (Asbóth & Polgár 1977; Azarkan *et al.* 2003). Horst *et al.* (2012) reported that the treatment of diatom *Phaeodactylum tricornutum* cells with papain facilitates lipid extraction and Kose & Oncel (2015) reported an increase in the digestibility of *Chlorella vulgaris* treated with pancreatin enzyme.

In this study, the anaerobic co-digestion of microalga-bacteria biomass, generated during municipal wastewater treatment, and papaya waste was evaluated under semi-continuous conditions. The performance of the reactor regarding the methane yield and the methane production rate were evaluated, assessing the effect of papain on microalgae-bacteria biomass.

METHODOLOGY

Microalga-bacteria biomass

Microalgae-bacteria aggregates were generated in a 50 L high rate algal pond (HRAP) operated with 10 days of hydraulic residence time (HRT) and artificially illuminated as described by Arcila & Buitrón (2016). Primary municipal wastewater from a wastewater treatment plant was used as the substrate. Under these conditions, the growth of agglomerates formed by Chlorophyta (mainly *Stigeoclonium* and *Scenedesmus*), diatoms and bacteria were developed (Figure 1). The effluent of the HRAP was settled. The microalgae-bacteria biomass was harvested from the settler tank and further concentrated to feed the anaerobic reactor.

Papaya waste

Papaya waste (includes skin, pulp and seed) was collected from the fruit and vegetable market. The waste was sliced (into pieces of about 0.5 cm of diameter) in a crusher equipped with stainless steel rotating blades. The waste was mixed thoroughly and then stored at −20 °C. Papaya waste batches were brought to room temperature and then mixed with microalgae-bacteria biomass to feed the reactor.

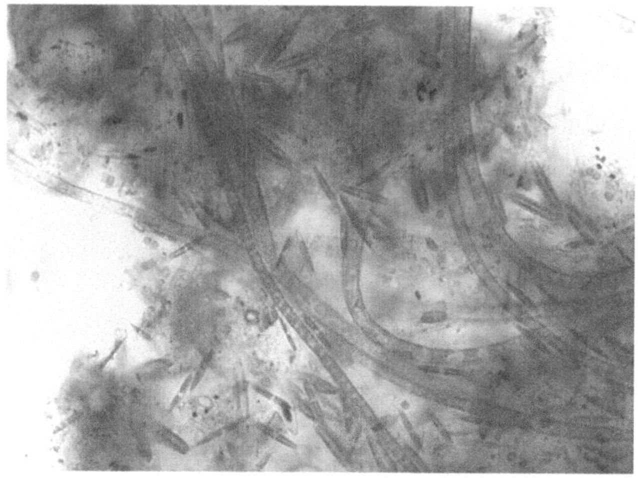

Figure 1 | Microscopic image of the microalgae-bacteria biomass cultivated in wastewater (magnification: 400X).

Anaerobic reactor and inoculum

The anaerobic reactor was fed with two microalga-bacteria to papaya waste (w/w) ratios (MAB/PW): 1:1 and 1:4, reaching a mean concentration for the mixture of 23 ± 1.7 gTS/L and 18.7 ± 1.5 gTVS/L. The 1:4 ratio was used at the start-up stage (first 20 days of operation) and the 1:1 ratio at the operational stage. Co-digestion was performed in a stirred lab-scale reactor of 5 L. The entire reactor was covered to reduce light penetration and hence prevent microalgae growth. The temperature was regulated at 35 °C and pH was monitored but not controlled. The produced biogas volume was continuously quantified (Ritter Drum-type Gas Meter TG-05). The reactor was fed twelve times per day just after the extraction of the digested sludge and was operated with 31 days of HRT and an OLR of 1.1 ± 0.1 gCOD/L/d. Once a week, the total and soluble phase of the feed and the digestate were characterized. The reactor was inoculated with granular anaerobic sludge from a brewery wastewater treatment plant operating at 35 °C. The initial solid content was 27 gTS/L and 19 gVS/L. The maximum specific methane activity of the sludge was 17 $mLCH_4/gVS/d$. Before the co-digestion assay, the inoculum was kept under endogenous conditions for 14 days to reduce non-specific biogas generation.

Enzymatic assay of papain

Enzymatic assays of papain (EC 3.4.22.2) were performed to determine the presence of papain in the feed mixture and its possible effect on microalgae-bacteria biomass. Assays were conducted according to the titrimetric rate determination method proposed by the Sigma-Aldrich Quality Control test procedure. The titrimetric analysis consists of determining the number of moles of reagent (titrant), required to react quantitatively with the substance being determined. In this case, the reaction is according to Equation (1):

$$\mathrm{BAEE} + \mathrm{H_2O} \xrightarrow{\mathrm{Papain}} \mathrm{N\alpha{-}Benzoyl - L - Arginine + Ethanol} \quad (1)$$

where one unit of papain hydrolyze 1.0 μmole of BAEE (Nα-Benzoyl-L-Arginine Ethyl Ester) per minute at pH 6.2 at 25 °C and the titrant (NaOH) reacts with the hydrolysis product (Nα-Benzoyl-L-Arginine). The calculation was made following Equation (2):

$$\frac{Units}{mL\ enzyme} = \frac{0.05N1000}{TV} \quad (2)$$

where 0.05 is the volume in milliliters of NaOH used to maintain the pH at 6.2., N is the normality of NaOH, 1,000 is the conversion factor from millimoles to micromoles, T is the time in minutes required to maintain the pH at 6.2 of a 50 μl aliquot sample and V is the volume in milliliter of enzyme used.

One mL of either papaya waste, feed mixture or microalgae-bacteria suspension were used as enzyme preparation. The results are expressed in Units/mL enzyme or Units/mg TVS. Additionally, solubilization kinetics of COD, carbohydrates and proteins of the feed mixture were analyzed by measuring the respective concentrations in the soluble phase during 7 days of storage at 4 °C to determine the effect of the papain on the solid phase (microalgae-bacteria biomass).

Analytical methods

Total solids (TS), total volatile solids (TVS) and total Kjeldahl nitrogen (TKN) were determined according to standard methods (APHA 2005). Total chemical oxygen demand (COD_T), soluble chemical oxygen demand (COD_S), total ammonium ($N\text{-}NH_4^+$) and total phosphate ($P\text{-}PO_4^{3-}$) were measured by a colorimetric method using Hach vials. Samples were centrifuged (600 × g, 10 min), followed by filtration at 1 μm (Whatman GF/A filter), to separate the particles from the soluble phase. The carbon dioxide and methane produced were periodically analyzed using a gas chromatograph (Agilent 6890 N) equipped with a thermal conductivity detector and a 30 m long (0.53 mm id) Carboxen 1010 Plot column. The temperature of the injection port, column and detector, were 200, 100 and 230 °C, respectively. Nitrogen was used as the carrier gas at a flow rate of 4 mL/min. Liquid samples were taken at the outlet for the analysis of volatile fatty acids (VFAs). 1 mL of sample was centrifuged at 600 g for 5 min. VFAs concentrations were determined using a chromatograph (Varian 3300) fitted with an FID detector and a 15 m long (0.53 mm id) Zebron ZB-FFAP column. Injector and detector temperatures were maintained at 190 and 210 °C, respectively. The temperature of the column was maintained at 45 °C for 1.5 min; then, it was increased to 135 °C at a rate of 8 °C/min. The carrier gas was nitrogen at 9.5 ml/min. Carbohydrate concentration was determined by the phenol-sulfuric acid method, using glucose as a standard according to Dubois *et al.* (1956), and the protein concentration was analyzed according to the Lowry *et al.* (1951) method using seroalbumin as standard.

RESULTS AND DISCUSSION

Methane production and content

The start-up of the bioreactor was carried out during the first 20 days of operation where the reactor was fed with a MAB/PW ratio of 1:4. This ratio was applied to promote the activity of methanogenic inoculum under co-digestion conditions. Subsequently, the reactor was fed with a MAB/PW ratio of 1:1. Figure 2 shows the reactor performance

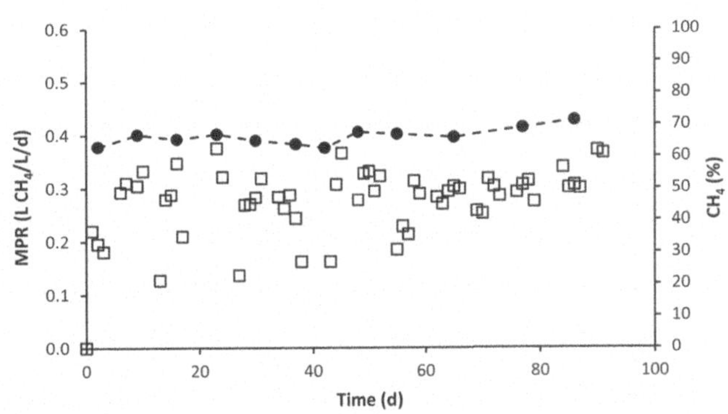

Figure 2 | MPR (squares) and methane content (circles) during the start-up of the reactor (0 to 20 days) and co-digestion period (20 to 95 days).

during the start-up and the co-digestion operation time. The MPR, expressed in L CH_4/L·d, indicates the quantity of methane that can be produced per unit reactor working volume. The improvement of this parameter is important in maximizing the efficiency of the anaerobic digestion process. Moreover, the methane content in the biogas is related to the process efficiency. A low methane content implies some form of inhibition that decreases the methanogenic activity within the microbial consortium.

A relatively elevated coefficient of variation of 34.3% for the MPR was observed (0.27 ± 0.09 L CH_4/L/d) during the first 58 days; however, the methane content was maintained around 63%. From day 59 forward, the coefficient of variation of the MPR decreased to 10.1%, indicating the stabilization of the reactor (0.30 ± 0.03 L CH_4/L/d). For this period, the methane content in the biogas increased up to 71% (Figure 2). Those values agree with the results obtained by Xie *et al.* (2012) who reported MPR values around 0.27 L CH_4/L/d using an OLR of 1 g TSV/L/d similar to the OLR used in the present study (0.6 g TSV/L/d). MPRs as high as 0.8 L CH_4/L/d were obtained at a higher OLR (2 g TSV/L/d) and using lipids as co-digestate (Park & Li 2012).

Effect of co-digestion on methane yield and digestate characteristics

MY (L CH_4/g VS_{fed}), expresses the substrate conversion efficiency into methane. Figure 3 shows the MY during the reactor operation. When the reactor was operated with a MAB/PW ratio of 1:1, an average MY of 0.55 ± 0.07 L CH_4/g VS_{feed} was obtained. Other studies (Olsson *et al.* 2014; Caporgno *et al.* 2015) have reported MY values of 0.4 L CH_4/g VS_{feed} for the co-digestion of saline microalgae and lipid waste under an OLR of 2 g TSV/L/d. When the OLR was increased to 3 g TSV/L/d, the MY was also increased (0.54 L CH_4/g VS_{feed}). Moreover, co-digestion studies of microalgae and sewage sludge in batch reactors report MY between 0.3 and 0.4 L CH_4/g VS.

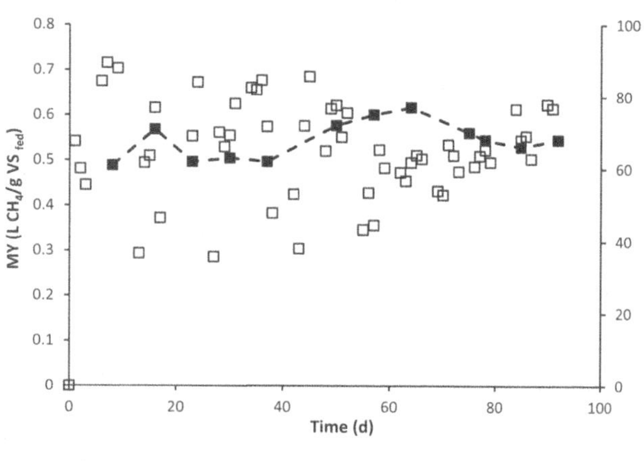

Figure 3 | MY (open squares) and TVS removal (filled squares) during the start-up of the reactor (0 to 20 days) and co-digestion period (20 to 95 days).

Figure 2 also presents the TVS removal. The TVS, COD_T and COD_S removals achieve 68%, 72% and 91%, respectively, evidencing the good reactor performance (Table 1). The results showed that about 8% of organic nitrogen has been mineralized (Table 1), which was evidenced by a decrease of TKN values in the digestate. The released of N-NH_4^+ in the digestate reached concentrations of 311 ± 50 mg/L. Important levels of N-NH_4^+ have been observed in the microalga digestate, ranging between 800 and 1,100 mg N-NH_4^+/L (Ras *et al.* 2011; Alzate *et al.* 2012). High levels of N-NH_4^+ increase the N-NH_3 toxicity. That behavior is related to the high protein content of microalgae, which can vary between 20% and 65% of the dry weight (Becker 2007; Alzate *et al.* 2012). The co-digestion of microalgae with papaya waste improved the N-NH_4^+ profile, maintaining the reactor at a non-inhibitory level of N-NH_4^+.

On the other hand, the digestion of fruit wastes at a high OLR becomes difficult because of the rapid acidification, resulting in the inhibition of methanogenic activity (Mata-Alvarez *et al.* 1992). The co-digestion of microalgae with papaya waste shows a synergistic effect, decreasing the ammonium to carbon ratio due to the papaya organic matter addition and the pH control levels due to the microalgal biomass. That resulted in an improvement in the performance of the fruit wastes digestion.

Evaluation of co-digestion performance

After 95 days of operation, the reactor was only fed with papaya waste at the same operation conditions (HRT and OLR) to determine the MY of this waste (data not shown). Table 2 summarizes the MY obtained in the present study

Table 1 | Feed and digestate physicochemical parameters for the co-digestion

	Feed	Digestate
COD_T (g O_2/L)	35 ± 3.2	9.8 ± 2
COD_S (g O_2/L)	–	0.92 ± 0.1
N-NH_4^+ (mg/L)	13 ± 9	311 ± 50
PO_4^{3-} (mg/L)	283 ± 110	162 ± 37
TKN (mg N-NH_4^+/L)	1,015 ± 10	938 ± 39
pH	7.0	7.15 ± 0.3

Table 2 | Comparison between the different MY and biodegradability percentage obtained for each substrate

Substrate	MY (L CH_4/gCOD$_{fed}$)	Biodegradability (%)
Papaya waste	0.312	89.1
Microalga-bacteria biomass[a]	0.144	41.1
Theoretical mixture of biomasses[b]	0.205	58.6
Co-digested MAB/PW[c]	0.230	65.7

[a]Arcila and Buitrón (2016).
[b]Calculated by using the data of papaya waste and microalga-bacteria biomasses alone under the condition applied to the reactor.
[c]Experimentally obtained during co-digestion.

and compared with the MY of microalga-bacteria biomass obtained in a previous study.

The MY (regarding COD) during the reactor operation was calculated by using Equation (3):

$$MY = \sum_{0}^{n} \frac{CH_{4TP}}{COD_{TF}} \tag{3}$$

where MY is methane yield in L/gCOD$_{fed}$, CH_{4TP} is the total production of methane during the operation time in L, COD_{TF} is the COD_T fed in g O_2, and n is reactor operation days.

The MY values were calculated for the co-digestion period (from day 20 to day 95) and the papaya waste feeding, after day 95 (Table 2). The waste conversion efficiency or biodegradability, expressed in percentage, was calculated considering the theoretical production yield of 0.35 L CH_4/$COD_{degraded}$ at 1 atm and 0 °C. Results indicate an increase in the biodegradability of the microalga-bacteria biomass under co-digestion from 41.1 to 65.7%. That represents an increment on the MY of 59.8% (0.230 versus 0.144 L CH_4/gCOD$_{fed}$). Similar co-digestion improvements have already been reported when microalgae and sewage sludge were co-digested (Olsson *et al.* 2014; Caporgno *et al.* 2015). It is interesting to note the synergistic effect of co-digestion: a higher biodegradability was observed in the experimental reactor than the one expected from theoretical calculations of the mixture of microalga-bacteria biomass with the papaya waste. That indicates that papaya waste improves the biodegradability.

The evidence that the co-digestion improves the MY of the microalgae-bacteria biomass, increasing its biodegradability, suggested a possible effect of pre-treatment by the papaya waste on the microalgae-bacteria biomass. On the other hand, the co-digestion improved the performance of the system due to a decrease in the concentration of VFAs of the digestate (101 mg acetate/L papaya waste in digestate against 5 mg acetate/L microalgae-bacteria/papaya waste in digestate), becoming an alternative to operating a higher OLR without an acidification of the system.

Effect of papaya waste on the feed mixture

The presence of papain activity in the feed mixture was determined to evaluate the effect of papaya waste on microalga-bacteria biomass degradability. Figure 4 shows the enzyme activity measured in the raw papaya waste, in the feed mixture at time 0 and after 3 days of storage at 4 °C. As a control sample, the activity was measured in the microalgae-biomass suspension to determine potential interferences of other substrates with the methodology. When the feed mixture is prepared, 80% of the initial enzyme activity was lost. Nevertheless, after three days of storage at 4 °C, the papain activity of the feed mixture remains (loss of 29% of activity). These results evidence the presence of active papain in the feed mixture and suggest a possible hydrolysis of the enzyme on the microalgae-bacteria biomass. The hydrolysis kinetics of the feed mixture was performed by measuring the COD_S, carbohydrates and proteins of the soluble phase over time (Figure 5). Proteins concentrations measured by Lowry protein assay increase during the incubation period. The Lowry protein assay can measure protein during enzyme fractionations, small peptides and free amino acids such as tyrosine and tryptophan (Lowry *et al.* 1951). However, the COD_S and carbohydrates kinetics are inconclusive. The carbohydrates consumption kinetics could be the result of the solubilization and consumption of carbohydrates present in the solid and liquid phase respectively, and COD_S kinetics could be the result of the solubilization and consumption of different macromolecules.

The main structures of the biomass that might be hydrolyzed by papain are microalgae cell walls and bacteria aggregates. Rupture of cell walls can result in the release

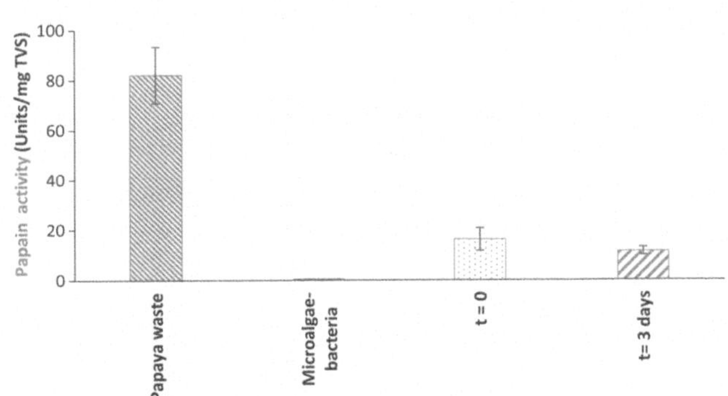

Figure 4 | Papain activities of papaya waste and feed mixture measured by titrimetric rate determination method using BAEE as substrate.

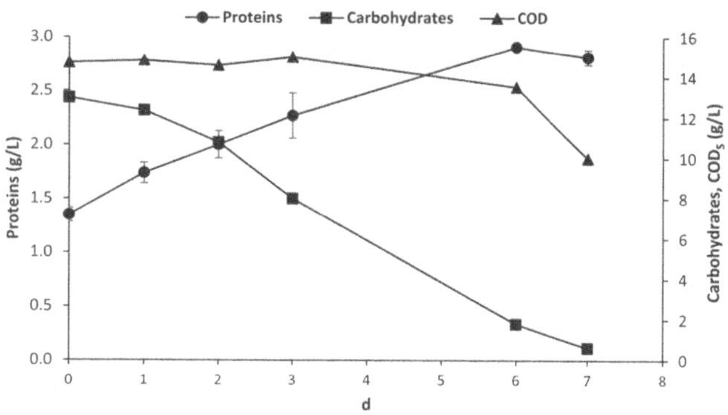

Figure 5 | COD_S, carbohydrates and proteins concentrations in the soluble phase of the feed mixture over seven days of storage at 4 °C.

into the medium of other cellular components, such as intracellular proteins, carbohydrates, chlorophyll and DNA. The microalgae from the Chlorophyceae class possess cell walls with assemblages of polymers containing hyp-rich glycoproteins and arabinogalactan proteins (Domozych *et al.* 2012). Papain can interact with the proteins contained in these structures.

CONCLUSIONS

Anaerobic co-digestion of microalgae-bacteria biomass and papaya waste was evaluated. Co-digestion improved the biodegradability of the microalga-bacteria biomass and overall performance of the system indicated by the increment of the MY of the microalgae-bacteria biomass from 0.144 to 0.230 L CH_4/gCOD_{fed} (0.55 ± 0.07 L CH_4/g VS_{feed}). The co-digestion poised the ammonium concentration of the microalga-bacteria biomass and decreased the risk of acidification of papaya waste. Furthermore, it was possible to determine the presence of papain activity in the feed mixture and its effect on the solid phase, showing solubilization of proteins. The results demonstrate that the co-digestion of microalga-bacteria biomass with a residue containing enzymatic activity, such as the papaya waste, is an effective strategy for an efficient methane production system from microalgae. However, more research and development is required to make this technology an economically viable process.

ACKNOWLEDGEMENTS

The authors gratefully acknowledge the financial support of the project SENER-CONACYT (249590) and DGAPA-UNAM (PAPIIT IN101716). Gloria Moreno is acknowledged for their technical assistance and DGAPA-UNAM for the postdoctoral fellowship to G. Cea Barcia.

REFERENCES

Alzate, M. E., Muñoz, R., Rogalla, F., Fdz-Polanco, F. & Pérez-Elvira, S. I. 2012 Biochemical methane potential of microalgae: influence of substrate to inoculum ratio, biomass concentration and pretreatment. *Bioresour. Technol.* **123**, 488–494.

APHA 2005 *Standard Methods for the Examination of Water and Wastewater*, 21st edn. American Public Health Association, American Water Works Association and Water Environment Federation, Washington, DC, USA.

Arcila, J. S. & Buitrón, G. 2016 Microalgae–bacteria aggregates: effect of the hydraulic retention time on the municipal wastewater treatment, biomass settleability and methane potential. *J. Chem. Technol. Biotechnol.* **91**, 2862–2870.

Asbóth, B. & Polgár, L. 1977 On the enhanced catalytic activity of papain towards amide substrates. *Acta Biochim. Biophys. Acad. Sci. Hung.* **12** (3), 223–230.

Azarkan, M., El Moussaoui, A., van Wuytswinkel, D., Dehon, G. & Looze, Y. 2003 Fractionation and purification of the enzymes stored in the latex of carica papaya. *J. Chromatogr. B Analyt. Technol. Biomed. Life Sci.* **790** (1–2), 229–238.

Bahadar, A. & Bilal Khan, M. 2013 Progress in energy from microalgae: a review. *Renew. Sust. Energ. Rev.* **27**, 128–148.

Becker, E. W. 2007 Micro-algae as a source of protein. *Biotechnol. Adv.* **25** (2), 207–210.

Caporgno, M. P., Trobajo, R., Caiola, N., Ibañez, C., Fabregat, A. & Bengoa, C. 2015 Biogas production from sewage sludge and microalgae co-digestion under mesophilic and thermophilic conditions. *Renew. Energ.* **75**, 374–380.

Chisti, Y. 2013 Constraints to commercialization of algal fuels. *J. Biotechnol.* **167**, 201–214.

Domozych, D. S., Ciancia, M., Fangel, J. U., Mikkelsen, M. D., Ulvskov, P. & Willats, W. G. T. 2012 The cell walls of green algae: a journey through evolution and diversity. *Front. Plant Sci.* **3** (82), 1–7.

Dubois, M., Gilles, K. A., Hamilton, J. K., Rebers, P. A. & Smith, F. 1956 Colorimetric method for determination of sugars and related substances. *Anal. Chem.* **28**, 350–356.

Ge, X., Matsumoto, T., Keith, L. & Li, Y. 2014 Biogas energy production from tropical biomass wastes by anaerobic digestion. *Bioresour. Technol.* **169**, 38–44.

Golueke, C. G. & Oswald, W. J. 1959 Biological conversion of light energy to the chemical energy of methane. *Appl. Microbiol.* **7** (4), 219–227.

González-Fernández, C., Molinuevo-Salces, B. & García-González, M. C. 2011 Evaluation of anaerobic codigestion of microalgal biomass and swine manure via response surface methodology. *Appl. Energ.* **88** (10), 3448–3453.

Horst, I., Parker, B. M., Dennis, J. S., Howe, C. J., Scott, S. A. & Smith, A. G. 2012 Treatment of *Phaeodactylum tricornutum* cells with papain facilitates lipid extraction. *J. Biotechnol.* **162**, 40–49.

Kose, A. & Oncel, S. S. 2015 Properties of microalgal enzymatic protein hydrolysates: biochemical composition, protein distribution and FTIR characteristics. *Biotechnol. Rep.* **6**, 137–143.

Lim, P. N., Wu, T. Y., Sim, E. Y. S. & Lim, S. L. 2011 The potential reuse of soybean husk as feedstock of Eudrilus eugeniae in vermicomposting. *J. Sci. Food Agric.* **91** (14), 2637–2642.

Lowry, O. H., Rosebrough, N. J., Farr, A. L. & Randall, R. J. 1951 Protein measurement with the folin phenol reagent. *J. Biol. Chem.* **193**, 265–275.

Mata-Alvarez, J., Cecchi, F., Llabrés, P. & Pavan, P. 1992 Anaerobic digestion of the Barcelona central food market organic wastes. Plant design and feasibility study. *Bioresour. Technol.* **42** (1), 33–42.

Mata-Alvarez, J., Dosta, J., Romero-Güiza, M. S., Fonoll, X., Peces, M. & Astals, S. 2014 A critical review on anaerobic co-digestion achievements between 2010 and 2013. *Renew. Sust. Energ. Rev.* **36**, 412–427.

Olsson, J., Feng, X. M., Ascue, J., Gentili, F. G., Shabiimam, M. A., Nehrenheim, E. & Thorin, E. 2014 Co-digestion of cultivated microalgae and sewage sludge from municipal waste water treatment. *Bioresour. Technol.* **171**, 203–210.

Park, S. & Li, Y. 2012 Evaluation of methane production and macronutrient degradation in the anaerobic co-digestion of algae biomass residue and lipid waste. *Bioresour. Technol.* **111**, 42–48.

Ras, M., Lardon, L., Sialve, B., Bernet, N. & Steyer, J. P. 2011 Experimental study on a coupled process of production and anaerobic digestion of *Chlorella vulgaris*. *Bioresour. Technol.* **102**, 200–206.

Sialve, B., Bernet, N. & Bernard, O. 2009 Anaerobic digestion of microalgae as a necessary step to make microalgal biodiesel sustainable. *Biotechnol. Adv.* **27**, 409–416.

Vargas, G., Donoso-Bravo, A., Vergara, C. & Ruiz-Filippi, G. 2016 Assessment of microalgae and nitrifiers activity in a consortium in a continuous operation and the effect of oxygen depletion. *Electronic Journal of Biotechnology* **23**, 63–68.

Xie, S., Wu, G., Lawlor, P. G., Frost, J. P. & Zhan, X. 2012 Methane production from anaerobic co-digestion of the separated solid fraction of pig manure with dried grass silage. *Bioresour. Technol.* **104**, 289–297.

Yen, H. W. & Brune, D. E. 2007 Anaerobic co-digestion of algal sludge and waste paper to produce methane. *Bioresour. Technol.* **98** (1), 130–134.

Zamalloa, C., Vulsteke, E., Albrecht, J. & Verstraete, W. 2011 The techno-economic potential of renewable energy through the anaerobic digestion of microalgae. *Bioresour. Technol.* **102** (2), 1149–1158.

First received 29 October 2017; accepted in revised form 5 July 2018. Available online 10 July 2018

Valorization of waste micro-algal biomass – collected from coke oven effluent treatment plant and evaluation of sorption potential for fluoride removal

Gargi Biswas, Philips Prince Pokkatt, Aratrika Ghosh, Biswajit Kamila, Kalyan Adhikari and Susmita Dutta

ABSTRACT

Fluoride contamination in groundwater is now becoming a global concern. In the present study, removal of fluoride using dry biomass (DBM) of a micro-algal consortium of *Chlorococcum infusionum* and *Leptolyngbya foveolaurum*, collected from a coke-oven effluent treatment plant, Durgapur, India, has been investigated. The large volume of algal bloom in the industrial effluent has created serious disposal issues and caused severe environmental concerns. A biosorption technique has been carried out to valorize the waste algae biomass into a potential adsorbent. Response Surface Methodology (RSM) is used to model and optimize fluoride removal. Maximum fluoride removal (72%) is obtained at pH 4, 5 mg/L initial fluoride concentration, 0.5 g/L adsorbent dose (AD), and 25 °C temperature during one-factor-at-a-time (OFAT) analysis. The optimum condition of removal as specified by RSM is – initial concentration of fluoride: 30 mg/L, pH: 4.5, AD: 3.5 g/L and temperature: 30 °C. FESEM-EDX, FTIR and BET isotherm studies are done to characterize raw and fluoride treated biomass. Lagergren first order kinetic model and Freundlich isotherm model, are found to analyze best kinetic and equilibrium data, respectively. Adsorption capacity of DBM has been found to be 34.36 mg/g. The kinetics of fluoride removal have been well described by COMSOL Multiphysics.

Key words | biosorption, COMSOL Multiphysics, defluoridation, micro-algae consortium, RSM, waste valorization

Gargi Biswas
Philips Prince Pokkatt
Aratrika Ghosh
Susmita Dutta (corresponding author)
Department of Chemical Engineering,
National Institute of Technology Durgapur,
Durgapur 713209,
India
E-mail: *susmita_che@yahoo.com*

Biswajit Kamila
Department of Chemical Engineering,
University of Calcutta,
Kolkata 700009,
India

Kalyan Adhikari
Department of Earth and Environmental Studies,
National Institute of Technology Durgapur,
Durgapur 713209,
India

NOMENCLATURE

A	IC (mg/L)
B	pH
b'	Temkin isotherm constant
B'	Temkin constant related heat of sorption (J/mol)
C	AD (g/L)
$C_{\exp,i}$	Experimental values of fluoride concentration (mg/L)
$C_{sim,i}$	Simulated values of fluoride concentration (mg/L)
C_o	Initial fluoride concentration (mg/L)
C_e	Residual liquid phase concentration of fluoride at equilibrium time (mg/L)
c	Concentration at any time (mg/L)
C''	Morris-Weber constant
D	Temperature (°C)
d	Diffusivity of fluoride element within the solution (m^2/s)
K_F	Freundlich isotherm constant [(mg/g) $(L/mg)^{1/n}$]
K_L	Langmuir isotherm constant (L/mg)
K_T	Temkin isotherm constant ()
K_{DB}	Dubinin-Radushkevich constant (g^2/kJ^2)
k_1	Adsorption rate constant as used in Lagergren first order adsorption rate equation (min^{-1})
k_2	Adsorption rate constant as used in pseudo second order adsorption rate equation (g mg^{-1} min^{-1})
k_3	Intra particle diffusion rate coefficient (g mg^{-1} $min^{-0.5}$)
N	Total number of experiments
n	Number of factors
n_c	Centre point
q_e	Solid phase concentration of fluoride at equilibrium (mg/g)

doi: 10.2166/wst.2017.638

q_t	Solid phase concentration of fluoride at time 't' (mg/g)
q_{max}	Maximum adsorption capacity of the adsorbent (mg/g)
$R1$	Removal of fluoride (%)
R	Universal gas constant (8.31 J/mol/K)
T	Temperature (K)
t	Contact time (min)
$\overline{U}$	Stirring velocity (m/s)
x_i, x_j	Factors (independent variables)
Y	Response of system
β_0	Constant coefficient
β_i	Linear coefficient
β_{ii}	Quadratic coefficient
β_{ij}	Interaction coefficient
ε'	Dubinin-Radushkevich isotherm constant
ε	Residual error
ρ	Density of contaminant solution (g/cc)

INTRODUCTION

The presence of fluoride ions in potable water has unique effects on human health (Venkata Mohan *et al.* 2007). For a concentration ranging from 0.5 to 1.5 mg/L, fluoride is essential for the development and formation of dental enamel (Hiremath & Binnal 2013). However, prolonged exposure to higher fluoride concentration (>1.5 mg/L) causes enamel-decay and osteosclerosis (Biswas *et al.* 2016). It has been observed that lower calcium and higher bicarbonate alkalinity favor high fluoride concentration (hydro-geochemical origin) in groundwater (Biswas *et al.* 2017). In India, approximately 60 million people in 19 states have been poisoned through consumption of well water contaminated with excessive fluoride in the range of 0.5–20 mg/L (Veeraputhiran & Alagumuthu 2011). Birbhum and Bankura districts of West Bengal, an eastern province of India, have been recorded to be fluoride contaminated districts with concentrations in the range of 1.5–13 mg/L (WBPHED 2006).

Several methods have been developed for the defluoridation of water such as membrane technology, ion exchange, chemical precipitation and adsorption (Biswas *et al.* 2017). Among different processes, adsorption is found to be an effective approach, and the application of biosorption for fluoride removal using bio-materials was studied by several researchers (Bhatnagar *et al.* 2002; Kumar *et al.* 2015). Biosorption, a physicochemical process, utilizes dry dead biomass to remove pollutants and deals with sorption of the chemicals on the biological matrix (Bhatnagar *et al.* 2002). This approach has a significant edge over the conventional processes, such as no production of chemical sludge, and effectiveness in the treatment of wastewater (Kumar *et al.* 2015). Furthermore, since the non-living biomass does not require any nutrient/minerals for growth or metabolic system, the operation costs become minimized. Application of dead algal biomass is a potent bioremediation technology due to their availability, cheapness and effectiveness for wastewater treatment (Bhatnagar *et al.* 2002; Kumar *et al.* 2015). Algal blooms are the overgrowth of different types of algae in clean/waste water, and they have a negative impact on several industries. Treated wastewater usually contains high concentration of nutrients (e.g. phosphate, nitrate, etc) that may cause algal blooms on the surface of storage ponds' water. The negative impact is not limited to the toxic effect of algae, but also includes their abundance, which may provoke the operational problems and increase of biochemical oxygen demand (BOD), pH and total suspended solids (TSS) (EPA 2017). Hence, the reuse or discharge of such water needs removal of algae. Such removed algal biomass thus can be valorized as a potential biosorbent. It is a newly devised substitute for the prevailing methods to remove pollutants from wastewater. The important functional groups of dead algal biomass are carboxyl, phosphate, hydroxyl, sulphate, etc. Ions can bind with biomass by processes including adsorption, ion exchange and chemisorption on surface pores (Hiremath & Binnal 2013). Usually micro-algal or cyanobacterial biomasses have been studied at a large scale for removal of various heavy metals (Sen *et al.* 2017). However, biosorption of anionic pollutants (e.g.: phosphate, nitrate, fluoride, etc.) is much lower than cationic contaminants such as heavy metals. Uptake of cationic pollutants is high when solution pH is higher than pH 7; however, better removal of anionic components can be obtained at low pH ($<pH_{zpc}$) (Hiremath & Binnal 2013). Application of dead micro-algal biomass for defluoridation of wastewater in batch contactors has been reported in few literatures (Bhatnagar *et al.* 2002; Venkata Mohan *et al.* 2007; Hiremath & Binnal 2013). However, application of algal biomass for treatment of fluoride-contaminated groundwater or industrial wastewater is very terse. Therefore, biosorption using inexpensive algal biomass for treatment of real wastewater is the thrust of the present work.

The present study investigates the removal of fluoride using dry biomass (DBM) of a micro-algal consortium of *Chlorococcum infusionum* and *Leptolyngbya foveolaurum*, as a potential biosorbent. The consortium was collected from the storage tank of a coke-oven wastewater effluent treatment plant after biological treatment. A comprehensive study starting from collection, preparation, characterization, kinetic and equilibrium studies of defluoridation from synthetic solution has been carried out. A large number of experiments are required to optimize multivariable systems to examine the best possible level, which is a time consuming process (Kushwaha & Dutta 2017). Such effort can be reduced by altering a number of variables at a time, and the main objective of Response Surface Methodology (RSM) is to establish an experiential model using statistical equation to attain a suitable result at a minimum effort (Myers *et al.* 2005). Therefore, RSM has been used for optimization of defluoridation parameters. So far as is known, no report on application of COMSOL Multiphysics to predict the kinetics of biosorption of a pollutant using algal biomass has been published. In the present article, COMSOL Multiphysics has been employed to predict the kinetics of biosorption of fluoride using DBM. Finally, defluoridation of groundwater, and Coal Bed Methane (CBM) produced water has been done using the prepared biosorbent and reported in the article.

MATERIALS AND METHODS

Collection, preparation and characterization of biosorbent

The algal bloom was collected from the water surface of the storage tank of a coke-oven effluent treatment plant, being used to hold wastewater after biological treatment, in Durgapur, India (Latitude: 23°50′92″N, Longitude: 87°32′10″E). After collection, the fresh micro-algal lump was washed with water carefully to remove dirt and chemicals. The washed biomass was then spread on a tray and dried under sunlight for one week. The biomass was then kept in a hot air oven at 60 °C for 24 h. The dried biomass was then ground to obtain 90 μ sized powdered biomass as final biosorbent. The Eppendorf tube containing an aqueous suspension of live micro-algal sample was sent to BSI (Botanical Survey of India, Ministry of Environment & Forests, Kolkata, India) for identification and the algal biomass was identified as a consortium of *Chlorococcum infusionum* (green alga) and *Leptolyngbya foveolaurum* (cyanobacterium).

Bulk density, specific gravity, pHzpc, BET surface area and pore volume of the native biosorbent were determined (Biswas *et al.* 2016). A specific sized particle (90 μ) was used in each case. BET (Brunauer Emmett Teller) of native DBM was determined by BET analyzer (Quantachrome, Poremaster 60, USA). The pH_{zpc} of the native biosorbent was determined by following standard protocol (Biswas *et al.* 2016). The test consortium was incubated in a fluoride loaded solution with 10 mg/L initial concentration (IC) of fluoride for 12 h. The cyanobacterial biomass was then centrifuged to separate it from solution, and dried. Field emission scanning electron microscopic (FESEM) image, Energy dispersive x-ray (EDX) analysis and Fourier transform infrared (FTIR) spectroscopy studies were performed to characterize the native and fluoride loaded DBM. A FESEMicroscope (ZEISS SIGMA VP, Japan) was used to take FESEM images of native and fluoride loaded DBM to obtain their topographical characteristics at specific magnification. The EDX study (ZEISS SIGMA VP, Japan) was performed to obtain the elemental composition of the samples. FTIR analyses of native and fluoride loaded DBM were done by FTIR spectrometer (Spectrum 100, Perkin-Elmer, USA) to determine the functional bends and stretches at specific frequencies. The wave number was kept between 400–4,000 cm^{-1}.

Biosorption of fluoride

Sodium fluoride (Merck) (2.21 g) was added in 1 L of deionized water (DW) to make 1,000 mg/L of fluoride stock. It was diluted with the requisite amount of distilled water to get the desired fluoride concentration. All the experiments were done three times. Several parameters such as adsorbent dose (AD) and pH were varied in the range of 0.04–0.5 g/L and 2–12, respectively. In the present study, the pH of the solution was adjusted by adding 4(N) hydrochloric acid or 1(N) sodium hydroxide. A definite amount of DBM (0.05 g in 100 mL) was contacted in 10 mg/L of fluoride solution, varying the initial pH (2–12). The samples were collected after 150 min of operation. The solution containing spent DBM was separated by centrifugation at 5,500 rpm for 5 min and a clear solution from the top section was experimented for remaining F^- by Ion-meter (4 Star, B 36531, Thermo-scientific) using TISAB (total ionic strain adjustment buffer) III-Thermo Scientific. Similarly, the effect of AD (0.04–0.5 g/L) was examined. The optimum pH as analyzed from the previous set of experiments was maintained and a solution with 10 mg/L fluoride concentration was used. The experimentation was conducted at an ambient temperature of 25 °C with shaking speed of 150 rpm.

Kinetic and equilibrium studies

A thorough kinetic study was done to observe the performance of DBM in removal of fluoride from simulated solution. Experiments were run for 150 min at 150 rpm and 25 °C. IC of fluoride was varied within 5–50 mg/L for both kinetics and equilibrium studies and other parameters, viz AD and pH, were kept constant at optimum condition.

Conductivity study of DBM

To assess the ionic characteristics of DBM samples (native and fluoride treated forms), conductivity analysis was done with both native biomass and fluoride loaded biomass. Fluoride loaded forms were prepared by contacting 0.05 g of DBM with fluoride solution of 10 mg/L concentration for 120 min. The biomass was separated from solution by centrifugation and dried in an air oven. An equal amount (0.05 g) of native and fluoride loaded DBM samples was added in a definite amount of DW individually, and the variation of conductivity was taken to 120 min using a multi-parameter tester (PCSTestr 35, EUTECH instrument, Thermo Scientific). The conductivity of distilled water was measured as control for same time span.

Optimization of fluoride biosorption using RSM

In the present study, the statistical design for optimization of fluoride sorption was done using the RSM, and the most popular Central Composite Design (CCD) was chosen to analyze the experimental data (Myers *et al.* 2005). RSM enumerates the relationship between responses and the fundamental independent input parameters (Myers *et al.* 2005):

$$Y = f(x_1, x_2 \ldots\ldots x_n) \tag{1}$$

Total number of experiments that are required to be carried out depends on the number of input factors. In the case of four input parameters, the total number of experiments can be obtained as follows:

$$N = 2^n + 2n + n_c = 2^4 + 2 \times 4 + 6 = 30 \tag{2}$$

The results of all sets were determined and an experiential second order polynomial equation has been obtained to correlate the results with three independent factors for removal of fluoride by DBM.

$$Y = \beta_0 + \sum_{i=1}^{n} \beta_i x_i + \sum_{i=1}^{n} \beta_{ii} x_i^2 + \sum_{i=1}^{n-1} \sum_{j=i+1}^{n} \beta_{ij} x_i x_j + \varepsilon \tag{3}$$

Four parameters, i.e., IC, pH, AD and temperature, were considered as input variables. They were varied as per the design prepared by Design Expert 8.0.6 software to obtain the final optimum conditions. The basic system was prepared by using analysis of variance (ANOVA), consisting of input and output responses and optimized by estimating the statistical conditions. The min (−1) and max (+1) point of IC of fluoride, pH, AD and temperature were 10 and 90 mg/L, 2 and 12, 0.5 and 4.5 g/L, and 25 and 45 °C, respectively.

Regeneration study and toxicity characteristics leaching procedure (TCLP) test

Regeneration of any adsorbent can be justified by adsorption-desorption experiment (Venkata Mohan *et al.* 2007; Kushwaha & Dutta 2017). Three sets of adsorption-desorption experiments were carried out in DW, 0.1 (N) hydrochloric acid, and 0.1 (N) sodium hydroxide solution individually, up to three rounds. After adsorption, the resultant fluoridated DBM was semi-dried at 60 °C for 5 min and reintroduced for desorption study. A single round consisted of 100 min run-time of adsorption and 50 min run-time of desorption, when IC, AD, and temperature were kept at optimum values, as obtained from RSM.

Fluoride-loaded DBM having solid phase fluoride concentration was subjected to a TCLP test (EPA 1992). The TCLP test was used to assess the risk of fluoride leaching to the environment (EPA 1992). An extraction solvent was prepared following the standard protocol of EPA (1992). The solvent was mixed with solid in the solvent-solid ratio of 10 L/Kg (EPA 2017) and then agitated at 150 rpm for 18 h at 30 °C. Thereafter, the mixture was filtered and tested for remaining F^- in the clear solution. The test was carried out at pH 3, 5, 7 and 11.

Removal of excess fluoride from groundwater and CBM produced water

Field survey by West Bengal Public Health Engineering Department (WBPHED) has shown that groundwater in some villages of Bankura district of West Bengal, India, are affected by excess fluoride (WBPHED 2006). Hence, groundwater samples were collected from several wells and tubewells of that affected area. Therefore, collected samples were characterized using the standard protocol (Clesceri *et al.* 2012).

CBM recovery is associated with production of a huge volume of underground water (Mendhe *et al.* 2017). The coal seams are depressurized by pumping out water for regular gas production (Liu *et al.* 2013). To maintain produced water quality, it is essential to manage it through some combinations of treatments (Mendhe *et al.* 2017). CBM produced water is mostly characterized by high TDS (total dissolved solids) and salinity (Mendhe *et al.* 2017). It consists mainly of sodium (Na), bicarbonate (HCO_3), chloride (Cl), iron (Fe), calcium (Ca), fluoride (F) and other trace elements. Before discharging the CBM produced water, several treatments are done based on the quality of the water. This work is focused on removal of excess fluoride using DBM from CBM produced water, collected from Raniganj Coalfield, in the area of West Bengal, India. Similar to groundwater, the CBM produced water was characterized (Clesceri *et al.* 2012).

After characterization, groundwater and CBM produced water were treated with 0.05 g of DBM in 100 mL for 120 min. Removal studies were performed at pH 4, the optimum value as analyzed by RSM, and at their native pH (7.8 and 8.8). After 120 min, the spent DBM was separated by centrifugation and the supernatant was analyzed for residual fluoride concentration.

THEORETICAL ANALYSIS

CFD modeling

COMSOL (4.3.a) Multiphysics was chosen to develop and solve a model for the above discussed batch mode adsorption process using equations of Lagrange's-first order model of adsorption (Augier *et al.* 2008). In order to solve the above equations, a suitable model of the reactor system was established with fine mesh size, mesh configuration and boundary conditions as maintained during experimentation. The model is mesh independent. Based on the real geometry of the reactor system, the model was developed. To simplify the model, a few assumptions were made, such as:

1. The adsorbent used in the experimentation was considered of equal size with uniform porosity.
2. An adsorbent bed is considered settled at the bottom and the system was stationary throughout the process.
3. The volume of the adsorbent bed was considered free of fluoride.
4. Mass transfer resistance in the system was very small and considered as negligible.
5. Biosorption resistance was encountered.
6. The whole system was isothermal as well as isobaric.

RESULTS AND DISCUSSION

Characterization of DBM

Physico-chemical characterization of DBM has been done in terms of the specific gravity, bulk density, surface area, pore volume and pH_{pzc}. The results are shown in Table 1. The specific gravity and bulk density of DBM are 1.0463 and 0.76 g/cc, respectively. The pH_{zpc} of DBM has been found to be 6.8. It indicates that the surface of DBM shows positive charge in a solution having a pH lower than 6.8, and thus binds with negatively charged ions (F^-) in the solution. The result is in agreement with Hiremath & Binnal (2013), for the algal biomass of *Spirulina platensis*. The BET surface area and pore volume of DBM have been found to be 2.365 m^2/g and 0.013 cc/g, respectively.

FESEM images were taken for both native and fluoride treated DBM, as shown in Figure 1(a) and 1(b), respectively. It can clearly be seen from the images that the surface of both adsorbents are not homogenous. However, before adsorption of fluoride on DBM, the rough surface textures along with many small pores are observed; whereas after adsorption, a comparatively smooth surface is observed. The elemental compositions of raw and fluoride loaded consortium biomass as obtained from the EDX study are shown in Figure 1(c) and 1(d). DBM contains 91% carbon (C), 6% nitrogen (N) and 3% phosphorus (P). After treatment with fluoride, DBM contains 2% fluoride (F), 90% carbon (C), 1% phosphorus (P) and 7% nitrogen (N), which clearly indicates the fluoride sorption on the DBM surface. Nitrogen adsorption and desorption isotherm studies of native DBM (Figure 1(e)) reveal the mesoporous nature of DBM, which reconfirms the surface morphology as obtained from FESEM images. The biosorption behaviour is dependent on the functional groups of the DBM surface, the primary sites of fluoride sorption. FTIR data of raw (Figure 1(f)) and fluoride treated (Figure 1(g)) DBM have been done to scrutinize the changes in spectra due to introduction of fluoride in the later case. The responsible bonds such as ≡C-N< (stretch), -C-O (bend), -O-H (stretch,), C = C (strong bend) and phenolic groups are observed at the wave numbers of 1,054, 1,269, 1,406, 3,433, 1,560, 1,648 and 2,922 cm^{-1}, respectively. However, an FTIR spectrogram of fluoride treated DBM shows some different spectra. The presence of $-CHF_3$ group (Tri-fluro methyl) at the wave number of

Table 1 | Physical characterization of DBM

Property	Value
Bulk density	0.76 g/cc
Specific gravity	1.0463
pH_{zpc}	6.8
Surface area	2.365 m^2/g
Pore volume	0.013 cc/g

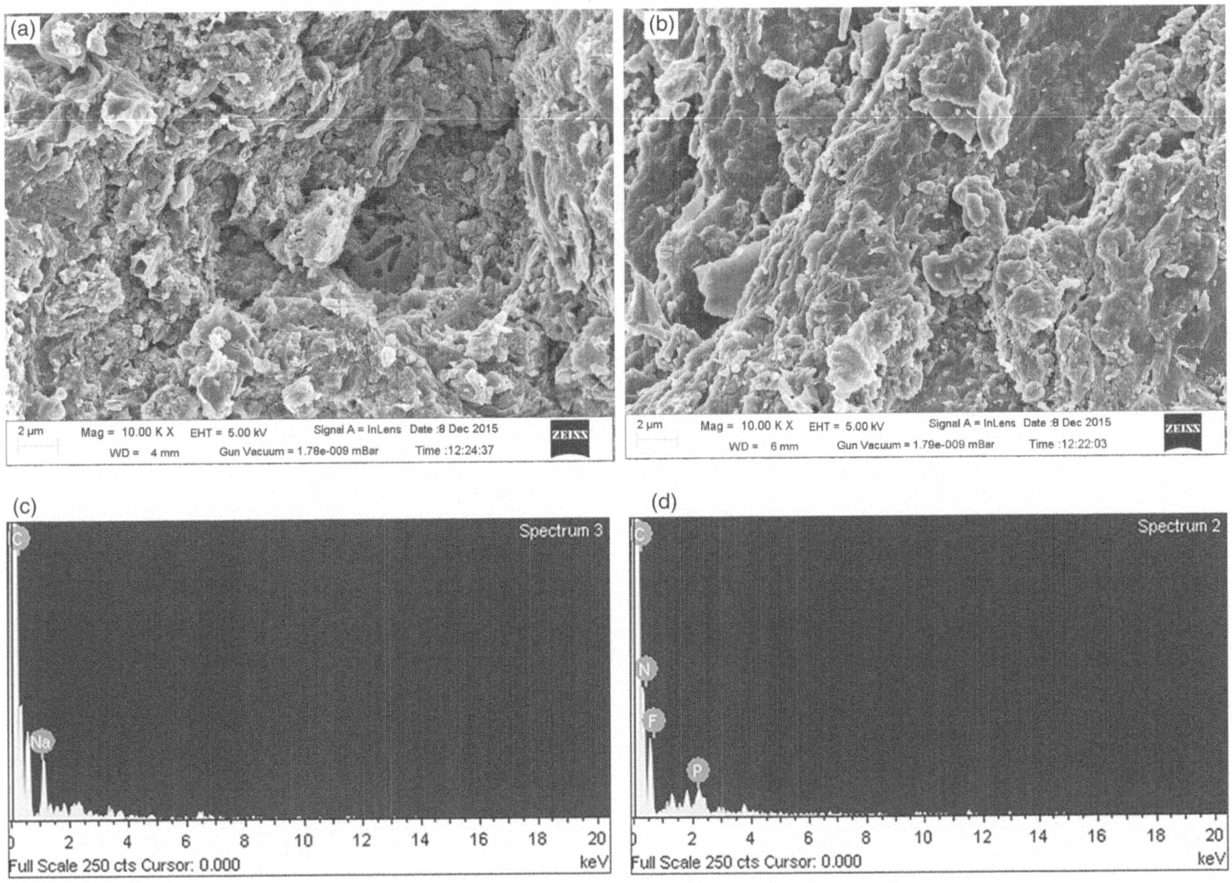

Figure 1 | (a) Morphological details (FESEM image) of native DBM, (b) morphological details (FESEM image) of fluoride loaded DBM, (c) elemental analysis (EDX study) of native DBM, (d) elemental analysis (EDX study) of fluoride loaded DBM, (e) nitrogen adsorption-desorption isotherm curve of native DBM, (f) FTIR study of native DBM, (g) FTIR study of fluoride loaded DBM. (*Continued*.)

1,205 cm^{-1} confirms the adsorption of fluoride on DBM. The other characteristic bonds such as ≡C-N< (1,050 cm^{-1}), -O-H (1,408 and 3,431 cm^{-1}), C = C (1,652 cm^{-1}) and phenolic group (2,924 cm^{-1}), are shifted as observed in the fluoride loaded form (Figure 1(g)). Hence, the fluoride adsorption onto DBM has initially been proved.

Biosorption of fluoride

One-factor-at-a-time (OFAT) analysis has been done to determine the optimum condition of biosorption of fluoride using DBM. From Figure 2(a), it is seen that, as the pH increases from 2 to 4, fluoride removal increases from 43 to 47.1%. However, when the pH increases from 4 to 12, fluoride removal decreases from 47.1 to 2.66% (Figure 2(a)). The surface of the algae contains a number of molecular groups, such as amino, phosphate, carboxyl, thiol, etc (Bhatnagar *et al.*, 2002). Protonation of the cell wall depends on the pH of the solution, i.e., the positive charged surface [H^+ (cationic) type] of DBM at pH< pH_{zpc} conditions (Hiremath & Binnal, 2013). Biswas *et al.* (2016) reported the maximum fluoride sorption capacity at pH lower than pH_{zpc}. Since the highest removal is obtained at pH 4, it has been chosen as the optimum pH for removal of fluoride. Figure 2(b) reveals that the maximum removal is achieved at 0.5 g/L of AD for the present biosorbent. At lower ADs (0.04–0.2 g/L), fluoride removal is lower because of the low availability of active sites, as reported in several literatures (Venkata Mohan *et al.* 2007). Although the removal increases with the increase in AD, 0.5 g/L is selected as the best AD from the economic point of the view (Venkata Mohan *et al.* 2007).

Kinetic and equilibrium studies on biosorption of fluoride

To appraise the fluoride biosorption function using DBM and to estimate the kinetic parameters, a detailed kinetic

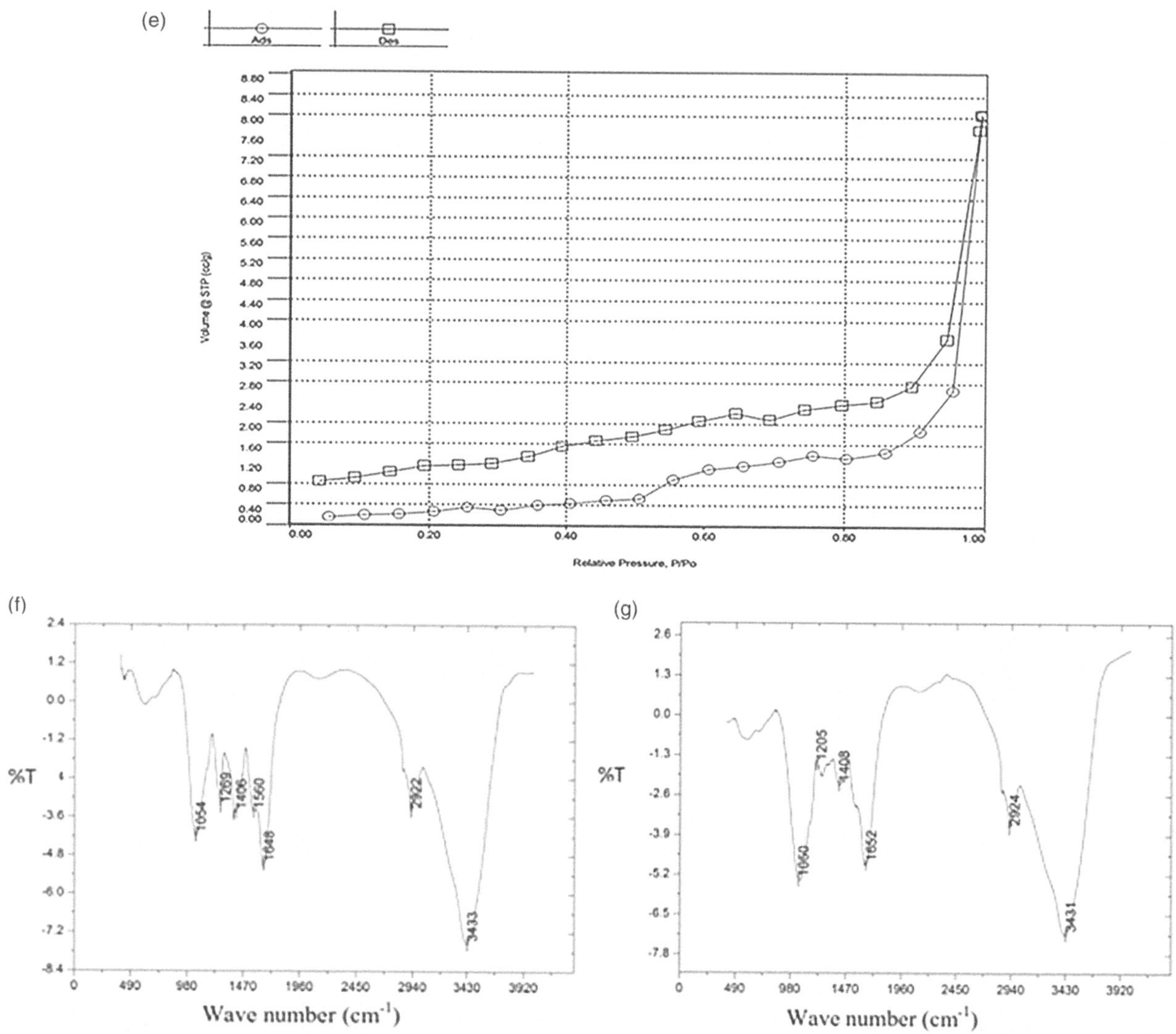

Figure 1 | Continued.

study has been done at optimum condition as obtained from OFAT analysis, such as: 0.5 g/L AD and pH 4. Volume of the solution, shaking speed and temperature are maintained at 100 ml, 150 rpm and 25 °C, respectively. The variation of fluoride removal (%) with time considering different IC of fluoride as a parameter is shown in Figure 3. It is seen that the removal achieves equilibrium condition gradually after 90 min. Maximum 72%, 67.4%, 42.5% and 36.1% have been removed from 5, 10, 25 and 50 mg/L of IC, respectively. The lower removal at higher IC may be due to the competitive nature of the overloaded F^- ions and the limitation of the active-sites on DBM (Bhatnagar *et al.* 2002). Furthermore, the higher removal at lower IC signifies the process as kinetically controlled rather than mass transfer limited one. Kinetic data have been fitted to three kinetic models viz., Lagergren first-order, Ho pseudo-second order and Morris-Weber model. It has been seen that the kinetic data fit best to the Lagergren first-order model with an R^2 value of 0.89 to 0.9. The experimental q_e matches quite well with simulated data for the Lagergren first-order model. The values of kinetic parameters are shown in Table 2. Lagergren (1898) presented a first-order rate equation to describe the kinetic process of liquid solid phase adsorption of oxalic acid and malonic acid onto charcoal. It can be presented as follows:

$$\frac{dq_t}{dt} = k_1(q_e - q_t) \tag{4}$$

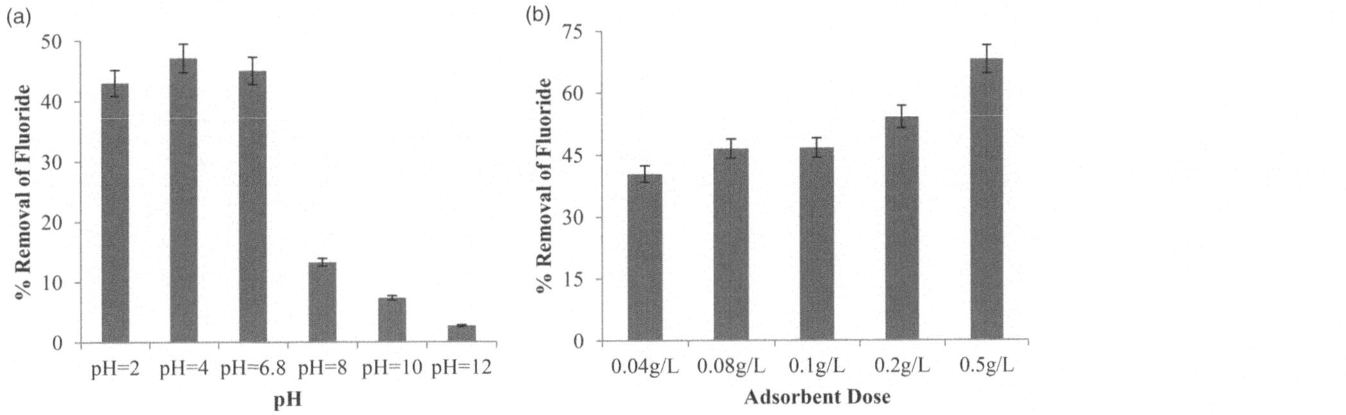

Figure 2 | (a) Effect of solution pH on fluoride removal using DBM (IC = 10 mg/L, AD = 0.04 g/L, Temp = 25 °C). (b) Effect of adsorbent doses on fluoride removal using DBM (IC = 10 mg/L, pH = 4, Temp = 25 °C).

The significance of fitting of kinetic data to the Lagergren first order model may be attributed to the dependence of binding of fluoride to the surface of DBM with first order relationship.

Equilibrium studies have been done at different IC of fluoride, ranging from 5 to 50 mg/L at an AD of 0.5 g/L, pH 4, shaking speed 150 rpm, contact time of 120 min and 25 °C temperature. Langmuir, Freundlich, Temkin and Dubinin-Radushkevich isotherm models have been used to fit the equilibrium data, and it is seen that the R^2 value (0.9894) of the Langmuir isotherm model holds good to predict equilibrium condition. This indicates the monolayer coverage of F^- on the DBM surface. Maximum adsorption capacity is obtained as 34.36 mg/g (Table 3). From kinetic and equilibrium studies, it can be concluded that DBM can be used as a good biosorbent for the purpose of defluoridation. A comparative study has been made based on the defluoridation works of other scientists (Table 4).

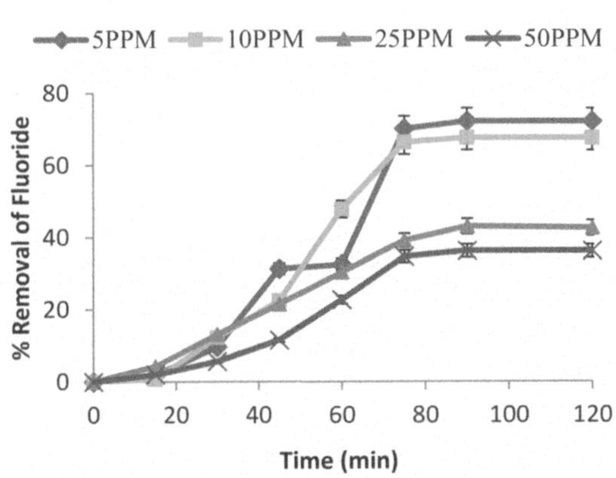

Figure 3 | Kinetics studies at different initial concentrations of fluoride (pH = 4, AD = 0.5 g/L, temp = 25 °C, shaking speed = 150 rpm).

Application of COMSOL multiphysics to predict kinetics of fluoride removal using DBM

A transient convection–diffusion equation was employed in COMSOL Multiphysics to study the experimental phenomena for removal of fluoride. The transient convection–diffusion equation is given in Equations (4) and (5):

$$\frac{\partial}{\partial t}(\rho C) + \nabla \cdot (\rho \overline{U} c) = \nabla \cdot (d \nabla \cdot C) + S \tag{5}$$

$$S = -q_e * (1 - \exp(-k_1 * t)) \tag{6}$$

In Equation (4), the first and second terms of the left hand side are accumulation and convective mass transport terms, respectively. In the right hand side, the first term is a diffusive term and the second is a source term. To validate and justify the mechanism of fluoride removal and the kinetic parameters obtained during experimentation of fluoride adsorption using dead biomass of the micro-algal consortium, such a model is proposed. The experimental system is shown in the schematic diagram (Figure 4(a)). Assuming the same kinetic parameters as obtained during experimentation, boundary conditions, Lagrange's-first order model of adsorption, and fine mesh size, the model predicts the removal of fluoride with time for different initial fluoride concentrations. The experimental data, when placed over the model-predicted data as shown in Figure 4(b), are found to be a good fit. It is seen that the model-predicted data show a similar nature of removal in terms of residual fluoride as found during experimentation. As the concentration of the solution increases, the biosorption rate decreases, i.e., there is no such mass transfer resistance

Table 2 | Values of kinetics parameters

Parameter		Lagergren first order model $\log(q_e - q_t) = \log\ q_e - k_1 t$			Pseudo second order model $\frac{t}{q_e} = \frac{1}{k_2} \times \frac{1}{q_e^2} + \frac{t}{q_e}$			Morris-Weber model $q_t = k_3 t^{1/2} + C''$	
IC (mg/L)	q_e(exp)	k_1	q_e	R^2	k_2	q_e	R^2	k_3	R^2
5	7.34	0.0113	7.795	0.8913	2.2*10^{-2}	−3.234	0.5017	1.0416	0.8726
10	12.48	0.0261	16.575	0.8648	3.9*10^{-3}	2.445	0.3781	1.7746	0.9212
25	24.9	0.0235	29.008	0.9003	1.7*10^{-6}	370.37	0.0432	3.028	0.968
50	37.6	0.0346	40.546	0.8648	1.0*10^{-4}	44.44	0.4656	5.3401	0.8798

Table 3 | Values of equilibrium parameters

Adsorption isotherm model	Expression	Constants
Langmuir model	$\frac{1}{q_e} = \frac{1}{q_{max} K_L C_e} + \frac{1}{q_{max}}$	R^2=0.9894 $q_{max} = 34.36$ K =0.186
Freundlich model	$lnq_e = lnK_F + \frac{1}{n} lnC_e$	$R^2 = 0.9796$ $1/n = 1.9425$ $K_F = 0.0289$
Temkin model	$q_e = \frac{RT}{b'} \ln(K_T) + \frac{RT}{b'} \ln(C_e)\ B' = \frac{RT}{b'}$	$R^2 = 0.9571$ $B' = 9.4165$ $K_T = 1.23$
Dubinin-Radushkevich model	$\ln(q_e) = \ln(q_{max}) - K_{DB} \varepsilon'^2$	$R^2 = 0.7634$ $K_{DB} = 0.7623$ $q_{max} = 25.41$

Table 4 | Adsorption capacities of different biomass derived adsorbent for removal of fluoride

Adsorbent (biomass)	Adsorption capacity (mg/g)	Reference
Spirogyra sp. IO2	1.272	Venkata Mohan *et al.* (2007)
Spirulina plantheis	34.0	Hiremath & Binnal (2013)
Anabaena fertilissima (Ca-treated)	7.0	Bhatnagar *et al.* (2002)
Cholorococcum humicola (Ca-treated)	4.5	Bhatnagar *et al.* (2002)
DBM	*34.34*	*Present study*

rather than biosorption resistance. So it may be concluded that the system is a biosorption controlled system. Thus, from the above observation, it can be said that the model developed is capable of predicting residual fluoride concentration vis-à-vis fluoride removal., To further justify the validity of the model, root mean square error (RMSE) and average error is calculated for all the cases using Equations (6) and (7), as shown in Table 5.

$$\mathrm{RMSE} = \frac{\sqrt{\sum_{i=1}^{N} (C_{\exp,i} - C_{sim,i})^2}}{N} \tag{7}$$

$$Error(\%) = \left[\frac{C_{\exp} - C_{sim}}{C_{\exp}}\right] \times 100 \tag{8}$$

Conductivity study of DBM

Conductivity analysis has been carried out to assess the ionic property of DBM. Figure 5 shows the change in conductivity of the solution with time under four different conditions: (1) distilled water (DW), (2) Simulated solution of fluoride (DW + F), (3) DBM in distilled water (DW + DBM), and (4) fluoride loaded DBM in distilled water (DW + DBMF).

(a)

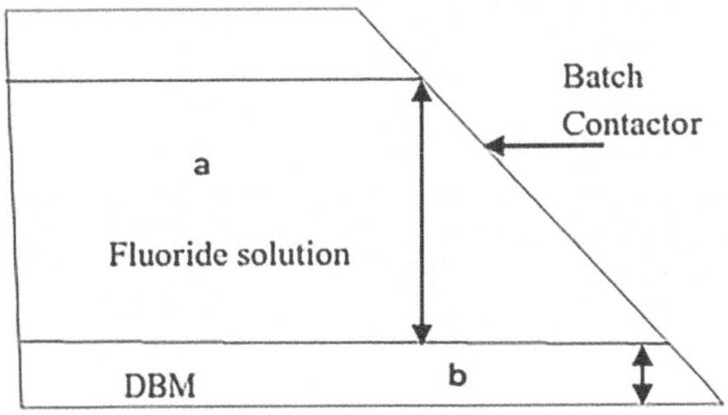

(b)

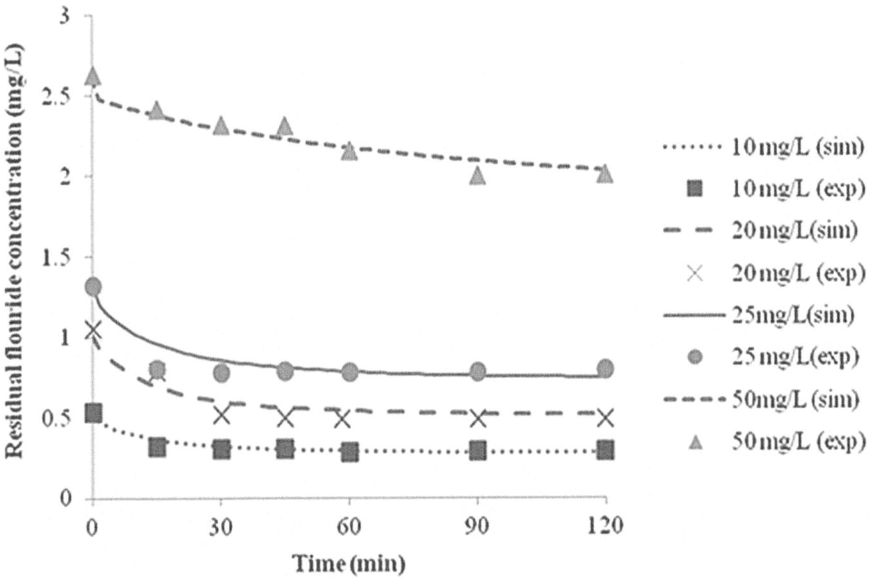

Figure 4 | (a) Schematic diagram of 1/4th of the 250 mL conical flask (a: fluoride solution; b: adsorbent DBM). (b) COMSOL Multiphysics predicted data (line) and the experimental results (point) of residual fluoride concentration for different initial fluoride concentration with time.

From the figure, it is seen that the conductivity of DBM in distilled water is maximum (143 μs/cm) and the conductivity is lowest for distilled water only. The conductivity of fluoride-loaded DBM in distilled water is 126 μs/cm, which is lower than DBM in distilled water (DW + DBM). The highest conductivity with native DBM confirms the presence of different ions in the solution, which may have come from algal biomass. The presence of ionic functional groups is already seen in the FTIR study (Figure 1(f)). The lower conductivity

Table 5 | RMSE and error (%) calculations for different initial concentrations

		Error (%)	
IC (mg/L)	RMSE	Avg. error	Max. error
10	0.0186	4.13	13.69
20	0.05	5.9	19.76
25	0.0048	8.9	15.98
50	0.051	1.87	4.66

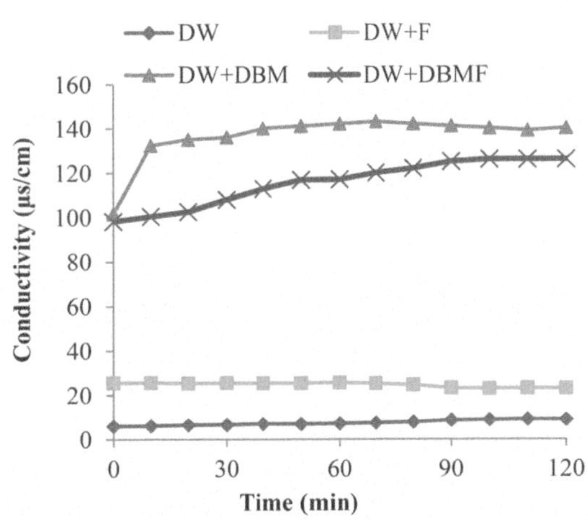

Figure 5 | Conductivity analysis of DBM with time.

of fluoride-loaded DBM may be due to the binding of fluoride with the ionic functional groups as present on the surface of DBM, and thus release of functional groups in solution is restricted. Lower conductivity of fluoride-loaded DBM in distilled water confirms that the adsorption is chemisorptive in nature (Oura *et al.* 2003).

Optimization using RSM

Four key parameters, i.e., pH, AD, IC and temperature, have been varied as the input condition according to CCD (Table 6). Defluoridation by DBM has been considered as final response, which have been analysed using ANOVA (Table 7). The F-value (22.49) obtained from ANOVA implies the significance of the model. As suggested by ANOVA, a quadratic model has been fitted to the responses. Equation (8) has been achieved in terms of coded factors has been given below:

$$R1 = 38 - 1.34^{*}A - 4.8^{*}B + 1.37^{*}C - 5.32^{*}D + 1.4^{*}AB - 0.09^{*}AC + 1.08^{*}AD - 0.68^{*}BC + 3.49^{*}BD - 1.25CD - 0.23^{*}A^2 - 1.63^{*}B^2 - 0.97^{*}C^2 + 0.94D^2 \quad (9)$$

R^2, R^2_{Adj} and R^2_{pred} have been obtained as 0.954, 0.912 and 0.738, respectively. The higher R^2 data indicates the good fitting of the model to the investigational data. Significant model terms are *B*, *D*, and *BD*. Moreover, the interactive effect of pH and temperature is significant. Figure 6 shows the interactive effect of pH and temperature on removal of fluoride using DBM. When the pH increases from 4.5 to 9.5 at a temperature of 30 °C, the removal decreases from 33.3% to 30.7%, and at a temperature of 40 °C, the removal decreases from 50.7% to 34.5%. The lower removal at higher pH condition has aready been

Table 6 | Experimental design and response of fluoride biosorption study

Sl no.	IC (mg/L)	pH	Dose (g/L)	Temperature (°C)	Removal (%)
19	50	2	2.5	35	38
18	90	7	2.5	35	36
13	30	4.5	3.5	40	35
3	30	9.5	1.5	30	31.2
1	30	4.5	1.5	30	51.2
28	50	7	2.5	35	38
23	50	7	2.5	25	54.5
16	70	9.5	3.5	40	28
22	50	7	4.5	35	38.5
5	30	4.5	3.5	30	59.2
21	50	7	0.5	35	30.6
9	30	4.5	1.5	40	33
27	50	7	2.5	35	38
17	10	7	2.5	35	39
24	50	7	2.5	45	30
4	70	9.5	1.5	30	29
10	70	4.5	1.5	40	32.5
8	70	9.5	3.5	30	33.8
2	70	4.5	1.5	30	42.5
12	70	9.5	1.5	40	29
6	70	4.5	3.5	30	47
11	30	9.5	1.5	40	30
14	70	4.5	3.5	40	32
20	50	12	2.5	35	25.8
26	50	7	2.5	35	38
29	50	7	2.5	35	38
25	50	7	2.5	35	38
7	30	9.5	3.5	30	32.5
15	30	9.5	3.5	40	28
30	50	7	2.5	35	38

Table 7 | ANOVA and statistical response of fluoride biosorption study

Source	Sum of Squares	df	Mean Square	F Value	*p*-value Prob > F
Model	1737.99	14	124.14	22.49	<0.0001 significant
A-IC	43.47	1	43.47	7.87	0.0133
B-pH	553.92	1	553.92	100.34	<0.0001
C-AD	45.1	1	45.1	8.17	0.012
D-T	681.6	1	681.6	123.47	<0.0001
AB	31.64	1	31.64	5.73	0.0302
AC	0.14	1	0.14	0.025	0.8753
AD	18.71	1	18.71	3.39	0.0855
BC	7.43	1	7.43	1.35	0.2643
BD	195.3	1	195.3	35.38	<0.0001
CD	25.25	1	25.25	4.57	0.0493
A^2	1.56	1	1.56	0.28	0.6027
B^2	73.64	1	73.64	13.34	0.0024
C^2	26.13	1	26.13	4.73	0.046
D^2	24.7	1	24.7	4.47	0.0515
Residual	82.8	15	5.52		
Lack of fit	82.8	10	8.28		
Pure error	0	5	0		
Cor total	1820.79	29			

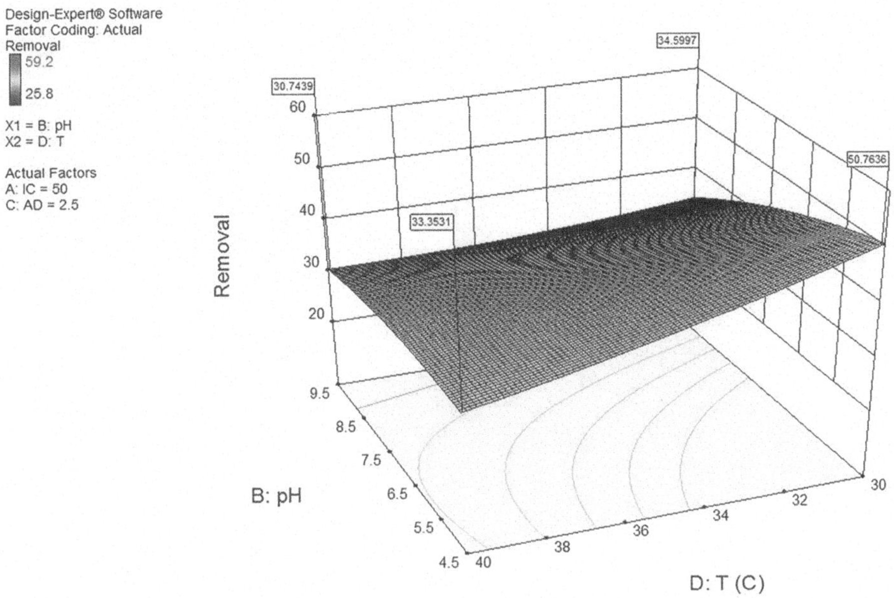

Figure 6 | Effect of pH and temperature on the removal of fluoride.

observed during OFAT analysis (Figure 2(a)). A pH higher than pH_{zpc} leads to a negative charge on the surface of the biosorbent and thus results in lower removal of fluoride. During selection of the optimum condition the following criteria have been set: IC of fluoride: in range, optimum values of IC of fluoride, pH of the solution, AD and temperature are obtained as 30 mg/L, 4.5, 3.5 g/L and 30 °C, respectively. The predicted removal of fluoride as analyzed by RSM is 56.9%, which is in accordance with the experimental removal (58%).

Regeneration study, TCLP test and CHNS study of DBMF

In Figure 7(a), three successive sorption/desorption rounds have been performed with three different desorption conditions. A consistent reduction in adsorption capacity has been noticed for each new cycle after a total adsorption-desorption series (Figure 7(a)). The adsorption capacities have been reduced from 5.71 mg/g (IC: 30 mg/L; AD: 3.5 g/L) after the first cycle of adsorption to 4.57 mg/g (distilled water), 4.28 mg/g (0.1 N HCl) and 2.85 mg/g (0.1 N NaOH) after the third cycle of adsorption. Furthermore, relatively more effective reusability has been noticed when the loaded biomass has been desorbed with 0.1 N HCl (Figure 7(a)). On the other hand, usage of acid incurs cost to the process. Therefore, from an economical point of view, distilled water can be used as the desorption medium for the present DBM system.

After a TCLP test at pH 3, the maximum 0.62 mg/L of fluoride was found (Figure 7(b)) in the supernatant which has been leached out from the fluoride-laden DBM, and such a concentration is far below the leachable permissible fluoride level (150 mg/L) as suggested by the Waste Classification Guidelines (2008). Therefore, the present study indicates that the fluoride laden DBM passes the TCLP tests successfully, and it can be disposed of safely or used as bio-fertilizer (Waste Classification Guidelines 2008).

CHNS study of fluoride-treated DBM is done for further assessment of used biomass (Figure 7(c)). From Figure 7(c), the carbon, nitrogen, hydrogen and sulphur contents of the fluoride-loaded DBM are obtained as 36.185, 3.965, 5.365 and 0.658%, respectively. The study shows that the used biomass can be used as feedstock for different types of bio-fuel production utilizing a large carbon source, which may come under the future scope of the study.

Comparison between removal of excess fluoride from groundwater and CBM produced water

The mineralogical properties of the groundwater sample have been determined. pH, TDS, total hardness, total alkalinity chloride, iron and fluoride are obtained as 7.4. 470 mg/L, 255 mg/L, 367 mg/L, 100 mg/L, 12.2 mg/L and 4.71 mg/L, respectively. The groundwater sample (100 mL) was then treated with DBM in batch mode with

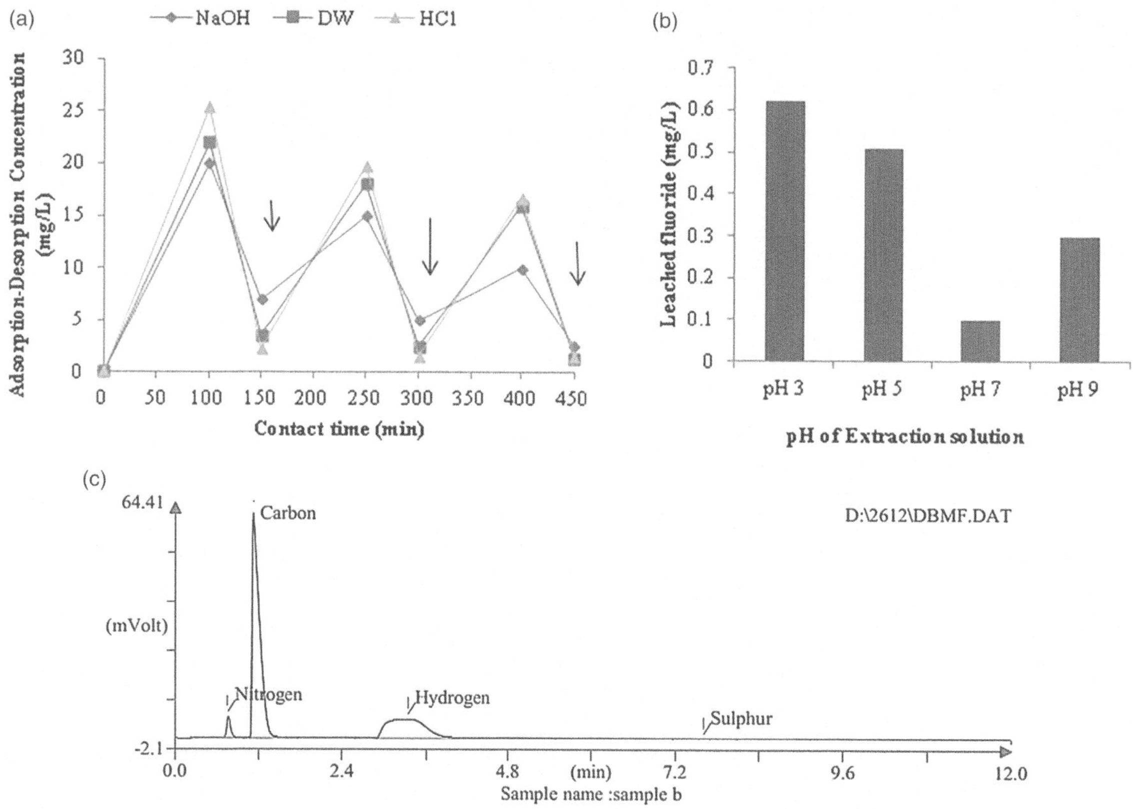

Figure 7 | (a) Pattern of adsorption-desorption of fluoride (represents desorption; represents adsorption). (b) TCLP test of fluoride loaded DBM. (c) CHNS study of fluoride laden DBM.

3.5 g/L AD at two pH conditions, i.e. at the groundwater's own pH (7.4), and optimum pH (4.5) as analyzed by RSM. From Figure 8, it is seen that after 120 min of contact time, 61.14% and 73.03% removal are obtained at pH 7.4 and pH 4.5, respectively. The final concentration of fluoride in treated groundwater sample at pH 4.5 is 1.27 mg/L, i.e. below the allowable limit as set by the Bureau of Indian Standard (BIS 2012).

The physico-chemical properties of raw CBM produced water are determined as: pH 8.8, TDS: 570 mg/L, salinity: 398 mg/L, carbonate: 209 mg/L, chloride: 287 mg/L, fluoride: 9.24 mg/L, nitrate: 18.6 mg/L, sulphate: 15.8 mg/L, sodium: 201 mg/L, potassium: 15.9 mg/L, calcium: 85.8 mg/L, magnesium: 7.5 mg/L and iron: 1.65 mg/L. After treatment with DBM, 66.45% and 69.16% removal of fluoride were obtained after 120 min of contact time from CBM produced water, at pH 8.8, and pH 4.5, respectively (Figure 8). The final concentration of fluoride at pH 4.5 was 2.85 mg/L, which is near the value of the industrial effluent acceptable limit (2 mg/L) as set by EPA (EPA 2017).

Therefore, it can be assured that a higher level of fluoride can be removed from both groundwater and CBM produced water using DBM under optimum conditions. Since in both cases the treated water meets the permissible limit, the biosorption of fluoride using the present algal consortium biomass may be considered as a potent technology for removal of fluoride.

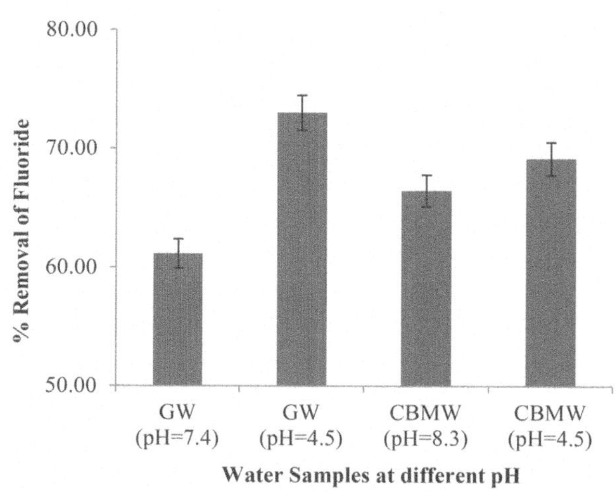

Figure 8 | Removal of fluoride from groundwater (GW) and CBM produced water (CBMW) at different pH.

CONCLUSION

Dead biomass (DBM) of micro-algal consortium (*Chlorococcum infusionum* and *Leptolyngbya foveolaurum*) has been used for biosorption of fluoride from both synthetic and real wastewater. Both kinetic and equilibrium studies have been performed to assess the efficacy of the present DBM in removal of fluoride. The percentage removal decreases from 72 to 36.1% when the initial fluoride concentration increases from 5 to 50 mg/L, at AD: 0.5 g/L, pH: 4, shaking speed: 150 rpm, volume of solution: 100 ml, contact time: 120 min and temperature: 25 °C. The higher removal at lower initial fluoride concentration indicates that the process is kinetically controlled rather than a mass transfer limited one. Equilibrium data have been found to fit well to the Langmuir isotherm model. The adsorption capacity of DBM has been found to be 34.36 mg/g. RSM has been employed to determine the optimum conditions for fluoride removal. The predicted removal of fluoride (56.9%) at optimum conditions (IC = 30 mg/L, AD = 3.5 g/L, pH = 4.5, temp = 30 °C) as analyzed by RSM matches quite well with the experimental one (58%). From the observation of COMSOL Multiphysics, it can be said that the model developed is capable of predicting residual fluoride present vis-à-vis fluoride removal., Conductivity analyses of native and fluoride-loaded biomass indicate the binding of fluoride onto biomass and restriction of release of the ionic group from biomass in the latter case. As the fluoride-bearing solid biomass passes the TCLP tests successfully, it can be classified as a non-toxic waste, and disposed of safely. Finally, the groundwater and CBM produced water have been treated at optimum pH (4.5) and at their native pH. In both cases, removal is greater at optimum pH and the final concentration of fluoride meets the permissible limit. Therefore, the present DBM, having good fluoride adsorption capacity, could be used for treatment of industrial wastewater as well as fluoride contaminated groundwater.

ACKNOWLEDGEMENT

These laboratory works are supported by Department of Chemical Engineering and Earth & Environmental Studies, National Institute of Technology Durgapur, West Bengal, India. The financial grant is funded by Department of Science and Technology, Govt. of West Bengal, India (Sanc. No. 891(Sanc.)/ST/P/S&T/9G-13/2012).

REFERENCES

Augier, F., Laroche, C. & Brehon, E. 2008 Application of computational fluid dynamics to fixed bed adsorption calculations: effect of hydrodynamics at laboratory and industrial scale. *Separation and Purification Technology* **63**, 466–474.

Bhatnagar, M., Bhatnagar, A. & Jha, S. 2002 Interactive biosorption by micro algal biomass as a tool for fluoride removal. *Biotechnol Letter.* **24**, 1079–1108.

Biswas, G., Dutta, M., Dutta, S. & Adhikari, K. 2016 A comparative study of removal of fluoride from contaminated water using shale collected from different coalmines in India. *Envl Sci Poll Res.* **23**, 9418–9431.

Biswas, G., Kumari, M., Adhikari, K. & Dutta, S. 2017 A critical review on occurrence of fluoride and its removal through adsorption with an emphasis on natural minerals. *Curr Pollution Rep.* **3**, 104–119.

Bureau of Indian Standard (BIS) 2012 Indian standard specification for drinking water. IS10500: 1–24. *Drinking Water [FAD25: Drinking Water]*. https://law.resource.org/pub/in/bis/S06/is.10500.2012.pdf (accessed 28 May 2017).

Clesceri, L. S., Greenberg, A. E. & Eaton, A. D. 2012 *Standard Methods for the Examination of Water and Wastewater.* 22nd edn. American Public Health Association (APHA), American Water Works Association (AWWA), Water Environment Federation (WEF), Washington, DC.

EPA 1992 Method 1311 Toxicity Characteristic Leaching Procedure. US Environmental Protection Agency Office of Ground Water and Drinking Water Standards and Risk Management Division. Washington, DC (1992). https://www.epa.gov/sites/production/files/2015-12/documents/1311.pdf (accessed 12 June 2017).

EPA 2017 Harmful Algal Blooms. US Environmental Protection Agency Office of Ground Water and Drinking Water Standards and Risk Management Division, Washington, DC. https://www.epa.gov/nutrientpollution/harmful-algal-blooms (accessed 10 June 2017).

Hiremath, P. G. & Binnal, P. 2013 First report on biosorption of fluoride on the microalga spirulina platensis: batch studies. *Asian-American Journal of Chemistry* **1**, 1–10.

Kumar, S. K., Dahms, H. U., Won, E. J., Lee, J. S. & Shin, K. H. 2015 Microalgae – a promising tool for heavy metal remediation. *Ecotoxicol. Env. Safety* **113**, 329–352.

Kushwaha, D. & Dutta, S. 2017 Experiment, modeling and optimization of liquid phase adsorption of Cu(II) using dried and carbonized biomass of *Lyngbya majuscula*. *Appl. Water Sci.* **7**, 935–949.

Lagergren, S. 1898 About the theory of so-called adsorption of soluble substances, K. *Svenska Vetenskapsakad. Handl.* **24** (4), 1–39.

Liu, X., Hu, Z., Zhu, C., Wen, G. & Meng, X. 2013 Removal of fluoride and total dissolved solids from coalbed methane produced water with a movable ultra-low pressure reverse osmosis system. *Desal. Water Treat.* **51**, 4359–4367.

Mendhe, V. A., Singh, A. P., Mishra, S. & Varma, A. K. 2017 Coalbed methane-produced water quality and its management options in Raniganj basin, West Bengal, India. *Appl. Water Sci.* **7**, 1359–1367.

Myers, R. H., Montgomery, D. C. & Anderson-Cook, C. M. 2005 *Response Surface Methodology: Process and Product Optimization Using Designed Experiments.* 3rd edn. John Wiley & Sons Inc., New Jersey, pp. 49–327.

Oura, K., Lifshits, V. G., Saranin, A. A., Zotov, A. V. & Katayama, M. 2003 *Surface Science, An Introduction.* Springer, Berlin, Germany.

Sen, S., Dutta, S., Guhathakurata, S., Chakrabarty, J., Nandi, S. & Dutta, A. 2017 Removal of Cr (VI) using a cyanobacterial consortium and assessment of biofuel production. *Int. Biodet. Biodeg.* **119**, 211–224.

Veeraputhiran, V. & Alagumuthu, G. 2011 Treatment of high fluoride drinking water using bioadsorbent. *Research Journal of Chemical Sciences* **1**, 49–54.

Venkata Mohan, S., Ramanaiah, S. V., Rajkumar, B. & Sarma, P. N. 2007 Biosorption of fluoride from aqueous phase onto algal *Spirogyra* IO1 and evaluation of adsorption kinetics. *Bioresource Technology* **98**, 1006–1011.

Waste Classification Guidelines 2008 *Part 1: Classifying Waste,* Department of Environment and Climate Change NSW, Sydney.

WBPHED 2006 Public Health Engineering Department, Government of West Bengal. http://www.wbphed.gov.in/main/images/pdf/Fluoride_Reports/FLVillagewise_A4.pdf (accessed 12 Jun 2017).

First received 9 July 2017; accepted in revised form 11 December 2017. Available online 18 December 2017

Removal of copper from an electroplating industrial effluent using the native and modified *spirogyra*

Nimra Ilyas, Sadia Ilyas, Sajjad-ur-Rahman, Sidra Yousaf, Aqsa Zia and Sidra Sattar

ABSTRACT

In the present study, biosorption behavior of a green filamentous alga, *spirogyra* in its native and modified states was investigated for copper removal from an electroplating industrial effluent. For this, the effluent containing 194 $mg{\cdot}L^{-1}$ Cu^{2+} in sulfate medium was contacted with both forms of *spirogyra*, under the parametric variations of effluent pH, adsorbent dosage, contact time, and sorption temperature. The study revealed *spirogyra* as a prominent candidate for removing contaminant metal cation; however, at the same condition, biosorption capacity of modified biomass in gel form was higher than the native *spirogyra*. At the optimized condition with 6 g sorbent dosage treated to 100 mL effluent for 30 min at pH 6.0 and temperature 20 °C, the maximum 82.8% and 96.4% copper could be adsorbed by the native and modified *spirogyra*, respectively. The batch sorption data using native biomass followed pseudo-first-order kinetic; exhibiting the multilayer sorption mechanism via surface diffusion could be defined by the Freundlich model. In contrast, the sulfuric acid treated modified *spirogyra* followed pseudo-second-order kinetics and intra particle diffusion as the rate-limiting step.

Key words | algae biomass, biosorption equilibria, cross-linked *spirogyra*, effluent treatment, electroplating waste solution

Nimra Ilyas
Sajjad-ur-Rahman
Institute of Microbiology,
University of Agriculture Faisalabad (UAF),
38040,
Pakistan

Sadia Ilyas (corresponding author)
Sidra Yousaf
Aqsa Zia
Sidra Sattar
Mineral and Material Chemistry Laboratory,
Department of Chemistry,
University of Agriculture Faisalabad (UAF),
38040,
Pakistan
E-mail: *sadiailyas1@yahoo.com*

INTRODUCTION

The availability of clean water is a global concern. In the recent past, much attention has been paid to the contamination of the aquatic environment, causing severe threat to aquatic life and human health. The industrial effluents, not limited, belong to the mining, metals, electroplating, fertilizer, leather, machining, boiler, etc. that contain heavy metals are primarily damaging the aquatic environment (Nayak & Pal 2017). The most adverse characteristics of heavy metals over other pollutants in water are the non-biodegradation and their unfavorable accumulation in living organisms. Notably, copper (used in this study as the contaminant metal in wastewater) is one of the vital used metals in demand, specifically as interconnecting material via surface electroplating, thus inevitably generating a large volume discharge by the electroplating industry (Jun *et al.* 2016). The rinse water generated after plating contains a significant amount of copper (up to 500 $mg{\cdot}L^{-1}$ Cu^{2+}) that cannot be directly discharged without treatment. Interestingly, the hazard quotient of copper is higher than the toxic metal, cadmium, and observed as: As > Cr > Pb > Cu > Cd (Yu *et al.* 2014). Due to this, the World Health Organization has restricted the maximum of 1.5 $mg{\cdot}L^{-1}$ copper in drinking water, or else it may cause severe damages of hepatic, renal and the central nervous systems, and necrotic changes in kidney and liver (Jun *et al.* 2016). A proper remedial action before discharging such effluent streams is therefore of great importance and often needed.

To eliminate the heavy metal contaminants, various physical, chemical and biological methods can be used, and some of them are already in practice (Volesky & Holan 1995). Such methodologies include flocculation, coagulation, precipitation, micro-filtration, resin ion-exchange, chemical/carbon adsorption, reverse osmosis, solvent extraction and electrochemical separation processes (Volesky & Holan 1995; Fu & Wang 2011; Jun *et al.* 2016). Nevertheless, each of them have their own constraints and may have proven unviable, especially when dealing with relatively less (<100 $mg{\cdot}L^{-1}$) of the contaminant metals

doi: 10.2166/wst.2018.226

(Volesky & Holan 1995). The high operating and reagent cost, and difficulty in treatment of the generated solid waste by the physical and chemical means, is problematic. Additionally, the proven toxicity of heavy metals exposure to living microbes is retardant to biological strategy in wastewater treatment (Fu & Wang 2011). In recent times, the sorption of heavy metals from waste streams onto microbial biomass has considered a new cost-effective technique of industrial relevance for removal and recovery of heavy metals, as well (Azizi *et al.* 2016). Previous studies revealed that metal uptake capacity by microorganisms usually follows passive adsorption of metal ions via binding with several functional groups (Chatterjee *et al.* 2010). To exhibit an efficient adsorption by biomass from a contaminant stream, several factors viz. pH, temperature, solubility of metal ions, their concentration and aqueous speciation have been identified to play key roles in the biosorption process (Aravindhan *et al.* 2009; Lesmana *et al.* 2009; Wang & Chen 2009; Chatterjee *et al.* 2010; Azizi *et al.* 2016). Using biomass of bacteria, fungi, algae and yeasts, the biosorption study for removal of various heavy metals such as cadmium, lead, chromium, uranium, arsenic, copper, nickel, iron, zinc, etc. from the divergent effluent streams has been performed (Lesmana *et al.* 2009; Bayramoglu *et al.* 2016). The ubiquitous eukaryotic organisms, containing chlorophyll and carrying oxygenic photosynthesis, named green algae (Wang & Chen 2009), fulfilling these criteria have been extensively studied for heavy metals removal or their recovery from waste effluent streams (Aravindhan *et al.* 2009). The presence of large amounts of proteins, lipids, and polysaccharides on the cell walls along with containing a number of functional groups (such as hydroxyl, amino, carboxyl, sulfate, and phosphate) can potentially bind the metal ions with algal biomass (Bayramoglu *et al.* 2018). Additionally, the studies on modification of adsorbents surface and functional groups have shown significant improvement in their performance and sorption capacity (Bayramoglu *et al.* 2015).

Therefore, in the present study, the treatment of a typical electroplating rinse water discharge containing 194 $mg{\cdot}L^{-1}$ copper in sulfate medium was investigated using biosorption technique. In which, the biomass of green filamentous algae, *spirogyra* as the potential sorbent was employed with aiming to assess the removal efficiency of copper from the effluent discharge. The sorption behaviour exhibited by *spirogyra* in its native and modified form as a function of parametric variation viz. effluent pH, contact time, adsorbent dosage, and temperature are being revealed and also used to determine the isotherm, kinetics and thermodynamics of the biosorption process. The present study can also be used for other metal-contaminant systems by employing the native and modified biomass of *spirogyra*.

EXPERIMENTAL

Industrial effluent

The copper bearing rinse water effluent, generated by a local electroplating industry, was used in the study as the sorbate feed solution. The effluent contains 194 $mg{\cdot}L^{-1}$ copper and also contains a lesser amount of <5 $mg{\cdot}L^{-1}$ nickel and zinc in sulfate medium. The characteristics of the industrial electroplating solution as supplied were: pH, 2.4; total dissolved solid, 56,800 $mg{\cdot}L^{-1}$; conductivity, 4.16 $\mu S{\cdot}cm^{-1}$; and chemical oxygen demand, 1,540 $mg{\cdot}L^{-1}$. When needed to vary the concentration of copper in solution, the synthetic solutions of desired copper content were prepared by dissolving the calculated amount of $CuSO_4{\cdot}5H_2O$ (supplied by Sigma Aldrich) in distilled water. The pH of the solution was maintained by 5 vol.% H_2SO_4 and NaOH (supplied by Riedel-deHaën and Lab Scan, respectively).

THE BIOMASS

Spirogyra belonging to the group of green filamentous algae and chlorophyceae family was used as the biosorbent in this study. The algal biomass was collected from the fish farm located at the University of Agriculture Faisalabad campus. The collected algae were water washed to remove the sand and other unwanted materials and squeezed to remove the water. Further, the biomass was sun-dried for 7 days and crushed to get 300 μm size of the biosorbent in its native form. In order to prepare the modified biomass as cross-linked gel, a 500 g algal sample was treated overnight with 1 L conc. H_2SO_4 (as the cross-linking agent) at 100 °C. It should be noted that the algal cell wall is basically composed of two layers: outer pectose layer and inner cellulose layer. By the treatment of native biomass with conc. H_2SO_4 (under the specific condition), cross-linked algal gel was obtained that have structural modification on cell wall that can improve the tendency of metal adsorption with the modified biomass. Subsequently, the filtered mass was neutralized with sodium bicarbonate and washed with distilled water until neutral pH was achieved. There after the biomass was dried at 50 °C for 3 h and crushed to 300 μm size that was used as the cross-linked modified *spirogyra* for biosorption tests of copper. The scheme of sample preparation and difference between their structures are presented in Figure 1. Further, the specific surface area of the native and modified *spirogyra* was determined by N_2 adsorption using the Brunauer–Emmett–Teller (BET) method, which was obtained to be 1.88 cm^2 and 3.42 cm^2, respectively.

Batch biosorption studies

All the batch sorption study was carried out at 100 mL scale in 250 mL conical flasks at room temperature (20 °C) and a fixed rotation speed 200 rpm in an orbital shaker (Kuhner-LT-X), until mentioned specifically. A predetermined amount of biomass (varied in the range of 1 g/L to 100 g/L pulp density) was introduced to the sorbate solution of 194 $mg{\cdot}L^{-1}$ Cu^{2+} at varied pH ranging from 2.0 to 8.0 and mixed for 30 min. However, the effect of contact time on sorption kinetics was later investigated separately for different duration from 2 min to 30 min. Whereas the thermodynamics parameters for using both native and modified *spirogyra* were studied under the varied temperature range of 20–50 °C, in which a fixed 6 g biomass in 100 mL sorbate solution of pH 6.0 was contacted for a duration of 30 min. After the completion of an individual set of sorption experiments, the samples were centrifuged (Kubota-2420) at 5,000 rpm for 10 min and the supernatant liquid was separated for analysis of the copper content in solution. An atomic absorption spectroscopy (AAS, Varian AA400) was used for this purpose. Based on the analytical results, the amount of copper adsorbed with the biomass was calculated using the equation:

$$q_e = (C_i - C_e) \times \frac{V}{W} \tag{1}$$

where q_e ($mg{\cdot}g^{-1}$) is the amount of metal adsorbed by biomass, C_i and C_e ($mg{\cdot}L^{-1}$) are the initial and equilibrium liquid phase concentrations of the metal, respectively, V (L), the initial volume of metal solution, and W (g), the weight of the biomass.

Regeneration and desorption studies

After the optimization of batch sorption study, desorption study of only modified algal biomass was carried out by contacting the metal-loaded biomass with 1.0 $mol{\cdot}L^{-1}$ HCl solutions. For this, the metal-loaded biomass separated after filtration was air dried overnight at room temperature and taken for desorption study. The biomass was introduced to acid solution (pre-heated at 50 °C) at a solid-to-liquid ratio of 1:4, and contacted for 60 min under a constant rotation speed of 200 rpm in an orbital shaker. After that the samples were centrifuged at 5,000 rpm for 10 min and the

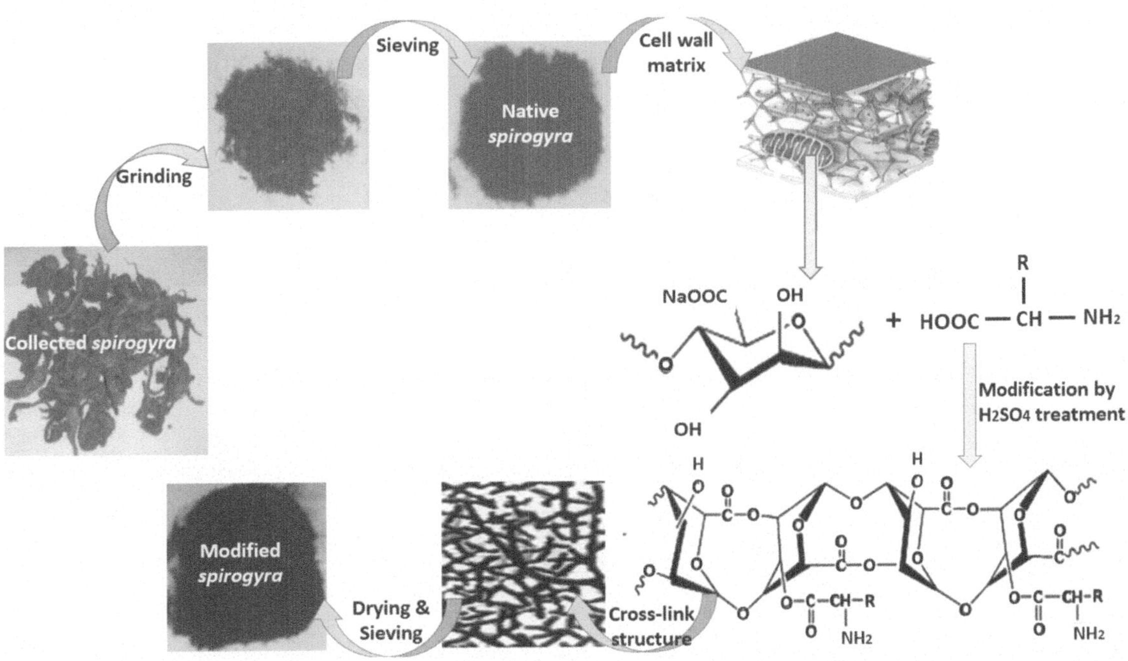

Figure 1 | Schematic preparation of biomass samples and representation of the structural changes in modified *spirogyra* after sulfuric acid treatment of the native *spirogyra*.

supernatant liquid was separated for analysis of the amount of copper eluted from the biomass, and the regeneration efficiency was determined. Four consecutive cycles of such sorption–desorption were performed with the same conditions; however, a washing with 10% H_2SO_4 was given each time of biomass recycling for adsorption.

RESULTS AND DISCUSSION

Effect of pH

By in-large the metal speciation and the active group attached on the cell wall sites primarily depends on pH of the sorption system (Hossain *et al.* 2014). Hence, the influence of pH on sorption behavior of copper onto the biomass of native and modified *spirogyra* was investigated in the range of 2.0 to 8.0. The other parameters viz. the biomass dosage 4 g/100 mL solution containing 194 mg·L^{-1} Cu^{2+} and contact time 30 min at room temperature 20 °C were invariably maintained. The results presented in Figure 2 shows a significant change in sorption capacity with respect to the sorption pH. The percentage sorption of copper was 30.1% and 32.5% by native and modified biomass, respectively, which surpasses above 50% at pH 4.0 and reached its maximum levels of 74.8% and 84.7% at a pH 6.0. A further increase in pH could decrease the sorption, declined to 64% and 72% at pH 8.0 using the biomass in native and modified form, respectively. Such results can directly be corroborated to the surface adsorption sites of the functional group attached with cell wall of biomass (Gupta & Rastogi 2008; Hossain *et al.* 2014; Celekli *et al.* 2016), hence a comparative Fourier transform infrared (FTIR) spectra of the biomasses before and after contacting the effluent were acquired and analyzed.

The spectra of native and modified *spirogyra* were acquired by FTIR photometer within the IR range and shown in Figure 3, which clearly depicts the presence of carboxyl, amino, hydroxyl and carbonyl groups in the biomasses. In both types of biomass, the band displayed at 3,639, 3,542, and stretching to 2,889 cm^{-1} can be attributed to –OH, –COOH, and –CH and –COOH groups, respectively (Gupta & Rastogi 2008; Celekli *et al.* 2016; Bayramoglu *et al.* 2018). The other common bands of the spectra could be assigned as: –N–H bending at 1,462 cm^{-1}; $-CH_3$ stretching at 1,379 cm^{-1}, overlapping of the stretching vibrational band for $-SO_3$ that represents the sulfonic acids of polysaccharides and NOO stretch at 1,274 cm^{-1}; >C=N; and >C=C stretching vibrations at 1,027 cm^{-1} (Gupta & Rastogi 2008; Celekli *et al.* 2016). However, the band appeared at 1,642 cm^{-1} (in spectra 2a and 2b) attributing the $-NH_2$ or, –C–N amide groups for the native *spirogyra* was found to be disappeared with the modified *spirogyra* (as shown in spectra 2c and 2d). A decrease in intensity and peak stretching at 2,889 cm^{-1} was also observed with modified *spirogyra*. This difference is certainly due to the treatment of native *spirogyra* with conc. sulfuric acid that could take out the amides via complexation. When comparing the spectra of before and after adsorption of Cu^{2+} ions with native and modified *spirogyra*, very significant changes in the band regions attributed to –OH, –COOH, and –CH groups are direct evidence for the adsorption of metal cation takes place via the binding with these active groups onto the cell walls (Bayramoglu *et al.* 2015).

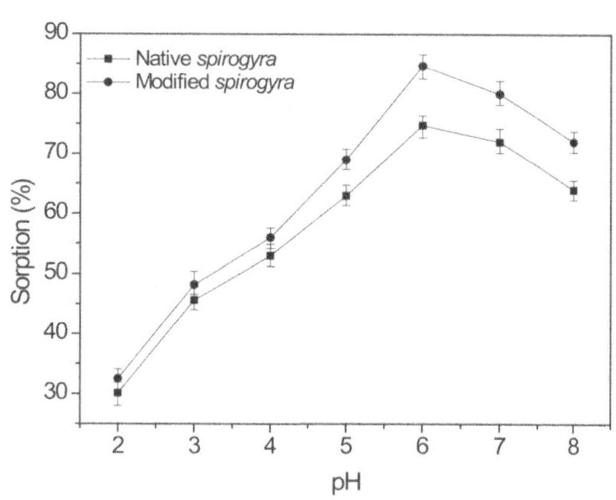

Figure 2 | Adsorption efficiency of copper using the native and modified biomass of *spirogyra* with respect to the pH of solution (conditions: biomass dosage 4 g/100 mL solution containing 194 mg·L^{-1} Cu^{2+}, contact time 30 min, temperature 20 °C, and agitation speed 200 rpm).

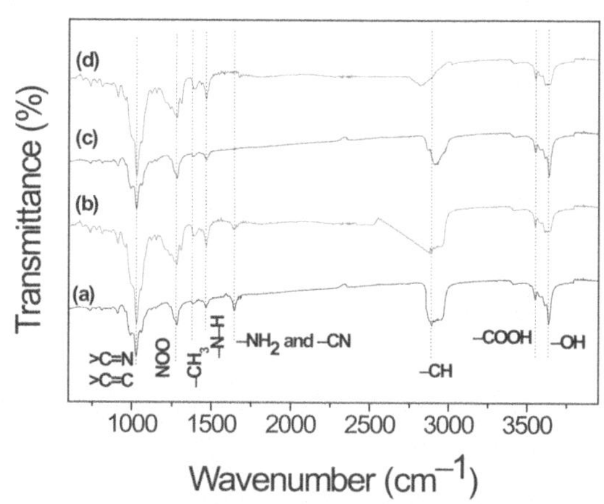

Figure 3 | FTIR spectra of (a) native *spirogyra* before adsorption, (b) native *spirogyra* after adsorption, (c) modified *spirogyra* before adsorption, (d) modified *spirogyra* after adsorption.

Thus, the sorption behavior exhibited as a function of pH suggests that at lower pH binding sites of the biomass cell becomes protonated, resulting in restriction for the metal cation Cu^{2+} to approach sorption sites (Jayakumar *et al.* 2015). With an increase in pH, more ligands with negative charges on active groups are expected to be exposed to attract the metal cation for its binding (Gupta & Rastogi 2008; Jayakumar *et al.* 2015; Celekli *et al.* 2016). However, the subsequent decline in sorption with pH above 6.0 can be attributed to the reduced solubility of Cu^{2+} and precipitation of metal hydroxide.

Effect of biosorbent dosage

The effect of biosorbent amounts on the removal efficiency of copper from the effluent stream was investigated under the range of 1–10 g biosorbent dosage added to a 100 mL solution. The other parameters viz. metal concentration 194 mg·L^{-1} Cu^{2+} in solution of pH 6.0, contact time 30 min, and temperature 20 °C were invariably maintained. The results presented in Figure 4 show that the sorption of copper increased with respect to the increasing amount of biosorbent up to 6 g/100 mL and further remains constant. The maximum sorption efficiency with native *spirogyra* could be achieved from 56% to ~83%; whereas it yielded 63% to more than 96% with the modified *spirogyra* when biosorbent dosage was increased from 1 g to ≥6 g. The behavior exhibited by both types of sorbent can be described by the increase of available sorption sites with increasing the amount of sorbent (Srivastava *et al.* 2012); however, at higher population of the biomass (>6 g) in the pulp may result in aggregation of biomass to decrease the surface area of interaction and thus no more improvement in metal adsorption (Jayakumar *et al.* 2015). At the chosen dosage of 6 g/100 mL, the sorption of copper with modified biomass was approximately 15% higher than that with the native *spirogyra* (82.8%). A higher biosorption with modified *spirogyra* can be attributed to the difference in its surface area with the native form, mainly due to oxidation of the green algae with H_2SO_4 which induces the reaction of phenolic –OH group present on the surface of *spirogyra*.

Effect of contact time

The batch sorption efficiency of the biomasses at different contact times was investigated from 2 to 30 min by exposing 100 mL effluent (194 mg·L^{-1} Cu^{2+}) to a 6 g sorbent at pH 6.0 and temperature of 20 °C. Figure 5 displays the copper sorption rate obtained using both native and modified biomasses of *spirogyra*. It can be seen that the removal of copper from effluent started to increase rapidly at the initial stage of the sorption process until 15 min, reaching near to 80% metal adsorption. Subsequently, it increased with a slow rate to attain equilibrium after a 20 min contact with native and 25 min contact time with the modified biomass that could remove approximately 82% and 96% of copper, respectively. For the engineering design of any sorption system, knowing the sorption rate is an important factor, hence the kinetics analysis of the copper removal rate was also investigated.

Biosorption kinetics

Based on the uptake capacity exhibited by the biomasses, biosorption kinetics can be described by pseudo-first-order

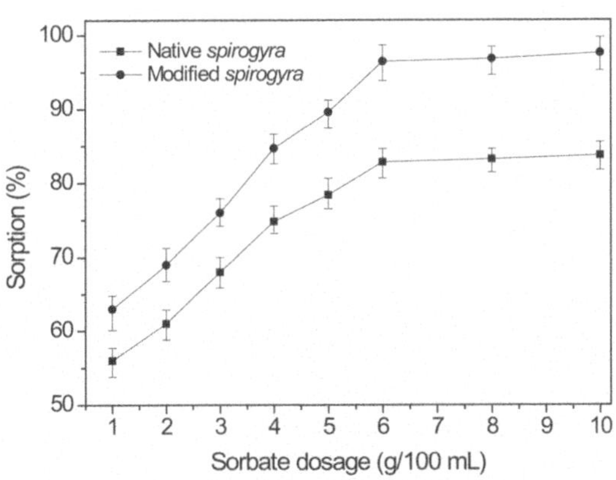

Figure 4 | Adsorption efficiency of copper using the native and modified biomass of *spirogyra* with respect to the sorbent dosage (conditions: solution containing 194 mg·L^{-1} Cu^{2+} of pH 6.0, contact time 30 min, temperature 20 °C and agitation speed 200 rpm).

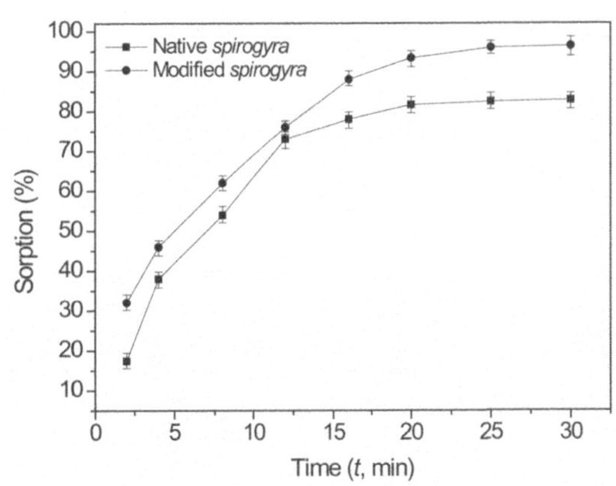

Figure 5 | Adsorption efficiency of copper using the native and modified biomass of *spirogyra* with respect to time (conditions: biomass dosage 6 g/100 mL solution containing 194 mg·L^{-1} Cu^{2+} of pH 6.0, temperature of 20 °C and agitation speed of 200 rpm).

and pseudo-second-order models (Lagergren & Svenska 1898; Blanchard *et al.* 1984), which can be expressed by Equations (2) and (3), respectively, as:

$$\log(q_e - q_t) = \log(q_e) - \frac{k_1}{2.303}t \quad (2)$$

$$\frac{t}{q_t} = \frac{1}{k_2 q_e^2} + \frac{1}{q_e}t \quad (3)$$

where q_t and q_e are the quantity of copper ($mg \cdot g^{-1}$) adsorbed onto biosorbents at any time, t (min) and at equilibrium, respectively. k_1 and k_2 are the pseudo-first order rate constant (min^{-1}) and pseudo-second order rate constant ($g \cdot mg^{-1} \cdot min^{-1}$), respectively that can be determined by the linear plots of $\log(q_e – q_t)$ *vs.* t and t/q_t *vs.* t, respectively using the adsorption data for native and modified *spirogyra*. As can be seen from Figure S1a and S1b (available with the online version of this paper), the adsorption data using the native form of the biomass was in good agreement with the pseudo-first-order model (with a higher correlation coefficient, $R^2 > 0.98$); whereas the modified biomass followed the pseudo-second-order model (with $R^2 > 0.99$). The rate constants and copper uptake values calculated for both types of biomass used in this study are given in Table 1, and revealed a higher metal sorption rate of the modified *spirogyra*.

Sorption mechanism

In an adsorption process, the solute transfer mechanism often may takes place either by external mass transfer or, intra particle diffusion; however, in some cases, both can also occur (Srivastava *et al.* 2013). To evaluate the sorption mechanism, the plot of q_t *vs.* $t^{0.5}$ for intra particle diffusion can be used, as given by Weber & Morris (1963):

$$q_t = k_{id} t^{0.5} \quad (4)$$

where k_{id} is the intra particle diffusion constant ($mg \cdot g^{-1} \cdot min^{0.5}$) that can be obtained from the slope value of q_t *vs.* $t^{0.5}$. As Figure 6 depicts that the liner plot using native *spirogyra* does not pass through the origin and has a lower R^2 value, hence disapproves the occurrence of intra-particle diffusion (Srivastava *et al.* 2013). In that case, the sorption follows external mass transfer through the surfaces of native *spirogyra*. Contrarily, the plot of modified *spirogyra* clearly passes through the origin, an indication for exhibiting the intra particle diffusion for copper adsorption. A difference in sorption mechanism with both forms of the biomass can be attributed to the obtainment of cross-link structural change after the sulfuric acid treatment of native *spirogyra*.

Effect of temperature and sorption thermodynamics

The role of temperature on copper adsorption with native and modified *spirogyra* was in the range of 20–50 °C and keeping other parameters constant as biosorbent dosage 6 g/100 mL, pH 6.0, contact time 30 min. The results presented in Figure 7(a) indicate that the adsorption was maximum (82.8% and 96.4%) at the lowest temperature of 20 °C which showed a decreasing trend with raising the temperature and declined to 51% and 59% at 50 °C by using the native and modified biomass of *spirogyra*, respectively. Based on the sorption data, the thermodynamic parameters of the adsorption system were also investigated. For this, the changes in enthalpy ($\Delta H°$) and entropy ($\Delta S°$) were determined by the slope and intercept values of the van't Hoff plot ($\ln K_{ad}$ *vs.* $1/T$) employing the following equation (Srivastava *et al.* 2013):

$$\ln K_{ad} = -\frac{\Delta H°}{RT} + \frac{\Delta S°}{R} \quad (5)$$

where R is the universal gas constant ($8.314\ J \cdot mol^{-1} \cdot K^{-1}$) and T is the absolute temperature (K). K_{ad} is the adsorption equilibrium constant, which can be calculated by Equation (6), where q_e and C_e are the amount of metal sorbed by biomass, and amount of metal in the liquid phase at

Table 1 | Investigated applicability of kinetic models for the adsorption of copper using native and modified *spirogyra*

		Pseudo-first-order kinetic			Pseudo-second-order kinetic		
Biosorbent used	$q_{e,exp}$ ($mg \cdot g^{-1}$)	k_1 (min^{-1})	$q_{e,cal}$ ($mg \cdot g^{-1}$)	R^2	k_2 ($g \cdot mg^{-1} \cdot min^{-1}$)	$q_{e,cal}$ ($mg \cdot g^{-1}$)	R^2
Native *spirogyra*	160.6	0.224	249.4	0.98	0.0004	232.5	0.96
Modified *spirogyra*	187.0	0.201	275.6	0.94	0.0007	238.1	0.99

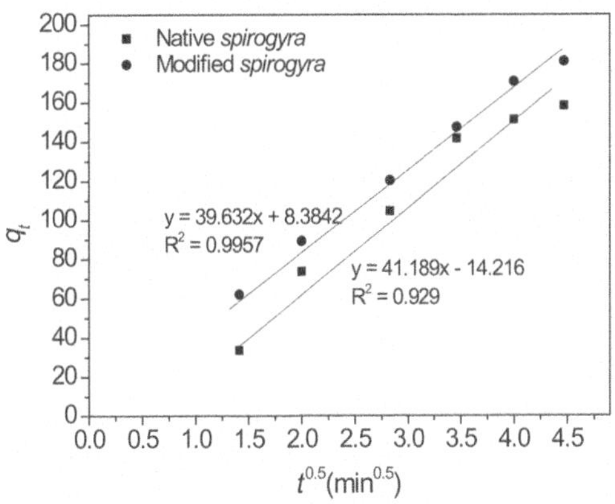

Figure 6 | Plot of q_t vs. $t^{0.5}$ for determining the sorption mechanism of copper using the native and modified biomass of *spirogyra*.

equilibrium, respectively.

$$K_{ad} = \frac{q_e}{C_e} \tag{6}$$

The calculated values of ΔH° (-41.29 kJ·mol^{-1} and -75.06 kJ·mol^{-1} for native and modified *spirogyra*, respectively)

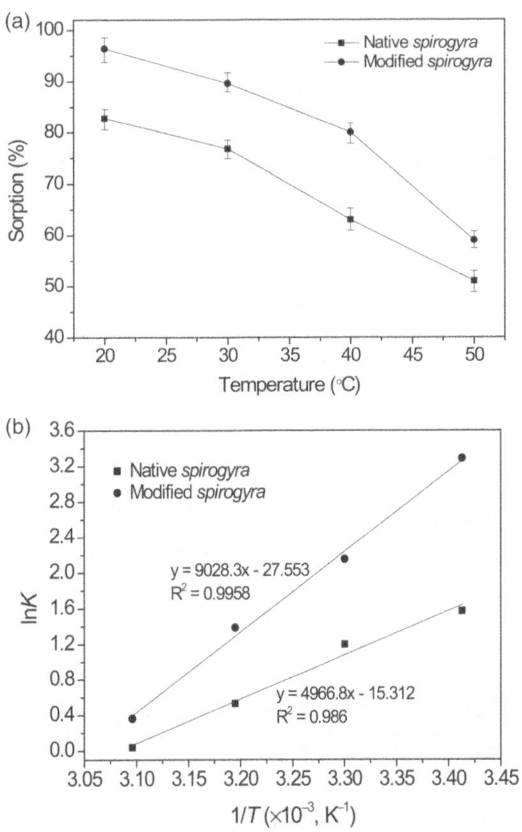

Figure 7 | Adsorption efficiency of copper using the native and modified biomass of *spirogyra* with respect to temperature, (b) van't Hoff plot for lnK_{ad} *vs.* 1/*T* (conditions: biomass dosage 6 g/100 mL solution containing 194 mg·L^{-1} Cu^{2+} of pH 6.0, contact time of 30 min, and agitation speed of 200 rpm).

and ΔS° (-0.13 kJ·mol^{-1}·K^{-1} and -0.23 kJ·mol^{-1}·K^{-1} for native and modified *spirogyra*, respectively) by the linear plot displayed in Figure 7(b) were further used to calculate the change in standard Gibbs free energy as follows:

$$\Delta G^\circ = \Delta H^\circ - T\Delta S^\circ \tag{7}$$

All the calculated thermodynamic parameters are given in Table S1 (available online). The negative values of the thermodynamic properties (ΔG°, ΔH° and ΔS°) indicate the feasibility of a spontaneous process, exhibiting the exothermic biosorption with structural changes on the cell wall surface of the biomasses. The enthalpy of physical and chemical adsorption falls into a range of 2.1–20.9 kJ·mol^{-1} and 80–200 kJ·mol^{-1}, respectively; therefore, the adsorption of copper in this study can be attributed to the physico-chemical adsorption process instead of being only physical or chemical adsorption.

Biosorption isotherm

The distribution of contaminant metal, copper between the effluent solution and biomasses at equilibrium were estimated by using the Langmuir and Freundlich isotherm models. For this, the copper content in the solution varied in the range of 56–308 mg·L^{-1} under the above optimized condition of 6 g biomass/100 mL solution, pH 6.0, contact time 30 min, and temperature 20 °C. The linear form of Langmuir (1918) model can be written as:

$$\frac{C_e}{q_e} = \frac{1}{k_L \cdot q_m} + \frac{C_e}{q_m} \tag{8}$$

where k_L (L·mg^{-1}) and q_m (mg·g^{-1}) are the Langmuir distribution coefficient and maximum sorption coefficient, respectively, which corresponds to the intercept and slope value of the linear plot of C_e/q_e *vs.* C_e.

Another isotherm model given by Freundlich can be expressed in logarithmic form as (Freundlich 1906):

$$\log q_e = \log k_F + \frac{1}{n}\log C_e \tag{9}$$

where k_F (mg$^{1-1/n}$·L$^{1/n}$·g^{-1}) and $1/n$ are Freundlich constants representing the distribution coefficient and heterogeneity constant, respectively, whose value can be obtained by the corresponding intercept and slope values of the linear plot log q_e *vs.* log C_e.

Using the experimental data, graphs were plotted for both models and shown in Figure S2a and S2b (available

Table 2 | Investigated applicability of the sorption isotherms of copper with the native and modified *spirogyra*

	Langmuir isotherm			Freundlich isotherm		
Biosorbent used	q_m (mg·g^{-1})	k_L (L·g^{-1})	R^2	1/*n*	k_F (mg$^{1-1/n}$·L$^{1/n}$·g^{-1})	R^2
Native *spirogyra*	434.7	3.8	0.96	1.5	15.3	0.99
Modified *spirogyra*	312.5	20.8	0.98	2.3	4.7	0.93

online). The analysis of plots indicate that the native biomass better fitted to Freundlich isotherm ($R^2 > 0.99$), revealing the heterogeneity of adsorption (Pathak 2017), whereas the modified biomass well fitted to the Langmuir model, indicating the homogeneous characteristics of the sorption process with cross-link structure of the modified *spirogyra* to exhibit the monolayer adsorption of copper. The values of the coefficients calculated for both isotherms are given in Table 2. Also, the obtained adsorption capacity in this study has been compared with the earlier reported different biomasses capacity (for sorption of Cu^{2+} from various solutions) and presented in Table S2 (available online). It indicates that the sorption capacity obtained with modified biomass in this study is the highest of all.

Desorption and reusability study

The batch sorption study established the higher sorption efficiency with modified algal biomass but in order to determine the reusability of this biomass further, the elution of adsorbed metal ion was evaluated in four consecutive cycles of sorption–desorption. The sorption condition was maintained as 6 g modified sorbent in 100 mL electroplating effluent, time 30 min, pH 6.0, and temperature 20 °C. Subsequently, the separated metal-loaded biomass was introduced to 1.0 mol·L^{-1} HCl solution (preheated at 50 °C) by maintaining a solid-to-liquid ratio of 1:4 and contacted for 60 min. The elution temperature (50 °C) was chosen because of the minimum adsorption observed at this temperature (as given in Figure 7(a)). By contacting the HCl solution, the coordination of Cu^{2+} ions with the active group of biomass gets disrupted to release the metal ions from the adsorbent surface (Bayramoglu *et al.* 2015). The prominent results (>98.5%) for elution of Cu^{2+} ions from the biomass was obtained with the fresh biomass, which was further used for the next three sorption–desorption cycles. In the second use of biomass, the sorption efficiency was slightly lower than fresh (~93.2%) but thereafter until the fourth cycle of study, the regeneration efficiency was good enough with ~90% efficiency.

CONCLUSIONS

In this study, the potential use of green algae *spirogyra* has been investigated for copper removal from an effluent generated by electroplating industry. A comparison between the biosorption capacity of biomasses in the native and modified forms after a sulfuric acid treatment revealed the attainment of cross-linked structural change could enhance the sorption efficiency as compared to the native *spirogyra*. The study showed a maximum sorption at pH 6.0, either increase or decrease in solution pH depicted a significant decline in adsorption of Cu^{2+} was directly affected by the surface adsorption sites of the functional group attached with the cell wall of biomass, which was also confirmed by FTIR spectral analyses. A 6 g sorbent mass treated to 100 mL effluent for 30 min at pH 6.0 and temperature of 20 °C were optimized to yield the maximum 82.8% and 96.4% adsorptive removal of copper from the effluent by the native and modified *spirogyra*, respectively. The sorption data exhibited pseudo-first-order kinetic for multilayer sorption via external surface diffusion onto the native biomass followed the Freundlich isotherm; whereas the modified *spirogyra* showed pseudo-second-order kinetics for monolayer intra particle diffusion followed the Langmuir model. The calculated ΔH° values (−41.29 kJ·mol^{-1} and −75.06 kJ·mol^{-1} for native and modified *spirogyra*, respectively) was attributed to the physico-chemical nature of the spontaneous adsorption with negative values of ΔG°. The sorbed metal was successfully eluted using 1.0 mol·L^{-1} HCl solution at solid-to-liquid ratio of 1:4, 50 °C, and 60 min, and good regeneration efficiency (>90%) was obtained after three stages of recycling.

ACKNOWLEDGEMENTS

This research did not receive any specific grant from funding agencies in the public, commercial, or not-for-profit sectors. The authors declare no competing financial interest.

REFERENCES

Aravindhan, R., Rao, J. R. & Nair, B. U. 2009 Application of chemically modified green macro alga as a biosorbent for phenol removal. *Journal of Environmental Management* **90**, 1877–1883.

Azizi, S., Kamika, I. & Tekere, M. 2016 Evaluation of heavy metal removal from wastewater in a modified packed bed biofilm reactor. *PLoS ONE*. doi.org/10.1371/journal.pone.0155462.

Bayramoglu, G., Akbulut, A. & Arica, M. Y. 2015 Study of polyethyleneimine- and amidoxime-functionalized hybrid biomass of *Spirulina (Arthrospira) platensis* for adsorption of uranium (VI) ion. *Environmental Science and Pollution Research* **22**, 17998–18010.

Bayramoglu, G., Akbulut, A. & Arica, M. Y. 2016 Aminopyridine modified Spirulina platensis biomass for chromium(VI) adsorption in aqueous solution. *Water Science & Technology* **74**, 914–926.

Bayramoglu, G., Akbulut, A., Acikgoz-Erkaya, I. & Arica, M. Y. 2018 Uranium sorption by native and nitrilotriacetate-modified *Bangia atropurpurea* biomass: kinetics and thermodynamics. *Journal of Applied Phycology* **30**, 649–661.

Blanchard, G., Maunaye, M. & Martin, G. 1984 Removal of heavymetals from waters by means of natural zeolites. *Water Research* **18**, 1501–1507.

Celekli, A., Gultekin, E. & Bozkurt, H. 2016 Morphological and biochemical responses of *Spirogyra setiformis* exposed to cadmium. *CLEAN* **44**, 256–262.

Chatterjee, S. K., Bhattacharjee, I. & Chandra, G. 2010 Biosorption of heavy metals from industrial waste water by *Geobacillus thermodenitrificans*. *Journal of Hazardous Materials* **175**, 117–125.

Freundlich, H. 1906 Uber die adsorption in Losungen. *Zeitschrift für Physikalische Chemie* **57A**, 385–470.

Fu, F. & Wang, Q. 2011 Removal of heavy metal ions from wastewaters: a review. *Journal of Environmental Management* **92**, 407–418.

Gupta, V. K. & Rastogi, A. 2008 Biosorption of lead from aqueous solutions by green algae *Spirogyra* species: kinetics and equilibrium studies. *Journal of Hazardous Materials* **152**, 407–414.

Hossain, A., Bhattacharyya, S. R. & Aditya, G. 2014 Biosorption of cadmium by waste shell dust of fresh water mussel *Lamellidens marginalis*: implications for metal bioremediation. *ACS Sustainable Chemistry & Engineering* **3**, 1–8.

Jayakumar, J., Rajasimman, M. & Karthikeyan, C. 2015 Optimization, equilibrium, kinetics, thermodynamics and desorption studied on the sorption of Cu(II) from an aqueous solution using marine green algae: *Halimeda gracilis*. *Ecotoxicology and Environmental Safety* **121**, 199 – 210.

Jun, M., Srivastava, R. R., Jeong, J., Lee, J.-C. & Kim, M.-S. 2016 Simple recycling of copper by the synergistic exploitation of industrial wastes: a step towards sustainability. *Green Chemistry* **18**, 3823–3834.

Lagergren, S. & Svenska, B. K. 1898 Zur theorie der sogenannten adsorption geloester stoffe. *Veternskapsakad Handlinkar* **24**, 1–39.

Langmuir, I. 1918 The adsorption of gases on plane surface of glass, mica and platinum. *Journal of the American Chemical Society* **40**, 1361–1403.

Lesmana, S. O., Febriana, N., Soetaredjo, F. E., Sunarso, J. & Ismadji, S. 2009 Studies on potential applications of biomass for the separation of heavy metals from water and wastewater. *Biochemical Engineering Journal* **44**, 19–41.

Nayak, A. K. & Pal, A. 2017 Green and efficient biosorptive removal of methylene blue by *Abelmoschus esculentus* seed: process optimization and multivariate modelling. *Journal of Environmental Management* **200**, 145–159.

Pathak, P. 2017 An assessment of strontium sorption onto bentonite buffer material in waste repository. *Environmental Science and Pollution Research* **24**, 8825–8836.

Srivastava, R. R., Mittal, N., Padh, B. & Reddy, B. R. 2012 Removal of tungsten and other impurities from spent HDS catalyst leach liquor by an adsorption route. *Hydrometallurgy* **127**, 77–83.

Srivastava, R. R., Kim, M.-S. & Lee, J.-C. 2013 Separation of Tungsten from Mo-Rich leach liquor by adsorption onto a typical Fe–Mn cake: kinetics, equilibrium, mechanism, and thermodynamics studies. *Industrial & Engineering Chemistry Research* **52**, 17591–17597.

Volesky, B. & Holan, Z. R. 1995 Biosorption of heavy metals. *Biotechnology Progress* **11**, 235–250.

Wang, J. & Chen, C. 2009 Biosorbents for heavy metals removal and their future. *Biotechnology Advances* **27**, 195–226.

Weber Jr, W. J. & Morris, J. C. 1963 Kinetics of adsorption on carbon from solution. *Journal of the Sanitary Engineering Division of American Society of Civil Engineers* **89**, 31–60.

Yu, B., Wang, Y. & Zhou, Q. 2014 Human health risk assessment based on toxicity characteristic leaching procedure and simple bioaccessibility extraction test of toxic metals in urban street dust of Tianjin, China. *PLoS ONE*. doi:10.1371/journal.pone.0092459.

First received 5 September 2017; accepted in revised form 8 May 2018. Available online 17 May 2018

Effective adsorption of nickel (II) with *Ulva lactuca* dried biomass: isotherms, kinetics and mechanisms

Jianyou Long, Xiaona Huang, Xiaoli Fan, Yan Peng and Jianrong Xia

ABSTRACT

This study aimed to evaluate the Ni^{2+} ions adsorption capability of *Ulva lactuca*. The isotherms, kinetics and mechanisms for the adsorption of Ni^{2+} from aqueous solution by *Ulva lactuca* were also investigated. Influencing factors including initial pH, initial Ni^{2+} concentration, biomass, contact time were examined. The results indicate that the maximum Ni^{2+} adsorption capacity of 38.28 mg/g was obtained at pH 5, initial Ni^{2+} concentration 250 mg/L, biomass dosage 0.5 g/L and contact time 30 min. The adsorption can be well fitted with Langmuir isotherm, and the kinetics were well described by the pseudo-second-order model. The parameters of thermodynamics verified that Ni^{2+} adsorption on *Ulva lactuca* was a spontaneous and endothermic process. Analyses of FT-IR, SEM-EDS and XPS indicate that carboxyl and hydroxyl groups on the surface of biomass are involved in Ni^{2+} adsorption. The dried biomass of *Ulva lactuca* can be a cost-effective and eco-friendly adsorbent for the removal of Ni^{2+} from wastewater.

Key words | isotherm, kinetics, mechanisms, Ni^{2+} adsorption, *Ulva lactuca*

Jianyou Long
Xiaona Huang
Xiaoli Fan
Yan Peng
Jianrong Xia (corresponding author)
School of Environmental Science and Engineering, Guangzhou University,
Guangzhou, 510006,
China
E-mail: *jrxia@gzhu.edu.cn*

Jianyou Long
Key Laboratory for Water Quality and Conservation of the Pearl River Delta, Ministry of Education,
Guangzhou University,
Guangzhou, 510006,
China

Yan Peng
Guangdong Provincial Key Laboratory of radionuclides pollution control and resources,
Guangzhou University,
Guangzhou, 510006,
China

INTRODUCTION

Heavy metal pollution has been a growing concern as a result of rapid industrialization worldwide (Tatsuki *et al.* 2017). Many heavy metals cause a great threat to human and other living organisms because of their carcinogenicity (Ghasemi *et al.* 2011), teratogenicity (Anastopoulos & Kyzas 2015), mutagenicity and biomagnification (He & Chen 2014; Kumar *et al.* 2015) in the food chain. Nickel (II) ions are frequently discharged into the environment from industries such as mineral processing, battery manufacturing, steam-electric power plants and electroplating (Aksu *et al.* 2002; Liu *et al.* 2009; Geetha *et al.* 2016). It can result in different types of disease, e.g. pulmonary fibrosis, renaledema, skin dermatitis, and gastrointestinal distress (Bulgariu & Bulgariu 2012). Therefore, it is necessary to remove Ni^{2+} from wastewater before discharge.

Conventional techniques for removing heavy metals from water include chemical precipitation (Tang *et al.* 2017), ion exchange (Jayakumar *et al.* 2015), solvent extraction, filtration (Sarı *et al.* 2011), membrane technologies (Oliveira *et al.* 2011) and electrochemical treatment (Tuzen & Sari 2010). But these proposed approaches are usually expensive and not eco-friendly, especially in the treatment of low concentration heavy metal effluents. In recent years, adsorption techniques with some natural biomaterials, including agricultural waste products and seaweeds, were reported to be utilized as adsorbents for the removal of heavy metals from the wastewaters (Vijayaraghavan *et al.* 2006). Among these materials, macroalgal biomass has been proven to be a promising adsorbent in removing heavy metals because of its high adsorption capacity and low cost (Akbari *et al.* 2015). The high adsorption capacity could be mainly attributed to its biochemical constitutions, especially its cell wall, which contains functional groups (such as amino, carboxyl and hydroxyl) that play an important role in the adsorption process (Daneshvar *et al.* 2017; Vijayaraghavan *et al.* 2017). Marine macroalgae can be divided into several sub-groups such as brown algae (Phaeophyta), red algae (Rhodophyta) and green algae (Chlorophyta). Different types of algae have been tested for removal of heavy metals with different effectiveness (Mohammad *et al.* 2014; Gutiérrez *et al.* 2015; Ahmad *et al.* 2016). It has been shown that the inactive biomass may be

doi: 10.2166/wst.2018.253

even more advantageous than active one in the removal of heavy metals because of no biotoxicity concerns and no requirements of growth media (Abdel-Aty *et al.* 2013). Therefore, non-living alga has become one of the popular choices among various adsorbents for disposing the heavy metals from effluents (Nath & Ray 2015).

Ulva lactuca, a green tide alga, widely grew in the southern coast of China with a large biomass. However, only a small portion of *Ulva lactuca* is used for food or forage. The biomass of *Ulva lactuca* was used for Cu^{2+}, Mn^{2+} and Cr^{3+} adsorption (Zeroual *et al.* 2003; Amany *et al.* 2007; Andrew *et al.* 2007). But little attention has been paid to use the algal material as adsorbent for Ni^{2+} removal (Hanan 2008). In the present study, we aimed to test the feasibility of *U. lactuca* as an adsorbent for removing Ni^{2+}. The influencing factors, including initial pH, initial Ni^{2+} concentration, sorbent biomass and contact time, on the adsorption performance were examined. In addition, to gain a better understanding of the adsorption mechanisms, the isotherms, kinetics and thermodynamics were also investigated. The surface structure and functional groups were characterized by scanning electron microscopy (SEM), Fourier transform infrared spectroscopy (FT-IR) and X-ray photoelectron spectroscopy (XPS) analyses.

MATERIALS AND METHODS

Chemicals and reagents

All chemicals and reagents used were of analytical grade and purchased from Guangzhou Huaxin Reagent Co. Ltd, Guangdong, China. Standard solution (1,000 mg/L) of Ni^{2+} was prepared with nickel nitrate ($Ni(NO_3)_2$). The working solution was serially diluted from standard solution daily. The deionized water used in the study was obtained from a deionized (DI) water generator (YL05-20 L 75G, Shenzhen, China).

Preparation of adsorbent

Biomass of *U. lactuca* was collected from Nan Ao Island of Shantou, Guangdong, China. The biomass was cleaned by washing repeatedly with deionized water for removing impurities, and then dried in oven at 80 °C for 6 h. Finally, the dried biomasses were crushed, sieved and stored at room temperature in an airtight container for future use.

Batch adsorption experiments

The batch experiments on Ni^{2+} adsorption were conducted in a temperature-controlled shaker. In order to determine the optimal adsorption conditions, initial pH (2–8), biomass dosage (0.5–5 g/L), initial Ni^{2+} concentration (0–400 mg/L), temperature (288 K, 298 K, 308 K) and contact time (0–250 min) were tested. Solutions of 1 M HCl and 1 M NaOH were used for pH adjustment, and all the experiments were conducted in triplicate.

The concentration of Ni^{2+} was detected by inductively coupled plasma-atomic emission spectroscopy (ICP-AES), and the adsorption capacity of Ni^{2+} was calculated according to the following equation (Lou *et al.* 2015):

$$q_e = \frac{(C_o - C_e)V}{W} \quad (1)$$

where q_e (mg/g) is the adsorption capacity of Ni^{2+} at equilibrium, C_0 and C_e (mg/L) are the initial and equilibrium concentration of Ni^{2+} in the solution, respectively. V (L) is the volume of the aqueous solution, and W (g) is the mass of adsorbent.

Adsorption isotherms

Adsorption isotherm is one important method to illuminate the mechanism of the adsorption systems, which describes how adsorbate interacts with the adsorbent under the equilibrium state (Yeslié *et al.* 2011). In this study, the two most often used adsorption isotherms, namely Langmuir and Freundlich models, were adopted to describe the adsorption process. The Langmuir model defines the monolayer sorption, and could be described by the following equation (Bakatula *et al.* 2014; Li *et al.* 2017a):

$$\frac{C_e}{q_e} = \frac{1}{q_{\max}b} + \frac{C_e}{q_{\max}} \quad (2)$$

where q_{max} is the maximum biosorption capacity (mg/g) and b is the constant of Langmuir isotherm which is closely related to the energy of adsorption.

The Freundlich model describes the nonideal adsorption on heterogeneous surface as well as multilayer adsorption; its equation is listed below (Zhang *et al.* 2015).

$$\mathrm{Inq_e} = \frac{1}{\mathrm{n}}\mathrm{InC_e} + \mathrm{InK_f} \quad (3)$$

where K_f (mg/g) is a constant related to the adsorption capacity, n is the constant related to the adsorption intensity.

Adsorption kinetics

In order to evaluate the adsorption rate of Ni^{2+} adsorption, two widely accepted kinetic models, pseudo-first-order and pseudo-second-order were applied to fit the adsorption capacity over the reaction time. The linear equation of pseudo-first-order model is expressed in the following formula (Shroff & Vaidya 2011).

$$\log(q_e - q_t) = \log q_e - \frac{k_1 t}{2.303} \quad (4)$$

where q_t (mg/g) is the adsorption capacity at time t, and k_1 (h^{-1}) is the rate constant. The formula of pseudo-second-order model can be expressed as the following equation (Daneshvar *et al.* 2012).

$$\frac{t}{q_t} = \frac{1}{k_2 q_e^2} + \frac{t}{q_e} \quad (5)$$

where k_2 (g/mg/h) is the second-order rate constant of the adsorption process.

Thermodynamic studies

The thermodynamics for the adsorption of Ni^{2+} onto *Ulva lactuca* were studied at 288 K, 298 K and 308 K, respectively. The thermodynamic parameters change of enthalpy (ΔH^0), entropy (ΔS^0) and Gibbs free energy (ΔG^0) can be calculated from the Equations (6)–(8). (Mitrogiannis *et al.* 2015; Ioannis & George 2016).

$$\Delta G^0 = -\text{RTIn}K_d \quad (6)$$

$$\ln K_d = \frac{\Delta S^0}{R} - \frac{\Delta H^0}{RT} \quad (7)$$

$$\Delta G^0 = \Delta H^0 - T\Delta S^0 \quad (8)$$

where K_d is the equilibrium adsorption distribution constant (K_d = qe/Ce), *R* is the gas constant (8.314 J/mol/K) and T is absolute temperature in Kelvin.

Characterization of *U. lactuca* adsorbent

FT-IR spectra (ALPHA-T, Bruker, Germany) were obtained to explore the surface functional groups presented in *U. lactuca* adsorbent in the range of 400–4,000 cm^{-1}. SEM-EDS (QUANTA200, USA) was used to investigate the morphological characteristics of adsorbent before and after adsorption of Ni^{2+} (Cho *et al.* 2013).

The XPS of adsorbent before and after Ni^{2+} adsorption were obtained by the Theta Probe Angle-Resolved XPS System (hv = 1,486.6 eV, Kratos Axis Ultra, Japan) which was used to identify the element and compounds of *U. lactuca*. Correction of the deviation of the binding energy (BE) owing to the relative surface charging was performed with the C 1 s level at BE of 284.6 eV as an internal standard. The fitting of all XPS spectra was conducted using the software XPSPEAK4.1 (Li *et al.* 2017a, 2017b).

RESULTS AND DISCUSSION

Influence of pH

Many studies have shown that the pH value of solution is one of the key parameters in the adsorption of metal ions which affects the metal binding sites of the sorbent and the metal ion chemistry in aqueous solution (Long *et al.* 2017; Yang *et al.* 2017; Zhang *et al.* 2018). In this study, the influence of initial pH on Ni^{2+} adsorption was conducted by varying the solution pH from 2 to 8 (Figure 1). The adsorption capacity for Ni^{2+} was low at pH less than 3; this could be due to the competition of H^+ over Ni^{2+} to bond the adsorption sites, causing the decrease in Ni^{2+} adsorption performance (Long *et al.* 2017). With the pH increase of the solution, more negatively charged on the surface of *U. lactuca* can be available for the ion-exchange of Ni^{2+}, resulting in the increase of adsorption for Ni^{2+}. As the pH increased the adsorption capacity reached a plateau and no longer increased. This may be explained by the

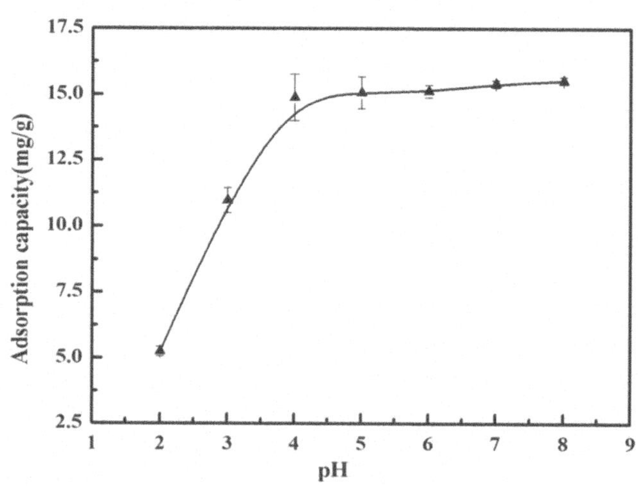

Figure 1 | Influence of solution pH on the adsorption of Ni^{2+} by *U. lactuca*.

formation of soluble hydroxylated complexes of the metal ions and their competition with the active sites on *U. lactuca*, and finally resulting in no increase upon adsorption capacity (Chen *et al.* 2008; Daneshvar *et al.* 2017).

Influence of initial Ni^{2+} concentration

The adsorption capacity increased with the increased Ni^{2+} concentration (Figure 2). This is due to the fact that the increasing initial concentration produces driving force to overcome the mass transfer resistance of Ni^{2+} between the aqueous and solid phase. Then, when the Ni^{2+} concentration further increased from 250 to 400 mg/L, there was no obvious variation on the adsorption efficiency, which may be attributed to the fact that maximum adsorption capacity was reached at the high Ni^{2+} concentration. Similar observations have been found by Tounsadi *et al.* (2015) during biosorption of Cd (II) and Co (II) on *Glebionis coronaria* L. biomasses.

Influence of biosorbent dosage

The biosorbent dosage was also an important parameter affecting the adsorption capacity as well as the removal efficiency. As shown in Figure S1 (available with the online version of this paper), the adsorption capacity decreased gradually as the biomass dosage increased, due to the fact that the adsorption became saturated with increasing adsorbent dosage. The available binding sites for limited adsorbate are excessive when the biomass is in high dosage, and thus leads to the decrease in adsorption capacity during the adsorption process (Christoforidis *et al.* 2015; Chen *et al.* 2008).

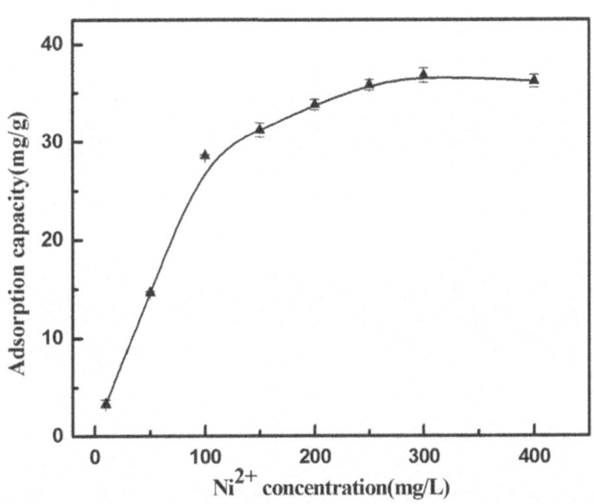

Figure 2 | Influence of initial concentration of Ni^{2+} on the adsorption by *U. Lactuca.*

Influence of contact time

The amount of adsorbed Ni^{2+} increased rapidly during the initial 30 min, and was kept constant when the contact time was further increased (Figure 3). This may be related to the availability of adsorption sites on the biomass surface. At first, adsorption capacity increased quickly owing to sufficient active sites in the beginning; afterward, these active binding sites were gradually occupied (Akbari *et al.* 2015). Therefore, adsorption capacity was kept invariable in the latter adsorption stage.

Adsorption isotherms

For all tested temperatures, the Langmuir model better described the adsorption process with a higher R^2 value of 0.9853, 0.9861 and 0.9819 than the Freundlich model (Table 1 and Figure S2, available online). The maximum adsorption capacity reached 38.28 mg/g at 308 K, and the constant b was 0.0347, respectively. It should be noted that the

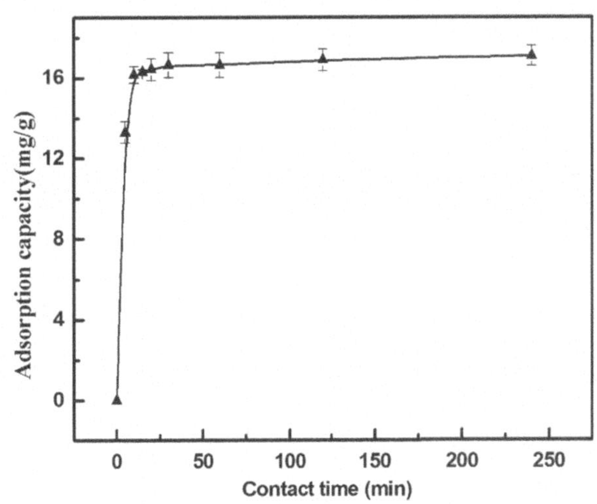

Figure 3 | Influence of contact time on the adsorption of Ni^{2+} by *U. lactuca.*

Table 1 | Isotherm constants for the adsorption of Ni^{2+} by *U. lactuca* at different temperatures

		Temperatures		
Isotherm model	**Parameter**	**288 K**	**298 K**	**308 K**
Langmuir	q_{max}(mg/g)	29.02	32.19	38.28
	b(L/mg)	0.0203	0.0211	0.0347
	R^2	0.9853	0.9861	0.9819
Freundlich	K_f (mg/g)	1.857	1.334	1.363
	n	2.399	1.745	1.722
	R^2	0.9669	0.9534	0.9569

data obtained by both isotherms demonstrate that the higher temperature was more beneficial to the adsorption under the tested conditions, which might indicate the endothermic nature during the adsorption process (Masoumeh *et al.* 2015).

Adsorption kinetics and thermodynamics

The R^2 value (0.999) of the pseudo-second-order kinetic model was higher than that ($R^2 = 0.926$) of pseudo-first-order (Table 2 and Figure S3, available online). This suggests that the pseudo-second-order kinetic model was better in describing the kinetics of Ni^{2+} adsorption. Similar results have been documented for various types of algae adsorbent reported by Pang *et al.* (2011).

The negative value of ΔG^0 at three temperatures (Table 3) indicated that the adsorption of Ni^{2+} was spontaneous (Henriques *et al.* 2015). The positive values of ΔH^0 showed the endothermic nature of the biosorption. The positive values of ΔS^0 suggest increased randomness at the solid/solution interface during the adsorption of Ni^{2+} on the biomass of *U. lactuca*.

Characterization of the adsorbent

FT-IR analysis

The broad and strong peaks around 3,404 cm^{-1} were assigned to the –OH groups (Xu *et al.* 2016) (Figure 4). The bands near

Table 2 | Kinetics constants for the adsorption of Ni^{2+} by *U. lactuca* at different temperatures

		Temperatures		
Kinetics model	**Parameter**	**288 K**	**298 K**	**308 K**
Pseudo-first-order	q_e (mg/g)	10.47	16.34	22.96
	K_1(1/min)	0.1798	0.1833	0.1864
	R^2	0.9142	0.9287	0.9368
Pseudo-second-order	qe (mg/g)	25.23	29.36	35.82
	K_2(g/mg/min)	0.0619	0.0648	0.0723
	R^2	0.9872	0.9913	0.9979

Table 3 | Thermodynamic parameters of Ni^{2+} adsorption by *U. lactuca* at different temperatures

T (K)	$\triangle G^0$(kJ/mol)	$\triangle H^0$(kJ/mol)	$\triangle S^0$(J/mol K)
288	−48.61		
298	−52.28	37.53	0.287
308	−88.86		

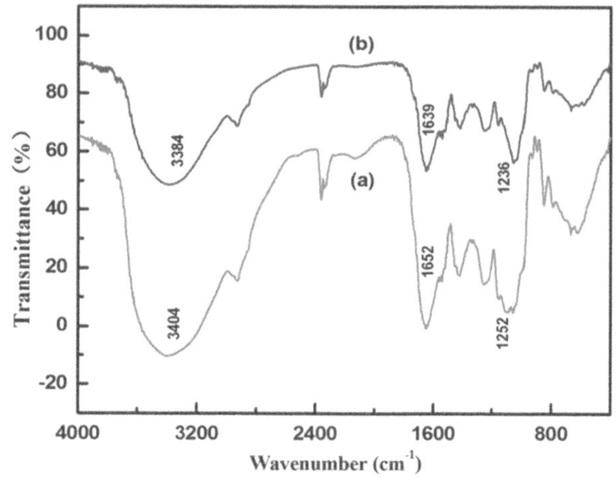

Figure 4 | FT-IR spectra of *U. lactuca* before (a) and after (b) adsorption of Ni^{2+}.

1,652 cm^{-1} could be attributed to the stretching vibration of C=O groups. The band near 1,252 cm^{-1} was assigned to C-O stretching (Castro *et al.* 2017). After Ni^{2+} adsorption, the stretching vibration bands of –OH groups were shifted to 3,384 cm^{-1} for Ni^{2+}-loaded biomass, the peaks of C=O and C-O groups were shifted to 1,639 cm^{-1} and 1,236 cm^{-1}, respectively. This indicates that the chemical interactions between Ni^{2+} and –OH or –NH, C=O and C-O groups of the biomass mainly occurred in the adsorption. The similar FT-IR results were reported for the adsorption of Ni^{2+} on other algae species reported by Castro *et al.* (2017).

SEM-EDS analysis

Before adsorption, the surface of the adsorbent is rough and uneven, surface ridge and void structure decreased after adsorption of Ni^{2+} (Figure 5), as element contents summarized in Table 4. After adsorption, a new peak appears with Ni^{2+} characterization, and the content Ni^{2+} increased by 0.09%, which indicates the successful capture of Ni^{2+} on to the adsorbent (Ramrakhiani *et al.* 2017).

XPS analysis

Figure 6 shows the new characteristic peaks of Ni 3 s and Ni 2P3, which further implies that Ni^{2+} was successfully adsorbed onto the surface of *U. lactuca*. To investigate the mechanism of Ni^{2+} adsorption on *U. lactuca*, the XPS spectra of the C1s, N1s and O1s are analyzed and summarized in Figure S4 (available online). Before adsorption, peaks appeared at 284.8 eV, 285.2 eV and 287.6 eV (Figure S4(a) and S4(b)), which correspond to the functional

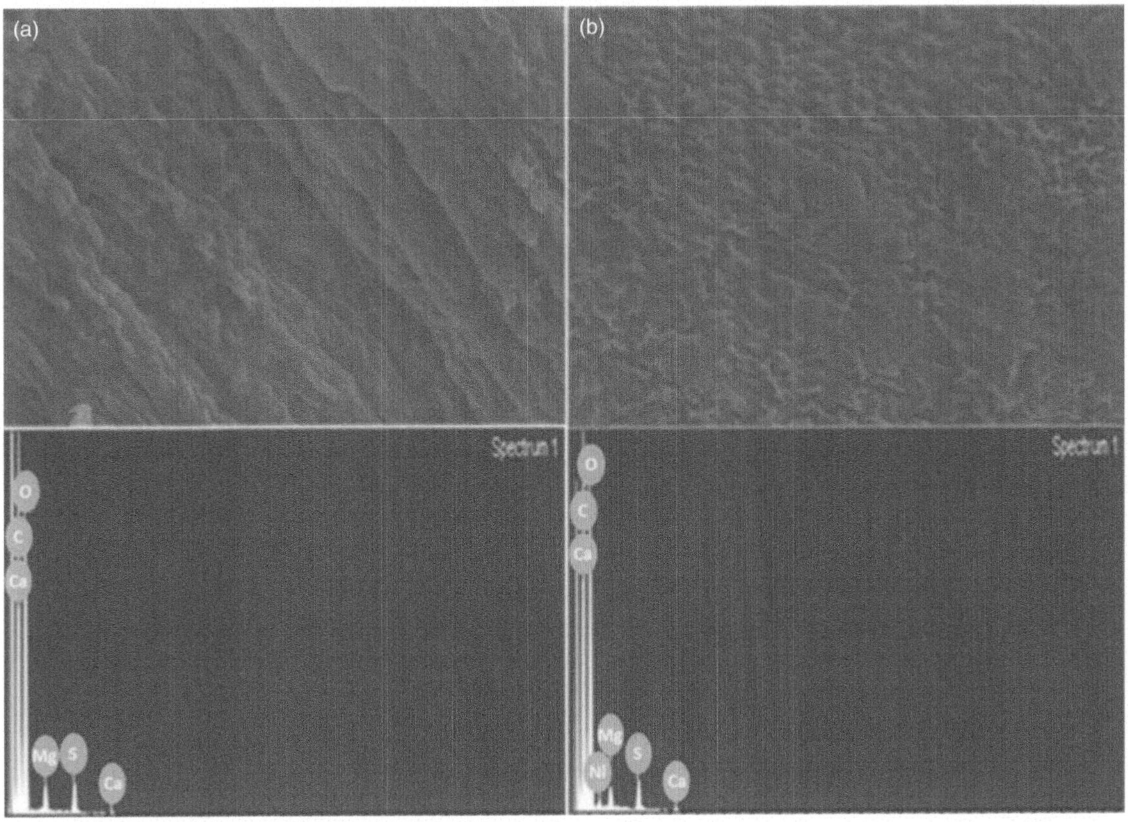

Figure 5 | SEM-EDS photos and spectra of *U. lactuca* before (a) and after (b) adsorption of Ni^{2+}.

groups C-C/C-H, C=O/O-C-O and C-O, respectively (Zhang *et al.* 2017). After adsorption, the peak area of the C-C/C-H and C=O/O-C-O decreased from 37.7% to 35.71%, and from 53.9% to 46.03%, respectively. The peak area of the C-O group increased, indicating that the three functional groups of C-C/C-H, C=O/O-C-O and C-O were involved in the adsorption of Ni^{2+}. The N1S spectra occurred with the characteristic peak of the group $N\text{-}C=O/NH_2$ at 400 eV before adsorption, and a new peak appeared at 400.6 eV after adsorption (Figure S4(c) and S4(d)). In addition, the peak area of the group $N\text{-}C=O/NH_3$ decreased to 91.1%, showing that the functional group of $N\text{-}C=O/NH_3$ is an active site for Ni^{2+} adsorption. Figure S4(e) and S4(f) show that the peak area of group O-H increased from

Table 4 | SEM-EDS quantitative elemental analysis of *U. lactuca* before and after the adsorption of Ni^{2+}

Before element	W%	A%	After element	W%	A%
C K	46.96	54.59	C K	45.79	53.38
O K	50.73	44.27	O K	52.10	45.60
Mg K	1.06	0.61	Mg K	0.86	0.49
S K	1.09	0.48	S K	1.02	0.44
Ca K	0.16	0.06	Ca K	0.15	0.05
Ni K	0	0	Ni K	0.09	0.02
Totals	100.00		Totals	100.00	

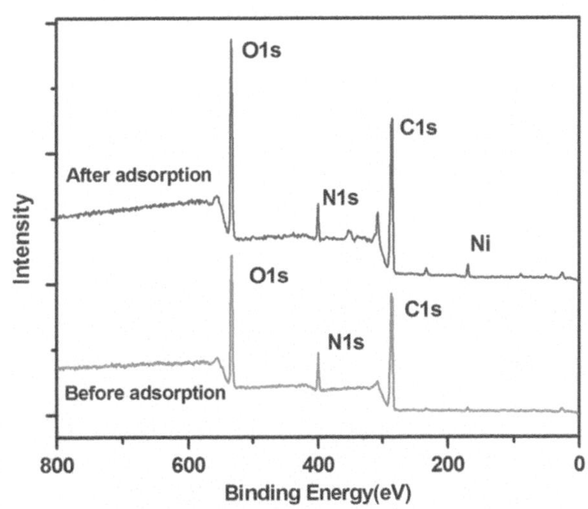

Figure 6 | XPS spectra survey of *U. lactuca* before and after adsorption of Ni^{2+}.

91.28% to 93.7% after adsorption, but the peak area of group C=O/C-O decreased from 8.72% to 6.3%. The group C=O/C-O had a small effect on Ni^{2+} adsorption, which is consistent with the report by Yang *et al.* (2017).

CONCLUSIONS

The dried biomass of *U. lactuca* is found to be an effective, cost-efficient and eco-friendly adsorbent for the removal of Ni^{2+} from aqueous solutions. Maximum adsorption capacity was obtained at pH 5, after 30 min of contact time.

The adsorption of Ni^{2+} can be well described by the Langmuir isotherm model, and the maximum adsorption capacity was 38.28 mg/g. The kinetic studies reveal that the adsorption process fitted the pseudo-second-order kinetic model better than the pseudo-first-order. The calculated values of ΔG and ΔH thermodynamic parameters imply that adsorption process of Ni^{2+} by *U. lactuca* is spontaneous and endothermic. In addition, higher temperature was more favorable to adsorption process. Characterization by FT-IR, SEM-EDS and XPS techniques demonstrates that the functional groups of –OH, N-H, C=O and C-O on the surface of the adsorbent were proven to take part in the adsorption of Ni^{2+}.

ACKNOWLEDGEMENTS

We gratefully acknowledge the financial support provided by National Natural Science Foundation of China (41376156), Guangzhou Science and Technology Program (201510010204, 201804010281), Guangdong innovation platform characteristic innovation project (2016KTSCX106), Guangzhou city science and technology project (201804010281), the Project of Education Bureau of Guangzhou City (1201620219) and the high level university construction project (regional water environment safety and water ecological protection).

REFERENCES

Abdel-Aty, A. M., Ammar, N. S., Ghafar, H. H. A. & Ali, R. K. 2013 Biosorption of cadmium and lead from aqueous solution by fresh water alga *Anabaena sphaerica* biomass. *Journal of Advanced Research* **4** (4), 367–374.

Ahmad, E., Seyedenayat, H., Samad, A. & Bahman, R. 2016 Modification of green algae harvested from the Persian Gulf by L-cysteine for enhancing copper adsorption from wastewater: experimental data. *Cheimical Date Collections* **2**, 36–42.

Akbari, M., Hallajisani, A., Keshtkar, A. R., Shahbeig, H. & Ghorbanian, S. A. 2015 Equilibrium and kinetic study and modeling of Cu(II) and Co(II) synergistic biosorption from Cu(II)-Co(II) single and binary mixtures on brown algae *C. indica*. *Journal of Environmental Chemical Engineering* **3** (1), 140–149.

Aksu, Z., Açıkel, Ü., Kabasakal, E. & Tezer, S. 2002 Equilibrium modelling of individual and simultaneous biosorption of chromium (VI) and nickel (II) onto dried activated sludge. *Water Research* **36** (12), 3063–3073.

Amany, E., Ahmed, E., Azza, K. & Ola, A. 2007 Removal of toxic chromium from wastewater using green alga *Ulva lactuca* and its activated carbon. *Journal of Hazardous Material* **148** (1–2), 216–228.

Anastopoulos, I. & Kyzas, G. Z. 2015 Progress in batch biosorption of heavy metals onto algae. *Journal of Molecular Liquids* **209**, 77–86.

Andrew, T., Marie, S. L., Leyla, S. & Murray, T. B. 2007 Uptake of platinum group elements by the marine macroalga, *Ulva lactuca*. *Marine Chemistry* **105** (3–4), 271–280.

Bakatula, E. N., Cukrowska, E. M., Weiersbye, I. M., Mihaly-Cozmuta, L., Peter, A. & Tutu, H. 2014 Biosorption of trace elements from aqueous systems in gold mining sites by the filamentous green algae (*Oedogonium sp.*). *Journal of Geochemical Exploration* **144**, 492–503.

Bulgariu, D. & Bulgariu, L. 2012 Equilibrium and kinetics studies of heavy metal ions biosorption on green algae waste biomass. *Bioresource Technology* **103** (1), 489–493.

Castro, L., Bonilla, L. A., González, F., Ballester, A., Blázquez, M. L. & Muñoz, J. A. 2017 Continuous metal biosorption applied to industrial effluents: a comparative study using an agricultural by-product and a marine alga. *Environmental Earth Science* **76**, 491–502.

Chen, Z., Ma, W. & Han, M. 2008 Biosorption of nickel and copper onto treated alga (*Undaria pinnatifida*): application of isotherm and kinetic models. *Journal of Hazardous Materials* **155**, 327–333.

Cho, H. J., Baek, K., Jeon, J. K., Park, S. K., Suh, D. J. & Park, Y. 2013 Removal characteristics of copper by marine macro-algae-derived chars. *Chemical Engineering Journal* **217**, 205–211.

Christoforidis, A. K., Orfanidis, S., Papageorgiou, S. K., Lazaridou, A. N., Favvas, E. P. & Ch, Mitropoulos A.,. 2015 Study of Cu(II) removal by *Cystoseira crinitophylla* biomass in batch and continuous flow biosorption. *Chemical Engineering Journal* **277**, 334–340.

Daneshvar, E., Kousha, M., Jokar, M., Koutahzadeh, N. & Guibal, E. 2012 Acidic dye biosorption onto marine brown macroalgae: isotherms, kinetic and thermodynamic studies. *Chemical Engineering Journal* **204**, 225–234.

Daneshvar, E., Vazirzadeh, A., Niazi, A., Sillanpää, M. & Bhatnagar, A. 2017 A comparative study of methylene blue biosorption using different modified brown, red and green macroalgae – effect of pretreatment. *Chemical Engineering Journal* **307**, 435–446.

Geetha, P., Latha, M. S., Pillai, S. S., Deepa, B., Kumar, K. S. & Koshy, M. 2016 Green synthesis and characterization of alginate nanoparticles and its role as a biosorbent for Cr(VI) ions. *Journal of Molecular Structure* **1105** (12), 54–60.

Ghasemi, M., Keshtkar, A. R., Dabbagh, R. & Safdari, S. J. 2011 Biosorption of uranium(VI) from aqueous solutions by Ca-pretreated *Cystoseira indica* alga: breakthrough curves studies and modeling. *Journal of Hazardous Materials* **189** (1–2), 141–149.

Gutiérrez, C., Hansen, H. K., Hernández, P. & Pinilla, C. 2015 Biosorption of cadmium with brown macroalgae. *Chemosphere* **138**, 164–169.

Hanan, H. O. 2008 Biosorption of copper, nickel and manganese using non-living biomass of marine alga, *Ulva lactuca*. *Journal of Biological Sciences* **11**, 964–973.

He, J. & Chen, J. P. 2014 A comprehensive review on biosorption of heavy metals by algal biomass: materials, performances, chemistry, and modeling simulation tools. *Bioresource Technology* **160** (6), 67–78.

Henriques, B., Rocha, L. S., Lopes, C. B., Figueira, P., Rui, J. R. M., Duarte, A. C., Pardal, M. A. & Pereira, E. 2015 Study on bioaccumulation and biosorption of mercury by living marine macroalgae: prospecting for a new remediation biotechnology applied to saline waters. *Chemical Engineering Journal* **281**, 759–770.

Ioannis, A. & George, Z. K. 2016 Are the thermodynamic parameters correctly estimated in liquid-phase adsorption phenomena? *Journal of Molecular Liquids* **218**, 174–185.

Jayakumar, R., Rajasimman, M. & Karthikeyan, C. 2015 Optimization, equilibrium, kinetic, thermodynamic and desorption studies on the sorption of Cu(II) from an aqueous solution using marine green algae: *Halimeda gracilis*. *Ecotoxicology and Environmental Safety* **121** (S12), 199–210.

Kumar, K. S., Dahms, H.-U., Won, E.-J., Lee, J.-S. & Shin, K.-H. 2015 Microalgae – A promising tool for heavy metal remediation. *Ecotoxicology and Environmental Safety* **113**, 329–352.

Li, H. S., Chen, Y. H., Long, J. Y., Jiang, D. Q., Liu, J., Li, S. J., Qi, J. Y., Zhang, P., Wang, J., Gong, J., Wu, Q. H. & Chen, D. 2017a Simultaneous removal of thallium and chloride from a highly saline industrial wastewater using modified anion exchange resins. *Journal of Hazardous Materials* **333**, 179–185.

Li, H. S., Chen, Y., Long, J. Y., Li, X. W., Jiang, D. Q., Zhang, P., Qi, J. Y., Huang, X. X., Liu, J., Xu, R. B. & Gong, J. 2017b Removal of thallium from aqueous solutions using Fe-Mn binary oxides. *Journal of Hazardous Materials* **338**, 296–305.

Liu, Y. H., Cao, Q. L., Luo, F. & Chen, J. 2009 Biosorption of Cd^{2+}, Cu^{2+}, Ni^{2+} and Zn^{2+} ions from aqueous solutions by pretreated biomass of brown algae. *Journal of Hazardous Materials* **163** (2–3), 931–938.

Long, J. Y., Li, H. S., Jiang, D. Q., Luo, D. G., Chen, Y. H., Xia, J. R. & Chen, D. Y. 2017 Biosorption of strontium(II) from aqueous solutions by *Bacillus cereus* isolated from strontium hyperaccumulator *Andropogon gayanus*. *Process Safety and Environmental Protection* **111**, 23–30.

Lou, Z. N., Wang, J., Jin, X. D., Wan, L., Wang, Y., Chen, H., Shan, W. J. & Xiong, Y. 2015 Brown algae based new sorption material for fractional recovery of molybdenum and rhenium from wastewater. *Chemical Engineering Journal* **273**, 231–239.

Masoumeh, A., Ahmad, H., Ali, R. K., Hossein, S. & Sohrab, A. G. 2015 Equilibrium and kinetic study and modeling of Cu(II) and Co(II) synergistic biosorption from Cu(II)-Co(II) single and binary mixtures on brown algae *C. indica*. *Journal of Environmental Chemical Engineering* **3**, 140–149.

Mitrogiannis, D., Markou, G., Çelekli, A. & Bozkurt, H. 2015 Biosorption of methylene blue onto *Arthrospira platensis* biomass: kinetic, equilibrium and thermodynamic studies. *Journal of Environmental Chemical Engineering* **3**, 670–680.

Mohammad, H., Hanafy, H. & Hassan, H. 2014 Remediation of lead by pretreated red algae: adsorption isotherm, kinetic, column modeling and simulation studies. *Green Chemistry Letters and Reviews* **7**, 342–358.

Nath, J. & Ray, L. 2015 Biosorption of Malachite green from aqueous solution by dry cells of *Bacillus cereus* m^1_{16} (MTCC 5521). *Journal of Environmental Chemical Engineering* **3** (1), 386–394.

Oliveira, R. C., Jouannin, C., Guibal, E. & Jr, O. G. 2011 Samarium (III) and praseodymium(III) biosorption on *Sargassum sp.*: batch study. *Process Biochemistry* **46** (3), 36–744.

Pang, C., Liu, Y. H., Cao, X. H., Li, M., Huang, G. L., Hua, R., Wang, C. X., Liu, Y. T. & An, X. F. 2011 Biosorption of uranium (VI) from aqueous solution by dead fungal biomass of *Penicillium citrinum*. *Chemical Engineering Journal* **170**, 1–6.

Ramrakhiani, L., Halder, A., Majumder, A., Mandal, A. K., Majumdar, S. & Ghosh, S. 2017 Industrial waste derived biosorbent for toxic metal remediation: mechanism studies and spent biosorbent management. *Chemical Engineering Journal* **308**, 1048–1064.

Sarı, A., Uluozlü, Ö. D. & Tüzen, M. 2011 Equilibrium, thermodynamic and kinetic investigations on biosorption of arsenic from aqueous solution by algae (*Maugeotia genuflexa*) biomass. *Chemical Engineering Journal* **167** (1), 155–161.

Shroff, K. A. & Vaidya, V. K. 2011 Kinetics and equilibrium studies on biosorption of nickel from aqueous solution by dead fungal biomass of *Mucor hiemalis*. *Chemical Engineering Journal* **171**, 1234–1245.

Tang, J., Li, Y., Wang, X. & Daroch, M. 2017 Effective adsorption of aqueous Pb^{2+} by dried biomass of Landoltia punctata and Spirodela polyrhiza. *Journal of Cleaner Production* **145**, 25–34.

Tatsuki, K., Makoto, N., Yoshihiro, K., Atsushi, I., Kota, A., Syuji, M., Katsunori, T., Hitoshi, K. & Jun, T. 2017 Biosorption of metal elements by exopolymer nanofibrils excreted from Leptothrix cells. *Water Research* **122**, 139–147.

Tounsadi, H., Khalidi, A., Abdennouri, M. & Barka, N., 2015 Biosorption potential of *Diplotaxis harra* and *Glebionis coronaria* L. biomasses for the removal of Cd(II) and Co(II) from aqueous solutions. *Journal of Environmental Chemical Engineering* **3**, 822–830.

Tuzen, M. & Sari, A. 2010 Biosorption of selenium from aqueous solution by green algae (*Cladophora hutchinsiae*) biomass: equilibrium, thermodynamic and kinetic studies. *Chemical Engineering Journal* **158** (2), 200–206.

Vijayaraghavan, K., Padmesh, T. V. N., Palanivelu, K. & Velan, M. 2006 Biosorption of nickel(II) ions onto *Sargassum wightii*: application of two-parameter and three parameter isotherm models. *Journal of Hazardous Materials* **133** (1–3), 304–308.

Vijayaraghavan, K., Rangabhashiyam, S., Ashokkumar, T. & Arockiaraj, J. 2017 Assessment of samarium biosorption from aqueous solution by brown macroalga *Turbinaria conoides*. *Journal of the Taiwan Institute of Chemical Engineers* **74**, 113–120.

Xu, L., Chen, X., Li, H. X., Hu, F. & Liang, M. X. 2016 Characterization of the biosorption and biodegradation properties of *Ensifer adhaerens*: a potential agent to remove polychlorinatedbiphenyls from contaminated water. *Journal of Hazardous Materials* **302**, 314–322.

Yang, W. C., Tian, S. Q., Tang, Q. Z., Chai, L. Y. & Wang, H. Y. 2017 Fungus hyphae-supported alumina: an efficient and reclaimable adsorbent for fluoride removal from water. *Journal of Colloid and Interface Science* **496**, 496–504.

Yeslié, G. B., Ivan, L. R., Omar, G. B. & Eric, G. 2011 Nickel biosorption using *Gracilaria caudata* and *Sargassum muticum*. *Chemical Engineering Journal* **166**, 122–131.

Zeroual, Y., Moutaouakkil, A., Dzairi, F. Z., Talbi, M., Chung, P. U., Lee, K. & Blaghen, M. 2003 Biosorption of mercury from aqueous solution by Ulva lactuca biomass. *Bioresource Technology* **90** (3), 349–351.

Zhang, D. N., Ran, Y., Cao, X. Y., Mao, J. D., Cui, J. F. & Schmidt-Rohr, K. 2015 Biosorption of nonylphenol by pure algae, field-collected planktons and their fractions. *Environmental Pollution* **198**, 61–69.

Zhang, B. P., Ma, Z. C., Yang, F., Liu, Y. & Guo, M. C. 2017 Adsorption properties of ion recognition rice straw lignin on $PdCl_4^{2-}$: equilibrium, kinetics and mechanism. *Colloids and Surfaces A: Physicochem. Eng. Aspects* **514**, 260–268.

Zhang, G., Fan, F., Li, X., Qi, J. & Chen, Y. 2018 Superior adsorption of thallium (I) on titanium peroxide: performance and mechanism. *Chemical Engineering Journal* **331**, 471–479.

First received 13 October 2017; accepted in revised form 24 May 2018. Available online 14 June 2018

Microalgal luxury uptake of phosphorus in waste stabilization ponds – frequency of occurrence and high performing genera

A. Crimp, N. Brown and A. Shilton

ABSTRACT

Microalgae commonly found in waste stabilization ponds (WSPs) are able to accumulate elevated phosphorus levels within their cells in a process known as luxury uptake. However, there are few studies focused on luxury uptake in full scale WSPs. In order to comprehensively quantify the occurrence of this phenomenon, eight different WSP sites comprising seven primary facultative, six maturation and two high rate algal ponds (HRAPs) spread over several climatic regions were monitored over four seasons. Of the 15 ponds studied, 13 of these exhibited elevated levels of biomass phosphorus content at some point; however, the occurrence in HRAPs was limited. More than half of the samples tested had elevated phosphorus contents and this occurred in all climatic zones surveyed. The phosphorus content of the biomass was significantly correlated to decreasing rainfall and increasing total dissolved phosphorus. Microscopic analysis revealed that nearly all the 17 microalgal and five cyanobacterial genera identified performed luxury uptake, but at varying frequencies. This is the first time that the genera of algae responsible for luxury uptake in full scale WSPs has been studied. *Chlamydomonas/Cryptomonas*, *Micractinium/Microcystis* and *Scenedesmus* were the only microalgal genera found to both commonly occur in WSPs and consistently perform luxury uptake.

Key words | luxury uptake, microalgae, phosphorus, waste stabilization pond, wastewater

A. Crimp
N. Brown (corresponding author)
A. Shilton
School of Engineering and Advanced Technology, Massey University,
Private Bag 11-222, Palmerston North,
New Zealand
E-mail: *n.brown@massey.ac.nz*

INTRODUCTION

One of the primary consequences of discharging high nutrient content effluent into waterways is eutrophication in the receiving waters, which can lead to depletion of the dissolved oxygen (DO) within the waterway, and in some cases, death of aquatic life.

To reduce the environmental effects of wastewater discharge, government authorities around the world are placing increasingly stringent requirements on wastewater treatment plants, requiring improved phosphorus removal (Hendriks & Langeveld 2017). Upgrading to meet these requirements is particularly difficult for smaller communities, as economies of scale diminish thereby making the relative 'per capita' capital and operating costs very high.

Due to their simple operation, waste stabilization ponds (WSPs) are used in tens of thousands of applications, serving millions of people around the world (Shilton & Walmsley 2005). However, while effective at removing biological oxygen demand and, when designed appropriately, pathogens, WSPs have been found to be inconsistent at removing phosphorus, with only approximately 15–50% removal achieved (Garcia *et al.* 2000). For communities employing WSPs, the two main upgrade options to improve their phosphorus removal are alum dosing and enhanced biological phosphorus removal. However, while these options have been proven to be effective (Tchobanoglous *et al.* 1991; Blackall *et al.* 2002) the significant capital and operational costs of both systems make them too expensive for most small communities.

As an alternative, increasing focus has gone into identifying potential mechanisms within the ponds themselves that could be utilized to improve phosphorus removal. Phosphorus removal within WSPs has been attributed to microalgal assimilation, phosphate precipitation and more recently direct adsorption onto the surface of microalgae. However, each of these mechanisms is known to have limitations, which is why we only see 15–50% phosphorus

doi: 10.2166/wst.2017.632

removal such as reported by Garcia *et al.* (2000). While microalgae assimilate phosphorus for growth, they normally contain no more than 1% phosphorus by dry weight (Xu *et al.* 2014) which means that at typical biomass concentrations this can only achieve removal of well under half of the total phosphorus (TP) that would be found in the pond effluent. Naturally occurring phosphate precipitation while known to be effective is, however, limited by pH and cation levels in the wastewater (Craggs 2005; Vendramelli *et al.* 2016). Adsorption of phosphorus to microalgae is also fickle, only making a contribution in specific ranges of light intensity, temperature and pH, outside of which this mechanism is considered largely ineffective (Martinez *et al.* 2000; Lu *et al.* 2014; Xu *et al.* 2014).

One further, but poorly researched, mechanism that has been found to occur within WSPs is the luxury uptake of phosphorus by microalgae. Luxury uptake within microalgae is the uptake of phosphorus beyond what is required for basic metabolism which is stored as polyphosphate (Powell *et al.* 2011). This leads to biomass phosphorus contents elevated above the 1%P/g volatile suspended solids (VSS) limit for microalgae undergoing normal assimilation. While this phenomenon has been known to occur in bacteria for some time, only limited studies have considered the ability of microalgae to perform this mechanism in the context of wastewater treatment. Powell *et al.* (2008) found that luxury uptake could occur within WSP microalgae grown in laboratory scale reactors operated continuously, with biomass phosphorus contents reaching over three times the standard value of 1%P/g VSS. Powell *et al.* (2011) then looked to see if microalgae also perform luxury uptake in full scale operational WSPs, and found that this did indeed also occur within the two WSPs studied with biomass phosphorus contents of up to 3.85%P/g VSS recorded. However, at that stage no further work was undertaken to link the incidences of high biomass phosphorus content to the environmental conditions.

In order to understand the luxury uptake mechanism further, more knowledge of the occurrence of luxury uptake in full scale systems is required. It is currently unknown how much the occurrence of luxury uptake varies in different pond designs or in response to different environmental conditions as experienced spatially across latitude and temporally across seasons. This paper reports on the findings of a comprehensive study of luxury uptake by microalgae in full scale WSPs. A range of full scale pond systems were monitored to determine how widespread the occurrence of luxury uptake is, which environmental variables affect its occurrence, and how primary facultative, maturation and high rate algal ponds (HRAPs) compare with regards to this phenomenon. In addition the performance of the microalgal genera identified within the ponds is reported in terms of their frequency of occurrence in ponds, and their propensity for performing luxury uptake of phosphorus.

METHODOLOGY

Site selection

Of the eight sites that were selected for analysis in the study, seven incorporated two separate ponds, thereby providing 15 ponds for monitoring. All sites treated wastewater from rural communities. Four sites were selected within and around the Manawatu region where the project was based, to assess any variation of biomass phosphorus content between WSPs within the same region. The four selected Manawatu sites were Halcombe, Sanson, Rongotea and Foxton Beach, each of which are two-pond systems featuring a primary facultative pond and a maturation pond. For the purposes of this paper these sites will be known as Manawatu 1, Manawatu 2, Manawatu 3 and Manawatu 4 respectively. An additional four sites from other regions of New Zealand were studied to evaluate the effect of the variation in the climatic environments on the ponds studied. The sites selected were Gore in the cool temperate south of the South Island, Ngaruawahia in the warm temperate central North Island, and Kaitaia in the sub-tropical north of the North Island. Of these sites, all except Ngaruawahia (single pond system) utilized two ponds.

To allow wider comparison against different WSP designs, two HRAPs were also studied, located in Hamilton, near Ngaruawahia. Both of these pilot scale HRAPs employed CO_2 addition to control the pH below 8. The locations of the sites selected are shown in Figure 1.

Due to the distance between each of the sites from other regions and the Manawatu region where the study was based, samples from Gore, Hamilton, Ngaruawahia and Kaitaia were preserved using Lugol's Iodine and sent via overnight courier. Lugol's Iodine was provided by microalgal biotechnologists at Cawthron Institute in Nelson, New Zealand.

Sampling methodology

At each site, WSP samples were taken using a 500 mL Duran bottle attached to a pole which was submerged to a

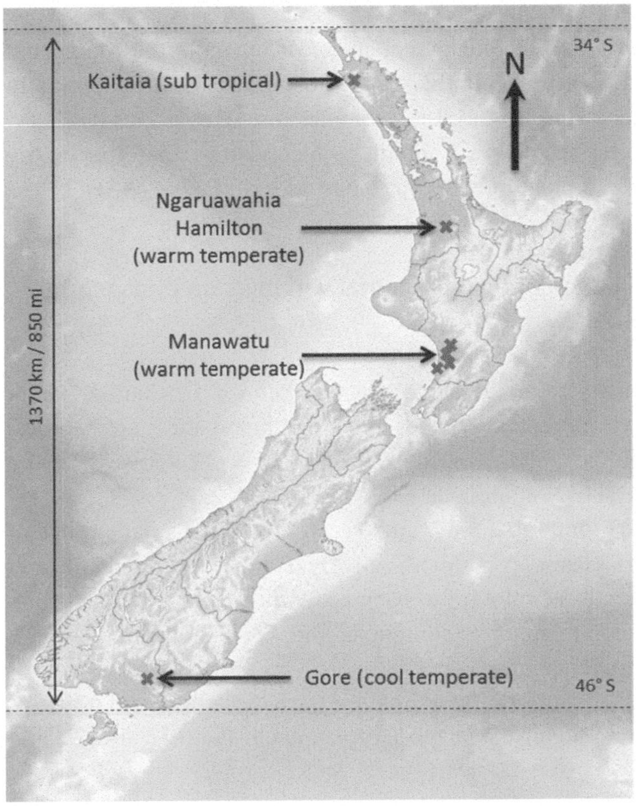

Figure 1 | Sampling locations for the study.

depth of 15 cm. This depth was chosen as it is within the microalgal band of typical WSPs (Pearson 2005). Using the sampling bottle, 500 mL samples were taken from each of the four sides of each pond studied, and combined into one container. This combined 2 L sample therefore represented an 'average sample' of the pond. Samples were taken from the sides of the pond rather than the corners to ensure the sample did not contain any stagnant material. All 15 ponds were monitored over a full year up to three times per season in duplicate for TP and total dissolved phosphorus (TDP), in triplicate for VSS and in quadruplicate for environmental parameters such as pH, pond temperature and DO.

Environmental variables

The pH, DO and pond temperature were measured using a Thermo Scientific Orion Star A326 pH-DO meter at the same points as where the pond samples were taken, with the results averaged for each pond. Monthly average rainfall, air temperature, wind speed and solar radiation were obtained from weather stations in close proximity to each site. All sites were monitored over the course of a full year. Statistical analysis of the effect of different variables on the biomass phosphorus content was carried out using the stepwise regression method with the statistical software MINITAB. Following the backward elimination approach (Devore & Farnum 2005), the environmental variables were put through a sequence of regression analyses, with the least significant variable removed after each analysis. The sequence was continued until the remaining variables all had *p*-values of less than 0.05, representing a 95% level of confidence.

Analysis

VSS measurements used the method outlined by APHA AWWA WEF (1995). TP analysis was carried out by first digesting the sample using the sulphuric acid-nitric acid digestion method then analysed using the ascorbic acid method (APHA AWWA WEF 1995). TDP samples were first filtered using a 0.45 μm filter to remove the biomass then digested and analysed using the same methodology as the TP samples. The VSS, TP and TDP values were then used to determine the biomass phosphorus content using Equation (1):

$$\frac{\%\mathrm{P}}{\mathrm{g\ VSS}} = \frac{\mathrm{TP} - \mathrm{TDP}}{\mathrm{VSS}} \times 100 \qquad (1)$$

Biomass phosphorus contents were compared between ponds using two-sample t-tests, again through MINITAB. This test compared the 95% confidence intervals around the means of two sample groups, with a significant difference determined if the confidence intervals do not overlap (Devore & Farnum 2005).

Powell *et al.* (2011) noted that values of biomass phosphorus content in WSP systems above 1%P/g VSS is indicative of luxury uptake occurring in the biomass. However, further confirmation of the occurrence of luxury uptake can also be obtained visually by staining to highlight polyphosphate granules within a microalgal cell. This was therefore also undertaken using 0.5 mL samples placed onto microscope slides. The samples were dried in an oven and then stained for polyphosphate granules following the method outlined by Bolier *et al.* (1992). This stain coloured the polyphosphate granules within the cells black, allowing them to be identified under the microscope. Microscopic analysis was carried out using a Leica light microscope with camera attachment at 1000× magnification. The different microalgal genera present in the sample was identified through comparison with other images of the genera utilizing the website AlgaeBase (Guiry & Guiry 1996). The presence of the stained polyphosphate granules within each algal cell was then assessed. Independent identification of microalgal genera was carried out by microalgal biotechnologists at Cawthron Institute of Nelson, New Zealand every three months for quality assurance. Their analysis also included determining the relative abundance of each genus identified, using an in-house method based on Hotzel & Croome (1999).

RESULTS AND DISCUSSION

Occurrence of elevated levels of biomass phosphorus content

The annual averages of the phosphorus content of the WSP biomass from the 15 different ponds are displayed in Table 1, which shows that 13 of the 15 ponds sampled exhibited elevated levels of phosphorus content at some point in the study and that it was found to be triggered at all eight sites over all the climatic zones. In some ponds, this occurred with high frequency, for example the Manawatu 3 primary facultative ponds exhibited luxury uptake in every sample. Of all the samples tested, over half had elevated phosphorus contents, which shows that the luxury uptake phenomenon often occurs in full scale WSPs.

Influence of environmental variables on phosphorus content

Statistical analysis was undertaken to determine which environmental variables had a significant effect on the phosphorus content of the WSP biomass. The results of the MINITAB analysis are shown in Table 2.

From Table 2, it can be seen that while a wide range of variables were studied, only TDP in the pond liquid and rainfall were found to have a significant impact on the biomass phosphorus content of the WSP biomass at a 95% confidence level.

It is interesting to note the variables which did not have a significant effect on luxury uptake. Ponds were selected

Table 1 | Annual average phosphorus content of biomass in the 15 ponds surveyed

	Primary facultative ponds						
	Man. 1	**Man. 2**	**Man. 3**	**Man. 4**	**Gore**	**Nga.**	**Kaitaia**
Avg (%P/g VSS)	1.4%	1.2%	1.8%	1.3%	0.9%	0.9%	1.0%
95% CI	0.4%	0.6%	0.3%	0.3%	0.2%	0.3%	0.3%
Max	2.5%	3.8%	3.3%	2.0%	1.6%	2.6%	1.9%
>1% P/g VSS	86%	63%	100%	67%	38%	19%	38%
	Maturation ponds						
	Man. 1	**Man. 2**	**Man. 3**	**Man. 4**	**Gore**		**Kaitaia**
Avg (%P/g VSS)	1.6%	1.4%	1.6%	1.0%	0.8%		0.7%
95% CI	0.3%	0.4%	0.5%	0.3%	0.2%		0.1%
Max	2.6%	2.5%	3.4%	2.3%	1.4%		0.9%
>1% P/g VSS	88%	57%	75%	67%	38%		0%
	HRAPs						
	HRAP 1	**HRAP 2**					
Avg (%P/g VSS)	0.8%	0.7%					
95% CI	0.3%	0.1%					
Max	1.3%	0.9%					
>1% P/g VSS	17%	0%					

(Man. = Manawatu; Nga. = Ngaruawahia; CI = confidence interval). Note that the row '> 1% P/g VSS' indicates the percentage of samples from each pond that contained biomass with phosphorus contents above 1% P/g VSS, indicating that luxury uptake is indeed occurring.

Table 2 | Results from MINITAB stepwise regression analysis of environmental variables tested against the biomass phosphorus content

Variable	Average	Range	p-value	Significant (95% confidence level)
VSS (mg/L)	114	28–687	0.881	No
pH	8.0	6.6–9.8	0.382	No
DO (mg/L)	9.7	0.2–20.3	0.406	No
Pond temperature (°C)	16.7	4.2–27.9	0.948	No
TDP (mg P/L)	3.2	0.5–8.6	0.009	Yes
Rainfall (mm)	81	10–328	0.024	Yes
Wind speed (km/h)	12.3	7.6–19.4	0.873	No
Air temperature (°C)	13.8	5.7–20.6	0.948	No
Solar radiation (W/m²)	172	51–299	0.654	No

Note that the *p*-values for each variable are for the last regression iteration before they were eliminated under the 'reverse elimination' method.

over a wide variety of climatic zones in order to enable environmental variables such as the temperature and solar radiation to be studied; however, neither of these variables were shown to have a significant effect (Table 2). This was surprising, particularly when these variables are known to affect luxury uptake in the laboratory under controlled conditions (Powell *et al.* 2008, 2009). It is possible that with some variables there is a delay in their subsequent effect on the phosphorus content of the biomass within the pond. The ponds studied did exhibit a wide range of temperatures and levels of solar radiation (Table 2) but it appears that under field conditions they were not statistically significant. In reality, in the field these conditions are constantly changing and the ponds would experience a large amount of variation in these variables throughout the day/night. This might mean that controlled conditions in the laboratory do not reflect those found in the field and it is therefore difficult to compare laboratory and field results.

To further examine the effects that were statistically significant, plots of the WSP biomass phosphorus content against TDP and rainfall are shown in Figures 2 and 3.

As can be seen, at low TDP concentrations the majority of samples were found to contain less than 1%P/g VSS and at a high TDP concentration the majority of samples exceeded 1%P/g VSS. This increase in biomass phosphorus content with TDP could be due to an increase in the phosphorus concentration gradient across the plasma membranes of the wastewater microalgae. This is known to positively affect phosphate transport via electrochemical potential gradient, passive diffusion, facilitated diffusion and active transport – the four known mechanisms for phosphorus uptake in microalgae (Cembella *et al.* 1982). Previous studies have shown that the phosphorus concentration is an important factor for luxury uptake in the laboratory (e.g. Aitchison & Butt 1973; Powell *et al.* 2008; Markou *et al.* 2012; Schmidt *et al.* 2016). However, this is the first time that the phosphorus concentration has been shown to influence luxury uptake in full scale WSPs. Also worth noting is the large variation in biomass phosphorus content across the same dissolved phosphorus concentration. For example, at a TDP concentration of 2.6 mg/L, one sample from Manawatu 4 was found to have a biomass phosphorus content of only 0.1%P/g VSS, while another sample from Manawatu 1 contained 2.5%P/g VSS, with biomass phosphorus contents of other sites varying between these values. This could be due to the biomass at each of the sites being at different stages in the cell life cycle, meaning that different levels of internal phosphorus would be maintained.

The effect of monthly total rainfall on the biomass phosphorus content is shown in Figure 3.

Figure 3 indicates that rainfall had a negative effect on the WSP biomass phosphorus content, an effect that

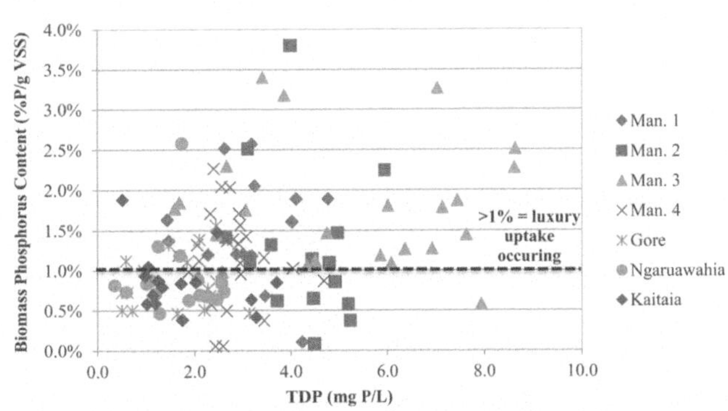

Figure 2 | Biomass phosphorus content vs. total dissolved phosphorus (TDP) within each of the pond systems studied. Note that the dashed black line at 1%P/g VSS indicates the limit for microalgae undergoing standard assimilation of phosphorus for growth. (Man = Manawatu)

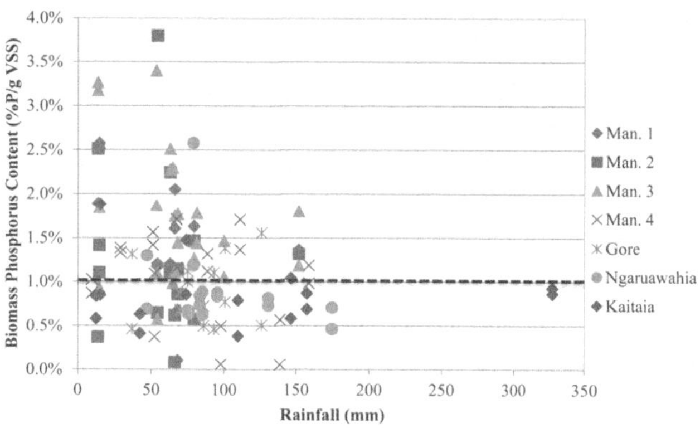

Figure 3 | Biomass phosphorus content vs. monthly total rainfall for each of the pond systems studied. Note that the dashed black line at 1%P/g VSS indicates the limit for microalgae undergoing standard assimilation of phosphorus for growth. (Man = Manawatu)

becomes particularly clear when considering the change in biomass phosphorus content at approximately 80 mm of total monthly rainfall. Of all the samples taken that had experienced less than 80 mm of rainfall, 67% of the samples exhibited elevated phosphorus contents above 1%P/g VSS. However, of the samples that had experienced greater than 80 mm rainfall, only 39% contained phosphorus contents above 1%P/g VSS. Most notably, of all of the samples measured with biomass phosphorus contents above 2%P/g VSS, none occurred above 80 mm of rainfall.

While the monthly rainfall had a statistically significant effect it is important to note that many environmental variables would be linked to rainfall. There is also the added complication that because the variable being tested is monthly rainfall, it would depend on when the rainfall occurred relative to when the sample was taken. However, while taking these complexities into account, it is possible to propose potential mechanisms responsible for the rainfall effect. Rainfall influences many variables known to affect microalgal growth and phosphorus uptake. For example due to infiltration, rainfall can cause a dilution in the phosphorus concentration of the wastewater and this has already been shown to influence the phosphorus content of the cells. Rainfall also influences the hydraulic retention time (HRT) as a result of additional flow due to infiltration. This reduced HRT would result in a lower biomass concentration due to washout, and after the rainfall event this could stimulate growth in preference to luxury uptake.

Comparison of pond types

To compare the relative phosphorus uptake achieved in the primary facultative ponds, maturation ponds and HRAPs, two sample t-tests were used to compare the different pond types. The results from this analysis are presented in Table 3.

Table 3 | Two sample t-test analysis comparing primary facultative WSPs, maturation WSPs and HRAPs

	Sample information		*p*-values		
	N	Mean (%P/g VSS)	Primary facultative	Maturation	HRAP
Primary facultative	71	1.22	N/A	0.881	0.000
Maturation	55	1.20	0.881	N/A	0.000
HRAP	13	0.71	0.000	0.000	N/A

Table 3 shows that there was no significant difference in biomass phosphorus content between primary facultative and maturation WSPs, with a *p*-value of 0.881. This is an interesting result, as the decreased organic loading experienced by maturation ponds is known to affect the pond ecology, for example resulting in different microalgal populations (Pearson 2005; Garcia *et al.* 2017). Table 3 also demonstrates that both primary facultative and maturation pond biomass had significantly higher phosphorus contents than HRAP biomass. To account for any variation in environmental variables that could have caused this discrepancy, the significant variables found for the study were compared between WSPs and HRAPs, as shown in Table 4.

The HRAPs had significantly lower TDP concentrations than the other WSPs (Table 4). HRAPs differ from WSPs as they are shallow, include paddlewheel mixing and in this case have CO_2 addition. All these properties of HRAPs result in elevated levels of microalgal growth and this increased growth consumes TDP, lowering its concentration and, as previously mentioned, makes build-up of intracellular phosphorus stores less likely. This suggests that microalgal phosphorus removal in HRAPs is likely to be dominated by growth rather than luxury uptake.

Table 4 | Comparison of TDP concentrations and rainfall levels for WSPs and HRAPs

	N	Mean	Standard deviation	Standard error	*p*-value
WSP TDP (mg P/L)	126	3.02	1.80	0.16	
HRAP TDP (mg P/L)	14	2.31	0.58	0.15	0.002
WSP rain (mm)	120	79	52	5	
HRAP rain (mm)	13	85	31	9	0.535

Luxury uptake of different microalgal genera

Microalgal genera identified in this study were assessed both based on how frequently they were identified in samples and the frequency at which they performed luxury uptake as verified by the presence of polyphosphate granules. These data are displayed in Table 5.

The most common microalgal and cyanobacterial genera were *Chlamydomonas/Cryptomonas*, *Microcystis/Micractinium*, *Monoraphidium* and *Scenedesmus*, all of which were found in over 70% of the samples (Table 5). While there were only a few genera that were frequently identified, luxury uptake across the range of genera was far more widespread, with 21 out of the 22 identified genera exhibiting luxury uptake at some point in the study.

Table 5 | Frequency of occurrence and frequency of luxury uptake of microalgal and cyanobacterial genera in WSPs

Genera	Luxury uptake observed	Frequency of occurrence within the pond	Frequency of luxury uptake
Actinastrum	Yes	47%	46%
Chlamydomonas/ Cryptomonas	Yes	76%	68%
Closterium	Yes	14%	65%
Crucigeniella	Yes	17%	65%
Cyclotella	Yes	5%	17%
*Dictyosphaerium**	No	3%	0%
Elakatothrix	Yes	2%	50%
Euglena	Yes	37%	73%
Kirchneriella	Yes	21%	14%
*Merismopedia**	Yes	24%	17%
*Micractinium/Microcystis**	Yes	88%	50%
Monoraphidium	Yes	71%	37%
Oocystis	Yes	14%	41%
Pandorina	Yes	2%	50%
Pediastrum	Yes	14%	82%
Phacus	Yes	34%	53%
*Planktothrix**	Yes	32%	84%
*Pseudanabaenaceae**	Yes	29%	23%
Scenedesmus	Yes	70%	73%
Schroederia	Yes	32%	79%
Tetraedron	Yes	8%	30%
Unknown colonial	Yes	2%	100%

Genera marked with a * are cyanobacteria. Note that the categories for *Chlamydomonas* and *Cryptomonas* as well as *Micractinium* and *Microcystis* were combined due to the difficulty of differentiating the two genera.

Aside from *Microcystis/Micractinium*, cyanobacteria were not commonly identified in samples taken from the WSPs, but typically represented a significant portion of the biomass when identified. Cyanobacteria were found to be capable of performing luxury uptake.

Nearly all of the identified genera were capable of performing luxury uptake; however, the frequency at which each genus stored polyphosphate as granules differed significantly. A large number of the microalgal and cyanobacterial genera were poor performers of luxury uptake, with *Kirchneriella*, *Merismopedia*, and *Pseudanabaenaceae* performing luxury uptake in less than 30% of the samples they were identified in. In contrast, despite being identified in less than 40% of the samples, when they were identified *Pediastrum*, *Planktothrix* and *Schroederia* exhibited luxury uptake at high rates: 82%, 84% and 79% respectively. Of significant interest with regards to the potential for improving phosphorus removal via this phenomenon, three of the most commonly identified genera, *Chlamydomonas/Cryptomonas*, *Micractinium/Microcystis* and *Scenedesmus*, all performed luxury uptake in more than 50% of the samples they were identified in. However, attempts to relate the presence or absence of a particular genus with the phosphorus content of the biomass in each pond found no relationship present. This could be due to a number of factors, including cell life cycles and the differing mass of polyphosphate stored within individual cells (Crimp 2015).

The results of these findings also provide one potential avenue for improving phosphorus removal from WSPs. If the growth of a particular microalgal genus with a high affinity for performing luxury uptake could be favoured in an engineered process, the average phosphorus content of the biomass found within WSPs could be increased significantly. Alternatively, favouring growth of a cyanobacterial genus with a high propensity for luxury uptake could also produce similar results. Subsequent removal of this biomass would result in greatly improved phosphorus removal achieved by this system, thus reducing the environmental impacts of the wastewater discharge.

CONCLUSIONS

The year-long monitoring WSPs and HRAPs over a range of climates showed for the first time that microalgal luxury uptake of phosphorus was both widespread and frequent in full scale pond systems.

Biomass phosphorus contents of up to 3.8%P/g VSS were measured within the biomass, nearly four times the maximum value for microalgae undergoing normal metabolism.

Analysis of environmental variables showed that the TDP concentration had a positive effect on the phosphorus content of the biomass, while monthly rainfall had a negative effect. In the range of climatic zones were studied, it was surprising that the temperature and solar radiation were not found to have a statistically significant effect on the phosphorus content of the biomass. While there were no significant differences between biomass phosphorus content in primary facultative and maturation ponds, both of these pond types had significantly higher biomass phosphorus contents than HRAPs. This was attributed to the low TDP concentrations experienced within the HRAPs which is likely to be due to elevated levels of microalgal growth in HRAP. This suggests that luxury uptake is less likely to occur in HRAP systems.

Microscopic analysis of the microalgae revealed that while luxury uptake occurred in nearly all of the genera observed, the frequency at which each genus performed luxury uptake varied significantly. There was a vast range in abilities to accumulate phosphorus within the microalgal genera present in ponds. Three of the most commonly found microalgae, *Chlamydomonas/Cryptomonas*, *Micractinium/Microcystis* and *Scenedesmus*, performed luxury uptake in more than 50% of the samples in which they were identified.

ACKNOWLEDGEMENTS

Acknowledgements are made to the Marsden Fund managed by the Royal Society of New Zealand for funding this work as part of the larger project titled: 'Discovering the trigger conditions for a new algal based phosphorus removal process'. Thanks to Steve Nind of Manawatu District Council and Marcus Coley of Horowhenua District Council for allowing access to their sites. Thanks are also given to Corey Reed and Ian Lowe of Gore District Council, Dr Jason Park at NIWA, David Kennington and Caleb Powell at Waikato District Council and Greg Timperley of Transfield Services Kaitaia for all their help with sampling. Thanks to Dr Nihal Jayamaha for his assistance with the statistical analysis. Lastly, thanks to Ramsay Huang for all of his help with laboratory analysis.

REFERENCES

Aitchison, P. A. & Butt, V. S. 1973 The relation between the synthesis of inorganic polyphosphate and phosphate uptake by *Chlorella vulgaris*. *Journal of Experimental Botany* **24** (80), 497–510.

APHA, AWWA, WEF 1995 *Standard Methods for the Examination of Water and Wastewater.* Water Environment Federation, Washington, DC.

Blackall, L. L., Crocetti, G. R., Saunders, A. M. & Bond, P. L. 2002 A review and update of the microbiology of enhanced biological phosphorus removal in wastewater treatment plants. *Antonie van Leeuwenhoek* **81**, 681–691.

Bolier, G., de Koningh, M. C. J., Schmale, J. C. & Donze, M. 1992 Differential luxury phosphate response to planktonic algae to phosphorus removal. *Hydrobiologia* **243/244**, 113–118.

Cembella, A. D., Antia, N. J. & Harrison, P. J. 1982 The utilization of inorganic and organic phosphorus compounds as nutrients by eukaryotic microalgae: a multidisciplinary perspective: part 1. *CRC Critical Reviews in Microbiology* **10** (4), 317–391.

Craggs, R. 2005 Nutrients. In: *Pond Treatment Technology* (A. Shilton, ed.). IWA Publishing, London, pp. 77–99.

Crimp, A. J. 2015 *Luxury Uptake of Phosphorus by Microalgae in New Zealand Waste Stabilisation Ponds.* Massey University, Palmerston North.

Devore, J. L. & Farnum, N. R. 2005 *Applied Statistics for Engineers and Scientists*, 2nd edn. Thomson Brooks/Cole, Belmont, CA, USA.

Garcia, J., Mujeriego, R., Bourrouet, A., Penuelas, G. & Freixes, A. 2000 Wastewater treatment by pond systems: experiences in Catalonia, Spain. *Water Science and Technology* **42** (10–11), 35–42.

Garcia, D., Posadas, E., Blanco, S., Acien, G., Garcia-Encina, P., Bolado, S. & Munoz, R. 2017 Evaluation of the dynamics of microalgae population structure and process performance during piggery wastewater treatment in algal-bacterial photobioreactors. *Bioresource Technology*. In press. doi.org/10.1016/j.biortech.2017.06.079.

Guiry, M. D. & Guiry, G. M. AlgaeBase. 20 March 1996. http://www.algaebase.org (accessed 1 June 2015.)

Hendriks, A. T. & Langeveld, J. G. 2017 Rethinking wastewater treatment plant effluent standards: nutrient reduction or nutrient control? *Environmental Science and Technology* **51**, 4735–4737.

Hotzel, G. & Croome, R. 1999 *A Phytoplankton Methods Manual for Australian Freshwaters.* Land and Water Resources Research and Development Corporation, Canberra.

Lu, H., Yang, L., Shabbir, S. & Wu, Y. 2014 The adsorption process during inorganic phosphorus removal by cultured periphyton. *Environmental Science and Pollution Research* **21** (14), 8782–8791.

Markou, G., Chatzipavlidis, I. & Georgakakis, D. 2012 Effects of phosphorus concentration and light intensity on the biomass composition of *Arthrospira (Spirulina) platensis. World Journal of Microbiology and Biotechnology* **28** (8), 2661–2670.

Martinez, M. E., Sanchez, S., Jimenez, J. M., El Yousfi, F. & Munoz, L. 2000 Nitrogen and phosphorus removal from urban wastewater by the microalga *Scenedesmus obliquus*. *Bioresource Technology* **73** (3), 263–272.

Pearson, H. 2005 Microbiology of waste stabilisation ponds. In: *Pond Treatment Technology* (A. Shilton, ed.). IWA Publishing, London, pp. 14–48.

Powell, N., Shilton, A. N., Pratt, S. & Chisti, Y. 2008 Factors influencing luxury uptake of phosphorus by microalgae in waste stabilization ponds. *Environmental Science Technology* **42** (16), 5958–5962.

Powell, N., Shilton, A., Chisti, Y. & Pratt, S. 2009 Towards a luxury uptake process via microalgae – defining the polyphosphate dynamics. *Water Research* **43**, 4207–4213.

Powell, N., Shilton, A., Pratt, S. & Chisti, Y. 2011 Luxury uptake of phosphorus by microalgae in full scale waste stabilisation ponds. *Water Science and Technology* **63** (4), 704–709.

Schmidt, J. J., Gagnon, G. A. & Jamieson, R. C. 2016 Microalgae growth and phosphorus uptake in wastewater under simulated cold region conditions. *Ecological Engineering* **95**, 588–593.

Shilton, A. & Walmsley, N. 2005 Introduction to pond treatment technology. In: *Pond Treatment Technology* (A. Shilton, ed.). IWA Publishing, London, pp. 1–13.

Tchobanoglous, G., Burton, F. L. & Stensel, H. D. 1991 *Wastewater Engineering: Treatment, Disposal and Reuse.* McGraw-Hill, New York.

Vendramelli, R. A., Vijay, S. & Yuan, Q. 2016 Phosphorus removal mechanisms in a facultative wastewater stabilisation pond. *Water, Air, & Soil Pollution* **227**, 417.

Xu, M., Bernards, M. & Hu, Z. 2014 Algae-facilitated chemical phosphorus removal during high density *Chlorella emersonii* cultivation in a membrane bioreactor. *Bioresource Technology* **153**, 383–387.

First received 3 September 2017; accepted in revised form 10 December 2017. Available online 21 December 2017

Enhanced nitrogen removal and energy saving in a microalgal–bacterial consortium treating real municipal wastewater

P. Foladori, S. Petrini, M. Nessenzia and G. Andreottola

ABSTRACT

The optimization of total nitrogen removal from municipal wastewater was investigated in a laboratory-scale photo-sequencing batch reactor (PSBR) operated with a mixed microalgal–bacterial consortium spontaneously acclimatized to real wastewater. No external aeration was provided in the PSBR to reduce energy consumption: oxygen was only supplied by the microalgal photosynthesis. The enhancement of total nitrogen removal was achieved through: (1) feeding of wastewater in the dark phase to provide readily biodegradable COD when oxygen was not produced, promoting denitrification; (2) intermittent use of the mixer to favor simultaneous nitrification–denitrification inside the dense flocs and to achieve 41% energy saving with respect to continuous mixing. Efficient COD removal (86 ± 2%) was observed, obtaining average effluent concentrations of 37 mg/L and 22 mg/L of total COD and soluble COD, respectively. TKN removal was 97 ± 3%, with an average effluent concentration of 0.5 ± 0.7 mg NH_4^+-N/L. Assimilation of nitrogen by heterotrophic bacteria accounted only for 20% of TKN removal, whilst the major part of TKN was nitrified. In particular, the nitrification rate was 1.9 mgN L^{-1} h^{-1} (specific rate 2.4 mgN $gTSS^{-1}$ h^{-1}), measured with dissolved oxygen near zero, when the oxygen demand was higher than the oxygen produced by photosynthesis. Total nitrogen of 6.3 ± 4.4 mgN/L was measured in the effluent after PSBR optimization.

Key words | microalgae, microalgal–bacterial consortia, nitrification–denitrification, nitrogen removal, photobioreactor, real wastewater

P. Foladori (corresponding author)
S. Petrini
M. Nessenzia
G. Andreottola
Department of Civil, Environmental and Mechanical Engineering,
University of Trento,
via Mesiano 77, Trento 38123,
Italy
E-mail: *paola.foladori@unitn.it*

INTRODUCTION

The treatment of real wastewater with microalgal–bacterial consortia has gained increasing attention in the last few years because the synergistic effects of microalgae and heterotrophic bacteria can be exploited for carbon and nitrogen removal through nitrification–denitrification, ensuring a significant energy saving (Su *et al.* 2012; Zhang *et al.* 2012; He *et al.* 2013; Abinandan & Shanthakumar 2015; Wang *et al.* 2016b; Arcila & Buitrón 2017; Quijano *et al.* 2017).

However, when ammonium removal occurs with nitrifiers in activated sludge, a huge amount of electrical energy is required to supply air in aerobic bioreactors. This energy consumption, represents up to half of the energy expenditure of a wastewater treatment plant (WWTP) based on activated sludge. In particular, considering the annual energy consumption by the air supply in aerated tanks in the range of 4.7–28 kWh PE^{-1} y^{-1} (Foladori *et al.* 2015), a medium-size WWTP (20,000 population equivalent, PE) may consume 94–560 MWh/y, which corresponds to an annual cost of 11,000–64,000 €/y (average EU electricity price for industrial consumers of 0.114 €/kWh; Eurostat 2017).

These figures emphasize the need to explore new alternatives aimed at reducing energy consumption and associated costs of wastewater treatment in WWTPs. In this context, the development of heterogeneous

doi: 10.2166/wst.2018.094

microalgal–bacterial consortia offers the possibility of moving towards quasi-zero-energy consumption; in particular, the oxygen produced by photosynthetic microorganisms may support the oxidation of organic matter and ammonium by heterotrophic and nitrifying bacteria (Van den Hende *et al.* 2014). Suspended-biomass reactors operating as photo-sequencing batch reactors (PSBR) represent a common configuration for the laboratory-scale implementation of microalgal–bacteria consortia (Arcila & Buitrón 2016; Wang *et al.* 2016b) because of the easy setup with small pumps and timers.

With regard to nitrogen removal in microalga–bacteria consortia, the main mechanisms are biomass assimilation, nitrification–denitrification and, when high pH values are reached, ammonia volatilization (Delgadillo-Mirquez *et al.* 2016; Wang *et al.* 2016b). González-Fernández *et al.* (2011) observed that microalgal assimilation of nitrogen is frequently overestimated. Su *et al.* (2011), treating municipal wastewater with a mixed population of microalgae and bacteria, observed that denitrification was considered to be a possible removal mechanism, while ammonia volatilization was excluded due to the pH being lower than 8.5.

In spite of the advantage of oxygen production by photosynthetic microorganisms, dissolved oxygen (DO) released in the bulk liquid may become excessive and may inhibit the transformation of nitrates into nitrogen gas through heterotrophic denitrification, which requires strictly anoxic conditions. Thus, it may be difficult to meet the legal requirements for total nitrogen in the effluent if the operational strategies in the microalgal system are not optimized. Wang *et al.* (2015) achieved 90% of nitrogen removal by alternating light and dark periods and without artificial aeration, but to promote denitrification during the dark period, an organic carbon source was added.

The present experimental study focuses on the optimization of total nitrogen removal in a laboratory-scale PSBR fed with real wastewater and operated with a mixed consortium of microalgae, cyanobacteria and bacteria. The phases in the PSBR were feeding + react + settling + draw, but no idle phase was included in the cycles. The length of the phases and the photoperiod (light/dark) were combined so as to affect the duration of the aerobic and anoxic periods and promote nitrification and denitrification, without the need for an additional external carbon source. This paper contributes to the optimization of a PSBR, in terms of enhancement of nitrogen removal, reduction of the time of treatment, and energy saving where possible, thus improving the global sustainability of the system.

MATERIALS AND METHODS

Influent wastewater

Influent municipal pre-settled wastewater was collected after the primary settler of the Trento Nord WWTP (100,000 PE) and fed in to a laboratory-scale photobioreactor. No filtration of the influent wastewater was performed before feeding in to the reactor. This was to allow the microorganisms naturally present in the influent wastewater to enter the reactor, thus significantly affecting the composition of the microalgal–bacterial consortium developed in the system. Table 1 shows the concentrations of the main parameters in the influent pre-settled wastewater: chemical oxygen demand (COD), soluble COD (sCOD), total Kjeldahl nitrogen (TKN), ammonium nitrogen (NH_4^+-N), nitrite nitrogen (NO_2^--N), nitrate nitrogen (NO_3^--N), TN, total phosphorus (TP), orthophosphate phosphorus (PO_4^{3-}-P) and total suspended solids (TSS). The average concentrations are in agreement with typical values expected in pre-settled wastewater (Tchobanoglous *et al.* 2003).

Photo-sequencing batch reactor

The PSBR setup consisted of a cylindrical bench-scale reactor made of Pyrex glass with diameter of 0.13 m, height of 0.29 m and a working volume of 2 L (Colaver, Italy) not sealed off from atmosphere (Figure 1(a)). The system was equipped with:

- peristaltic pumps (Kronos Seko, Italy) for pumping the influent and discharging the effluent; 0.7 L of influent wastewater was fed per cycle;
- cool-white lamp, 8 LEDs × 0.5 W, 0.18 m high and 0.065 m wide (Orion, Italy), arranged on one side of the PSBR;

Table 1 | Characterization of influent pre-settled wastewater (average ± standard deviation; number of samples = 30)

Parameter	Concentration (mg/L)
COD	257 ± 91
sCOD	120 ± 42
TKN	55 ± 20
NH_4^+-N	50 ± 14
NO_2^--N	0.1 ± 0.1
NO_3^--N	1.2 ± 0.9
TN	54 ± 22
TP	4.8 ± 1.5
PO_4^{3-}-P	2.8 ± 0.9
TSS	184 ± 122

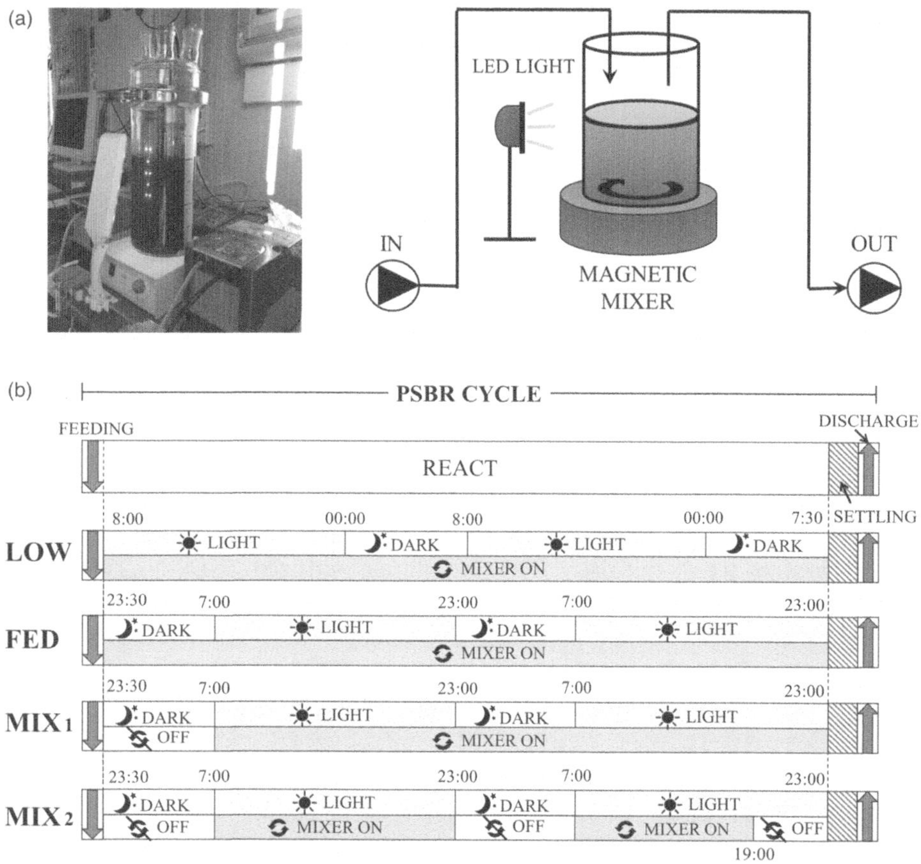

Figure 1 | (a) Laboratory-scale PSBR; (b) operating cycles during the four regimes (LOW, FED, MIX_1, MIX_2).

- magnetic mixer set at 200 rpm (AGE, Velp, Italy) used to maintain the biomass in suspension during the react phase, but avoiding excessive turbulence and reoxygenation from the air;
- timers (TR 612 top2, Theben, Germany) to control on/off of pumps, lamp and mixer.

A light intensity of 30 μmol m^{-2} s^{-1} (expressed as photosynthetically active radiation, PAR) was measured inside the reactor near the top of the liquid surface, at the most exposed point. Light originated from solar light penetration in to the laboratory and the lamp, but the reactor was never exposed to direct sunlight. This light intensity is in the same order of magnitude as other studies in the literature (for example 63 μmol m^{-2} s^{-1} in Karya *et al.* 2013; 90 μmol m^{-2} s^{-1}, in Mennaa *et al.* 2015) but lower than the PAR in outdoor plants which surpasses 200 μmol m^{-2} s^{-1} (light irradiances that saturate photosynthesis). It is worth noting that no external aeration was provided in the PSBR (in order to reduce energy consumption and move towards a zero-energy system); the oxygen available in the reactor was supplied only by microalgal photosynthesis.

The temperature of the mixed liquor during the experiment was 22.8 °C on average.

An acclimatized microalgal–bacterial consortium that had spontaneously developed over more than a year in the PSBR was employed in this study, which was carried out from January to July 2017. Microscopic observations of the biomass showed the presence of *Chlorella*, diatoms, many filamentous cyanobacteria and heterotrophic bacteria embedded in flocs.

Operational strategy in the PSBR

The typical PSBR cycle was based on the following phases: (1) feeding, with duration of 5 min; (2) react, 47.5 hours; (3) settling, 30 min; (4) draw, 5 min. The photoperiod was 16 hours of light + 8 hours of dark and the sequence of the light and dark periods is summarized in Figure 1(b). A constant photoperiod using artificial light allowed us to control the photoperiod and understand the process clearly. The sequence of light and dark causes the alternation of high and low oxygen in the bulk liquid and thus the alternation of aerobic and anoxic conditions, which can be advantageously exploited for optimizing nitrogen removal from wastewater.

In order to enhance total nitrogen removal and save energy in the PSBR, four different operational regimes were applied during the experiment (Figure 1(b)):

- LOW: feeding at the beginning of the light phase (at 8:00) and continuous mixing during the react phase. This basic operational regime was used as the control/starting point, but it did not meet the EU discharge limits in sensitive areas for total nitrogen (15 mg N/L for 10,000 < PE < 100,000; 10 mg N/L for PE > 100,000), requiring progressive optimizations.
- FED: feeding at the beginning of the dark phase (at 23:30) and continuous mixing in the react phase; the aim was to create an anoxic phase at the beginning of the cycle (to exploit readily biodegradable COD, RBCOD, and the anoxic conditions of the influent wastewater) in order to promote denitrification and thus to enhance total nitrogen removal.
- MIX_1: feeding at the beginning of the dark phase (at 23:30) with the mixer off until the beginning of the light phase (at 07:00); the aim was to create an anoxic phase at the beginning of the cycle and to reduce the energy consumption by stopping the mixing for 7.5 hours in the cycle (16% energy saving).
- MIX_2: feeding at the beginning of the dark phase (at 23:30) with the mixer off during all the dark phases and before the end of the react phase (at 19:00); the aim was to create longer anoxic conditions to enhance denitrification and to reduce the energy consumption by stopping the mixing for 19.5 hours in the cycle (41% energy saving).

Due to the good settleability of the biomass in all the operational regimes, the duration of the settling phase (30 min) was appropriate and remained unchanged.

Analytical methods

The chemical parameters COD, sCOD, TKN, NH_4^+-N, NO_2^--N, NO_3^--N, TP, PO_4^{3-}-P and TSS were analyzed in the influent and effluent wastewater according to *Standard Methods* (APHA 2012). Influent TN was estimated as the sum of TKN and NO_x-N. The parameter sCOD was measured after filtration of the sample through a 0.45 μm membrane. Fractionation of COD in the influent wastewater aimed at distinguishing soluble and particulate biodegradable COD was performed by respirometry (Vanrolleghem *et al.* 1999; Ziglio *et al.* 2001). TSS in the mixed liquor (APHA 2012) was measured to determine the concentration of the microalgal–bacterial consortium in the PSBR. DO and temperature were measured with a Multi3410 meter and an FDO®925 sensor (WTW, Germany). Oxidation–reduction potential (ORP) and pH were measured continuously (every 10 min) with a pH3310 meter coupled with the electrodes Sentix®ORP and Sentix®41, respectively (all from WTW, Germany). PAR was measured with a quantum sensor SQ-520 (Apogee Instruments, USA). Microscopic observations were performed using a Nikon Optiphot EFD-3 Microscope (Nikon, Japan) for the morphological characterization of the microalgal–bacterial consortium.

Track studies

Track studies were performed in the PSBR to measure the dynamics of the N-forms during a typical cycle. Samples were collected every hour, then filtered and finally analyzed for NH_4^+-N, NO_2^--N and NO_3^--N. The volumetric removal of NH_4^+-N (or the production of NO_3^--N) expressed as mg N L^{-1} h^{-1} was estimated using the slope of the straight line that interpolates the experimental concentrations over time during the track study. The corresponding specific rate, expressed as mg N g TSS^{-1} h^{-1}, was obtained by dividing the volumetric rate by the TSS concentration in the mixed liquor. The removal rate was also expressed per unit of surface, which was useful for appreciating the footprint of the system, taking into account the height of mixed liquor in the reactor.

Statistical analysis

A Student's t test, producing a p value, was used to evaluate the significant difference in results. Significance was assumed at $p < 0.05$.

RESULTS AND DISCUSSION

PSBR ensures robust and efficient COD removal (>85%) in all the operational regimes

The profiles of COD and sCOD concentrations in the influent and effluent wastewater during the 6-month experiment are shown in Figure 2. Although large variations of COD concentration in real wastewater were common (257 ± 91 mg/L), the average concentrations in the effluent were 37 ± 7 mg COD/L (min–max range equal to 21–54 mg COD/L), without statistically significant variations among the operational regimes. These results largely met

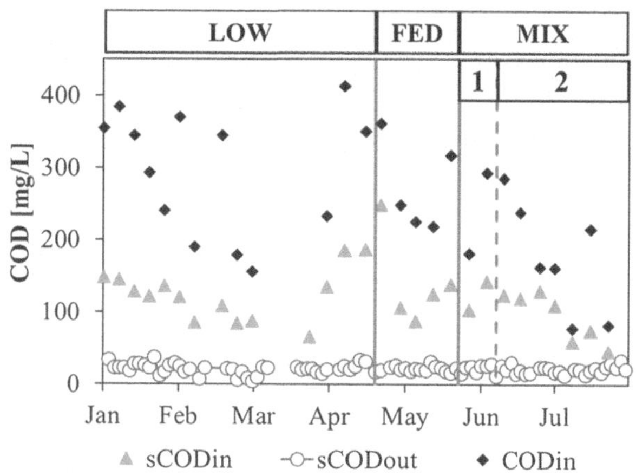

Figure 2 | Profiles of COD and sCOD in the influent and effluent wastewater.

the EU limit for discharge (125 mg COD/L, according to EU Directive 91/271/EEC). The effluent total COD was measured analytically or calculated by summing the soluble COD (22 mg sCOD/L on average) and the particulate COD estimated from effluent TSS concentration, VSS/TSS ratio of mixed liquor (that was 0.80 on average) and the conversion factor 1.42 g COD/gVSS (volatile suspended solids).

High removal efficiency of COD was achieved in all the PSBR operational regimes (86 ± 2% on average), indicating that the variations of influent load and stopping the mixing did not significantly affect organic matter removal. The PSBR appeared to be a very robust process to treat real municipal wastewater. For comparison, COD removal >70% was found by Arango *et al.* (2016) in the treatment of municipal wastewater with an average influent sCOD of 410 mg/L in a PSBR inoculated with microalgae and activated sludge. COD removal efficiency >80% was found in a PSBR for the treatment of synthetic wastewater with 420 mg COD/L (Gutzeit *et al.* 2005). The removal of COD from municipal wastewater by microalgal–bacterial consortia has been reported to be a very effective process (Gutzeit *et al.* 2005; Su *et al.* 2012; Wang *et al.* 2016a).

The fractionation of influent COD permits an understanding of the composition of the effluent COD (Henze *et al.* 2008):

- particulate inert COD (X_I = 46 ± 16 mg/L; 18% of COD) and particulate biodegradable COD (X_S = 100 ± 36 mg/L; 39%) can be flocculated on suspended solids;
- soluble biodegradable COD (S_S = 95 ± 34 mg/L; 37%) is rapidly biodegraded;
- soluble inert COD (S_I = 16 ± 5 mg/L; 6% of COD) cannot be removed and thus it leaves the system in the effluent.

This COD fractionation suggests that the effluent soluble COD (22 ± 6 mg/L) was mainly composed of the soluble inert fraction of the influent wastewater (S_I) which passed unchanged through the PSBR, while the rest may have been soluble microbial products (SMPs) originating from substrate metabolism and biomass decay (Azami *et al.* 2012).

The particulate COD in the effluent was associated with some suspended solids floating and leaving the system during discharge. The effluent TSS concentration was around 13 mg TSS/L during the whole experiment, indicating very good sedimentation of the biological flocs developed in the PSBR.

With regard to the removal of nutrients in the whole experiment, average P concentration decreased from 4.8 ± 1.5 mg/L in the influent to 2.5 ± 0.8 mg/L in the effluent (average removal efficiency of 47 ± 19%), while N removal is discussed in depth in the following sections.

TKN removal >95% occurs in the PSBR at near-zero DO

The influent TKN (55 ± 20 mg TKN/L) was almost entirely removed in the PSBR, leading to effluent concentrations of 1.5 ± 2.1 mg TKN/L, 0.5 ± 0.7 mgNH_4^+-N/L and 0.1 ± 0.4 mg NO_2^--N/L.

There are three ways in which TKN and ammonium can be transformed by the microalgal–bacterial consortium: ammonia volatilization, assimilation by biomass and nitrification. In particular, volatilization was negligible in this PSBR because the pH in the reactor was 7.6 ± 0.3 on average during the whole experiment, and ammonia volatilization requires relatively high pH values. Nitrogen is also assimilated and incorporated into biomass through cell synthesis during the growth of heterotrophs and, to a lesser extent, autotrophs (Tchobanoglous *et al.* 2003). The real amount of N removed by net microbial biomass growth was estimated taking into account sludge production, which was 0.03 g TSS/d. Considering the specific content of N in biomass (0.10 g N/g VSS), an amount of 2.4 mg N/L was used daily for the net growth. It accounts for about 4% of the TKN transformed in the system. This highlights that sludge production and nutrient uptake in PSBRs applied to the highly efficient treatment of real municipal wastewater may be significantly lower than that expected in PSBRs specifically designed to maximize algal growth.

The N mass balance reveals that the major role in TKN removal in the PSBR was played by autotrophic nitrification. Karya *et al.* (2013) also observed that nitrification accounted for more than 80% of ammonium removal during the treatment of synthetic artificial wastewater in an open photobioreactor inoculated with nitrifying activated sludge and a pure culture of *Scenedesmus* sp.

The very low concentration of NH_4^+-N in the effluent wastewater (<1.0 mg/L in 89 samples, with the exception of only three cases) demonstrates that nitrification was complete during the 6-month experiment and not negatively affected by the different operational regimes implemented in the PSBR. As shown in Figure 3, the average TKN removal efficiency was 97 ± 3% and very stable during the whole period, even when the influent TKN concentration reached very high values of around 100 mg TKN/L (Figure 3).

On the basis of these considerations, the nitrification kinetic in the PSBR reactor was investigated in depth. Track studies were carried out in the reactor after the addition of influent wastewater with a known amount of ammonium and the time profiles of NH_4^+-N, NO_2^--N, NO_3^--N were measured. The results obtained in a typical track study carried out during the light phase of the LOW operational regime (used as control) are shown in Figure 4. The decrease of ammonium occurred rapidly and the initial concentration of 14 mg NH_4^+-N/L was completely removed in less than 7 hours, corresponding to a nitrification rate of 1.9 mg N L^{-1} h^{-1}. Considering that the biomass in the PSBR was 0.8 g TSS/L, a specific nitrification rate of 2.4 mg NH_4^+-N g TSS^{-1} h^{-1} was calculated. Similar results for nitrification kinetics were found during the other operational regimes (FED, MIX_1, MIX_2), which did not significantly affect ammonium removal.

Considering the hourly nitrification rate, a daily rate of 30 mg NH_4^+-N L^{-1} d^{-1} can be calculated for a daily

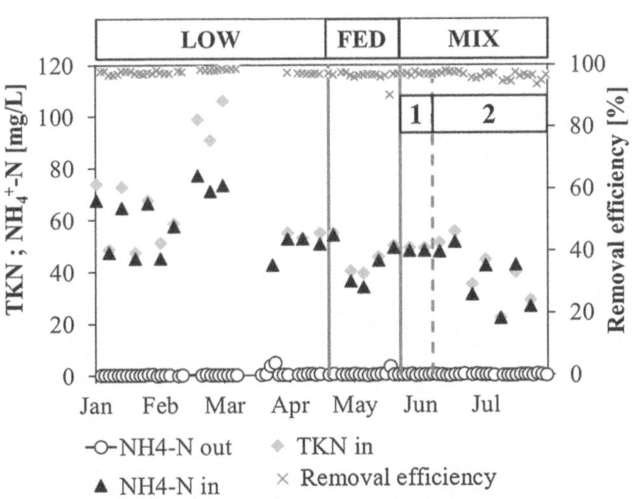

Figure 3 | Profiles of TKN and NH_4^+-N in the influent and effluent wastewater, and removal efficiency of TKN.

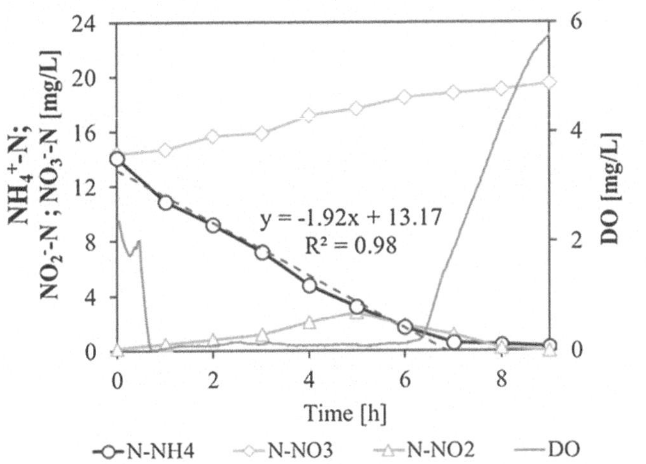

Figure 4 | Track study of NH_4^+-N, NO_3-N and NO_2-N in the PSBR to evaluate nitrification kinetics during the light phase.

photoperiod of 16/24 hours, which corresponds to a surface removal rate of 5.4 g NH_4^+-N m^{-2} d^{-1} (considering the reactor's diameter of 0.13 m, top surface area of 0.0133 m^2 and volume of 0.0024 m^3). This specific value gives an approximate estimate of the footprint. For a comparison, 10.2 g NH_4^+-N m^{-2} d^{-1} was found in a flat PSBR with irradiance an order of magnitude higher than in this experiment (Rada-Ariza *et al.* 2017).

The specific value of 2.4 mg N $gTSS^{-1}$ h^{-1} found for the microalgal–bacterial consortium is similar to the specific nitrification rate reported in the literature for activated sludge systems (Tchobanoglous *et al.* 2003). However, the important difference is that the microalgal–bacterial consortium works with a limited TSS concentration (around 1 g TSS/L) to allow light to penetrate in the biomass, while activated sludge works at 4 g TSS/L. This results in a significant difference in the volumetric nitrification rate of microalgal consortia and activated sludge. However, a further investigation is needed at a real scale to evaluate the volumetric nitrification rate of microalgal–bacteria consortia under more realistic conditions.

From Figure 4 it can be observed that the decrease of ammonium occurred steadily over a long period (>5 hours) with DO concentrations near zero. Since no external aeration was provided to the PSBR, only photosynthetic activity was responsible for oxygen production during the light phase. In the presence of NH_4^+, the oxygen demand required for nitrification surpassed the amount produced by photosynthesis, resulting in a near-zero DO concentration. Then, when ammonium was completely depleted, the DO concentration increased gradually, reaching values above 5 mg O_2/L.

Concerning the profile of NO_3^--N, the initial concentration of 14 mg N/L increased to 19.5 mg N/L at the end of the track study, indicating that only 40% of the ammonium removed was found as nitrates. The explanation of this low increase in nitrates is that simultaneous nitrification–denitrification occurred in this PSBR. The availability of soluble COD, which decreased from 121 mg/L in the influent wastewater to 17 mg/L at the end of the track study, may support denitrification. Analogous observations were found by Tiron *et al.* (2015) in activated algae granules treating low-strength wastewater in a PSBR. In the present reactor, dense clusters of microalgae, bacteria and abiotic solids were also observed, and this may contribute to initiating conditions for partial denitrification.

Although the near-zero DO phase may suggest the absence of oxygen, it must not be confused with a completely anoxic phase, because photosynthesis occurred and oxygen was continuously produced in the system, but there was not enough oxygen to satisfy the requirements of the biomass. The advantage of nitrification in municipal wastewater by exploiting the oxygen produced in a microalgal–bacterial consortium, without the need of artificial aeration, may offer very powerful opportunities in the evolution of WWTPs. However, further research is needed to prove the feasibility of the process under real conditions in order to evaluate the influence of the main parameters' variations (e.g. light supply, mass transfer and temperature).

Total nitrogen <10 mgN/L in the effluent through the enhancement of denitrification in the PSBR

The LOW operational regime, used as the control, produced a concentration of TN in the effluent wastewater in the range 6.8–40.0 mg N/L, which was over the limits of 10–15 mg N/L (as an annual average) for discharge in sensitive areas imposed by EU Directive 91/271/EEC (Figure 5). The high TN concentration in the effluent, largely composed by nitrates, was the result of: (1) slight assimilation of N by biomass and (2) only partial denitrification.

To enhance TN removal, efforts aimed at increasing the assimilation of N by pure microalgae do not seem to be so effective, because assimilation permits moderate N removal rates, often requires a longer reaction time and thus larger volumes for the treatment of wastewater.

The great advantage of using microalgal–bacterial consortia instead of pure microalgal strains is that bacterial nitrification can be exploited to speed up ammonium removal. Then a higher TN removal can be achieved by ensuring a denitrification step (Rada-Ariza *et al.* 2017).

In our case, the microalgal–bacterial consortium removed 30 mg N L^{-1} d^{-1} for a daily photoperiod of 16/24 hours, which was noticeably higher than those found by Álvarez-Díaz *et al.* (2017) (N removal of 6.6 mg N L^{-1} d^{-1}) treating secondary effluents under 14/24 hours of light with *Chlorella sorokiniana*. A similar N removal (6.63 mg N L^{-1} d^{-1}) was reported by Cabanelas *et al.* (2013) in the case of *Chlorella vulgaris* treating urban wastewater under 14/24 hours of light.

Concerning denitrification, there could still be room for improvement, when suitable conditions were applied in the PSBR. In the LOW operational regime the ORP profile (data not shown) assumed negative values only for 2–4 hours per cycle after the feeding phase, not long enough to create the conditions for full denitrification in the system. Therefore, the enhancement of TN removal was investigated through the optimization of the denitrification process, but without worsening the effluent quality in terms of COD and NH_4^+-N. This was achieved with the following modifications:

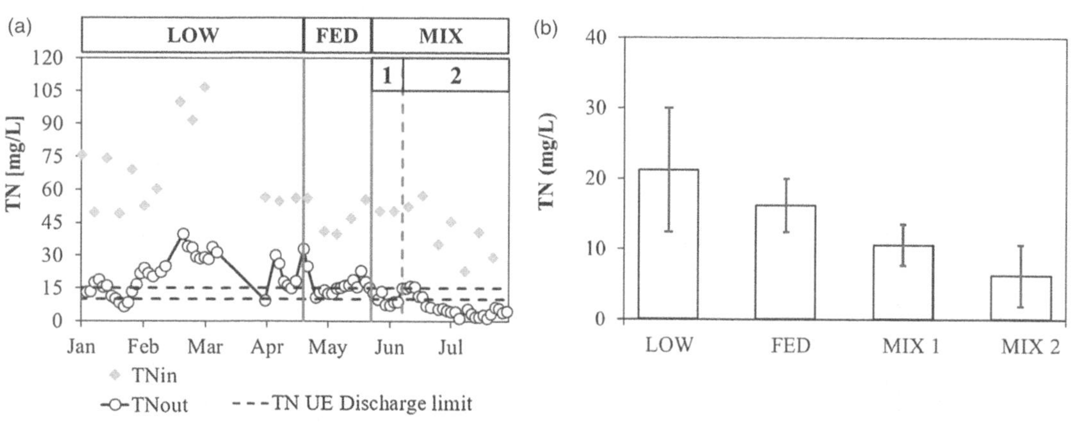

Figure 5 | (a) Profiles of total nitrogen in the influent and effluent wastewater compared with the discharge limit (10–15 mgN/L; EU Directive 91/271/EEC); (b) average effluent concentration and standard deviation for each operational regime.

1. The time of feeding (FED operational regime): wastewater was fed during the dark period instead of the light period, in order to provide readily biodegradable COD, required for denitrification, during the dark period when oxygen is consumed but not produced.
2. The intermittent use of the mixer (MIX operational regime): switched on during the light phases, when the biomass requires complete exposure to the light for photosynthesis, and switched off during the night, to avoid turbulence, favor denitrification inside flocs and energy saving; in this phase the biomass settled but the process was not negatively affected.

As shown in Figure 5, the application of the FED and MIX operational regimes resulted in a gradually decreasing profile of TN. The average TN concentration in the effluent was 21 ± 8.8 mg N/L in the LOW operational regime (not optimized), 16.2 ± 3.8 mg N/L in the FED regime, and 10.6 ± 2.9 mg N/L in the MIX_1 regime, reaching the lowest average value of 6.3 ± 4.4 mg N/L in the MIX_2 regime, which met the EU regulation limits. The enhancement of total N removal due to the progressive optimization of the PSBR was evident only when the MIX regimes were applied. In fact, no statistically significant difference was observed between the LOW and FED regimes, while effluent TN decreased significantly in the MIX_1 and MIX_2 regimes ($p < 0.05$).

The decrease of TN occurred along with the extension of the period with negative ORP in the cycle. ORP values in the range −50 to − 200 mV, which are favorable for denitrification, were measured for periods of 7–8 hours during the night in the MIX_2 regime. In this case, the mixer was switched off in the light phase (at 19:00, for 4 hours before settling and discharge), in order to move toward anoxic conditions before feeding, thus favoring denitrification at the beginning of the subsequent cycle.

CONCLUSIONS

The total nitrogen removal in a PSBR fed with real wastewater and operated with a mixed microalgal–bacterial consortium was investigated and optimized to achieve effluent TN concentrations in accordance with EU requirements for discharge in sensitive areas.

COD and TKN removal in the whole experiment were 86 ± 2% and 97 ± 3%, producing an average effluent concentration of 37 mg COD/L and 0.5 ± 0.7 mg NH_4^+-N/L. In spite of long periods with DO near zero during the PSBR cycle, due to an oxygen demand higher than photosynthesis production, the nitrification rate was 1.9 mg N L^{-1} h^{-1}, corresponding to a specific rate of 2.4 mg N g TSS^{-1} h^{-1}. PSBR optimization was aimed at enhancing denitrification through the feeding during the dark and the intermittent use of the mixer; in this way, the total nitrogen in the effluent during the MIX_2 regime reached 6.3 ± 4.4 mg N/L.

In conclusion, the PSBR allowed the nitrification of municipal wastewater under laboratory conditions without the need for artificial aeration and with an important energy saving, while simultaneous denitrification was exploited to enhance total nitrogen removal. However, considering the ability of microalgae or associated bacteria to synthesize N_2O, further research is needed to couple the nitrogen balance with a greenhouse gas mass balance.

ACKNOWLEDGEMENTS

The authors thanks Giovanna Flaim (Fondazione Edmund Mach, Italy) for the cooperation in the investigation of microalgae in the photobioreactor.

REFERENCES

Abinandan, S. & Shanthakumar, S. 2015 Challenges and opportunities in application of microalgae (*Chlorophyta*) for wastewater treatment: a review. *Renewable and Sustainable Energy Reviews* **52**, 123–132. http://dx.doi.org/10.1016/j.rser.2015.07.086.

Álvarez-Díaz, P. D., Ruiz, J., Arbib, Z., Barragán, J., Garrido-Pérez, M. C. & Perales, J. A. 2017 Freshwater microalgae selection for simultaneous wastewater nutrient removal and lipid production. *Algal Research* **24**, 477–485. http://dx.doi.org/10.1016/j.algal.2017.02.006.

APHA 2012 *Standard Methods for the Examination of Water and Wastewater*, 22nd edn. American Public Health Association/American Water Works Association/Water Environment Federation, Washington, DC, USA.

Arango, L., Cuervo, F. M., González-Sánchez, A. & Buitrón, G. 2016 Effect of microalgae inoculation on the start-up of microalgae–bacteria systems treating municipal, piggery and digestate wastewaters. *Water Science & Technology* **73** (3), 687–696. doi:10.2166/wst.2015.544.

Arcila, J. S. & Buitrón, G. 2016 Microalgae–bacteria aggregates: effect of the hydraulic retention time on the municipal wastewater treatment, biomass settleability and methane potential. *Journal of Chemical Technology and Biotechnology* **91**, 2862–2870. doi 10.1002/jctb.4901.

Arcila, J. S. & Buitrón, G. 2017 Influence of solar irradiance levels on the formation of microalgae-bacteria aggregates for municipal wastewater treatment. *Algal Research* **27**, 190–197. https://doi.org/10.1016/j.algal.2017.09.011.

Azami, H., Sarrafzadeh, M. H. & Mehrnia, M. R. 2012 Soluble microbial products (SMPs) release in activated sludge systems: a review. *Iranian Journal of Environmental Health Science and Engineering* **9** (1), 30. doi:10.1186/1735-2746-9-30.

Cabanelas, I. T. D., Ruiz, J., Arbib, Z., Chinalia, F. A., Garrido-Pérez, C., Rogalla, F., Nascimento, I. A. & Perales, J. A. 2013 Comparing the use of different domestic wastewaters for coupling microalgal production and nutrient removal. *Bioresource Technology* **131**, 429–436. http://dx.doi.org/10.1016/j.biortech.2012.12.152.

Delgadillo-Mirquez, L., Lopes, F., Taidi, B. & Pareau, D. 2016 Nitrogen and phosphate removal from wastewater with a mixed microalgae and bacteria culture. *Biotechnology Reports* **11**, 18–26. http://dx.doi.org/10.1016/j.btre.2016.04.003.

EU Directive 91/271/EEC 1991 Council Directive of 21 May 1991 concerning urban waste water treatment.

Eurostat, Statistics Explained: Electricity price statistics. http://ec.europa.eu/eurostat/statistics-explained/index.php/Electricity_price_statistics (accessed 26 September 2017).

Foladori, P., Vaccari, M. & Vitali, F. 2015 Energy audit in small wastewater treatment plants: methodology, energy consumption indicators, and lessons learned. *Water Science & Technology* **72** (6), 1007–1015. doi:10.2166/wst.2015.306.

González-Fernández, C., Molinuevo-Salces, B. & Cruz García-González, M. 2011 Nitrogen transformations under different conditions in open ponds by means of microalgae–bacteria consortium treating pig slurry. *Bioresource Technology* **102**, 960–966. doi:10.1016/j.biortech.2010.09.052.

Gutzeit, G., Lorch, D., Weber, A., Engels, M. & Neis, U. 2005 Bioflocculent algal-bacterial biomass improves low-cost wastewater treatment. *Water Science & Technology* **52** (12), 9–18.

He, P. J., Mao, B., Lü, F., Shao, L. M., Lee, D. J. & Chang, J. S. 2013 The combined effect of bacteria and *Chlorella vulgaris* on the treatment of municipal wastewaters. *Bioresource Technology* **146**, 562–568. http://dx.doi.org/10.1016/j.biortech.2013.07.111.

Henze, M., van Loosdrecht, M. C. M., Ekama, G. A. & Brdjanovic, D. 2008 *Biological Wastewater Treatment: Principles, Design and Modelling*. IWA Publishing, London, UK, pp. 526.

Karya, N. G. A. I., van der Steen, N. P. & Lens, P. N. L. 2013 Photo-oxygenation to support nitrification in an algal–bacterial consortium treating artificial wastewater. *Bioresource Technology* **134**, 244–250. http://dx.doi.org/10.1016/j.biortech.2013.02.005.

Mennaa, F. Z., Arbib, Z. & Perales, J. A. 2015 Urban wastewater treatment by seven species of microalgae and an algal bloom: biomass production, N and P removal kinetics and harvestability. *Water Research* **83**, 42–51. http://dx.doi.org/10.1016/j.watres.2015.06.007.

Quijano, G., Arcila, J. S. & Buitrón, G. 2017 Microalgal-bacterial aggregates: applications and perspectives for wastewater treatment. *Biotechnology Advances* **35**, 772–781. http://dx.doi.org/10.1016/j.biotechadv.2017.07.003.

Rada-Ariza, A. M., Lopez-Vazquez, C. M., van der Steen, N. P. & Lens, P. N. L. 2017 Nitrification by microalgal-bacterial consortia for ammonium removal in flat panel sequencing batch photo-bioreactors. *Bioresource Technology* **245**, 81–89.

Su, Y., Mennerich, A. & Urban, B. 2011 Municipal wastewater treatment and biomass accumulation with a wastewater-born and settleable algal-bacterial culture. *Water Research* **45**, 3351–3358. doi:10.1016/j.watres.2011.03.046.

Su, Y., Mennerich, A. & Urban, B. 2012 Synergistic cooperation between wastewater-born algae and activated sludge for wastewater treatment: influence of algae and sludge inoculation ratios. *Bioresource Technology* **150**, 67–73. doi:10.1016/j.biortech.2011.11.113.

Tchobanoglous, G., Burton, F. L. & Stensel, H. D. 2003 *Wastewater Engineering: Treatment and Reuse*, 4th edn. McGraw-Hill Metcalf & Eddy Inc., Boston.

Tiron, O., Bumbac, C., Patroescu, I. V., Badescu, V. R. & Postolache, C. 2015 Granular activated algae for wastewater treatment. *Water Science & Technology* **71** (6), 832–839. doi:10.2166/wst.2015.010.

Van Den Hende, S., Carré, E., Cocaud, E., Beelen, V., Boon, N. & Vervaeren, H. 2014 Treatment of industrial wastewaters by microalgal bacterial flocs in sequencing batch reactors. *Bioresource Technology* **161**, 245–254. http://dx.doi.org/10.1016/j.biortech.2014.03.057.

Vanrolleghem, P. A., Spanjers, H., Petersen, B., Ginestet, P. & Takacs, I. 1999 Estimating (combinations of) activated sludge Model No. 1 parameters and components by respirometry. *Water Science & Technology* **39** (1), 195–214.

Wang, M., Yang, H., Ergas, S. J. & van der Steen, P. 2015 A novel shortcut nitrogen removal process using an algal–bacterial consortium in a photo-sequencing batch reactor (PSBR). *Water Research* **87**, 38–48.

Wang, L., Liu, J., Zhao, Q., Wei, W. & Sun, Y. 2016a Comparative study of wastewater treatment and nutrient recycle via activated sludge, microalgae and combination systems. *Bioresource Technology* **211**, 1–5. http://dx.doi.org/10.1016/j.biortech.2016.03.048.

Wang, Y., Ho, S., Cheng, C., Guo, W., Nagarajan, D., Ren, N., Lee, D. & Chang, J. 2016b Perspectives on the feasibility of using microalgae for industrial wastewater treatment. *Bioresource Technology* **222**, 485–497. http://dx.doi.org/10.1016/j.biortech.2016.09.106.

Zhang, Y., Su, H., Zhong, Y., Zhang, C., Shen, Z., Sang, W., Yan, G. & Zhou, X. 2012 The effect of bacterial contamination on the heterotrophic cultivation of *Chlorella pyrenoidosa* in wastewater from the production of soybean products. *Water Research* **46**, 5509–5516. http://dx.doi.org/10.1016/j.watres.2012.07.025.

Ziglio, G., Andreottola, G., Foladori, P. & Ragazzi, M. 2001 Experimental validation of a single-OUR method for wastewater RBCOD characterisation. *Water Science & Technology* **43** (11), 119–126.

First received 31 October 2017; accepted in revised form 19 February 2018. Available online 11 April 2018

Photosynthetic oxygenation for urine nitrification

Maarten Muys, Joeri Coppens, Nico Boon and Siegfried E. Vlaeminck

ABSTRACT

Human urine accounts for only a fraction of the sewage volume, but it contains the majority of valuable nutrient load in wastewater. In this study, synthetic urine was nitrified in a closed photo-bioreactor through photosynthetic oxygenation by means of a consortium of microalgae and nitrifying bacteria. *In situ* production of oxygen by photosynthetic organisms has the potential to reduce the energy costs linked to conventional aeration. This energy-efficient strategy results in stable urine for further nutrient recovery, while part of the nutrients are biologically recovered in the form of valuable biomass. In this study, urine was nitrified for the first time without conventional aeration at a maximum photosynthetic oxygenation rate of 160 mg O_2 gVSS^{-1} d^{-1} (VSS: volatile suspended solids). A maximum volumetric nitrification rate of 67 mg N L^{-1} d^{-1} was achieved on 12% diluted synthetic urine. Chemical oxygen demand (COD) removal efficiencies were situated between 44% and 83% at a removal rate of 24 mg COD gVSS^{-1} d^{-1}. After 180 days, microscopic observations revealed that *Scenedesmus* sp. was the dominant microalga. Overall, photosynthetic oxygenation for urine nitrification is promising as a highly electricity efficient approach for further nutrient recovery.

Key words | activated sludge, high-rate algal ponds, microalgal-bacterial consortium, photo-aeration, resource recovery, source separation

Maarten Muys
Siegfried E. Vlaeminck (corresponding author)
Research Group of Sustainable Energy, Air and Water Technology, Department of Bioscience Engineering,
University of Antwerp,
Groenenborgerlaan 171, 2020 Antwerpen,
Belgium
E-mail: *siegfried.vlaeminck@uantwerpen.be*

Joeri Coppens
Nico Boon
Siegfried E. Vlaeminck
Center for Microbial Ecology and Technology (CMET),
Ghent University,
Coupure Links 653, 9000 Gent,
Belgium

INTRODUCTION

In the context of sustainable resource management, the transition towards energy-efficient and resource-recovery focused wastewater treatment is pivotal. For nutrients such as nitrogen (N) and phosphorus (P), the diluted characteristics of domestic wastewater make direct nutrient recovery technologically challenging and resource intensive. Nutrient recovery is therefore recommended for more concentrated waste streams, such as manure, digestate or urine (Verstraete *et al.* 2016). Urine accounts for only 1% of the total domestic wastewater volume, while it contains approximately 40% of the phosphorus load, 69% of the nitrogen load, and 60% of the potassium load arriving at municipal wastewater treatment plants (Kujawa-Roeleveld & Zeeman 2006). This makes it a valuable target stream for nutrient recovery.

While direct application of urine as fertilizer has been common practice in many rural societies around the world, the high water content makes transportation and application costly, while the high urine salinity causes soil salinization (Basakcilardan-Kabakci *et al.* 2007). In addition, potential presence of pathogens and pharmaceuticals further increases health concerns towards farmers and consumers (Udert *et al.* 2006). More advanced chemical and biological nutrient recovery strategies have therefore been demonstrated, such as ammonia stripping, struvite precipitation and microalgae cultivation (Maurer *et al.* 2006; Tuantet *et al.* 2013). Urine is, however, highly unstable, as urea hydrolysis through bacterial urease results in the production of ammonia and bicarbonate. The subsequent pH rise induces nutrient losses through unwanted phosphate precipitation and ammonia volatilization and concomitant odour and toxicity issues (Udert *et al.* 2003).

Chemical stabilization, for example through acid dosage, or biological stabilization through nitrification, has therefore been suggested as a pre-treatment step prior to nutrient recovery (Maurer *et al.* 2006; Feng *et al.* 2008). Biological urine stabilization through nitrification converts volatile ammonia to nitrate, thereby allowing for long-term storage and further use as agricultural fertilizer or for microalgae cultivation. However, oxygen requirements associated with the nitrification process result in large energy demands due to energy-intensive conventional aeration. Based on an oxygenation

doi: 10.2166/wst.2018.200

efficiency of 1 kg O_2 kWh^{-1} (Metcalf & Eddy 2002) and a urine chemical oxygen demand (COD)/N ratio of 0.8, the electricity need for nitrification and COD oxidation with conventional aeration is 31 kWh m^{-3} urine (solids retention time (SRT): 6.66 days; 25 °C).

An alternative for conventional aeration is photosynthetic oxygenation, or *in situ* oxygen production by photosynthetic organisms such as microalgae. By providing the oxygen demand for autotrophic nitrification and heterotrophic carbon oxidation through *in situ* photosynthesis while consuming heterotrophically produced carbon dioxide, electricity costs and greenhouse gas emissions affiliated to conventional aeration and wastewater treatment can be reduced (Praveen & Loh 2015). In addition, the produced microalgal-bacterial biomass can be used for energy production (anaerobic digestion), as a resource for the production of high-value biochemical and biofuels or it can be applied as slow-release fertilizer or microbial protein (Tuantet *et al.* 2013; Coppens *et al.* 2016).

The combination of photosynthetic oxygenation with nitrification or nitrification–denitrification has been described in both high-rate algal ponds (HRAPs) and in photo-bioreactors (PBRs), for various waste streams (Table 1). Karya *et al.* (2013) achieved a volumetric nitrification rate of 43 mg N L^{-1} d^{-1} and photo-oxygenation rate of 140 mg O_2 $gVSS^{-1}$ d^{-1} (VSS: volatile suspended solids), while van der Steen *et al.* (2015) reported a volumetric nitrification rate of 46 mg N L^{-1} d^{-1} and photo-oxygenation rate of 234 mg O_2 $gVSS^{-1}$ d^{-1}. Furthermore, direct microalgae cultivation on urine has been established (Adamsson 2000; Yang *et al.* 2008; Tuantet *et al.* 2013) and superior microalgal growth on nitrified urine compared to untreated urine has been demonstrated (Coppens *et al.* 2016).

Photosynthetic oxygenation for urine nitrification has so far not been documented. The goal of this study was to develop a microalgal-bacterial consortium for urine nitrification through photosynthetic oxygenation. The influence of salinity and ammonium concentration on microalgal growth was determined and subsequently, a PBR was operated for urine nitrification in the absence of external aeration.

METHODS

Nitrifying sludge and microalgae

The inoculum comprised commercially available nitrifying activated sludge (Avecom, Belgium), the microalgal species *Chlorella* sp., *Haematococcus* sp., *Desmodesmus* sp., *Ankistrodesmus* sp., *Pediastrum duplex*, *Chlorella vulgaris* and *Nannochloropsis* sp. (Marine Biology research group, Ghent University) and microalgae acquired from a grassland pond (Ghent, Belgium). The initial biomass concentration was set at 0.7 g VSS L^{-1} of which half was microalgal biomass, the other half activated sludge. Prior to inoculation, all microalgae were cultivated at 25 ± 1 °C in enriched seawater artificial water (ESAW) medium made with artificial seawater (Harrison *et al.* 1980), to provide a salinity level in the range of that of urine. The cultures were aerated with sterile air and continuously illuminated (Philips TL-D 90 De Lux 36 W).

Influence of ammonium and salt concentration on microalgal growth

A mixture of all microalgae was exposed to ESAW medium containing 1, 2, 3.5 and 5 g L^{-1} NaCl and 50, 100, 200 and 1,000 mg NH_4^+-N L^{-1}, brought to pH 6 with a phosphate buffer. The experiment was realized in quadruplicate in 96 well microtiter plates (300 μL per well). The optical density (OD) was measured daily at 620 nm with a plate reader (Tecan Infinite), after shaking. Each well was inoculated between 0.10 and 0.17 (OD620). The plates were incubated in an orbital shaking incubator at 25 °C for 160 hours.

PBR design and operation

A Plexiglas, gas-tight, bubble column PBR with a working volume of 4 L (ø 12 cm; height 50 cm), was operated in semi-continuous mode (Figure 1). The illuminated surface area was 0.16 m^2 resulting in a volume to surface ratio of 0.025 m^3 m^{-2}. Three fluorescent growth lamps (Grolux T5, 24 W, Sylvania) provided a photon flux density of 300 μmol PAR m^{-2} s^{-1} (PAR: photosynthetically active radiation), measured in a straight angle from the lamp at the outer reactor wall (Fieldscout quantum light meter, USA). Liquid recirculation at 0.5 L min^{-1}, to facilitate pH and dissolved oxygen (DO) measurements and mixed liquor sampling, and headspace gas recirculation at 6 L min^{-1} took place. CO_2 and N_2 gas were dosed in a 10/90 (v/v) ratio (Bronkhorst high-tech EL-FLOW select mass flow meter/controller) as 1% of the gas recirculation flow. Where mentioned, co-aeration with compressed air was applied at 5 L h^{-1}. The pH was controlled between 6.5 and 7 (Prominent dulcometer; pH electrode dulcotest PHEP-112). The temperature was kept at 25 ± 1 °C by operating in a temperature controlled room.

Initially, 10% diluted synthetic urine medium (Brooks & Keevil 1997) with additional trace element solutions

Table 1 | Application of photosynthetic oxygenation by means of algal-bacterial consortia for the treatment of various liquid waste streams (HRAP: high-rate algal pond; PBR: photo-bioreactor; SBR: sequencing batch reactor; SRT: solids retention time; HRT: hydraulic retention time; CSTR: continuously stirred tank reactor)

Waste stream	Consortium	Set-up and operation	N-loading rate (mg N L^{-1} d^{-1})	Volumetric nitrification rate (mg N L^{-1} d^{-1})	Biomass specific nitrification rate (mg N $gVSS^{-1}$ d^{-1})	Nitrification efficiency (%)	COD oxidation rate (mg COD $gVSS^{-1}$ d^{-1})	Photo-oxygenation rate (mg O_2 $gVSS^{-1}$ d^{-1})	Reference
10% swine manure	Inoculum from stabilization pond treating domestic wastewater	Outdoor HRAP; HRT: 10 d	30	9	10	30	73	110	de Godos *et al.* (2009)
Modified BG-11 medium	*S. quadricauda*, nitrifiers	PBR as SBR; SRT: 15–30 d; 60 μmol m^{-2} s^{-1}	50	43	31	85	0	140	Karya *et al.* (2013)
Pre-treated sewage	*C. vulgaris*, activated sludge	Continuous PBR; SRT: 15 d; HRT: 1.5 d; 2,000 μmol m^{-2} s^{-1}	50	2	0.97	2.9	184	154	Gutzeit *et al.* (2005)
10% (v/v) molasses wastewater	Municipal activated sludge	PBR as SBR; HRT: 5 d; 17 W LED	43	36	28	83	47	168	Tsioptsias *et al.* (2017)
Modified BG-11 medium	Microalgae, activated sludge	PBR as CSTR; HRT: 1 d; SRT: 15 d; 66 μmol m^{-2} s^{-1}	66	46	48	70	0	234	van der Steen *et al.* (2015)
12% synthetic urine	Microalgae, activated sludge	Semi-continuous PBR; HRT: 6.67 d; SRT = HRT; 300 μmol m^{-2} s^{-1}	97	67	31	64	24	160	This study

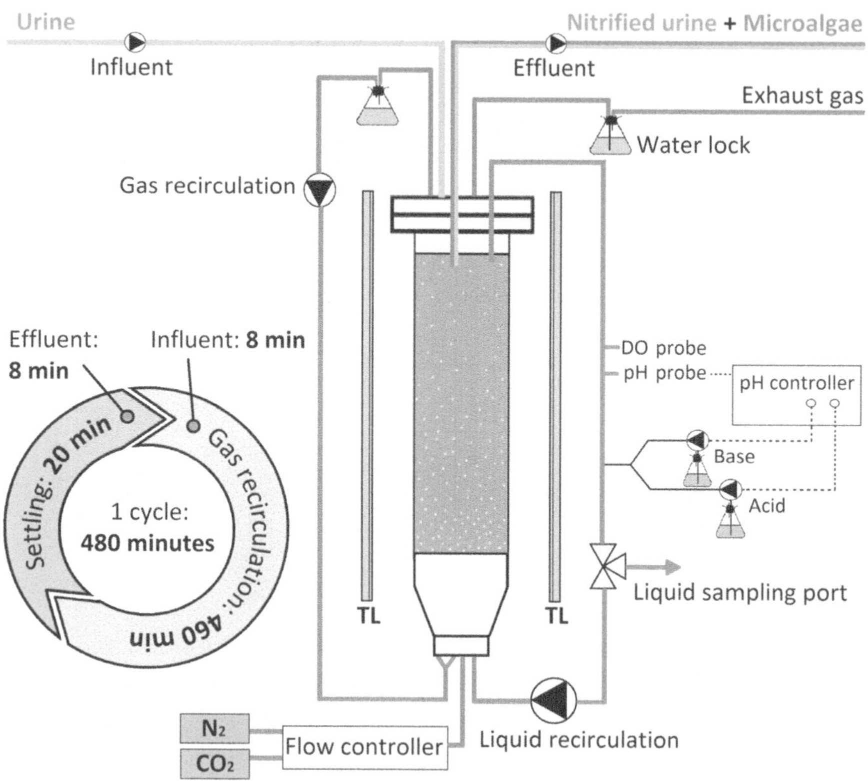

Figure 1 | Schematic overview of the PBR bubble column and the timeline for one operation cycle of 8 hours. Full lines: liquid; dotted lines: electric signal; blue = gas flow; green = bioreactor content. Please refer to the online version of this paper to see this figure in colour: http://dx.doi.org/10.2166/wst.2018.200.

(Kuai & Verstraete 1998) was fed, implying a nitrogen loading rate of 50 mg N L^{-1} d^{-1}. The influent electrical conductivity was altered to that of a 33% dilution of urine (6.66 mS cm^{-1}) using NaCl. The nitrogen loading was distributed over time using three cycles of 8 hours per day. Influent was dosed in the first 8 minutes of each cycle during gas recirculation, while effluent was extracted in the last 8 minutes of each cycle, at the end of the settling phase.

A nitrogen conversion balance was made to calculate how much of the incoming nitrogen was nitrified and how much was assimilated by the microalgae, nitrifiers and heterotrophs. This was done over six periods (I to VI) of each 7 days, just before an operational change took place (Table S1, Supplementary material, available with the online version of this paper). Incoming nitrogen was defined as nitrogen available for nitrification and for biomass growth, meaning the average total influent nitrogen together with the average Kjeldahl and nitrite nitrogen present in the reactor. Since ammonium is the preferred microalgal nitrogen source, it was assumed that nitrate was not assimilated. Incoming nitrogen can accumulate in the reactor unconverted (as urea or ammonium) or in the form of biomass, or it can be removed with the effluent as such or in the form of biomass. The amount of biomass-assimilated nitrogen removed with the effluent was calculated based on the mean biomass concentration in the reactor and the assumption that effluent equalled reactor biomass concentration. The biomass nitrogen content was determined by analysing filtered (0.45 μm) versus non-filtered reactor content for Kjeldahl-N. Nitrogen assimilation in nitrifying biomass was calculated based on the stoichiometry of nitrification (Equation (1); Metcalf & Eddy (2002)) and effluent nitrate concentrations.

$$NH_4^+ + 1.863O_2 + 0.098CO_2 \rightarrow 0.0196C_5H_7NO_2 + 0.98NO_3^- + 0.0941H_2O + 1.98H^+ \quad (1)$$

Nitrogen assimilation by heterotrophic bacteria was calculated based on COD removal and observed heterotrophic biomass yield (Metcalf & Eddy 2002). By subtracting the nitrogen assimilated in nitrifier and heterotrophic biomass from the total effluent biomass nitrogen, microalgal nitrogen assimilation was calculated.

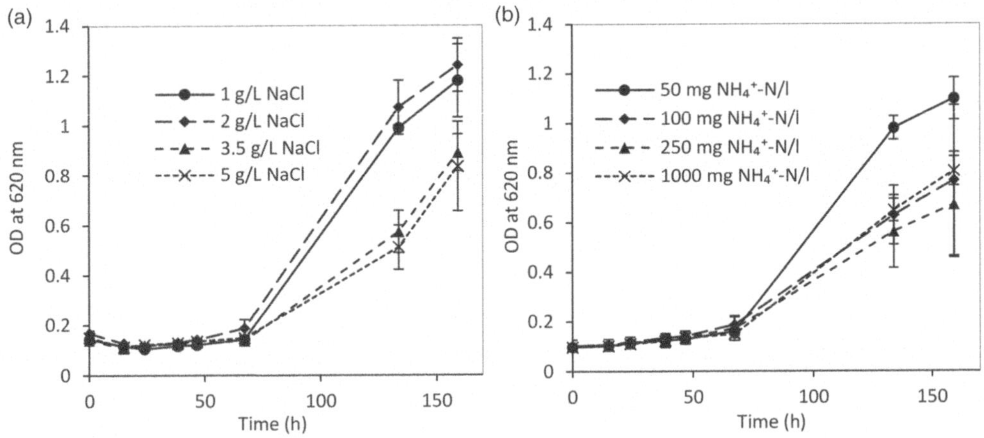

Figure 2 | Growth of a mixture of all microalgal species exposed to (a) different salinities (1, 2, 3.5 and 5 g L^{-1}) and (b) ammonium concentrations (50, 100, 250 and 1,000 mg NH_4^+-N L^{-1}) at 1 g L^{-1} NaCl.

Sampling and analytical methods

Every 2 days, samples of influent and bulk reactor solution at the end of a cycle were filtered (0.45 μm Chromafil Xtra, Machery-Nagel, PA, USA) and stored at 4 °C. The concentration of ammonium was determined following the standard method of Nessler (APHA 1992) or according to the standard methods (APHA 1992). Kjeldahl nitrogen was also analysed according to the standard methods (APHA 1992). Organic nitrogen was determined as the difference between Kjeldahl nitrogen and ammonium nitrogen. Nitrite and nitrate were analysed with ion chromatography (IC 761, Compact, Methrom AG, Switzerland). COD was determined photometrically (Nanocolor COD 160; Machery-Nagel, PA, USA). Total suspended solids (TSS) and VSS were measured according the standard methods (APHA 1992). The dissolved oxygen (DO) concentration was measured daily (Hach HQ40d) and electrical conductivity weekly (Consort C833). Reactor biomass was analysed with light microscopy (Zeiss Axioskop 2 plus). During PBR operation, ammonia-oxidizing bacteria (AOB) and nitrite-oxidizing bacteria (NOB) nitrification activities were determined regularly following a standardized protocol (Coppens *et al.* 2016).

RESULTS AND DISCUSSION

Influence of ammonium and salt concentration on algal growth

Since fresh unhydrolysed urine contains a high salinity (±20 mS cm^{-1}) and urea nitrogen concentration (5 g N L^{-1} for synthetic urine), the influence of different salt and ammonium concentrations was investigated on the growth of a mixture of all microalgal species. Four different salinities (1, 2, 3.5 and 5 g L^{-1} NaCl, corresponding to 4.54, 6.55, 9.95 and 13.60 mS cm^{-1}, respectively) and ammonium concentrations (50, 100, 250 and 1,000 mg NH_4^+-N L^{-1}), were applied (Figure 2). A longer lag phase was observed at salt concentrations equal to or higher than 3.5 g NaCl L^{-1}. This might be due to the predominance of fresh water microalgal species. Results indicate that a concentration of 50 mg NH_4^+-N L^{-1} is preferred. Since no significant change in growth was observed for the mixed culture up to 2 g L^{-1} NaCl, corresponding to a salinity of 6.55 mS cm^{-1}, these tests indicate that the salinity of roughly 2 mS cm^{-1}, for a 10% dilution of synthetic urine, will not inhibit microalgal growth. In contrast, ammonium concentrations (500 mg NH_4^+-N L^{-1} for a 10% diluted urine solution) could result in slower microalgal growth if ammonium accumulates in the PBR, depending on the nitrogen loading rate and nitrification activity.

Nitrogen evolution in the PBR

Start-up phase

During the first 15 days of reactor operation, nitrogen removal efficiencies between 60% and 95% (Figure 3(a)) were achieved, but no nitrate accumulation was observed. COD removal efficiencies fluctuated between 61% and 93% (Figure 3(d)) with a maximum COD removal rate of 45 mg COD g VSS^{-1} d^{-1} at day 15. The DO concentration fluctuated around 3.6 ± 2.6 mg L^{-1}, indicating sufficient

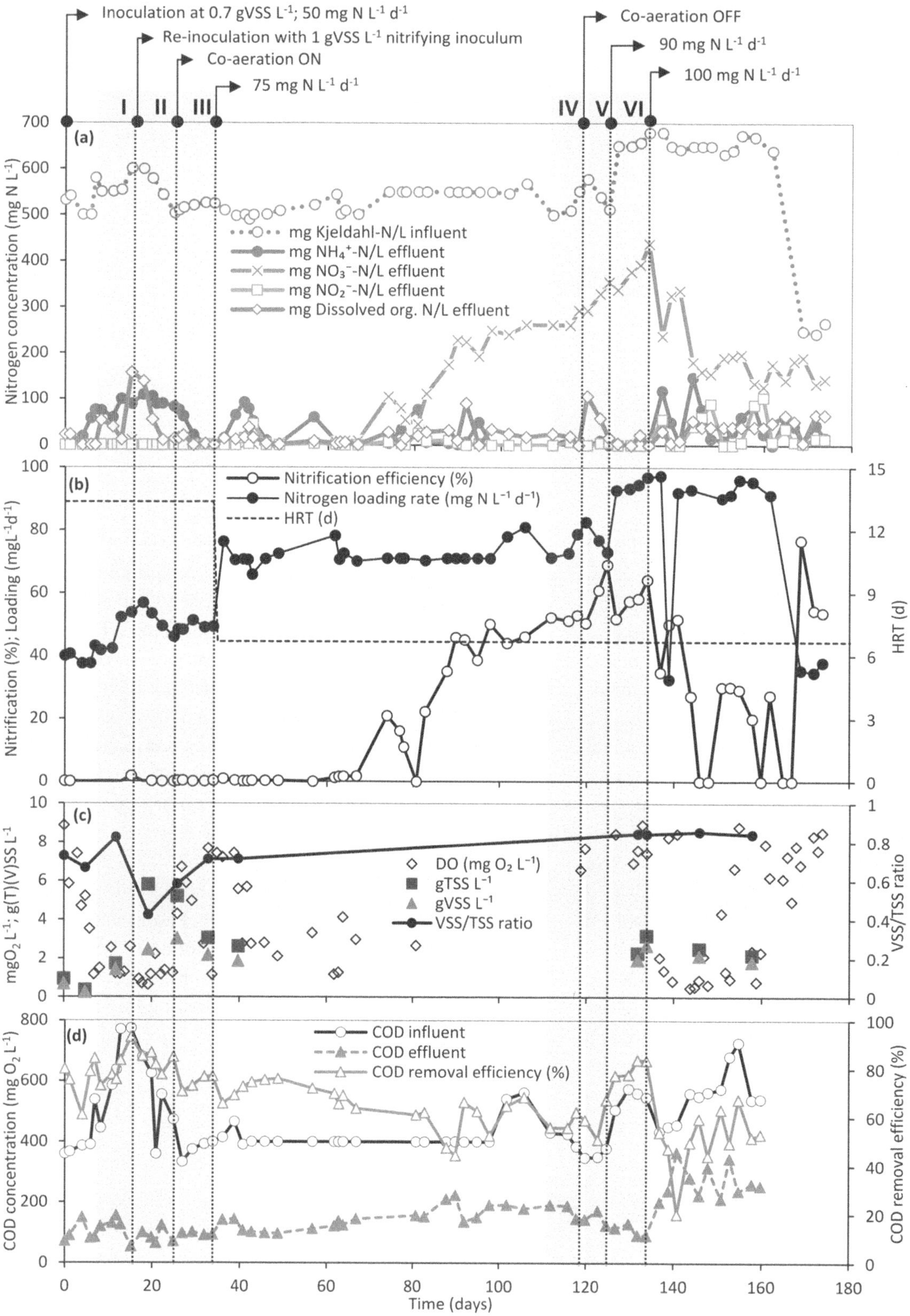

Figure 3 | (a) PBR influent and effluent nitrogen species; (b) nitrogen loading rate, nitrification efficiency and hydraulic retention time (HRT); (c) DO, TSS and VSS concentration; (d) influent and effluent COD concentrations and COD removal efficiency, during 174 days of PBR operation. Changes in reactor operation are indicated with arrows while the zones from I to VI indicate the periods just before a change in reactor operation, over which a 7 day nitrogen conversion balance was made.

photo-oxygenation. An estimated 22% of the incoming nitrogen was assimilated, 51% accumulated in the reactor or was unconverted and removed with the effluent (Figure 4). The remaining 27% of incoming nitrogen was assumed to be lost mainly due to ammonia stripping. Indeed, free ammonia concentrations were the highest (around 0.6 mg N L^{-1} as calculated from 100 mg NH_4^+-N L^{-1} at pH 7 and 25 °C). It is unlikely that denitrification contributed to the nitrogen loss, as no consistent nitrification was observed, and as average DO concentrations were rather high (around 2 mg O_2 L^{-1}), limiting anoxic zones.

To stimulate nitrification, at day 15, the reactor was re-inoculated with 1 g VSS L^{-1} of commercially available nitrifying inoculum (Avecom, Belgium). The high calcium carbonate content of the inoculum resulted in a decrease in VSS/TSS ratio from 0.83 to 0.42. Although DO concentrations between 0.6 and 2 mg L^{-1} indicated successful photo-oxygenation, re-inoculation did not result in nitrification. However, activity tests demonstrated the low nitrification potential of the PBR biomass in both fresh medium and reactor supernatant. Additionally, no inhibitory effect was observed of reactor supernatant on new nitrifying inoculum (Figure 5(a); Figure S1, Supplementary material, available with the online version of this paper).

To increase oxygen concentrations and stimulate nitrification, co-aeration with compressed air was started at day

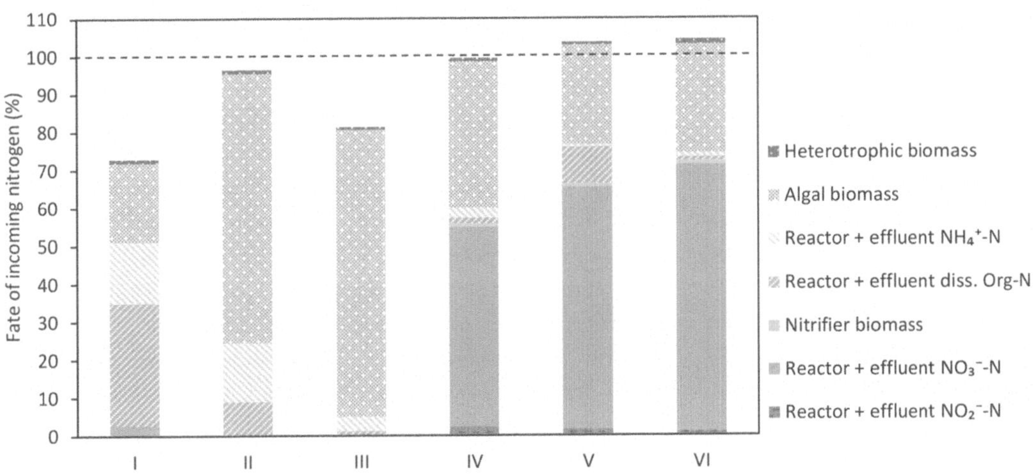

Figure 4 | Nitrogen (N) conversion balance over the periods I to VI. Incoming N is defined as the average total influent N together with the average Kjeldahl- and nitrite-N present in the reactor. Incoming N can accumulate in the PBR or can be removed with the effluent.

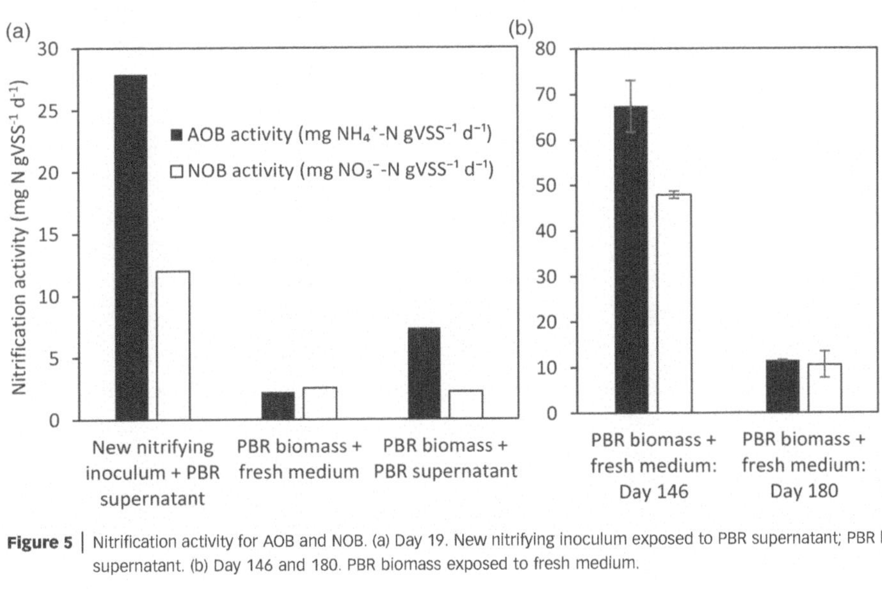

Figure 5 | Nitrification activity for AOB and NOB. (a) Day 19. New nitrifying inoculum exposed to PBR supernatant; PBR biomass exposed to fresh medium; PBR biomass exposed to PBR supernatant. (b) Day 146 and 180. PBR biomass exposed to fresh medium.

25 at a flow rate of 0.25 L O_2 $L_{reactor}^{-1}$ h^{-1}. Complete nitrogen removal was obtained (Figure 3(a)) and nitrogen assimilation (77%) in microbial biomass remained the dominant nitrogen removal pathway, with 76% assimilated as microalgal and 1% as chemoheterotrophic biomass (Figure 4).

Nitrification phase

From day 55 onwards, accumulation of nitrate indicated successful nitrification, while the COD removal efficiency was situated between 44% and 83%. Starting from day 110, nitrification efficiencies higher than 50% were obtained. Therefore, at day 118, external aeration was stopped and since oxygen levels remained above 6 mg O_2 L^{-1}, photosynthetic oxygenation was sufficient to sustain nitrification and organic carbon oxidation. At day 134, nitrification efficiency reached 64% at a nitrogen loading rate of 97 mg N L^{-1} d^{-1} (12% synthetic urine) (Figure 3(b)). Furthermore, well-settling microbial biomass was obtained, as indicated by the low sludge volume index (SVI_{30}) of 58.06 mL g^{-1}.

The increase of nitrogen loading rate to 100 mg N L^{-1} d^{-1} (15% dilution of synthetic urine) at day 134 resulted in a nitrification efficiency decline to values lower than 30% and reduction in biomass concentration to 2.14 gVSS L^{-1} at day 145. Nitrification batch activity tests demonstrated maximum biomass specific ammonium and nitrite oxidation rates of 67 and 48 mg N $gVSS^{-1}$ d^{-1} (Figure 5(b); Figure S2, Supplementary material, available online) (or volumetric oxidation rates of 122 and 87 mg N L^{-1} d^{-1}, respectively), which are higher than the nitrogen loading rate. A DO concentration below 1 mg L^{-1} indicated a drop in photo-aeration rate.

Between day 146 and 167, nitrification efficiency fluctuated between 0 and 30% and, at day 168, the nitrogen loading rate was decreased to 35 mg N L^{-1} d^{-1}. From day 168 until 180, nitrification efficiencies remained between 52% and 53%. At day 180, a last activity test was performed (Figure 5(b); Figure S2, Supplementary material), demonstrating that both AOB and NOB activity decreased to 12 mg NH_4^+-N $gVSS^{-1}$ d^{-1} and 11 mg NH_4^+-N $gVSS^{-1}$ d^{-1}, respectively.

PBR nitrification efficiency and photosynthetic oxygenation

The maximum achieved volumetric nitrification rate was 67 mg N L^{-1} d^{-1} at a loading rate of 97 mg N L^{-1} d^{-1}, which is higher than in the outdoor HRAP described by de Godos *et al.* (2009), where a volumetric nitrification rate of 9 mg N L^{-1} d^{-1} was achieved at a loading rate of 30 mg N L^{-1} d^{-1}. Karya *et al.* (2013) nitrified artificial wastewater supported by photo-oxygenation in an algal-bacterial consortium at a nitrogen loading rate of 50 mg N L^{-1} d^{-1} and maximum volumetric nitrification efficiency of 43 mg N L^{-1} d^{-1}. This artificial wastewater did, however, not contain COD and all nitrogen was present as ammonium. Van der Steen *et al.* (2015) reached a comparable volumetric nitrification rate of modified BG-11 medium of 46 mg N L^{-1} d^{-1} (Table 1).

The observed sudden nitrification activity in this study could be the consequence of adaptation of the nitrifying community to light, since it is known that AOB and NOB are inhibited by light at an intensity of 500 and 1,250 μmol m^{-2} s^{-1} (Vergara *et al.* 2016). Guerrero & Jones (1996) studied light inhibition on marine AOB and NOB and concluded that photo-inhibition is species-specific and dependent on light intensity, lighting period and wavelength. AOB were found to be more sensitive to blue light than NOB, while cool-white fluorescent light inhibited AOB activity but did not influence NOB. Abeliovich & Vonshak (1993) observed complete nitrification inhibition during 4 days of exponentially growing *Nitrosomonas europeana* after 1 hour light exposure, while ammonia presence provided some protection. Alleman *et al.* (1987) observed that light with a wavelength in the range of 410–415 nm is responsible for *Nitrosomonas* inhibition, during a period without respiration and nutrient absence. The basis for light sensitivity is assumed to be damage to the many cytochromes of AOB and NOB, involved in the nitrification energy transduction pathways (Ward 2011). No records were found on a longer period of light irradiation and potential adaptation to light-inhibiting conditions. Since in this study, no light–dark cycle was used to enable recovery from light inhibition and, additionally, the selected light source to stimulate microalgal growth was rich in the inhibitory wavelengths, conditions were unfavourable for nitrifiers.

Theoretically, nitrifying bacteria require 4.57 g oxygen to convert 1 g NH_4^+-N to 0.95 g NO_3^--N and 0.05 g biomass-N, while heterotrophic bacteria consume 0.69 g oxygen to oxidize 1 g of organic carbon to CO_2 (SRT: 6.67 days; 25 °C). At day 134, the moment of optimal nitrification, the oxygen consumption and thus production rate was 160 mg O_2 $gVSS^{-1}$ d^{-1}, calculated based on the stoichiometry of nitrogen and organic carbon oxidation. This value is comparable to the photo-oxygenation rates calculated based on the results in other studies (Table 1). Additionally, in this study, the light supply rate was 1.04 mol photons L^{-1} d^{-1}. Together with the volumetric oxygen production rate of 429 mg O_2 L^{-1} d^{-1} (recalculated from 160 mg O_2 gVSS^{-1} d^{-1}), this results in an oxygen quantum yield of 0.013 mol O_2 mol photons^{-1}. This is lower than the experimental maximal quantum yield of 0.1 mol O_2 mol photons^{-1} under low light (Zijffers *et al.* 2010), indicating room for improvement in photosynthetic efficiency. However, the real oxygen quantum yield could be slightly higher due to a lower photon flux density than the measured 300 μmol photons m^{-2} s^{-1}, since perfect homogenous light distribution was lacking.

To check if photosynthetic oxygenation was sufficient, DO was monitored during one cycle at day 134 (Figure S3, Supplementary material, available online). To assess the oxygen consumption rate without algal oxygen production, the lights were turned off during the first 2 hours. When the DO reached values lower than 3 mg O_2 L^{-1}, lights were turned on. Initially the DO remained constant while nitrification was still ongoing, indicating a sufficient photo-oxygenation rate due to the simultaneous oxygen production and consumption. When ammonium and nitrite were fully oxidized (measured with ammonium and nitrite test strips), DO increased at a rate equal to the oxygen production rate minus the heterotrophic oxidation rate. The measured oxygen production rate was 4.24 mg O_2 L^{-1} h^{-1} taking into account the ongoing organic carbon oxidation. This oxygen production rate is within the range of other microalgal-bacterial systems (Karya *et al.* 2013). However, 95 mg O_2 L^{-1} was consumed for nitrification and 23 mg O_2 L^{-1} was consumed for COD oxidation leading to a total theoretical oxygen consumption rate of 15 mg O_2 L^{-1} h^{-1}. The difference between this rate and the oxygen consumption rate based on the DO profile might be due to the direct consumption of oxygen within the floc entity.

The optimal nitrifying PBR

Since each involved type of micro-organism has its own function, nutritional requirement and metabolic rate, the potential nitrification rate can be calculated. Taking into account the stoichiometry of photosynthesis, nitrification and heterotrophic COD oxidation, ideally without nitrogen losses, 23% of incoming nitrogen should be assimilated by the microalgae, resulting in a sufficient amount of oxygen to oxidize the leftover 77% of the ammonium nitrogen and present COD (Figure 6). Considering algal oxygen production rates in an optimal designed PBR between 128 and 192 mg O_2 L^{-1} h^{-1} (Javanmardian & Palsson 1992), it is theoretically possible to nitrify urine at a nitrogen influent loading rate of 1.14 g N L^{-1} d^{-1} and 0.91 g COD L^{-1} d^{-1} (urine COD/N ratio of 0.8). In cheaper HRAPs, however,

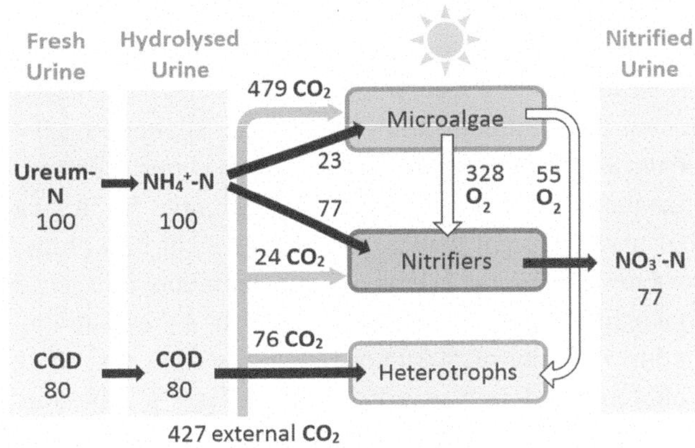

Figure 6 | Theoretical behaviour of the ideal consortium of microalgae, nitrifiers and heterotrophs in a nitrifying bioreactor, starting from 100 mass units of ureum nitrogen and 80 mass units of COD (average COD/N mass ratio in urine).

optimal oxygen production rates up to 9.55 mg O_2 L^{-1} h^{-1} are reported (Arbib *et al.* 2017), indicating a maximum influent loading rate of 57 mg N L^{-1} d^{-1}.

Several operational parameters influence the activity of the consortium organisms. The specific growth rate and activity ratio between microalgae and nitrifying bacteria is influenced by parameters such as available nutrients and light and is related to the substrate affinity constants (K_s) for ammonium nitrogen and inorganic carbon. Due to the higher nitrifier K_s values for inorganic carbon (21.4 mg C L^{-1}; Guisasola *et al.* (2007)) in comparison to microalgae (0.58 mg C L^{-1}; Shelp & Canvin (1980)), higher CO_2 concentrations favour nitrifiers. The same can be assumed in terms of ammonium concentration due to the lower K_s value for microalgae (0.01 mg N L^{-1}; Hein *et al.* (1995)) compared to AOB (0.42 to 4.47 mg N L^{-1}; Kayee *et al.* (2016)).

Next to high substrate concentrations, an SRT increase favours slow growing nitrifiers and increases nitrification activity. To obtain better settling characteristics, dense floc formation could be stimulated by a low hydraulic retention time or high volumetric exchange ratio (VER), causing poor-settling biomass to wash out. The microalgae could form the outer layer of the granule while the nitrifying bacteria are protected from light inhibition in the inner layers. Nevertheless, granule formation is a complex process which depends on many parameters. Additionally, the anoxic core could facilitate denitrification, resulting in an unwanted nitrogen loss. With an increase in VER from 15% to 50%, by reducing the number of cycles from three to two and increasing the influent flow rate to 2 L d^{-1} of 10% dilution of urine, a nitrogen loading rate of 250 mg N L^{-1} d^{-1} could be reached. Attention has to be paid to potential free ammonia formation when the same nitrogen loading rate is applied with fewer cycles.

Since parameters such as pH, nutrient quantity and quality, stirring and temperature were optimal for most microalgae, higher oxygen production rates may be achieved by increasing the illuminated volume. Light attenuates exponentially as it penetrates into the culture medium, estimated by Lambert–Beer's law (Lee 1999). Ogbonna & Tanaka (2000) observed light penetration of only 2 cm in a photobioreactor containing a biomass concentration of 1 g L^{-1} with a light absorption coefficient of 200 $m^2\ s^{-1}$ and an illumination intensity of 500 μmol PAR $m^{-2}\ s^{-1}$. In this study the photon flux density was 300 μmol PAR $m^{-2}\ s^{-1}$ and biomass density fluctuated around 2 g VSS L^{-1}, indicating a light penetration depth less than 1 cm or an illuminated volume of less than 30%.

Microscopy on PBR biomass

Microscopic observation after 180 days of reactor operation revealed that microalgal-bacterial flocs were formed with a size of 50–100 μm (Figure 7). *Scenedesmus* sp. was the dominant green microalga present, but also *Chlorella* sp. and the cyanobacteria *Synechocystis* sp. and *Leptolyngbya* sp. were detected. Between the large microalgal cells, the bacteria were present in low numbers. The electrical conductivity in the PBR was fluctuating around 8.5 mS cm^{-1} corresponding to 100 mM NaCl. Among the inoculated microalgae, only *Nannochloropsis* sp. was a distinct salt water species, although all fresh water species were initially cultivated in a salt water medium (ESAW) and, depending on species, microalgae can be halotolerant and show adaptation (Hart *et al.* 1991). *Chlorella* sp. presents optimal growth at a salinity level of 100 mM (Abdel-Rahman *et al.* 2005), indicating feasible conditions in the PBR. In contrast, for *Scenedesmus obliquus* optimal growth was observed in BBM medium supplemented with 25 mM NaCl (Salama *et al.* 2013). This shift towards *Scenedesmus* sp. in a concentrated wastewater solution was observed before by Koreiviene *et al.* (2014), suggesting the larger surface area to volume ratio enabled quicker nutrient uptake. Karya *et al.* (2013) inoculated with *Scenedesmus quadricauda* and selection took place towards cyanobacteria; however, reactor salinity was not mentioned.

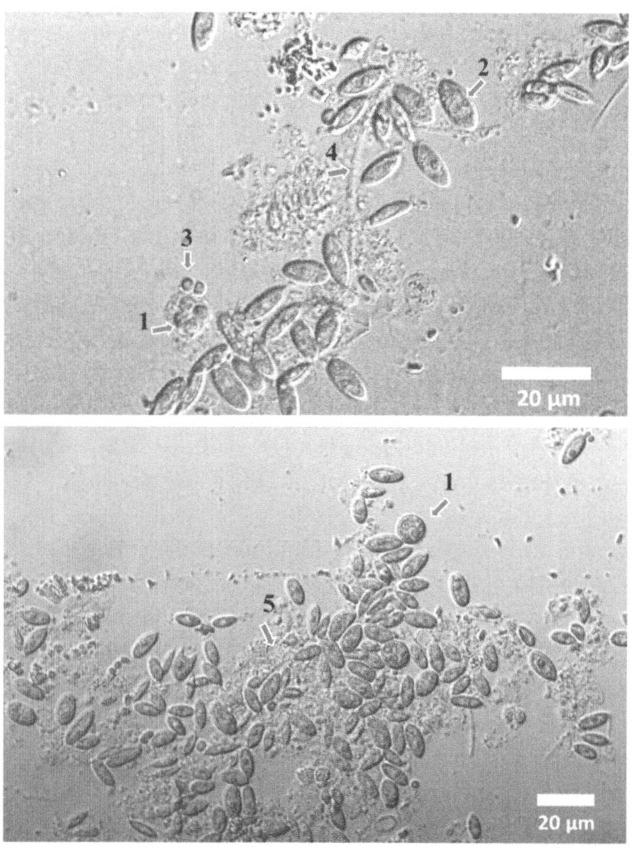

Figure 7 | Microscopy on the PBR biomass after 180 days of operation. (1) *Chlorella* sp.; (2) *Scenedesmus* sp.; (3) *Synechocystis* sp.; (4) *Leptolyngbya* sp.; (5) Bacteria.

Potential applications for photosynthetic oxygenation

Since the energy gain of photosynthetic oxygenation would completely disappear in a PBR with synthetic light supply, HRAPs and PBRs in natural sunlight are more suitable. Although electricity costs and associated carbon footprint are reduced and electricity independence allows application in developing countries, the volumetric nitrification rate of 67 mg N $L^{-1}\ d^{-1}$ achieved in this study is low compared to the urine nitrification rate of 450 mg N $L^{-1}\ d^{-1}$ achieved with conventional aeration in a membrane bioreactor (Coppens *et al.* 2016). Due to this lower nitrification rate and outdoor day–night light regime, the required surface area and reactor volume are higher compared to systems relying on conventional aeration. Outdoor experiments on a larger scale should be performed to get a better insight on the true energy gain of photosynthetic oxygenation.

CONCLUSION

It was experimentally demonstrated that biological oxidation of all nitrogen present in urine to nitrate is a promising pre-treatment stabilization step to substitute for expensive conventional aeration, before further nutrient recovery. A consortium of microalgae and nitrifying bacteria was successfully developed in which the photosynthetic oxygenation rate was sufficiently high to nitrify urine at a volumetric nitrification rate of 67 mg N L^{-1} d^{-1}. Additionally, a maximum biomass specific photo-oxygenation rate of 160 mg O_2 $gVSS^{-1}$ d^{-1} was achieved and microscopic observations revealed that *Scenedesmus* sp. was the dominant microalga after 180 days of reactor operation. Finally, outdoor experiments on a larger scale should be performed to estimate the true energy gain of photosynthetic oxygenation.

ACKNOWLEDGEMENT

The authors would like to thank Dr Claudio Sili from the ISE-CNR, Firenze for determination of the microalgae and Dr Marc Spiller and Dr ir. Erik Van Eynde for the critical discussions. This study was also supported by the European Space Agency (ESA) and the Belgian Science Policy (BELSPO) in the framework of the MELiSSA project.

REFERENCES

Abdel-Rahman, M. H. M., Ali, R. M. & Said, H. A. 2005 Alleviation of NaCl-induced effects on *Chlorella vulgaris* and *Chlorococcum humicola* by riboflavin application. *International Journal of Agriculture & Biology* **7** (1), 58–62.

Abeliovich, A. & Vonshak, A. 1993 Factors inhibiting nitrification of ammonia in deep waste-water reservoirs. *Water Research* **27** (10), 1585–1590.

Adamsson, M. 2000 Potential use of human urine by greenhouse culturing of microalgae (*Scenedesmus acuminatus*), zooplankton (*Daphnia magna*) and tomatoes (*Lycopersicon*). *Ecological Engineering* **16** (2), 243–254.

Alleman, J. E., Keramida, V. & Panteakiser, L. 1987 Light-induced *Nitrosomonas* inhibition. *Water Research* **21** (4), 499–501.

APHA 1992 *Standard Methods for the Examination of Water and Wastewater*, 18th edn. American Public Health Association/American Water Works Association/Water Environment Federation, Washington, DC, USA.

Arbib, Z., Crespo, I. D., Corona, E. L. & Rogalla, F. 2017 Understanding the biological activity of high rate algae ponds through the calculation of oxygen balances. *Applied Microbiology and Biotechnology* **101** (12), 5189–5198.

Basakcilardan-Kabakci, S., Ipekoglu, A. N. & Talini, I. 2007 Recovery of ammonia from human urine by stripping and absorption. *Environmental Engineering Science* **24** (5), 615–624.

Brooks, T. & Keevil, C. W. 1997 A simple artificial urine for the growth of urinary pathogens. *Letters in Applied Microbiology* **24** (3), 203–206.

Coppens, J., Lindeboom, R., Muys, M., Coessens, W., Alloul, A., Meerbergen, K., Lievens, B., Clauwaert, P., Boon, N. & Vlaeminck, S. E. 2016 Nitrification and microalgae cultivation for two-stage biological nutrient valorization from source separated urine. *Bioresource Technology* **211**, 41–50.

de Godos, I., Blanco, S., Garcia-Encina, P. A., Becares, E. & Munoz, R. 2009 Long-term operation of high rate algal ponds for the bioremediation of piggery wastewaters at high loading rates. *Bioresource Technology* **100** (19), 4332–4339.

Feng, D. L., Wu, Z. C. & Xu, S. H. 2008 Nitrification of human urine for its stabilization and nutrient recycling. *Bioresource Technology* **99** (14), 6299–6304.

Greenberg, A. E., Clesceri, L. S. & Eaton, A. D. 1992 *Standard Methods for the Examination of Waste and Wastewater.* American Public Health Association, Washington.

Guerrero, M. A. & Jones, R. D. 1996 Photoinhibition of marine nitrifying bacteria. 2. Dark recovery after monochromatic or polychromatic irradiation. *Marine Ecology Progress Series* **141** (1–3), 193–198.

Guisasola, A., Petzet, S., Baeza, J. A., Carrera, J. & Lafuente, J. 2007 Inorganic carbon limitations on nitrification: experimental assessment and modelling. *Water Research* **41** (2), 277–286.

Gutzeit, G., Lorch, D., Weber, A., Engels, M. & Neis, U. 2005 Bioflocculent algal-bacterial biomass improves low-cost wastewater treatment. *Water Science and Technology* **52** (12), 9–18.

Harrison, P. J., Waters, R. E. & Taylor, F. J. R. 1980 A broad-spectrum artificial seawater medium for coastal and open ocean phytoplankton. *Journal of Phycology* **16** (1), 28–35.

Hart, B. T., Bailey, P., Edwards, R., Hortle, K., James, K., Mcmahon, A., Meredith, C. & Swadling, K. 1991 A review of the salt sensitivity of the Australian fresh-water biota. *Hydrobiologia* **210** (1–2), 105–144.

Hein, M., Pedersen, M. F. & Sandjensen, K. 1995 Size-dependent nitrogen uptake in micro- and macroalgae. *Marine Ecology Progress Series* **118** (1–3), 247–253.

Javanmardian, M. & Palsson, B. O. 1992 Design and operation of an algal photobioreactor system. *Life Sciences and Space Research Xxiv (4): Natural and Artificial Ecosystems* **12**, 231–235.

Karya, N. G. A. I., van der Steen, N. P. & Lens, P. N. L. 2013 Photo-oxygenation to support nitrification in an algal-bacterial consortium treating artificial wastewater. *Bioresource Technology* **134**, 244–250.

Kayee, P., Rongsayamanont, C., Kunapongkiti, P. & Limpiyakorn, T. 2016 Ammonia half-saturation constants of sludge with different community compositions of ammonia-oxidizing bacteria. *Environmental Engineering Research* **21** (2), 140–144.

Koreiviene, J., Valciukas, R., Karosiene, J. & Baltrenas, P. 2014 Testing of *Chlorella/Scenedesmus* microalgae consortia for remediation of wastewater, CO_2 mitigation and algae biomass feasibility for lipid production. *Journal of Environmental Engineering and Landscape Management* **22** (2), 105–114.

Kuai, L. P. & Verstraete, W. 1998 Ammonium removal by the oxygen-limited autotrophic nitrification-denitrification system. *Applied and Environmental Microbiology* **64** (11), 4500–4506.

Kujawa-Roeleveld, K. & Zeeman, G. 2006 Anaerobic treatment in decentralised and source-separation-based sanitation concepts. *Reviews in Environmental Science and Biotechnology* **5** (1), 115–139.

Lee, C.-G. 1999 Calculation of light penetration depth in photobioreactors. *Biotechnology and Bioprocess Engineering* **4**, 78–81.

Maurer, M., Pronk, W. & Larsen, T. A. 2006 Treatment processes for source-separated urine. *Water Research* **40** (17), 3151–3166.

Metcalf & Eddy 2002 *Wastewater Engineering: Treatment and Reuse*. McGraw-Hill, Boston, MA, USA.

Ogbonna, J. C. & Tanaka, H. 2000 Light requirement and photosynthetic cell cultivation – development of processes for efficient light utilization in photobioreactors. *Journal of Applied Phycology* **12** (3–5), 207–218.

Praveen, P. & Loh, K. C. 2015 Photosynthetic aeration in biological wastewater treatment using immobilized microalgae-bacteria symbiosis. *Applied Microbiology and Biotechnology* **99** (23), 10345–10354.

Salama, E. S., Kim, H. C., Abou-Shanab, R. A. I., Ji, M. K., Oh, Y. K., Kim, S. H. & Jeon, B. H. 2013 Biomass, lipid content, and fatty acid composition of freshwater *Chlamydomonas mexicana* and *Scenedesmus obliquus* grown under salt stress. *Bioprocess and Biosystems Engineering* **36** (6), 827–833.

Shelp, B. J. & Canvin, D. T. 1980 Utilization of exogenous inorganic carbon species in photosynthesis by *Chlorella pyrenoidosa*. *Plant Physiology* **65** (5), 774–779.

Tsioptsias, C., Lionta, G. & Samaras, P. 2017 Microalgae-activated sludge treatment of molasses wastewater in sequencing batch photo-bioreactor. *Environmental Technology* **38** (9), 1120–1126.

Tuantet, K., Janssen, M., Temmink, H., Zeeman, G., Wijffels, R. H. & Buisman, C. J. N. 2013 Microalgae growth on concentrated human urine. *Journal of Applied Phycology* **26** (1), 287–297.

Udert, K. M., Larsen, T. A., Biebow, M. & Gujer, W. 2003 Urea hydrolysis and precipitation dynamics in a urine-collecting system. *Water Research* **37** (11), 2571–2582.

Udert, K. M., Larsen, T. A. & Gujer, W. 2006 Fate of major compounds in source-separated urine. *Water Science and Technology* **54** (11–12), 413–420.

van der Steen, P., Rahsilawati, K., Rada-Ariza, A. M., Lopez-Vazquez, C. M. & Lens, P. N. L. 2015 A new photo-activated sludge system for nitrification by an algal-bacterial consortium in a photo-bioreactor with biomass recycle. *Water Science and Technology* **72** (3), 443–450.

Vergara, C., Munoz, R., Campos, J. L., Seeger, M. & Jeison, D. 2016 Influence of light intensity on bacterial nitrifying activity in algal-bacterial photobioreactors and its implications for microalgae-based wastewater treatment. *International Biodeterioration & Biodegradation* **114**, 116–121.

Verstraete, W., Clauwaert, P. & Vlaeminck, S. E. 2016 Used water and nutrients: recovery perspectives in a 'panta rhei' context. *Bioresource Technology* **215**, 199–208.

Ward, B. B. 2011 Measurement and distribution of nitrification rates in the oceans. *Methods in Enzymology: Research on Nitrification and Related Processes* **486**, 307–323.

Yang, C. L., Liu, H., Li, M., Yu, C. Y. & Yu, G. 2008 Treating urine by *Spirulina platensis*. *Acta Astronautica* **63** (7–10), 1049–1054.

Zijffers, J. W. F., Schippers, K. J., Zheng, K., Janssen, M., Tramper, J. & Wijffels, R. H. 2010 Maximum photosynthetic yield of green microalgae in photobioreactors. *Marine Biotechnology* **12** (6), 708–718.

First received 24 October 2017; accepted in revised form 20 April 2018. Available online 9 May 2018

Outdoor flat-panel membrane photobioreactor to treat the effluent of an anaerobic membrane bioreactor. Influence of operating, design, and environmental conditions

J. González-Camejo, R. Barat, M. V. Ruano, A. Seco and J. Ferrer

ABSTRACT

As microalgae have the ability to simultaneously remove nutrients from wastewater streams while producing valuable biomass, microalgae-based wastewater treatment is a win–win strategy. Although recent advances have been made in this field in lab conditions, the transition to outdoor conditions on an industrial scale must be further investigated. In this work an outdoor pilot-scale membrane photobioreactor plant was operated for tertiary sewage treatment. The effects of different parameters on microalgae performance were studied including: temperature, light irradiance (solar and artificial irradiance), hydraulic retention time (HRT), biomass retention time (BRT), air sparging system and influent nutrient concentration. In addition the competition between microalgae and ammonium oxidising bacteria for ammonium was also evaluated. Maximum nitrogen and phosphorus removal rates of 12.5 ± 4.2 mgN·L^{-1}·d^{-1} and 1.5 ± 0.4 mgP·L^{-1}·d^{-1}, respectively, were achieved at a BRT of 4.5 days and HRT of 2.5 days, while a maximum biomass productivity of 78 ± 13 mgVSS·L^{-1}·d^{-1} (VSS: volatile suspended solids) was reached. While the results obtained so far are promising, they need to be improved to make the transition to industrial scale operations feasible.

Key words | flat-panel, membrane photobioreactor, microalgae, outdoor, pilot plant, wastewater treatment

J. González-Camejo
R. Barat (corresponding author)
J. Ferrer
CALAGUA – Unidad Mixta UV-UPV, Institut Universitari d'Investigació d'Enginyeria de l'Aigua i Medi Ambient – IIAMA, Universitat Politècnica de Valencia,
Camí de Vera s/n, 46022 Valencia,
Spain
E-mail: *rababa@dihma.upv.es*

M. V. Ruano
A. Seco
CALAGUA – Unidad Mixta UV-UPV, Departament d'Enginyeria Química,
Universitat de València,
Avinguda de la Universitat s/n, 46100 Burjassot, Valencia,
Spain

INTRODUCTION

Microalgae are microorganisms that carry out photosynthesis and thus require inorganic carbon and light (energy source) to grow. They also require nutrients (mainly nitrogen and phosphorus), which can be obtained from wastewater streams (Ledda *et al.* 2015), avoiding eutrophication of natural water bodies.

Algae-based wastewater treatment has some interesting advantages over other classical technologies: it (i) produces valuable biomass; (ii) reduces chemicals, and (iii) reduces sludge production (Gao *et al.* 2016). Green microalgae seem to be more appropriate for wastewater treatment than other types of microalgae such as cyanobacteria (Arias *et al.* 2017). In this respect, green algae *Chlorella* and *Scenedesmus* have been extensively reported as ideal for wastewater treatment because of their adaptability to such media (Xu *et al.* 2015; Wu *et al.* 2017).

Many authors have studied pure microalgae cultures in highly controlled lab conditions looking for fast-growth strains. However, single-genus cultures are difficult to maintain on a large scale under outdoor conditions. On the other hand, polycultures can increase microalgae performance, since they are more robust before contamination by other microorganisms (Gouveia *et al.* 2016).

Microalgae can be used to treat different types of wastewater streams: urban (raw wastewater, primary and secondary effluents, centrate), aquaculture, etc. Each type has different characteristics which can affect microalgae growth positively or negatively. In this regard, Ledda *et al.* (2015) reported that the organic matter was the main factor affecting microalgae growth, as it was directly related with turbidity, and that nutrient content did not affect the microalgae process, while Gao *et al.* (2016) found that high nutrient concentrations are needed to maintain high microalgae growth rates.

There are two main groups of microalgae cultivation systems: open ponds and closed photobioreactors (PBRs). Open

doi: 10.2166/wst.2018.259

ponds allow CO_2 uptake by microalgae directly from the atmosphere, but CO_2 can also be supplied by an aerator. Although they have lower investment and operational costs than PBRs, they also have disadvantages: large surface areas are required; contamination by predators; high CO_2 diffusion to the atmosphere; ineffective light distribution from the surface to the bottom of the reactor and high evaporative losses. PBRs are designed to improve photosynthesis efficiency by increasing the light available to the microalgae culture. While they are perfectly mixed to avoid wall fouling and enable light and nutrient homogenisation, their investment and maintenance costs are high. Moreover, photoinhibition, overheating, biofouling and oxygen accumulation can cause microalgae growth inhibition (Arbib *et al.* 2013). Table 1 summarises the results of different microalgae cultivation systems which treated wastewater under outdoor conditions.

Generally, closed PBRs obtained high nitrogen (NRE) and phosphorus removal efficiencies (PRE) (around 80–100%), while open ponds are less efficient. Moreover, Table 1 shows that the highest productivities and nutrient removal efficiencies were obtained in batch experiments. However, both batch and high hydraulic retention time (HRT) operations would imply considerably high surface areas to treat wastewater at industrial scale. Thus, algae-based wastewater treatment technologies must operate at minimum HRT. In this respect, membrane photobioreactors (MPBRs), which are the combination of PBRs and membrane technology, appear as an ideal solution for microalgae cultivation to treat wastewater. Membranes separate the microalgae biomass from the water effluent, so that high nutrient loads can be maintained while microalgae biomass wash-out is avoided (Gao *et al.* 2016).

This paper summarises the results obtained from an outdoor MPBR pilot plant under different environmental, design, and operating conditions. This plant was fed by the effluent of an anaerobic membrane bioreactor (AnMBR) treating sewage. The aim of the MPBR plant was to simultaneously reduce the nutrient load in the AnMBR effluent and to produce microalgae biomass.

MATERIAL AND METHODS

The substrate

The microalgae substrate consisted of the nutrient-rich effluent from an AnMBR plant that treated real sewage (Giménez *et al.* 2011). Its nutrient concentration varied in the range of 40–80 $mgN{\cdot}L^{-1}$ and 4–10 $mgP{\cdot}L^{-1}$ due to variations in wastewater characteristics and AnMBR performance. The substrate also contained large amounts of sulphide (around 100–120 $mgS{\cdot}L^{-1}$), which inhibit microalgae growth (González-Camejo *et al.* 2017). The substrate was therefore aerated before feeding the PBRs to oxidise the sulphide to sulphate (González-Camejo *et al.* 2017). Moreover, the AnMBR effluent presented a chemical oxygen demand (COD) concentration of 72 ± 37 $mg{\cdot}L^{-1}$ (mostly non-biodegradable) and an alkalinity of 370 ± 67 $mg\,CaCO_3{\cdot}L^{-1}$.

Pilot plant

The MPBR pilot plant was located in the Carraixet wastewater treatment plant (WWTP) (Valencia, Spain), and consisted of two outdoor flat-panel PBRs connected to a filtration system. Each PBR had a working volume of 550 L: 2.00 m long × 1.10 m high × 0.25 m wide. The aeration system consisted of two perforated pipes (5 mm diameter) placed on the bottom of the PBRs, which continuously

Table 1 | Results of algae-based wastewater treatment studies under outdoor conditions

Type of PBR	Type of wastewater	HRT (d)	N-Feed ($mgN{\cdot}L^{-1}$)	P-Feed ($mgP{\cdot}L^{-1}$)	Productivity ($mgVSS{\cdot}L^{-1}{\cdot}d^{-1}$)	NRE (%)	PRE (%)	Reference
Vertical PBR	Primary effluent	13[a]	133	8.3	100	84	95	Gouveia *et al.* (2016)
High rate algal pond	Secondary effluent	8	25.7	2.2	30	56.3	86.5	Arbib *et al.* (2013)
Rectangular PBR	Municipal wastewater	15[a]	30.5	2.6	–	96	99	Woertz *et al.* (2009)
Rectangular PBR	ADAS[b] + secondary effluent	21[a]	259.7	42.6	109	73.3	66.5	Tan *et al.* (2016)
Flat-panel PBR	AnMBR effluent	8	44.7	5.2	23.4	41.6	36.1	Viruela *et al.* (2016)
Flat-panel PBR	AnMBR effluent	14	81.5	9.2	13.8	50.9	50.9	Viruela *et al.* (2016)

[a]Batch operation. HRT indicates the length of the study.
[b]ADAS: anaerobically digested activated sludge.

introduced air at a flow rate of 0.09 vvm. This way, microalgae settling and wall fouling were minimised. Whenever the pH value of the culture was over 7.5 (set point), pure CO_2 (99.9%) was introduced into the air system, reaching a maximum percentage of CO_2 in the air flow of 4%. This way, phenomena such as ammonia volatilisation and phosphorus precipitation were considered negligible (Whitton *et al.* 2016).

Both PBRs had 12 white LED lamps (Unique Led IP65 WS-TP4S-40 W-ME) installed at the back, offering a continuous light irradiance of 300 μmol·m^{-2}·s^{-1} (light:dark cycle of 24:0 h).

Both PBRs were connected to a filtration system, which mainly consisted of two membrane tanks which included industrial hollow-fibre ultrafiltration membrane units (PURON® Koch Membrane Systems (PUR-PSH31), 0.03 μm pore size), with a working volume of 38 L and filtering area of 6.8 m^2. They were stirred by the same CO_2-enriched air flow as the PBRs to reduce cake formation and avoid undesirable phenomena.

During the experiments with inhibition of nitrification, a concentration of 5 mg·L^{-1} of allylthiourea (ATU) was maintained in the PBRs to inhibit AOB growth (Table 2).

Experimental periods

Before each operating period, the MPBR plant went through a start-up phase, consisting of: (i) adding 10% of the working volume with microalgae biomass (300–500 mgVSS·L^{-1} (VSS: volatile suspended solids); mainly *Scenedesmus* and *Chlorella*; although bacteria and cyanobacteria were also present) and 90% of the working volume with the aforementioned substrate; (ii) batch mode until reaching a biomass concentration of around 250–400 mgVSS·L^{-1} (data not shown); and (iii) continuous feeding maintaining the desired biomass retention time (BRT) and HRT.

The experimental set-up consisted of four periods in which the MPBR was operated under different environmental (temperature, solar irradiance and influent nutrient concentration), operating (BRT and HRT) and design (bubble size of the air sparging system and operating the MPBR plant without membrane filtration, i.e. as a PBR system) conditions. Moreover, artificial light and ATU addition were also modified (Table 2).

Period 1 was operated without microalgae biomass filtration so that BRT was equal to HRT (PBR system). No additional artificial light source was used. It was divided into four sub-periods: (1) 1A was operated at HRT of 8 days and ATU was continuously added; (2) 1B was operated at the same HRT without ATU; (3) in sub-period 1C, HRT was increased to 14 days without ATU; (4) in 1D, an initial ATU dose of 5 mg·L^{-1} was added. The rest of the sub-period was operated at HRT of 14 days without further ATU addition.

In period 2, the pilot plant was also operated as a PBR system (without membranes), maintaining HRT (i.e. BRT) at 8 days. A neoprene diffuser with 0.5 mm pore size was installed in PBR1. In PBR2, the same air sparging system (5 mm pore size) was maintained. The rest of the operating and outdoor conditions were the same for both PBRs. Thus, only in this period, PBR1 and PBR2 were operated separately in order to compare the effect of different bubble size of the air sparging system.

In period 3, the plant was operated as an MPBR system at BRT of 4.5 days and variable HRT: 2.5, 2 and 3 days, for sub-periods 3A, 3B and 3C, respectively.

Table 2 | Operation and outdoor conditions of each period

Sub-period	Days of operation	Daily average solar PAR[a] (μmol·m^{-2}·s^{-1})	Average artificial PAR (μmol·m^{-2}·s^{-1})	Temperature (°C)	BRT (d)	HRT (d)	NLR[a] (gN·d^{-1})	PLR[a] (gP·d^{-1})	ATU (mg·L^{-1})
1A	17	171 ± 55	0	28.0 ± 1.5	8	8	2.5 ± 0.2	0.3 ± 0.0	5
1B	13	164 ± 34	0	25.4 ± 1.9	8	8	3.0 ± 0.2	0.4 ± 0.0	0
1C	21	294 ± 100	0	24.4 ± 2.2	14	14	1.7 ± 0.3	0.2 ± 0.0	0
1D	33	249 ± 111	0	16.8 ± 2.3	14	14	2.2 ± 0.5	0.3 ± 0.1	5[b]
2[c]	24	119 ± 32	300	23.0 ± 1.1	8	8	3.9 ± 0.3	0.4 ± 0.1	5
3A	20	234 ± 19	300	23.5 ± 0.3	4.5	2.5	9.7 ± 2.3	1.3 ± 0.2	5
3B	22	259 ± 43	300	26.9 ± 4.0	4.5	2	14.4 ± 1.8	1.8 ± 0.1	5
3C	47	283 ± 75	300	24.8 ± 1.3	4.5	3	8.4 ± 1.1	1.1 ± 0.2	5
4	40	357 ± 105	300	23.2 ± 2.1	4.5	2.5	13.6 ± 2.0	1.4 ± 0.2	5

[a]PAR: photosynthetically active radiation; NLR, PLR: nutrient (N, P) loading rate to each PBR.
[b]single ATU dosage.
[c]Smaller bubble size in PBR1 than PBR2.

Period 4 was operated as an MPBR system at a BRT and HRT of 4.5 days and 2.5 days, respectively, but the period started with a microalgae biomass concentration of 160 mgVSS·L^{-1} (lower than the other periods).

Analytical methods

Grab samples were collected in duplicate from the influent and effluent streams of the MPBR pilot plant three times a week. Ammonium, nitrite, nitrate, and phosphate were analysed in a Smartchem 200 automatic analyser (Westco Scientific Instruments), according to standard methods (APHA *et al.* 2005). VSS was also analysed following APHA *et al.* (2005).

Fifty microlitres of sample were measured twice a week according to Pachés *et al.* (2012) to count (in duplicate) the total eukaryotic cells (TEC).

Calculations

Biomass productivity (mgVSS·L^{-1}·d^{-1}), nitrogen removal rate (NRR) (mgN·L^{-1}·d^{-1}), phosphorus removal rate (PRR) (mgP·L^{-1}·d^{-1}), NRE (%) and PRE (%) were calculated by the Equations (1)–(5), respectively:

$$\text{Biomass productivity} = \frac{X_{VSS}}{\text{BRT}} \tag{1}$$

$$\text{NRR} = \frac{Q \cdot (N_i - N_e)}{V_{PBR}} \tag{2}$$

$$\text{PRR} = \frac{Q \cdot (P_i - P_e)}{V_{PBR}} \tag{3}$$

$$\text{NRE} = \frac{(N_i - N_e)}{N_i} \cdot 100 \tag{4}$$

$$\text{PRE} = \frac{(P_i - P_e)}{P_i} \cdot 100 \tag{5}$$

where X_{VSS} (mg VSS·L^{-1}) is the VSS concentration in the PBRs, BRT is the biomass retention time (d), Q is the wastewater flow rate (L·d^{-1}), N_i is the nitrogen concentration of the influent (mgN·L^{-1}), N_e is the nitrogen concentration of the effluent (mgN·L^{-1}), P_i is the phosphorus concentration of the influent (mgP·L^{-1}), P_e is the phosphorus concentration of the effluent (mgP·L^{-1}) and V_{PBR} is the total volume of the PBRs (L).

In order to compare different operating periods with variations in solar irradiances, the NRR:light irradiance ratio (NRR:I) (mgN·mol photons^{-1}), and PRR:light irradiance ratio (PRR:I) (mgP·mol photons^{-1}) were calculated by Equations (6) and (7), respectively:

$$\text{NRR:I} = \frac{\text{NRR} \cdot V_{PBR} \cdot 10^6}{I \cdot S \cdot 24 \cdot 3600} \tag{6}$$

$$\text{PRR:I} = \frac{\text{PRR} \cdot V_{PBR} \cdot 10^6}{I \cdot S \cdot 24 \cdot 3600} \tag{7}$$

where I is the total light photosynthetically active radiation. (PAR) irradiance on the PBR surface, i.e. the 24-hour average solar irradiance plus the light from the LED lamps (µmol photons·m^{-2}·s^{-1}) and S is the illuminated PBR surface (m^2).

RESULTS

Period 1

In this period, in which *Scenedesmus* remained the main microalgae genus (>99% of TEC), the effect of different BRT (i.e. HRT) under different environmental conditions and the competition of microalgae and ammonium oxidising bacteria (AOB) for ammonium was evaluated.

In sub-period 1A, AOB growth was inhibited by ATU addition, so that nitrite and nitrate concentration remained at negligible values, although ammonium and phosphate stayed at high values during the entire sub-period (Figure 1(a)).

In sub-period 1B, ATU was not added, but nitrite and nitrate concentrations remained at negligible concentrations, which suggested that no nitrifying bacteria proliferation occurred. Biomass concentration dropped from 301 ± 15 mgVSS·L^{-1} in sub-period 1A to 213 ± 28 mgVSS·L^{-1} in 1B. Since temperature has been shown to have a direct effect on biomass productivity (Viruela *et al.* 2016), the biomass reduction was assumed to be due to the temperature dropping from 28.0 ± 1.5 °C in sub-period 1A to 25.4 ± 1.9 °C in 1B. This temperature reduction could have also favoured microalgae over AOB (González-Camejo *et al.* 2018).

In sub-period 1C, the HRT (i.e. BRT) was raised from 8 to 14 days. In consequence, VSS concentration achieved a maximum concentration of 304 mg·L^{-1} (Figure 1(b)). However, this increased biomass concentration could also have been related to a solar PAR increase from 164 ± 34 µmol·m^{-2}·s^{-1} in sub-period 1B to 294 ± 100 µmol·m^{-2}·s^{-1} in 1C. On the other hand, by the end of sub-period 1C, nitrite concentration reached a maximum value of 18.5 mgN·L^{-1} (Figure 1(a)), which indicated that an AOB proliferation occurred.

Lastly, a single ATU dose was added at the beginning of sub-period 1D to inhibit AOB growth. Consequently, nitrite

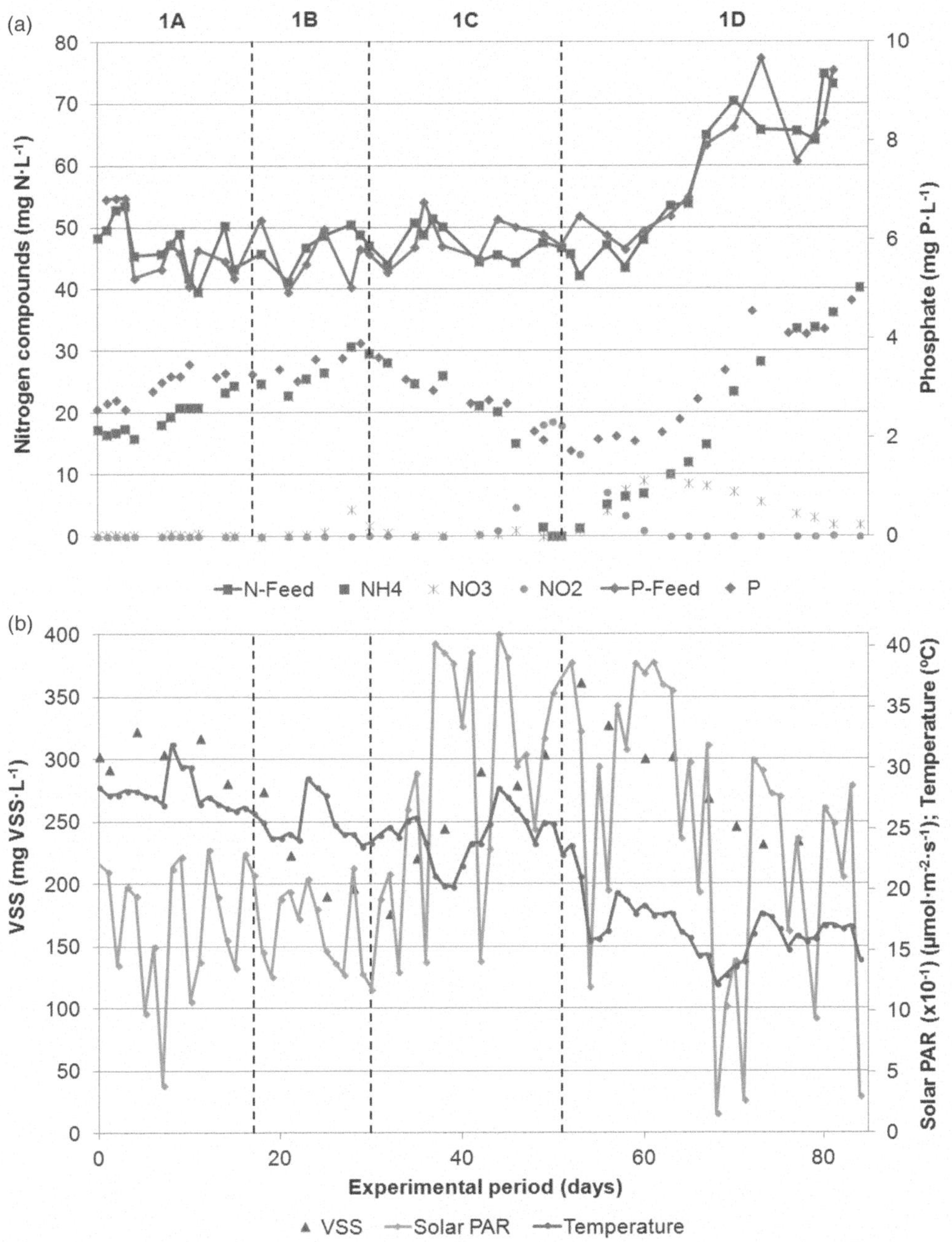

Figure 1 | Evolution during period 1 of (a) effluent concentration of ammonium (NH4), nitrite (NO2), nitrate (NO3) and soluble phosphorus (P); and feed concentration of nitrogen (N-Feed) and phosphorus (P-Feed); (b) VSS concentration, solar PAR and culture temperature.

concentration dropped due to the nitrate oxidising bacteria (NOB) proliferation, which oxidised nitrite to nitrate (Figure 1(a)). When the nitrite was exhausted, the NOB could no longer grow and nitrate concentration declined due to wash-out.

In terms of microalgae biomass, sub-period 1D started with a concentration of 360 $mgVSS{\cdot}L^{-1}$, but steadily decreased mainly due to a significant reduction in the culture temperature (Figure 1(b)).

It is worth mentioning that HRT was not directly related to nutrient loading rates due to both WWTP intake dynamics and AnMBR plant performance. For instance, sub-period 1A (HRT of 8 days) had a similar nitrogen loading rate (NLR) and phophorus loading rate (PLR) to 1D (HRT of 14 days) (Table 2). Hence, NLR and PLR must also be considered as controlling parameters.

In this period, the highest biomass productivities were achieved in sub-periods 1A and 1B (Table 3), probably because the temperature was higher (Table 2). Similar results were obtained by Viruela *et al.* (2016). Moreover, the nutrient removal rates in terms of NRR:I and PRR:I were also higher in sub-periods 1A and 1B, although the solar irradiances were considerably lower than in periods 1C and 1D (Table 2). Since nutrient removal rates have been reported to be directly related to light irradiance (Viruela *et al.* 2016), these results suggested that the culture could have been nutrient-limited during sub-periods 1C and 1D. In fact, the ammonium concentration remained under 10 mgN·L^{-1} during days 49–63, i.e., in sub-periods 1C and 1D (Figure 1(a)). In this respect, ammonium values below 10 mgN·L^{-1} have been reported to limit ammonium absorption by microalgae (Ruiz-Martinez *et al.* 2014). This low ammonium concentration in sub-periods 1C and 1D was mainly due to an AOB proliferation, which competed with microalgae for ammonium (González-Camejo *et al.* 2018). Hence, the proliferation of AOB did not seem to be desirable, as the system can get nutrient-limited. Further research on this topic must be developed in order to better understand the operating conditions which favour microalgae growth over AOB.

When the system was non-nutrient-limited, the effluent nutrient concentration followed approximately the same trend as the influent (Figure 1(a)). This tendency was in agreement with Arbib *et al.* (2013), who reported higher effluent nutrient concentrations at higher influent nutrient concentrations in outdoor microalgae cultivation.

Period 2

The effect of the bubble size of the air sparging system was studied in this period. Bubble diameter in PBR1 was reduced to 0.5 mm, while it remained at 5 mm in PBR2. PBR1 and PBR2 showed similar behaviour (Figure 2), reaching no significant differences between nutrient removal rates and biomass productivity (Table 3).

However, the genera distribution in the cultures was different; PBR1 had 40% *Scenedesmus* and 55% *Chlorella*, while PBR2 had 85% *Scenedesmus* and 10% *Chlorella*. Moreover, by the end of the period, the phosphorus concentration in PBR2 was slightly lower than in PBR1. These differences could have been related to a cyanobacteria proliferation observed in PBR1 at the end of period 2 (Figure 3). This agrees with Kin *et al.* (2014), who reported that small bubble size favours cyanobacteria growth over green algae. The proliferation of cyanobacteria is not desirable, as they have been reported to excrete some allelopathic substances that can damage green microalgae (Leão *et al.* 2009).

The results obtained in this period showed that nutrient removal rates and nutrient removal efficiencies were higher in period 2 than in period 1 (Table 3), mainly due to an additional light source that had not been used in period 1. Increasing the light irradiance on the PBRs was therefore considered beneficial for nutrient removal in outdoor conditions.

Period 3

The use of the membrane system in this period enhanced the treatment capacity of the MPBR plant: HRT was significantly reduced from 8 (period 2) to 2.5 days (sub-period 3A). This means that nutrient loading rates were considerably higher during this period (Table 2), which has been

Table 3 | Results obtained in each sub-period

Sub-period	Biomass productivity (mgVSS·L^{-1}·d^{-1})	NRE (%)	PRE (%)	NRR (mgN·L^{-1}·d^{-1})	PRR (mgP·L^{-1}·d^{-1})	NRR:I (mgN·mol^{-1})	PRR:I (mgP·mol^{-1})
1A	38 ± 2	56 ± 9	46 ± 8	2.8 ± 1.0	0.3 ± 0.1	36.4 ± 9.5	4.0 ± 2.1
1B	27 ± 4	40 ± 6	38 ± 6	1.9 ± 1.4	0.2 ± 0.1	31.3 ± 25.8	3.8 ± 2.0
1C	19 ± 3	49 ± 7	52 ± 10	2.3 ± 0.5	0.3 ± 0.1	25.8 ± 12.7	3.5 ± 1.5
1D	20 ± 3	57 ± 8	60 ± 8	1.6 ± 0.7	0.2 ± 0.1	24.6 ± 13.7	3.6 ± 1.8
2-PBR1[a]	28 ± 6	57 ± 4	76 ± 7	3.3 ± 2.0	0.3 ± 0.2	37.1 ± 32.2	3.7 ± 2.3
2-PBR2	28 ± 6	56 ± 7	87 ± 10	3.1 ± 1.2	0.4 ± 0.1	31.7 ± 28.1	4.3 ± 3.4
3A	72 ± 8	67 ± 11	69 ± 9	12.5 ± 4.2	1.5 ± 0.4	64.2 ± 22.5	12.7 ± 3.4
3B	69 ± 5	43 ± 11	43 ± 10	11.5 ± 2.9	1.4 ± 0.3	56.4 ± 15.4	11.8 ± 2.9
3C	78 ± 13	50 ± 15	56 ± 12	7.5 ± 1.8	1.1 ± 0.3	36.3 ± 9.5	9.6 ± 2.5
4	53 ± 15	33 ± 7	49 ± 12	7.8 ± 2.5	1.2 ± 0.3	33.5 ± 9.9	9.5 ± 2.4

[a]Bubble diameter: 0.5 mm.

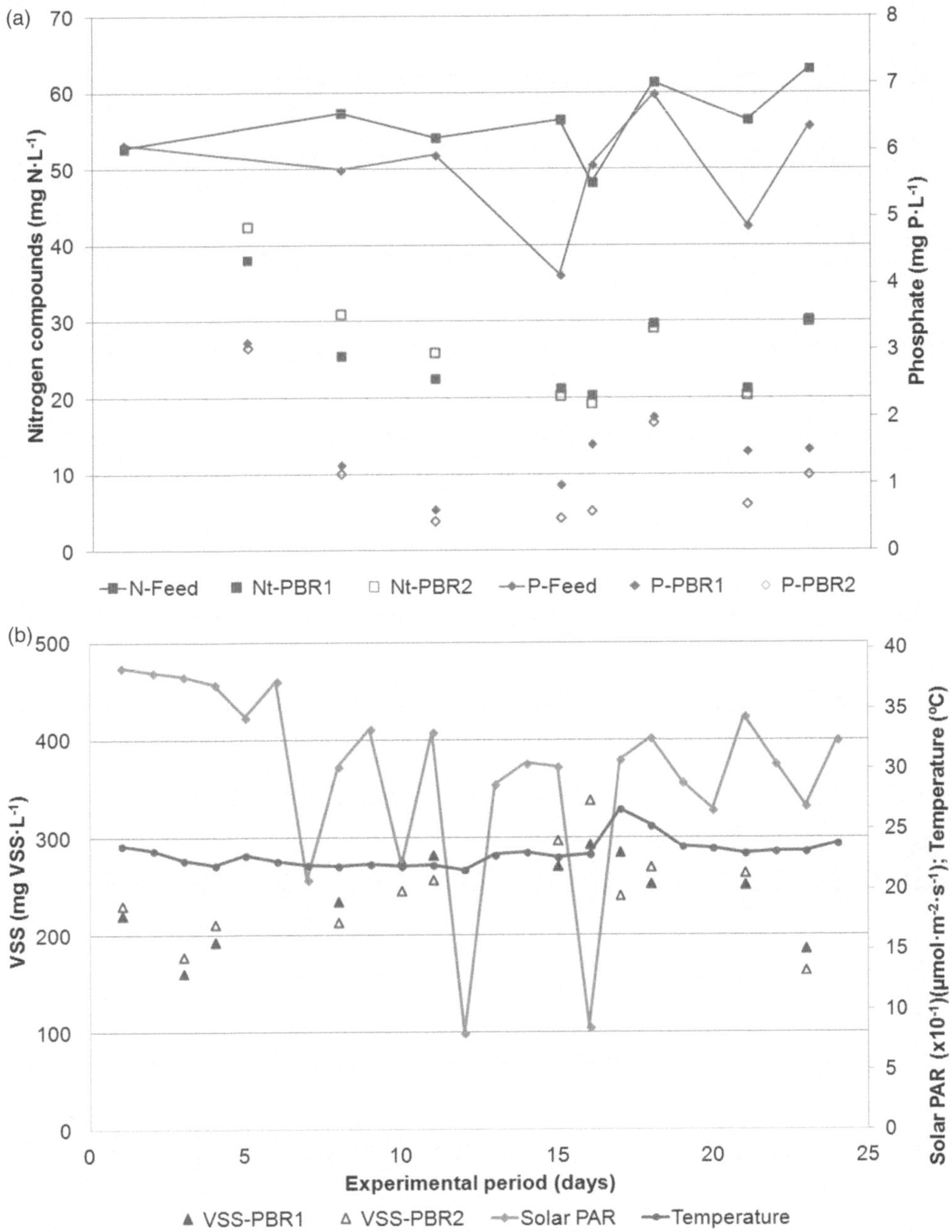

Figure 2 | Evolution during period 2 in PBR1 and PBR2 of (a) effluent concentration of soluble nitrogen (Nt) and soluble phosphorus (P), and feed concentration of nitrogen (N-Feed) and phosphorus (P-Feed); (b) solar PAR, culture temperature and VSS concentration.

reported to favour microalgae growth (Gao *et al.* 2016). In consequence, nutrient removal rates and biomass productivity were considerably higher in period 3 than in the previous periods (Table 3), reaching maximum NRR, PRR and biomass productivity in sub-period 3A: 12.5 ± 4.2 $mgN{\cdot}L^{-1}{\cdot}d^{-1}$, 1.5 ± 0.4 $mgP{\cdot}L^{-1}{\cdot}d^{-1}$ and 78 ± 13 $mgVSS{\cdot}L^{-1}{\cdot}d^{-1}$, respectively. The light use efficiency of the microalgae improved in this period (operating as an MPBR system), since NRR:I and PRR:I values were around twofold and threefold higher than in the previous periods, in which the system operated as a PBR (Table 3).

In sub-period 3B nutrient removal rates started at values around 15 $mgN{\cdot}L^{-1}{\cdot}d^{-1}$ and 1.7 $mgP{\cdot}L^{-1}{\cdot}d^{-1}$, but after day 30 they suddenly dropped to 7–10 $mgN{\cdot}L^{-1}{\cdot}d^{-1}$ and 1.0 $mgP{\cdot}L^{-1}{\cdot}d^{-1}$ and did not recover their high initial values (Figure 4(c)). This reduced nutrient removal rates could have been due to a significant increase in the culture temperature from around 25 to 33 °C in days 30–35 (Figure 4(b)).

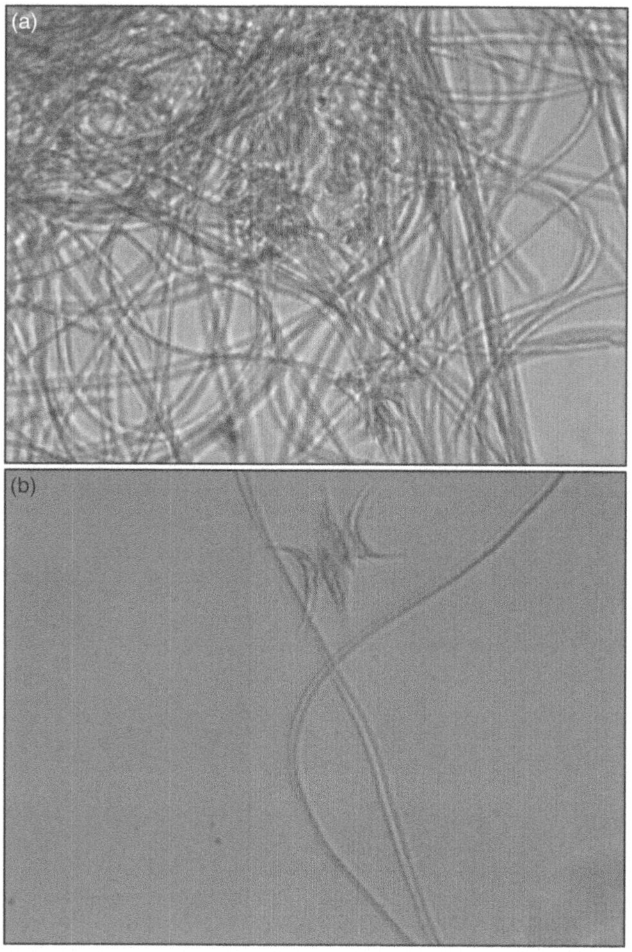

Figure 3 | Samples observed under epifluorescence microscope (Leica DM2500/DFC420c digital camera, 63×) in period 2 (day 21). (a) PBR1: cyanobacteria and green algae (mainly *Scenedesmus* and *Chlorella*) floc; (b) PBR2: *Scenedesmus* in four-cell coenobia and a small amount of cyanobacteria.

These high temperatures could have affected biomass productivity. Indeed, the biomass concentration dropped from around 400 to 300 mgVSS·L^{-1}. Consequently, nutrient removal capacity also decreased.

In sub-period 3C, temperature stabilised and NRR was solar PAR-dependent (Figure 4(c)), which was in agreement with Viruela *et al.* (2016). However, NRR and PRR were lower in sub-period 3C than in sub-periods 3A and 3B, which could be explained by the following. i) After the high temperatures in sub-period 3B, the system took around 2 weeks to recover the initial microalgae biomass (Figure 4(b)), so that its nutrients removal capacity was reduced. ii) Sub-period 3C had the lowest nutrient loading rates of period 3 (Table 2). Consequently, effluent nitrogen concentration (which was mainly ammonium) was reduced to values of 10–15 mgN·L^{-1} during days 55–68 (Figure 4(a)). Ruiz-Martinez *et al.* (2014) reported that NRR decreased whenever ammonium concentration in the culture was below 10–13 mgN·L^{-1}. Hence, in sub-period 3C, the culture was considered to be nutrient-limited. iii) In spite of having received a higher solar PAR in sub-period 3C (Table 3), this irradiance was more variable than in sub-periods 3A and 3B (Figure 4(c)). This means that the alternation of very sunny days, in which photoinhibition could have occurred, with photo-limited days could have negatively affected microalgae growth.

Throughout period 3, *Scenedesmus* remained as dominant genus (80–95% of TEC) and *Chlorella* only reached 5–20% of TEC.

The best efficiencies of this period (67 ± 11% and 69 ± 9%, for nitrogen and phosphorus, respectively) were obtained at an HRT of 2.5 days, even though solar PAR in sub-period 3A was the lowest of the period (Table 2). On the other hand, in sub-period 3C, with the lowest nutrient loading rates, the culture could be nutrient-limited and therefore nutrient removal efficiencies were lower than in 3A (Table 3). NLR and PLR thus appear to be key parameters in assessing MPBR performance.

Period 4

The same HRT and BRT were used in period 4 as in sub-period 3A, but at higher nutrient loading rates (Table 2); however, the results obtained were significantly different (Table 3).

As Figure 5(b) shows, microalgae biomass concentration was under 250 mgVSS·L^{-1} for the entire period, while in sub-period 3A it always remained over 250 mgVSS·L^{-1} (Figure 4(b)), so that the nutrient removal capacity of the system diminished and nutrient removal rates were not as high as in sub-period 3A (Table 3). This lower biomass concentration could have been influenced by the lower initial microalgae concentration in the start-up period: 160 mgVSS·L^{-1} in period 4, while sub-period 3A started at 270 mgVSS·L^{-1}. Su *et al.* (2012) also obtained higher NRR and PRR in the culture with a higher initial biomass concentration. Moreover, Feng *et al.* (2011) reported that cultures with denser initial biomass concentration achieved higher biomass productivity and adapted quickly to outdoor conditions.

Solar PAR, in spite of being higher than in sub-period 3A (Table 2), was quite variable in period 4 (Figure 5(c)) and, as in Period 3, could have negatively affected microalgae growth.

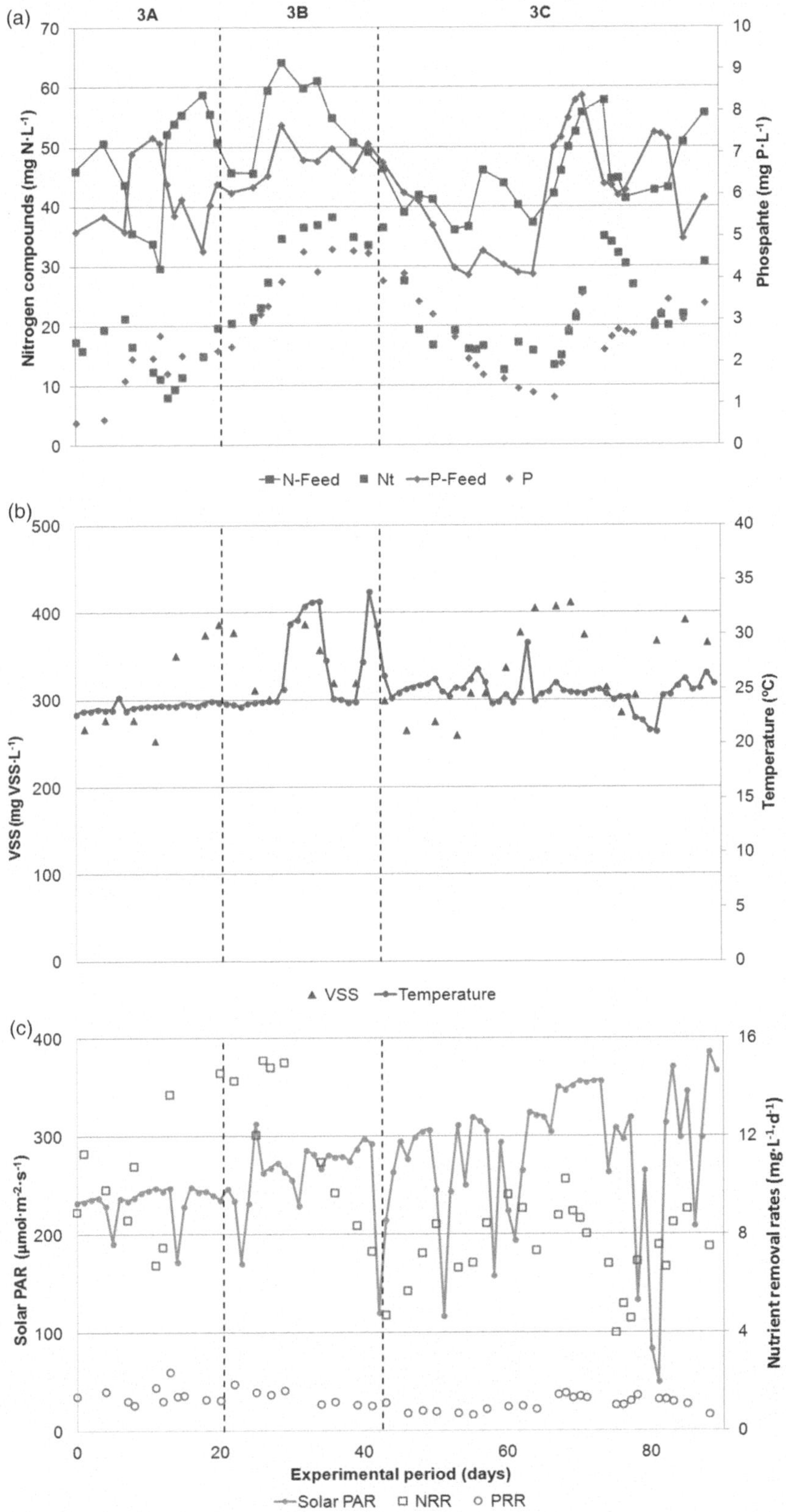

Figure 4 | Evolution during period 3 of (a) effluent concentration of soluble nitrogen (Nt) and soluble phosphorus (P), and feed concentration of nitrogen (N-Feed) and phosphorus (P-Feed); (b) culture temperature and VSS concentration; (c) solar PAR and nutrient removal rates.

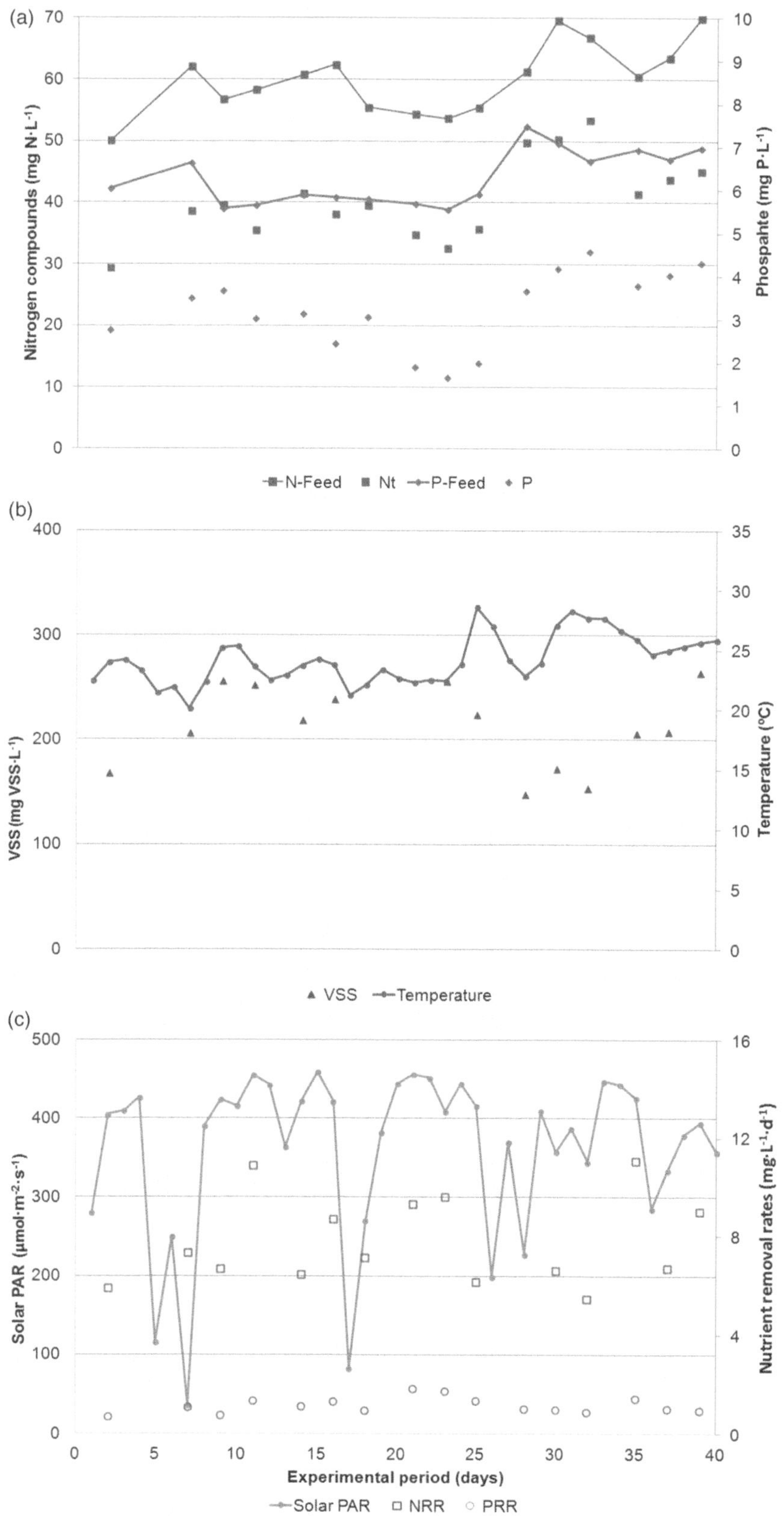

Figure 5 | Evolution during period 4 of (a) effluent concentration of soluble nitrogen (Nt) and soluble phosphorus (P); and feed concentration of nitrogen (N-Feed) and phosphorus (P-Feed); (b) culture temperature and VSS concentration; (c) solar PAR and nutrient removal rates.

Nutrient removal rates could also have been influenced by a shift in the microalgae culture. In period 4 there was a proliferation of *Monoraphidium* (45% TEC) which co-habited with *Scenedesmus* (50% TEC). No significant amount of *Chlorella* was present.

As happened in period 1, in period 4 the effluent nutrient concentrations followed the same trend as the influent nutrient concentrations (Figure 5(a)), since the system was not nutrient-limited. According to Arbib *et al.* (2013), in these conditions, microalgae are mainly limited by outdoor conditions.

DISCUSSION

The performance of this outdoor MPBR pilot plant treating AnMBR effluent within a wide range of environmental, design, and operating conditions produced some interesting results, which deserve to be commented on.

When the plant was operated as a PBR system without membrane filtration (periods 1 and 2), the highest values in terms of nutrient removal and biomass productivity were obtained when HRT was 8 days (Table 3). When the plant was operated as an MPBR system (periods 3 and 4), the best results were achieved at a BRT and HRT of 4.5 days and 2.5 days, respectively (sub-period 3A, Table 3). In this respect, optimum BRT and HRT must be assessed to further improve MPBR performance.

Comparing PBR and MPBR performance, nutrient removal rates and biomass productivity were significantly higher in MPBR as the use of membranes to separate microalgae from water enabled operation at lower HRT (i.e. higher nutrient loading rates), avoiding microalgae wash-out.

Generally, the plant performance was strongly dependent on outdoor conditions; solar irradiance seemed to be one of the main factors affecting nutrient removal, while temperature variations had a major impact on biomass productivity. The plant performance yields were reduced when the culture was nutrient-limited, which meant that high nutrient loading rates were required to reach high nutrient removal rates. In this respect, the proliferation of AOB in the culture can worsen PBR performance since they compete with microalgae for ammonium consumption.

Increasing the light supply to the microalgae seemed to be beneficial for nutrient removal as nutrient removal rates were lower in period 1 with no artificial lighting (Table 3).

Small bubble size (0.5 mm diameter) in the air sparging system was not found to be suitable, as it favoured the proliferation of filamentous cyanobacteria, which could hinder green microalgae growth.

The initial biomass concentration appeared to have some influence on the plant performance, since higher biomass concentrations attained better results at quite similar operating conditions.

Overall, as the nutrient removal efficiencies achieved in this continuously operated MPBR under outdoor conditions and using real anaerobically treated sewage were not particularly high, some improvements need to be made to comply with legal requirements. Special efforts should be focused on increasing the efficiency of the light applied to the PBRs, lowering the plant HRT to further increase its treatment capacity, controlling BRT (and HRT when treatment capacity can be variable) to optimise microalgae productivity and nutrient removal, avoiding AOB growth without using chemical inhibitors, and reducing operating costs.

CONCLUSIONS

In this study, an MPBR plant was operated outdoors under different conditions: BRT, HRT, temperature, light irradiance, influent nutrient concentration, ATU addition, and bubble size of the air sparging system; reaching maximum biomass productivity and nitrogen and phosphorus removal rates of 78 ± 13 mgVSS·L^{-1}·d^{-1}, 12.5 ± 4.2 mgN·L^{-1}·d^{-1} and 1.5 ± 0.4 mgP·L^{-1}·d^{-1}, respectively. Although these values are promising, further research needs to be carried out to make this technology feasible on an industrial scale. The main challenges to overcome include increasing the efficiency of the light supplied to the PBRs, avoiding AOB growth, improving the plant's treatment capacity and reducing its operating costs.

ACKNOWLEDGEMENTS

This research work has been supported by the Spanish Ministry of Economy and Competitiveness (MINECO, CTM2011-28595-C02-01 and CTM2011-28595-C02-02) jointly with the European Regional Development Fund (ERDF), both of which are gratefully acknowledged. It was also supported by the Spanish Ministry of Education, Culture and Sport via a pre-doctoral FPU fellowship to author J. González-Camejo (FPU14/05082).

REFERENCES

APHA, AWWA & WEF 2005 *Standard Methods for the Examination of Water and Wastewater*, 21st edn. American

Public Health Association (APHA)/American Water Works Association (AWWA)/Water Environment Federation (WEF), Washington, DC, USA.

Arbib, Z., Ruiz, J., Álvarez-Díaz, P., Garrido-Pérez, C., Barragan, J. & Perales, J. A. 2013 Long term outdoor operation of a tubular airlift pilot photobioreactor and a high rate algal pond as tertiary treatment of urban wastewater. *Ecol. Eng.* **52**, 143–153.

Arias, D. M., Uggetti, E., García-Galán, M. J. & García, J. 2017 Cultivation and selection of cyanobacteria in a closed photobioreactor used for secondary effluent and digestate treatment. *Science of the Total Environment* **587–588**, 157–167.

Feng, P., Deng, Z., Hu, Z. & Fan, L. 2011 Lipid accumulation and growth of *Chlorella zofingiensis* in flat plate photobioreactors outdoors. *Bioresour. Technol.* **102** (22), 10577–10584.

Gao, F., Li, C., Yang, Z., Zeng, G., Feng, L., Liu, J., Liu, M. & Cai, H. 2016 Continuous microalgae cultivation in aquaculture wastewater by a membrane photobioreactor for biomass production and nutrients removal. *Ecol. Eng.* **92**, 55–61.

Giménez, J. B., Robles, A., Carretero, L., Durán, F., Ruano, M. V., Gatti, M. N., Ribes, J., Ferrer, J. & Seco, A. 2011 Experimental study of the anaerobic urban wastewater treatment in a submerged hollow-fibre membrane bioreactor at pilot scale. *Bioresour. Technol.* **102**, 8799–8806.

González-Camejo, J., Serna-García, R., Viruela, A., Pachés, M., Durán, F., Robles, A., Ruano, M. V., Barat, R. & Seco, A. 2017 Short and long-term experiments on the effect of sulphide on microalgae cultivation in tertiary sewage treatment. *Bioresour. Technol.* **244**, 15–22.

González-Camejo, J., Barat, R., Pachés, M., Murgui, M., Ferrer, J. & Seco, A. 2018 Wastewater nutrient removal in a mixed microalgae-bacteria culture: effect of light and temperature on the microalgae-bacteria competition. *Environ. Technol.* **39** (4), 503–515.

Gouveia, L., Graça, S., Sousa, C., Ambrosano, L., Ribeiro, B., Botrel, E. P., Neto, P. C., Ferreira, A. F. & Silva, C. M. 2016 Microalgae biomass production using wastewater: treatment and costs: scale-up considerations. *Algal Res.* **16**, 167–176.

Kin, K., Choi, J., Ji, Y., Park, S., Do, H., Hwang, C. & Lee, B. 2014 Impact of bubble size on growth and CO_2 uptake of *Artrhospira (Spirulina) platensis* KMMCC CY-007. *Bioresour. Technol.* **170**, 310–315.

Leão, P. N., Teresa, M., Vasconcelos, S. D. & Vasconcelos, V. M. 2009 Allelopathy in freshwater cyanobacteria. *Crit. Rev. Microbiol.* **35**, 271–282.

Ledda, C., Idà, A., Allemand, D., Mariani, P. & Adan, F. 2015 Production of wild *Chlorella* sp. cultivated in digested and membrane-pretreated swine manure derived from a full-scale operation plant. *Algal Res.* **12**, 68–73.

Pachés, M., Romero, I., Hermosilla, Z. & Martinez-Guijarro, R. 2012 PHYMED: an ecological classification system for the Water Framework Directive based on phytoplankton community composition. *Ecol. Indic.* **19**, 15–23.

Ruiz-Martinez, A., Serralta, J., Pachés, M., Seco, A. & Ferrer, J. 2014 Mixed microalgae culture for ammonium removal in the absence of phosphorus: effect of phosphorus supplementation and process modeling. *Process Biochem.* **49**, 2249–2257.

Su, Y., Mennerich, A. & Urban, B. 2012 Coupled nutrient removal and biomass production with mixed algal culture: impact of biotic and abiotic factors. *Bioresour. Technol.* **118**, 469–476.

Tan, X. B., Zhang, Y. L., Yang, L. B., Chu, H. Q. & Guo, J. 2016 Outdoor cultures of *Chlorella pyrenoidosa* in the effluent of anaerobically digested activated sludge: the effects of pH and free ammonia. *Bioresour. Technol.* **200**, 606–615.

Viruela, A., Murgui, M., Gómez-Gil, T., Durán, F., Robles, A., Ruano, M. V., Ferrer, J. & Seco, A. 2016 Water resource recovery by means of microalgae cultivation in outdoor photobioreactors using the effluent from an anaerobic membrane bioreactor fed with pre-treated sewage. *Bioresour. Technol.* **218**, 447–454.

Whitton, R., Le Mével, A., Pidou, M., Ometto, F., Villa, R. & Jefferson, B. 2016 Influence of microalgal N and P composition on wastewater nutrient remediation. *Water Res.* **91**, 371–378.

Woertz, I., Feffer, A., Lundquist, T. & Nelson, Y. 2009 Algae grown on dairy and municipal wastewater for simultaneous nutrient removal and lipid production for biofuel feedstock. *J. Environ. Eng.* **135** (11), 1115–1122.

Wu, Y. H., Zhu, S. F., Yu, Y., Shi, X. J., Wu, G. X. & Hu, H. Y. 2017 Mixed cultivation as an effective approach to enhance microalgal biomass and triacylglycerol production in domestic secondary effluent. *Chem. Eng. J.* **328**, 665–672.

Xu, M., Li, P., Tang, T. & Hu, Z. 2015 Roles of SRT and HRT of an algal membrane bioreactor system with a tanks-in-series configuration for secondary wastewater effluent polishing. *Ecol. Eng.* **85**, 257–264.

First received 21 November 2017; accepted in revised form 29 May 2018. Available online 11 June 2018

Activity assessment of microalgal-bacterial consortia based on respirometric tests

S. Rossi, M. Bellucci, F. Marazzi, V. Mezzanotte and E. Ficara

ABSTRACT

Respirometric techniques are useful tools to evaluate bacterial activities in activated sludge processes due to their fast execution and the possibility to obtain several kinetic parameters from a single test. Using such techniques in microalgae-bacteria consortia treating wastewater could allow a better understanding of mutual interactions between the microbial populations as a function of environmental parameters. This work aims at developing and testing a novel experimental respirometric protocol to determine oxygen uptake rates and oxygen production rates by a microalgae-bacteria consortium. The defined protocol is characterized by alternating light/dark regimes and by dosing substrates/inhibitors to selectively activate/inactivate microalgal and bacterial metabolisms. The protocol was then applied on microalgal and bacterial consortia, which were grown on the liquid fraction of black water from biogas plants fed on agricultural and municipal waste sludge. Results elucidate the presence and activity of microalgae and nitrifying bacteria in the tested systems, suggesting that the respirometric tests could be included into monitoring procedures of photobioreactors/algal ponds.

Key words | activity assessment, microalgae-bacteria consortium, nitrification inhibition, oxygen mass transfer, respirometry, wastewater treatment

S. Rossi
M. Bellucci
E. Ficara (corresponding author)
Dipartimento di Ingegneria Civile e Ambientale (DICA),
Politecnico di Milano,
P.zza L. da Vinci 32, Milano 20133,
Italy
E-mail: *elena.ficara@polimi.it*

F. Marazzi
V. Mezzanotte
Dipartimento di Scienze dell'Ambiente e della Terra (DISAT),
Università degli Studi di Milano – Bicocca,
P.zza della Scienza 1, Milano 20126,
Italy

INTRODUCTION

Biological wastewater treatment with microalgae has gained an increasing interest in the last decades, thanks to the capacity of microalgae to uptake nutrients (ammonia, nitrate and phosphate), to produce bio-fuels and to fix inorganic carbon (Chisti 2007). One of the major advantages of using microalgal-based wastewater treatment is that microalgae produce oxygen with their photosynthetic activity, and the produced oxygen is used by autotrophic bacteria to oxidize ammonia nitrogen. Indeed, the synergism between nitrifying bacteria and microalgae was reported by several authors (Godos *et al.* 2009; González-Fernández *et al.* 2011). The use of microalgal-bacterial (MB) consortia could decrease the operational costs connected to aeration in aerobic wastewater treatment.

Respirometric and titrimetric techniques are useful tools to determine the optimal growth conditions for bacteria in activated sludge systems and have been applied to various biological systems (Spanjers & Vanrolleghem 1995; Munz *et al.* 2009). By means of these techniques, the oxygen consumption of bacterial suspensions can be quantified, from which bacterial activity is then computed. In addition, respirometry can be successfully applied to characterize influent wastewaters, to identify potential inhibitory effects and to evaluate kinetic and stoichiometric parameters (Vanrolleghem 2002). The knowledge of these parameters is essential for the calibration and validation of mathematical models, describing the behaviour of biological processes, such as the activated sludge models (Henze *et al.* 2000).

The application of respirometric tests to algal-bacterial suspensions could represent a powerful alternative to determine the activity and the suspended solids apportioning among the mass of relevant microbial components. This information allows for a better knowledge and understanding of wastewater-treating photobioreactors/ponds. Nonetheless, to date, the application of respirometric/titrimetric techniques for the characterization of MB suspensions is scarce. Some authors used respirometry for evaluating the effect of different environmental factors, such as light regimes and intensity (Brindley *et al.* 2010), pH, bicarbonates or nutrient sources (Decostere *et al.* 2013, 2016; Tang *et al.* 2014). Respiratory activity tests were also performed to assess the photosynthetic activity or the dark respiration of microalgal cultures (Kliphuis

doi: 10.2166/wst.2018.078

et al. 2011; Ruiz-Martinez *et al.* 2016). However, a lack of uniformity in units and procedures makes the comparison of data and protocols quite difficult. Previous works generally refer to different test conditions or experimental devices; significant discordance can be particularly found regarding: light availability (light intensity and light/dark phase durations), biomass characteristics (microalgae strains and initial biomass concentrations), environmental conditions (nutrient and inorganic carbon sources and availability, pH and temperature range) and respirometric devices (volume and geometry of bioreactors). Moreover, except for a few cases (Vargas *et al.* 2016), respirometric procedures are generally applied to pure microalgal cultures, in the absence of bacteria, which makes those procedures not directly applicable to MB reactors. This work aims at developing a respirometric device and a standardized procedure to study MB suspensions, by evaluating both microalgal and nitrifying activity during the test. To the extent of the authors' knowledge, this combined approach has never been applied before to MB consortia. Specifically, the respirometric protocol was tested on MB consortia grown on agro-zootechnical and municipal black waters.

MATERIALS AND METHODS

Microalgae cultivation

MB suspensions were collected from three microalgal culturing systems (a laboratory-scale system, a pilot-scale bubble-column, a pilot-scale raceway), briefly described below.

The laboratory-scale cultivation system was composed of four glass cylinders (hereafter referred to as 'Lab-C'), with a working volume of 2 L each. Lab-cylinders were operated semi-continuously, with an average hydraulic retention time (HRT) of 20 d and were fed on the liquid fraction of anaerobically digested agro-wastewaters (hereafter referred to as 'black water', BW). During Lab-C cultivation, BW was diluted twice with tap water, due to high nutrient contents. Every cylinder was equipped with an aeration system for mixing and CO_2-supply. Lab-cylinders were exposed to artificial fluorescent lights (OSRAM Fluora, 2 × 30 W), providing a photosynthetically active radiation (PAR) of 68 ± 17 mE m^{-2} s^{-1} with a 12 h/12 h light/dark cycle. The average room temperature was about 24 °C. The MB culture was dominated by *Chlorella* and *Scenedesmus* spp., as confirmed by optical microscopy observations.

The pilot-scale bubble-column (hereafter referred to as 'Pilot-C') operated outdoors and was characterized by a working volume of 75 L and a diameter of 29 cm. The reactor was fed on the same BW described above, without applying dilutions. Feeding was continuous with an HRT of 20 d and CO_2 was supplied to the culture by means of an air compressor, bubbling air at the bottom of the column. Air bubbling also provided adequate mixing to the suspension. The MB culture was dominated by *Chlorella* and *Scenedesmus* spp.

The pilot-scale raceway was an open pond reactor (hereafter referred to as 'Pilot-RW'), operating in outdoor conditions. The system was characterized by a working volume of about 1 m^3, a surface of 5.8 m^2 and a water depth of 18 cm. Mixing was provided by a paddle-wheel, and a pH-controlled injection of off-gas from the combined heat and power unit was used to transfer CO_2 to the culture. No temperature regulation was implemented. The reactor was fed first in batch and then continuously, with an HRT of 10 d. The feed was the BW of the anaerobically digested waste sludge of a municipal wastewater treatment plant. MB samples observed under the microscope contained *Chlorella, Chlamydomonas* and *Scenedesmus* spp., but the dominant algal species in the culture were *Chlorella*.

The average composition of influent wastewaters is reported in Table 1.

Data in Table 1 indicate that all cultivation systems were fed with quite concentrated wastewater, especially in ammonia nitrogen and organic matter (although the latter is expected to consist of recalcitrant organics that the anaerobic process could not convert into biogas). For this reason, bacteria were also expected to develop together with

Table 1 | Composition of the black waters used as influent to the cultivation systems (values are expressed as mean ± standard deviation, N is the number of measurements, Pilot-C is the pilot-scale bubble-column bioreactor, Lab-C is the laboratory-scale cultivation system, Pilot-RW is the pilot-scale raceway reactor)

		Cultivation system		
Parameter		**Pilot-C**	**Lab-C**	**Pilot-RW**
NH_4^+	mg N L^{-1}	251 ± 86	134 ± 47	244 ± 79
NO_3^-	mg N L^{-1}	8.6 ± 2.9	4.3 ± 1.1	0.2 ± 0.3
PO_4^{3-}	mg P L^{-1}	19.8 ± 8.2	9.1 ± 4.4	5.7 ± 0.8
Soluble chemical oxygen demand	mg O_2 L^{-1}	362 ± 153	271 ± 216	122 ± 54
Turbidity	FAU	137 ± 176	76 ± 91	68 ± 48
TSS	mg TSS L^{-1}	126 ± 119	54 ± 62	80 ± 40
OD_{680}	–	0.16 ± 0.15	0.09 ± 0.06	0.07 ± 0.14
pH	–	7.9 ± 0.2	7.8 ± 0.5	8.1 ± 0.3
Conductivity	mS cm^{-1}	3.2 ± 0.3	2.7 ± 0.7	1.5 ± 0.3
N	–	36	12	31

microalgae. Indeed, nitrifying consortia easily developed in the bioreactors, as confirmed by high concentrations of oxidized N-compounds (NO_3^- and NO_2^-) measured in the algal suspensions (data not shown). The presence of high concentrations of heterotrophic bacteria (HB) was not expected, since the organic load in BW is generally constituted by recalcitrant or scarcely biodegradable substances, as reported by Akhiar *et al.* (2017). High turbidities, associated with the presence of suspended solids, were sporadically measured in the BW, due to malfunctioning in the dewatering step obtained by centrifuging.

Respirometric equipment and procedure

A simple *respirometer* was set up (Figure 1) to perform respirometric assays, consisting og: a bioreactor (250 mL transparent glass flask) which was covered in aluminium foils during dark phases; a dissolved oxygen (DO) probe (Hach-Lange, LDO101) connected to a multi-meter device (Hach-Lange, HQ40D) recording both DO and temperature data (2 data min^{-1}); an aeration system including an air pump (KW, Mouse 2-Beta, 110 NL h^{-1}) and a porous diffuser; a lighting system made of fluorescent lights (OSRAM Fluora, 2 × 30 W); and a magnetic mixing system (VELP Scientifica, Microstirrer), operated at a fixed revolution speed of 150 rpm to minimize the oxygen mass transfer.

During the respirometric test, the following *solutions* were used:

- Substrates solution (SS): 9.5 g NH_4Cl L^{-1}, corresponding to 3.2 g NH_4^+-N L^{-1}; 12.3 g $NaNO_2$ L^{-1} corresponding to 8.2 g NO_2^--N L^{-1}, which was dosed in order to achieve an ammonium and nitrate concentration of 10 mg N L^{-1}.

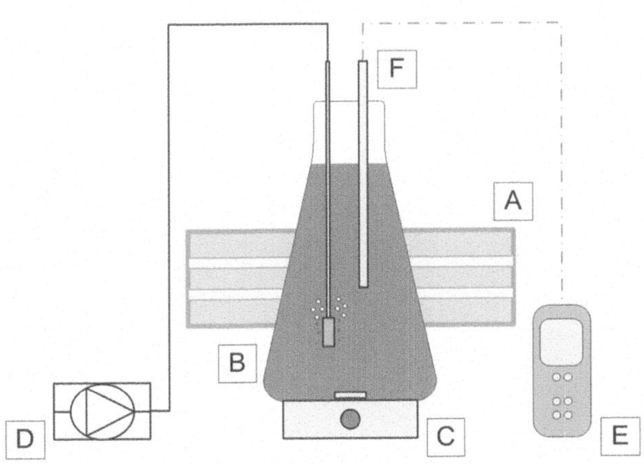

Figure 1 | Lay-out of the respirometer (A: lights, B: 250 mL glass flask, C: magnetic mixer, D: air pump, E: multi-meter and data logger, F: DO probe).

- Inhibitors solutions (IS): allylthiourea (ATU) solution (IS_1: 2.5 g ATU L^{-1}), chlorate solution (IS_2: 60 g $KClO_3$ L^{-1}), which were dosed in order to achieve ATU concentration of 10 mg L^{-1} and a chlorate concentration of 1.2 g L^{-1}, known to cause the complete inhibition of ammonia-oxidizing bacteria (AOB) and nitrite-oxidizing bacteria (NOB), respectively (Hynes & Knowles 1983; Ginestet *et al.* 1998). Inhibitors of AOB and NOB were dosed simultaneously because preliminary analysis demonstrated that reliable results on the sole NOB activity were difficult to be obtained, probably due to both their limited occurrence in the MB suspensions (as suggested by frequent NO_2^- accumulation) and their limited specific oxygen request compared to AOB.

The sum of injected volumes (10.5 mL) represents an increase lower than 5% in the final respirometer volume. Therefore, dilution correction factors were not considered during calculations, since the volume increase was considered negligible.

To perform respirometric tests, the following *protocol* was set up. First, an aliquot (250 mL) of MB suspension was sampled from the cultivation system. The optical density at the wavelength of 680 nm (OD_{680}) was measured before each test and adjusted to a fixed interval (from 0.4 to 0.6) to obtain comparable light penetration. Therefore, when necessary, a mineral medium was used to dilute the original suspension. The composition of the mineral medium was defined in order to mimic the ionic composition of the original MB samples, thus avoiding the exposure of MB consortia to osmotic shocks. The diluted MB suspension was transferred into the respirometer bioreactor.

Then DO data collection was started. Along the test duration, environmental conditions (light conditions and the presence/absence of specific substrates and/or inhibitors) were modified, identifying different phases (L_i are light phases, D_i are dark phases).

Phase durations were set to 15 min for light phases and 10 min for dark phases, according to Brindley *et al.* (2010). The duration of the first light phase was set to 30 min to let the consortium acclimate to respirometer conditions. Hereafter each phase is described, including environmental conditions and expected biological reactions affecting the DO concentration:

- L_1: Light on, addition of SS. Oxygen production is expected due to active microalgal photosynthesis; a contribution to oxygen consumption is also expected due to the aerobic activity of nitrifying bacteria (hereafter referred to as 'NIT', including both AOB and NOB), of

HB and of microalgae (MA) respiration; therefore, the net oxygen production rate of this phase (OPR_{L1}) is made of the following contributions:

$$OPR_{L1} = OPR_{MA,P} + OPR_{MA,R} + OPR_{HB} + OPR_{NIT}$$

where $OPR_{MA,P}$ is the photosynthetic MA oxygen production, $OPR_{MA,R}$ is the respiration by microalgae, OPR_{HB} is the oxygen production by heterotrophs (negative), OPR_{NIT} is the oxygen production by nitrifiers (negative).

- D_1: Light off. Oxygen consumption by MA respiration, NIT activity and HB activity; therefore:

$$OPR_{D1} = OPR_{MA,R} + OPR_{HB} + OPR_{NIT}$$

- D_2: Light off, addition of IS_1 and IS_2. NIT are inhibited and the oxygen consumption is only attributed to MA respiration and HB activity; therefore:

$$OPR_{D2} = OPR_{MA,R} + OPR_{HB}$$

- L_2: Light on. Oxygen production due to MA photosynthetic activity is expected and simultaneous oxygen consumption by MA respiration and HB activity; therefore:

$$OPR_{L1} = OPR_{MA,P} + OPR_{MA,R} + OPR_{HB}$$

- D_3: Same conditions as D_2. Oxygen consumption by MA respiration and by HB activity; therefore:

$$OPR_{D3} = OPR_{MA,R} + OPR_{HB}$$

The typical output of a respirometric test and its phases are reported in Figure 2, in which the DO profile and the calculated OPR for each phase are plotted against time. Under the graph, a simplified representation of the activity/inactivity of the main microbial populations is reported.

Then the test was ended and DO data were retrieved and used to compute the OPR (positive/negative when a net production/consumption was observed) in each phase, according to the following *data processing* procedure. Experimental DO data were fitted by a model considering the coexistence of oxygen mass transfer, due to mixing, and the biological DO consumption/production rates, as follows:

$$\frac{dDO}{dt} = k_La*(DO_{SAT} - DO) + OPR \qquad (1)$$

where: dDO/dt is the variation of DO over time [mg O_2 L^{-1} h^{-1}], k_La is the global oxygen mass transfer coefficient [h^{-1}], DO_{SAT} is the oxygen saturation concentration at the actual temperature [mg O_2 L^{-1}], DO is the actual DO concentration in the bioreactor [mg O_2 L^{-1}], OPR is a constant oxygen production rate [mg O_2 L^{-1} h^{-1}]. The minimum mean square error criterion was used assess the optimal fit. The k_La coefficient was evaluated during abiotic tests in clean water (data not shown), implementing the log-deficit procedure as described in ASCE (1993). A final value of 0.9 h^{-1} was calculated and used for OPR determinations.

OPR values of each phase (OPR_{L1}, OPR_{L2} for the two light phases; OPR_{D1}, OPR_{D2}, OPR_{D3} for the three dark phases) were determined, which depend on the microbial

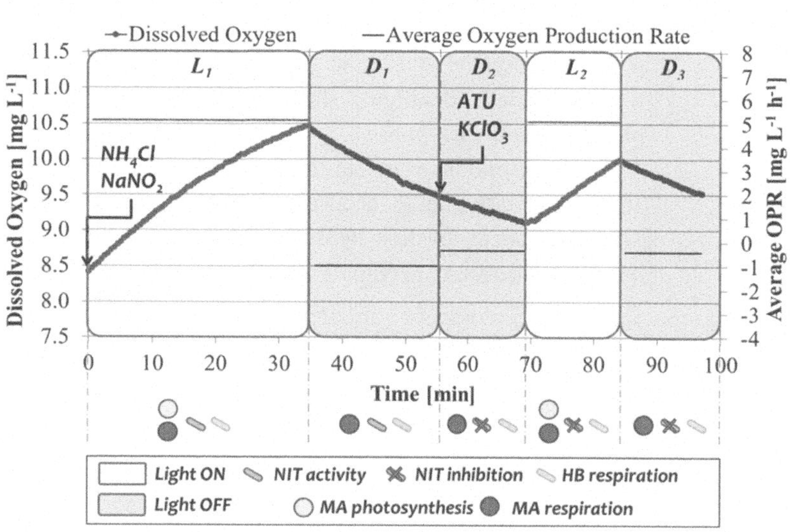

Figure 2 | Typical result of a respirometric test. Phases are reported together with the addition of substrates or inhibitors and the active or inactive populations.

oxygen uptake/consumption as follows:

$$OPR_{MA,P\_1} = OPR_{L1} - OPR_{D1} \quad (2)$$

$$OPR_{MA,P\_2} = OPR_{L2} - OPR_{D2} \quad (3)$$

$$OPR_{MA,P\_3} = OPR_{L2} - OPR_{D3} \quad (4)$$

$$OPR_{NIT\_1} = OPR_{D1} - OPR_{D2} \quad (5)$$

$$OPR_{NIT\_2} = OPR_{L1} - OPR_{L2} \quad (6)$$

All the OPR values are expressed as mg O_2 L^{-1} h^{-1}. From Equations (2)–(6), various estimates were obtained for both microalgal photosynthetic oxygen production (OPR_{MA,P_1}, OPR_{MA,P_2}; OPR_{MA,P_3}) and the nitrifiers' oxygen consumption (OPR_{NIT_1}, OPR_{NIT_2}). Mean values were then calculated ($OPR_{MA,P,mean}$, OPR_{NIT_mean}). OPR_{MA,P_1} and the mean value of OPR_{MA,P_2} and OPR_{MA,P_3} were compared by means of heteroscedastic T-tests (computed using data analysis toolbox in Microsoft Office Excel 2016) to exclude the possibility of short-term inhibition effects of ATU and chlorate on MA. The dark oxygen consumption including MA and HB respiration was also obtained, although it gives an aggregated information (OPR_{R*}). OPR values were corrected by using a temperature-dependent function to compare results obtained under laboratory conditions with results calculated for the pilot-scale bioreactor. OPRs obtained for NIT were corrected with an Arrhenius-type function, as described in the activated sludge models (Henze *et al.* 2000). The OPR for MA was corrected using the temperature correction suggested in Solimeno (2017).

Finally, specific OPRs (sOPRs), expressed as mg O_2 g TSS^{-1} h^{-1}, could be calculated by dividing the OPR values by the concentration of total suspended solids (TSS) in the suspension:

$$sOPR\ (t) = \frac{OPR\ (t)}{TSS\ (t)} \quad (7)$$

Experimental plan

Preliminary tests were conducted on samples taken from Lab-C and Pilot-C over three consecutive days to assess the repeatability of the described methodology. Pilot-C samples were transported from the pilot plant in white plastic bottles and tested within 2–3 hours, while Lab-C samples were tested immediately after sampling.

After preliminary tests, the described procedure was used to monitor the Pilot-RW system during a 3-month continuous operation (June–August). A respirometric test was performed every second week (seven tests) to follow the activity of the MB consortium and to evaluate the robustness of the method, by comparing the respirometric results to the parameters (nitrification rates and microalgal growth rates) calculated for the pilot plant from conventional monitoring analyses (nitrogen forms, OD_{680} and TSS trends, data not shown).

Analytical methods

Soluble chemical oxygen demand, ammonia, nitrate- and nitrite-nitrogen and phosphate-phosphorus were determined using spectrophotometric test kits (Hach-Lange, LCK 314, LCK 303, LCK 339, LCK 342 and LCK 348, respectively) on filtered samples (0.45 μm). TSS were determined according to standard methods (APHA 2005). Optical density at 680 nm and turbidity were measured in 1 cm and 5 cm cuvettes respectively, by a spectrophotometer (Hach-Lange, DR 3900). pH and DO were measured by a portable multi-meter (Hach-Lange, HQ40D). Microalgae were counted using an optical microscope (Optika, B 350) at 40 × magnification.

Fluorescence *in situ* hybridization (FISH)

Fluorescence *in situ* hybridization (FISH) is a molecular technique that uses fluorescently labelled oligonucleotide probes, which target the 16S rRNA, to detect specific organisms directly into their environmental matrix (Nielsen *et al.* 2009). In this study, it was used to confirm the presence of AOB in MB suspensions. FISH was carried out on samples collected and stored in ethanol (sample/ethanol ratio 1:1 [v/v]) at 20 °C prior to fixation, which was performed using 4% paraformaldehyde fixative solution as described by Amann *et al.* (1990) and Bellucci & Curtis (2011). Hybridization was performed with a mix of Cy3 labelled probes and competitors targeting AOB (Nso 1225, NEU, CTE, 6a192, c6a192), according to a protocol described in Bellucci & Curtis (2011). Hybridized MB samples were then visualized under a fluorescent microscope (Zeiss, Axioskop HBO 50) and recorded.

RESULTS AND DISCUSSION

Preliminary tests

Test conditions and main results obtained for Lab-C and for Pilot-C are reported in Table 2. Samples were collected from

Table 2 | sOPR results obtained in preliminary tests on samples collected in Lab-C and Pilot-C (tests were performed in triplicate, results are expressed as mean ± standard deviation; PAR is the photosynthetically active radiation, sOPR is the specific oxygen production rate). Italic results represent the main output of the respirometric tests

sOPR [mg O_2 g TSS^{-1} h^{-1}]	Lab-C OD_{680} = 0.43 ± 0.02 PAR = 59.2 ± 3.0 μE m^{-2} s^{-1}	Pilot-C OD_{680} = 0.58 ± 0.01 PAR = 40.8 ± 6.7 μE m^{-2} s^{-1}
$sOPR_{NIT\_1}$	−4.8 ± 4.1	−7.2 ± 8.8
$sOPR_{NIT\_2}$	−6.0 ± 2.2	−7.8 ± 5.9
$sOPR_{NIT\_mean}$	*−5.4 ± 2.9*	*−7.5 ± 7.0*
$sOPR_{MA,P\_1}$	11.7 ± 3.0	14.1 ± 2.7
$sOPR_{MA,P\_2}$	12.9 ± 2.1	14.7 ± 5.3
$sOPR_{MA,P\_3}$	12.4 ± 2.2	10.0 ± 0.6
$sOPR_{MA,P\_mean}$	*12.3 ± 1.9*	*12.9 ± 2.0*
$sOPR_{R*}$	*−4.3 ± 0.7*	*−4.1 ± 0.9*

each plant within 3 days to minimize variations in the MB community and used to perform respirometric tests.

The nitrifying activity could be calculated by both Equation (2) and (3). The results of the two systems are comparable and show that NIT activity is present at a similar level, though slightly higher in the pilot column (7.5 mg O_2 g TSS^{-1} h^{-1}) than in laboratory reactors (5.4 mg O_2 g TSS^{-1} h^{-1}). Nonetheless, by taking into account the quite high standard deviation, these two values cannot be considered statistically different. This was verified by applying a heteroscedastic T-test including both $sOUR_{NIT\_1}$ and $sOUR_{NIT\_2}$ (6 data per set: sOUR: specific oxygen uptake rate). A *p*-value of 0.16 was obtained, suggesting that the difference in the NIT activity in the two culturing systems was not significant at a significance level of 0.05. As stated before, the presence of NIT in the MB community of both Lab-C and Pilot-C culturing systems was confirmed by the presence of oxidized nitrogen forms (both as nitrite and nitrate) in the algal suspension. In contrast, higher variability was found for results from the Pilot-C, probably associated with possible slight alterations (considering the short time interval) in the outdoor conditions, which may have affected the microbial consortium.

The gross photosynthetic activity was very similar for the Lab-C samples (12.3 mg O_2 g TSS^{-1} h^{-1}) and the Pilot-C (12.9 mg O_2 g TSS^{-1} h^{-1}), corresponding to 5.1 and 6.4 mg O_2 L^{-1} h^{-1} respectively, which is comparable to the results obtained by Decostere *et al.* (2013) and Decostere *et al.* (2016) in which the gross OPR was found to be lower than 10.4 mg O_2 L^{-1} h^{-1}. By comparing OPR_{MA,P_1} to the mean value of OPR_{MA,P_2} and OPR_{MA,P_3}, the potential inhibition of ATU and chlorate to microalgae photosynthetic activity could be assessed. No significant difference was found by applying a paired T-test (*p*-level = 0.39); therefore no short-term inhibition could be demonstrated. Therefore, nitrification inhibitors (allylthiourea and potassium chlorate) used during the test were proven to not affect the microalgal photosynthetic activity.

Results about algal respiration during the dark phases (OPR_{R*}) were very similar in the two sets of experiments (4.3 and 4.1 mg O_2 g TSS^{-1} h^{-1} for Lab-C and Pilot-C, respectively). Microalgal dark respiration rates were consistent with those indicated by Ruiz-Martinez *et al.* (2016) (0.9–5.1 mg O_2 g TSS^{-1} h^{-1}). The variability was generally lower than that recorded for bacterial activity. This may be due to the stronger effect of photosynthesis compared to nitrification on DO data, which makes the photosynthetic activity easier to discriminate from the oxygen mass transfer (see Equation (1)). Reducing the relevance of the oxygen mass transfer compared to the biological oxygen production/consumption is therefore a desirable improvement.

As previously mentioned, the endogenous respiration of heterotrophic biomass cannot be computed, and it is eventually embedded in the measurement of the OPR_{R*}.

Overall, the similar feed and HRT applied to the two culturing systems promoted the development of similar MB communities in terms of $sOPR_{NIT}$, $sOPR_{MA}$, and $sOPR_{R*}$.

Monitoring tests

In this second experimental campaign, the respirometric protocol was applied to monitor the Pilot-RW. A selection of operational data for the Pilot-RW is reported in Table 3. As suggested by Marazzi *et al.* (2017), average values of the 4 days preceding the day of test were considered for irradiance and temperature. Relevant variations of OD_{680} were observed, as a consequence of the changing environmental conditions. Similarly, a quite high variability was observed

Table 3 | Environmental conditions and operating parameters in the Pilot-RW in relation to the respirometric tests

Test	Date	OD_{680} [-]	Temperature [°C]	PAR [μE m^{-2} s^{-1}]	NH_4^+ [mg N L^{-1}]	NO_2^- [mg N L^{-1}]	NO_3^- [mg N L^{-1}]
1	31/05/17	0.62	24.2	1520	33.0	10.9	3.1
2	14/06/17	0.71	25.5	1000	80.9	16.4	2.0
3	28/06/17	1.05	21.9	672	60.9	107	12.0
4	12/07/17	1.37	24.9	1290	43.2	109	7.1
5	26/07/17	1.41	24.1	1320	65.9	145	12.0
6	04/08/17	1.68	30.7	1340	60.9	157	12.4
7	25/08/17	1.79	25.9	1120	40.2	146	13.2

in the oxidized nitrogen forms, though constant ammonium concentrations were observed.

Respirometric tests were performed on the MB suspension after dilution (applied dilution factors: 1–2.9) to a standardized OD_{680} of 0.63 in order to improve repeatability.

The sOPRs detected by respirometric tests are summarized in Figure 3. As for the MA, $sOPR_{MA,P,mean}$ is reported together with the standard deviation for the three measures (i.e. computed from OPR_{MA,P_1}, OPR_{MA,P_2}, OPR_{MA,P_3}). A generally satisfactory reproducibility was obtained, especially for tests 2–6, with an average coefficient of variation of 10.7%. The precision of the method for the mentioned experiments is similar to that calculated by Brindley *et al.* (2010) for determining microalgal photosynthetic rates, where values of 2.8–7.6% were obtained during repeatability tests. Average standard deviation of 8–10% was observed by Rozzi *et al.* (2003) while applying respirometric tests on HB in activated sludge samples using three different respirometric protocols. Results are also consistent with the study of Ricco *et al.* (2004), in which coefficients of variation of 2–10% were determined during toxicity assessments for xenobiotic compounds in a closed respirometer. The specific oxygen production is quite constant during the whole experimental period with a significant decrease in test 7. The test was executed at the end of the summer, when the outdoor irradiance and temperature were lower than in July and August.

The specific oxygen consumption during the dark phase ($sOPR_{R^*}$) varied significantly over time (2–19 mg O_2 g TSS^{-1} h^{-1}). During the first two tests the highest results were recorded, later microalgal dark respiration decreases reaching stability around 2 to 5 mg O_2 g TSS^{-1} h^{-1}. This is consistent with the values obtained by Ruiz-Martinez *et al.* (2016). The initial higher respiratory activity during test 1 and 2 might be explained with a fast microalgal growth rate, resulting in an increased energy requirement, as suggested by Kliphuis *et al.* (2011). It should be considered that algal respiration may also include heterotrophic oxygen uptake, although its contribution to the overall oxygen consumption is not expected to be relevant because of the recalcitrant nature of the organic matter in the BW that was used to feed the raceway.

The OPR of microalgae during light phases was 10–25 mg O_2 L^{-1} h^{-1}, slightly higher than that obtained by Decostere *et al.* (2013), probably because of the different operating conditions applied. However, results are consistent with that obtained for *Chlorella vulgaris* by Tang *et al.* (2014), under similar conditions. The gross specific oxygen production rate ($sOPR_{MA,P,mean}$) faced an initial increase (from about 33 to 44 mg O_2 g TSS^{-1} h^{-1}), then decreased in the last test. The photosynthetic activity of MA detected by the respirometric assay could not be directly linked to any calculated parameter in the open pond. However, comparisons of the obtained data with the OD_{680} variation rate ($rOD_{680} = \Delta OD/\Delta t$, h^{-1}), which is commonly related to the concentration of microalgal biomass (Figure 4), suggest that the respirometric assay is a reliable tool to follow the evolutionary trend of microalgae activity in the photobioreactor. The few discrepancies could be justified by the fact that the microalgae growth rate and OPR values are influenced by other operating and environmental parameters that varied during the course of the experimentation, such as the light/dark daily cycle.

The sOPR for nitrifiers reported in Figure 5 suggests that the activity of nitrifying bacteria was substantially stable in Pilot-RW from test 3 onward. This is in agreement with the stable concentration of NO_X ($NO_2^- + NO_3^-$) measured in the Pilot-RW (see Table 3).

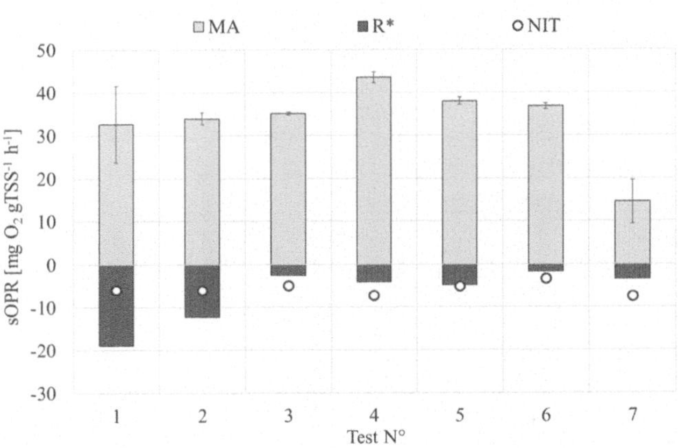

Figure 3 | Average sOPR measured during Pilot-RW monitoring tests.

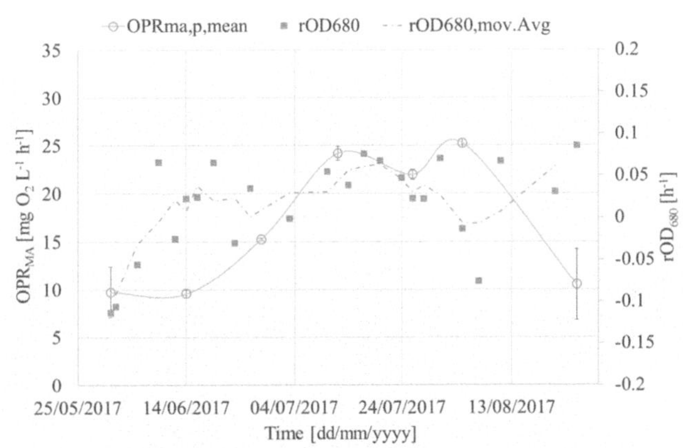

Figure 4 | Calculated microalgal oxygen production rates using the respirometric protocol ($OPR_{MA,P,mean}$) and optical density variation rates (rOD_{680}) from RW monitoring data. Dotted line represents the 3-points moving average on rOD_{680}.

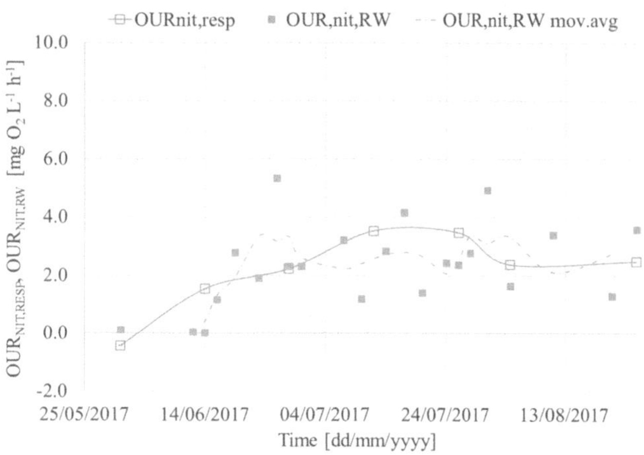

Figure 5 | Comparison between calculated oxygen uptake rate for nitrification from Pilot-RW monitoring data ($OUR_{NIT,RW}$) and its 3-points moving average ($OUR_{NIT,RWmov.avg}$) and nitrifying activity obtained by respirometric tests ($OUR_{NIT,RESP}$).

From the trend of the NO_X concentrations assessed in the Pilot-RW, and assuming a conventional stoichiometric oxygen request for nitrification, the oxygen consumption rate by nitrifiers could be estimated and compared to the oxygen uptake rate evaluated with the respirometric assay. The oxygen uptake rate, OUR, detected in the respirometric assays (Figure 5) seemed to be well correlated to the average level of the oxygen uptake estimated for the Pilot-RW. Only the first OUR value was slightly negative, indicating that the nitrifying activity was scarce and the low oxygen consumption was poorly detected with the respirometric procedure. These results suggest that respirometric tests could provide a rough estimation of the potential nitrifying activity of the consortium.

FISH analyses also confirmed a stable occurrence of AOB in the MB suspensions (Figure 6).

CONCLUSIONS

This study demonstrated that respirometry is a promising technique to gain insights into the oxygen dynamics in microalgae-bacteria systems. The proposed protocol involved the implementation of changing environmental conditions (in terms of light availability and presence of inhibitors) and successfully distinguished specific oxygen production/consumption rates deriving from the different guilds of microbes involved. Nevertheless, the experimental setup needs to be further improved to be routinely applied. Suggestions are: (i) the reduction in the oxygen mass transfer during the test; (ii) the search for specific inhibitors of algal dark respiration, which would help to discriminate between algal dark respiration and oxygen consumption due to HB in mixed MB consortia. Once optimized, the protocol could become a very simple and effective tool for monitoring outdoor and indoor MB bioreactors treating industrial and municipal wastewaters as well in the study of optimal operational conditions to be applied to these bioreactors.

ACKNOWLEDGEMENTS

The authors wish to thank Fondazione Cariplo for their funding in support of this research ('IMAP – Integration of microalgal based processes in wastewater treatment' project, Grant 2015-0063, and 'Il polo delle microalghe' project), and Nadia Margariti, Arianna Catenacci, Marco Mantovani and Tania Fantasia for their valuable work and help in laboratory. The authors are also thankful to the reviewers, whose contributions allowed us to improve the quality of the submitted paper.

Figure 6 | Microalgae (dark grey; dark red in online version) and AOB (light grey; light red in online version) colonies observed in MB samples collected in Pilot-RW using a fluorescent microscope. Please refer to the online version of this paper to see this figure in colour: http://dx.doi.org/10.2166/wst.2018.078.

REFERENCES

Akhiar, A., Battimelli, A., Torrijos, M. & Carrere, H. 2017 Comprehensive characterization of the liquid fraction of digestates from full-scale anaerobic co-digestion. *Waste Management* **59**, 118–128. DOI: 10.1016/j.wasman.2016.11.005.

Amann, R. I., Binder, B. J., Olson, R. J., Chisholm, S. W., Devereux, R. & Stahl, D. A. 1990 Combination of rRNA-targeted oligonucleotide probes with flow cytometry for analysing mixed microbial populations. *Applied and Environmental Microbiology* **56** (6), 1919–1925.

APHA 2005 *Standard Methods for the Examination of Water and Wastewater*. American Public Health Association/American Water Works Association/Water Environment Federation, Washington, DC, USA.

ASCE 1993 *Measurement of Oxygen Transfer in Clean Water*, 2nd edn. American Society of Civil Engineers, New York, USA.

Bellucci, M. & Curtis, T. P. 2011 Ammonia-oxidizing bacteria in wastewater. In: *Methods in Enzymology* (M. G. Klotz & L. Y. Stein, eds), Burlington Academic Press, Burlington, MA, USA. pp. 269–286. DOI: 10.1016/B978-0-12-386489-5.00011-7.

Brindley, C., Acién, F. G. & Fernandéz-Sevilla, J. M. 2010 The oxygen evolution methodology affects photosynthetic rate measurements of microalgae in well-defined light regimes. *Biotechnology and Bioengineering* **106** (2), 228–237.

Chisti, Y. 2007 Biodiesel from microalgae. *Biotechnology Advances* **25**, 294–306. DOI: 10.1002/bit.22676.

Decostere, B., Janssens, N., Alvarado, A., Maere, T., Goethals, P., Van Hulle, S. W. H. & Nopens, I. 2013 A combined respirometer-titrimeter for the determination of microalgae kinetics: experimental data collection and modelling. *Chemical Engineering Journal* **222**, 85–93. DOI: 10.1016/j.cej.2013.01.103.

Decostere, B., Van Hulle, S. W. H., Duyck, M., Maere, T., Vervaeren, H. & Nopens, I. 2016 The use of a combined respirometric-titrimetric setup to assess the effect of environmental conditions on micro-algae growth rate. *Journal of Chemical Technology & Biotechnology* **91**, 248–256. DOI: 10.1002/jctb.4574.

Ginestet, P., Audic, J. M., Urbain, V. & Block, J. C. 1998 Estimation of nitrifying bacterial activities by measuring oxygen uptake in the presence of the metabolic inhibitors allylthiourea and azide. *Applied and Environmental Microbiology* **64** (6), 2266–2268.

Godos, I. D., Blanco, S., García-Encina, P. A., Becares, E. & Muñoz, R. 2009 Long-term operation of high rate algal ponds for the bioremediation of piggery wastewaters at high loading rates. *Bioresource Technology* **100** (19), 4332–4339. DOI: 10.1016/j.biortech.2009.04.016.

González-Fernández, C., Molinuevo-Salces, B. & García-González, M. C. 2011 Nitrogen transformations under different conditions in open ponds by means of microalgae-bacteria consortium treating pig slurry. *Bioresource Technology* **102** (2), 960–966. DOI: 10.1016/j.biortech.2010.09.052.

Henze, M., Gujer, W., Mino, T. & Van Loosdrecht, M. 2000 *Activated Sludge Models ASM1, ASM2, ASM2D and ASM3*. Scientific and Technical Report No. 9. IWA Publishing, London, UK. DOI: 10.2166/9781780402369.

Hynes, R. K. & Knowles, R. 1983 Inhibition of chemoautrophic nitrification by sodium chlorate and sodium chlorite: a reexamination. *Applied and Environmental Microbiology* **45** (4), 1178–1182.

Kliphuis, A. M. J., Janssen, M., van den End, E. J., Martens, D. E. & Wijffels, R. H. 2011 Light respiration in *Chlorella sorokiniana*. *Journal of Applied Phycology* **23**, 935–947. DOI: 10.1007/s10811-010-9614-7.

Marazzi, F., Ficara, E., Fornaroli, R. & Mezzanotte, V. 2017 Factors affecting the growth of microalgae on blackwater from biosolid dewatering. *Water, Air & Soil Pollution* **228** (2), 68. DOI: 10.1007/s11270-017-3248-1.

Munz, G., Gori, R., Mori, G. & Lubello, C. 2009 Monitoring biological sulphide oxidation processes using combined respirometric and titrimetric techniques. *Chemosphere* **76** (5), 644–650. DOI: 10.1016/j.chemosphere.2009.04.039.

Nielsen, P. H., Dams, H., Lemmer, H., Arslan-Alaton, I. & Olmez-Hanci, T. 2009 *FISH Handbook for Biological Wastewater Treatment*. IWA Publishing, London.

Ricco, G., Tomei, M. C., Ramadori, R. & Laera, G. 2004 Toxicity assessment of common xenobiotic compounds on municipal activated sludge: comparison between respirometry and Microtox. *Water Research* **38**, 2103–2110. DOI: 10.1016/j.watres.2004.01.020.

Rozzi, A., Ficara, E. & Rocco, A. 2003 Dissolved oxygen-Stat titration respirometry: principle of operation and validation. *Journal of Environmental Engineering* **129**, 602–609. DOI: 10.1061/(ASCE)0733-9372(2003)129:7(602).

Ruiz-Martinez, A., Serralta, J., Seco, A. & Ferrer, J. 2016 Behaviour of mixed Chlorophyceae cultures under prolonged dark exposure. Respiration rates modelling. *Ecological Engineering* **91**, 265–269. DOI: 10.1016/j.ecoleng.2016.02.025.

Solimeno, A. 2017 *Numerical Modelling of Microalgae Systems for Wastewater Treatment*. PhD thesis, Universitat Politecnica de Catalunya, Barcelona, Spain.

Spanjers, H. & Vanrolleghem, P. 1995 Respirometry as a tool for rapid characterization of wastewater and activated sludge. *Water Science and Technology* **31** (2), 105–114.

Tang, T., Fadaei, H. & Hu, Z. 2014 Rapid evaluation of algal and cyanobacterial activities through specific oxygen production rate measurement. *Ecological Engineering* **73**, 439–445. DOI: 10.1016/j.ecoleng.2014.09.095.

Vanrolleghem, P. 2002 *Principles of Respirometry in Activated Sludge Wastewater Treatment*. PhD thesis, University of Ghent, Ghent, Belgium.

Vargas, G., Donoso-Bravo, A., Vergara, C. & Ruiz-Filippi, G. 2016 Assessment of microalgae and nitrifiers activity in a consortium in a continuous operation and the effect of oxygen depletion. *Electronic Journal of Biotechnology* **23**, 63–68. DOI: 10.1016/j.ejbt.2016.08.002.

First received 31 October 2017; accepted in revised form 12 February 2018. Available online 22 February 2018

Exploiting symbiotic interactions between *Chlorella protothecoides* and *Brevundimonas diminuta* for an efficient single-step urban wastewater treatment

Martina Pastore and Eleonora Sforza

ABSTRACT

The application of microalgal bacteria consortia to the treatment of wastewater is receiving increasing attention, meeting the demand for new green and efficient technologies for water remediation. The specificity of the consortium, however, may strongly affect the performance of the treatment. In fact, even though a general exploitation of the O_2/CO_2 exchange between microalgae and bacteria is effective, some specific interactions may increase the pollutant removal. With this aim, the co-cultivation of *Chlorella protothecoides* and *Brevundimonas diminuta* was tested, with particular attention to the removal capability of nitrogen, phosphorus and chemical oxygen demand (COD) from wastewater. Batch experiments were carried out both for the consortium and, separately, for the bacteria and microalgae alone, in order to compare their performances. *B. diminuta* showed a remarkable capability for removing organic substances and transforming organic nitrogen to ammonium. *C. protothecoides* efficiently removed nitrogen and phosphorus. As the specific growth rates of the two organisms are different, the co-cultivation was also carried out also in a continuous system, and the effect of hydraulic retention time (HRT) on the steady-state biomass concentration and nutrient removal efficiency was verified. Residence time was found as the main operating variable for obtaining a significant reduction of pollutants from wastewater.

Key words | consortium, continuous reactor, microalgae, mixotrophy, steady state

Martina Pastore
Eleonora Sforza (corresponding author)
Interdepartmental Centre Giorgio Levi Cases,
Via Marzolo 9, 35131 Padova,
Italy
E-mail: *eleonora.sforza@unipd.it*

Eleonora Sforza
Department of Industrial Engineering DII,
University of Padova,
Via Marzolo 9, 35131 Padova,
Italy

INTRODUCTION

In the last few years it has become imperative to study more efficient and environmental friendly methods to recycle fresh water and recover nutrients from wastewaters. Even though conventional biological wastewater treatment plants (WWTPs), based on activated sludge technology, have been set up and constantly improved, some issues still remain unresolved. In this type of plant, the nitrogen and phosphorous recovery efficiency is usually low and requires additional steps that often involve the addition of chemicals, with a consequent increase of costs and environmental impact. The most energy-consuming step is the use of blowers for the oxygenation of the aerobic tank, which constitutes almost 50% of the total operative cost of conventional WWTPs. In addition, from a circular economy perspective, more efficient and green technologies for resource recovery are being investigated. Many researchers have recently been focusing on the use of algal–bacteria consortia: it is known that gas exchange between microalgae and bacteria may act as a growth-promoting agent for both cultures. Microalgae are also able to reduce nitrogen and phosphorous concentrations in wastewaters, thus avoiding any other chemical steps (Beltràn-Rocha *et al.* 2017). In the natural environment, most microalgae are associated with either aerobic or anaerobic microorganisms and, in an industrial setting, using a system that is self-sustaining in terms of oxygen has potential not only for wastewater bioremediation but also for the production of commercially valuable by-products (Subashchandrabose *et al.* 2011; Beltràn-Rocha *et al.* 2017).

Interactions between microorganisms may be either positive or negative (Fukami *et al.* 1997) and this may strongly affect treatment performance: for example, *Azospirillum brasilense* produces indole-3-acetic acid (IAA), among other compounds, which is a plant hormone able to act as a growth-enhancing compound for *Chlorella vulgaris* (Gonzalez-Bashan *et al.* 2000). On the other hand, other bacterial

doi: 10.2166/wst.2018.155

strains, such as *Pseudomonas putida*, are used as biological control agents against algal blooms (Doucette *et al.* 1999; Kang *et al.* 2005). Bacteria can also affect microalgal growth by limiting the access to light of these photosynthetic microorganisms (Tricolici *et al.* 2014). Even microalgae may have an influence on bacterial growth, in both negative and positive ways. As an example, several microalgal species secrete compounds with specific antibacterial activity (Natrah *et al.* 2014), or that are able to interfere with bacterial quorum sensing (Natrah *et al.* 2011). In summary, the specificity of the consortium should be carefully considered in biological processes and, possibly, controlled.

Some examples of natural and synthetic consortia with plant-growth-promoting bacteria have been already reported in the literature, as in the case of *A. brasilense* (de-Bashan *et al.* 2002, 2004; Palacios *et al.* 2016b) and *Brevundimonas diminuta* (Park *et al.* 2008). Biochemical mechanisms exploited by these bacteria to promote microalgal growth are still unknown, despite many authors reporting that *B. diminuta* may protect plants from the detrimental effects of heavy metal pollution (Ji *et al.* 2016), arsenic stress (Singh *et al.* 2016) and the ability to degrade pesticides based on organophosphates (Parthasarathy *et al.* 2017). *B. diminuta* is also able to solubilize phosphates, produce siderophore and plant growth hormones such as IAA (Singh *et al.* 2016). In particular, it is known that the *Chlorella* genus and *B. diminuta* may have symbiotic interactions in the aquatic environment (Park *et al.* 2008) and that *Chlorella* sp. is able to secrete thiamine, which is an important cofactor of the enzyme responsible for IAA biosynthesis and in many other metabolic ways (Palacios *et al.* 2016b). *Chlorella* sp. is also shown to secrete tryptophan, an amino acid used as a precursor of IAA by bacteria (Palacios *et al.* 2016a, 2016b).

Aside from the mechanisms of interaction between the two species, which are still under investigation, the application of this consortium for wastewater treatment has not yet been tested. Accordingly, in this work the co-cultivation of *C. protothecoides* and *B. diminuta* was assessed, with particular attention to the simultaneous removal capability of nitrogen, phosphorus and chemical oxygen demand (COD) from wastewater. Both batch and continuous experiments were carried out: batch experiments were performed separately for bacteria and microalgae, in order to compare the performance of the consortium. As the specific growth rates of the two organisms are different, the co-cultivation was also carried out in a continuous system, and the effect of residence time (hydraulic retention time; HRT) on the steady-state biomass concentration and nutrient removal efficiency was investigated.

METHODS

Microalgal and bacterial species, growth medium and wastewater

Chlorella protothecoides 33.80 (obtained from SAG Goettingen, Germany) was maintained in liquid BG11 medium for the inoculum ($Na_2MgEDTA$ 1 mg L^{-1}, ferric ammonium citrate 6 mg L^{-1}, citric acid·H_2O 6 mg L^{-1}, $CaCl_2{\cdot}2H_2O$ 36 mg L^{-1}, $MgSO_4$ 75 mg L^{-1}, K_2HPO_4 30.5 mg L^{-1}, H_3BO_3 2.86 mg L^{-1}, $MnCl_2{\cdot}4H_2O$ 1.81 mg L^{-1}, $ZnSO_4{\cdot}7H_2O$ 0.222 mg L^{-1}, $CuSO_4{\cdot}5H_2O$ 0.70 mg L^{-1}, $COCl_2{\cdot}6H_2O$ 0.050 mg L^{-1}, $Na_2MoO_4{\cdot}2H_2O$ 0.391 mg L^{-1}, Na_2CO_3 20 mg L^{-1}, $NaNO_3$ 0.943 mg L^{-1}, HEPES 1M pH = 8).

Brevundimonas diminuta LMG 02089 (obtained from Belgian Coordinated Collections of Microorganisms, BCCM) was maintained in liquid flasks with Luria broth, and inoculated in fresh medium at 30 C° one day before each experiment, as a preinoculum.

To avoid the influence of the environmental changes of real wastewater on the performances of the consortium, a synthetic formula was used to obtain a medium with constant quality. A standard synthetic wastewater (http://www.oecd.org/chemicalsafety/testing/) was slightly modified in order to have an average characterization similar to local urban wastewater (peptone 80 mg L^{-1}, meat extract 110 mg L^{-1}, NH_4Cl 40 mg L^{-1}, CH_3COONa 159 mg L^{-1}, K_2HPO_4 23 mg L^{-1}, NaCl 7 mg L^{-1}, $CaCl_2{\cdot}2H_2O$ 4 mg L^{-1} and $Mg_2SO_4{\cdot}7H_2O$ 2 mg L^{-1}). The final characterization of the synthetic sewage was the following: ammonium (NH_4^+-N) 13 mg L^{-1}, total nitrogen (N_{tot}) 29.5 mg L^{-1}, orthophosphate (PO_4^{3-}-P) 3.75 mg L^{-1}, COD 310 mg L^{-1}.

Experimental set-up and procedures

Both batch and continuous experiments with *C. protothecoides* and *B. diminuta* grown in synthetic wastewater were carried out in order to investigate their growth and their nutrient removal capacity under continuous light.

Batch growth curves were carried out in Drechsel bottles of 250 mL, with 5 cm diameter, continuously mixed by a magnetic stirrer. Each run started with an initial biomass inoculation of OD_{750} = 0.6. Before starting each experiment, the volume of pre-inoculum was centrifuged for 5 min at 1,500 rpm in order to remove the preinoculum medium.

Four conditions were studied in order to better understand the effect of non-limiting gas supply and the possible exploitation of gas produced by the two populations in the case of consortium:

(a) *C. protothecoides* alone with air–CO_2 bubbling;
(b) *C. protothecoides* alone without bubbling;
(c) *C. protothecoides* and *B. diminuta* cultivated together without bubbling;
(d) *B. diminuta* alone without bubbling.

Aeration (air enriched with 5% v/v of CO_2) was provided to the microalgae as a control, at a total flow rate of $1\ L\ h^{-1}$. The temperature was controlled at 24 °C in an incubator (Frigomeccanica Andreaus, Padova, Italy) and artificial light (white neon lamps OSRAM) was supplied continuously at an intensity of 30 µmol photons $m^{-2}\ s^{-1}$ of photosynthetic active radiation (PAR), measured by a photoradiometer (Model LI-189, LI-COR, USA).

Continuous experiments were carried out in a 300 mL flattened glass tank, with 6.5 cm thickness, under 150 $\mu mol\ m^{-2}\ s^{-1}$ of irradiation. The reactor was initially inoculated with the consortium and it was first run in batch mode, allowing cells to enter an exponential phase of growth, so as to reach a sufficient biomass concentration and avoid the occurrence of washout when the operation mode was shifted to continuous mode. After activating the peristaltic pump (Sci-Q 400, Watson Marlow, USA), continuous operating conditions were achieved. Fresh medium was continuously supplied from an external sterilized stirred bottle by means of the peristaltic pump, at a constant volumetric flow-rate Q ($mL\ d^{-1}$). The culture volume was kept constant by the presence of an overflow tube, placed at the opposite side with respect to the fresh medium inlet, from which biomass was continuously collected at the same volumetric flow-rate. The HRT of the reactor is calculated as:

$$\tau = \frac{V_R}{Q} \tag{1}$$

where V_R is the reactor volume (mL). The flow-rate was regulated in order to obtain the desired residence time. As no biomass separation and recirculation was done, the solid retention time (SRT) was equal to the HRT.

After a transition period of acclimatization, steady state was reached, with a constant biomass concentration, which depends on residence time and nutrients in the inlet (de Farias Silva *et al.* 2017). Subsequently, flow-rate was changed, resulting in a different residence time. This resulted in a new transition period followed by a further steady state.

Analytical methods

The microalgal and bacterial concentration was indirectly measured as absorbance, by spectrophotometric analysis of the optical density (OD; measured at 750 nm, by double beam spectrophotometer UV-Visible UV 500 from Spectronic Unicam, UK), which however did not allow discrimination between bacteria and microalgae, as it is just a measure of light scattering. Accordingly, microalgal cell concentration was also counted with a Bürker Chamber (HBG, Germany). Microalgal growth rates in the batch experiments were measured by linear regression of 6–8 experimental points of cell concentration during the logarithmic phase of growth, from two or three independent biological replicates. The specific growth rate was calculated as the slope on graph of logarithm of cells as a function of time.

At the beginning and the end of the growth curve the final biomass concentration of each experiment was also measured as dry weight (DW) in terms of $g\ L^{-1}$. This was done gravimetrically on biomass filtered (0.22 µm) and dried at 100 °C in a laboratory oven for 2 hours. Nitrate (NO_3^--N), nitrites (NO_2^--N), NH_4^+-N, N_{tot}, PO_4^{3-}-P and COD were determined at time 0 and at the end of each growth curve, in order to verify the consumption by microalgae, bacteria or consortium, after biomass removal by filtration, to measure dissolved compounds only. Nitrate, nitrites, ammonium and orthophosphate were analysed using Hydrocheck Spectratest kits by Reasol®, while an Aquanal® kit by Sigma-Aldrich was used for COD. N_{tot} was measured as nitrate, after persulfate digestion in an autoclave for 1 hour.

Dissolved oxygen (DO) concentration was measured by means of a Handylab Ox 12 SCHOTT® oximeter connected to a PC by using Multi/ACHAT II software provided by WTW.

In continuous experiments, biomass was sampled at least five times at steady state, and all analytical measures reported before were carried out, including N, P and COD in the outlet stream.

Statistical analysis

Student's *t*-tests were applied to ascertain meaningful differences in nutrient consumption and growth curves (both OD values and cell numbers) in the continuous system at three different HRT values. The level of statistical significance was assumed for $p < 0.05$, and significantly different results are highlighted with letters in the figures. Biomass samples were collected at least five times for each residence time and were found to be normally distributed.

RESULTS AND DISCUSSION

Batch experiments of co-cultivation of the consortium

To ascertain the possible mutual interactions between *C. protothecoides* and *B. diminuta* the two species were cultivated in batches, both individually and in consortium, and the nutrient removal was measured and compared. As shown in Figure 1, the presence of bacteria increased the performance of microalgal growth. In fact (Figure 1(a)) a higher specific growth rate of microalgae can be calculated by the trend of cell concentration, when bacteria are present: the specific growth rate of *C. protothecoides* increased from 0.689 d^{-1} to 0.995 d^{-1} in the presence of *B. diminuta*.

A number of authors have suggested that one of the mechanisms behind the higher performance of the bacterial–microalgal consortium is related to gas exchange (Subashchandrabose *et al.* 2011; Boelee *et al.* 2014), where oxygen produced by photosynthesis is supplied to the bacteria, while CO_2 can be exploited as carbon source by the microalgae. To find out if gas exchange is the main reason of the increased performances of the consortium, some other experiments were carried out by bubbling air–CO_2 (5% v/v) to the axenic culture of *C. protothecoides* (Figure 1), in order to compare non-limiting gas conditions with the performance of the mixed culture. Even though the algal biomass concentration reached higher values when external CO_2 was provided, the specific growth rate was lower than in the co-culturing condition (0.808 and 0.995 d^{-1}, respectively). This supports the conclusion that the mutual positive interaction between the two populations is not limited to a simple gas exchange, but is the result of an actual symbiotic interaction. The positive interactions between microalgae and bacteria may occur with a wide range of mechanisms (Fukami *et al.* 1997; Subashchandrabose *et al.* 2011; Unnithan *et al.* 2014), and they may involve nutrient exchange as well (Park *et al.* 2008; Kouzuma & Watanabe 2015; Palacios *et al.* 2016b), facilitating, for instance, nitrogen assimilation by microalgae (Le Chevanton *et al.* 2013).

Nutrient consumption data are shown in Figure 2, for all the conditions tested. One interesting result relates to phosphorus reduction, which reached 100% when microalgae were present. In the case of co-cultivation, about 94% of the P was removed, which is remarkable considering that, in current wastewater treatment, the conventional method to reduce this pollutant requires a chemical precipitation step. This P uptake capability is well known and described for many algal species, including luxury uptake (Solovchenko *et al.* 2016). *B. diminuta*, along with other bacterial species, is not able itself to consume all the phosphorus present in wastewaters: the reduction was only about 34% in the axenic culture of this species.

On the other hand, the bacterial species showed a remarkable capability for organic carbon removal, higher than the mixotrophic capability of microalgae. From Figure 2(b) it is clear that microalgae can consume organic carbon by mixotrophy, as already demonstrated for this species in previous

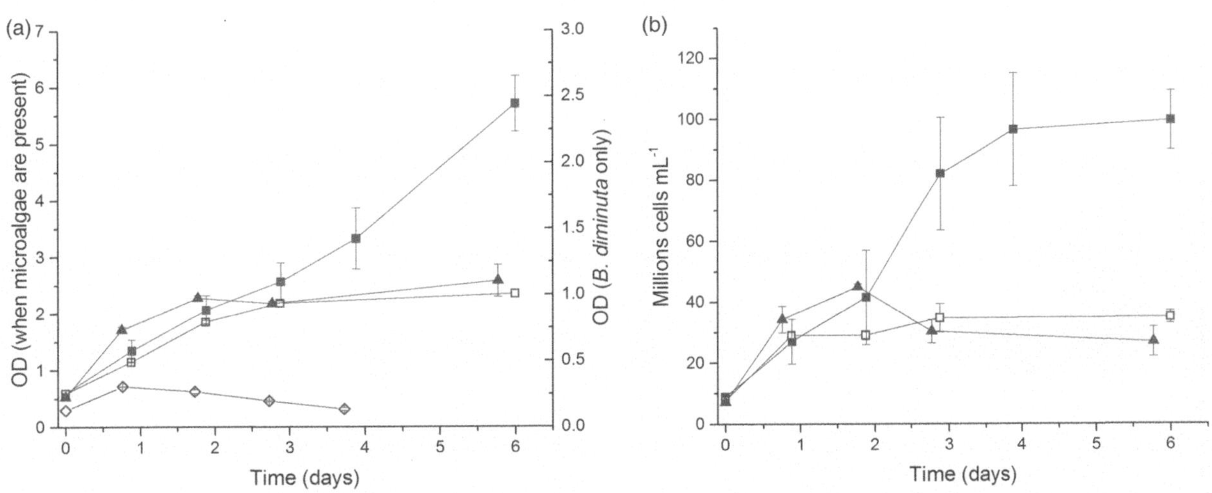

Figure 1 | Growth curve (OD data in (a), cell counts in (b)) of *C. protothecoides–B. diminuta* consortium (dark triangles), microalgae alone without bubbling (open squares) and with non-limiting CO_2 supply (dark squares) and *B. diminuta* alone (open diamonds) in synthetic wastewater. In (a), data of cultures with bacteria only (open diamonds) relate to the right Y axis, for graphical reasons only.

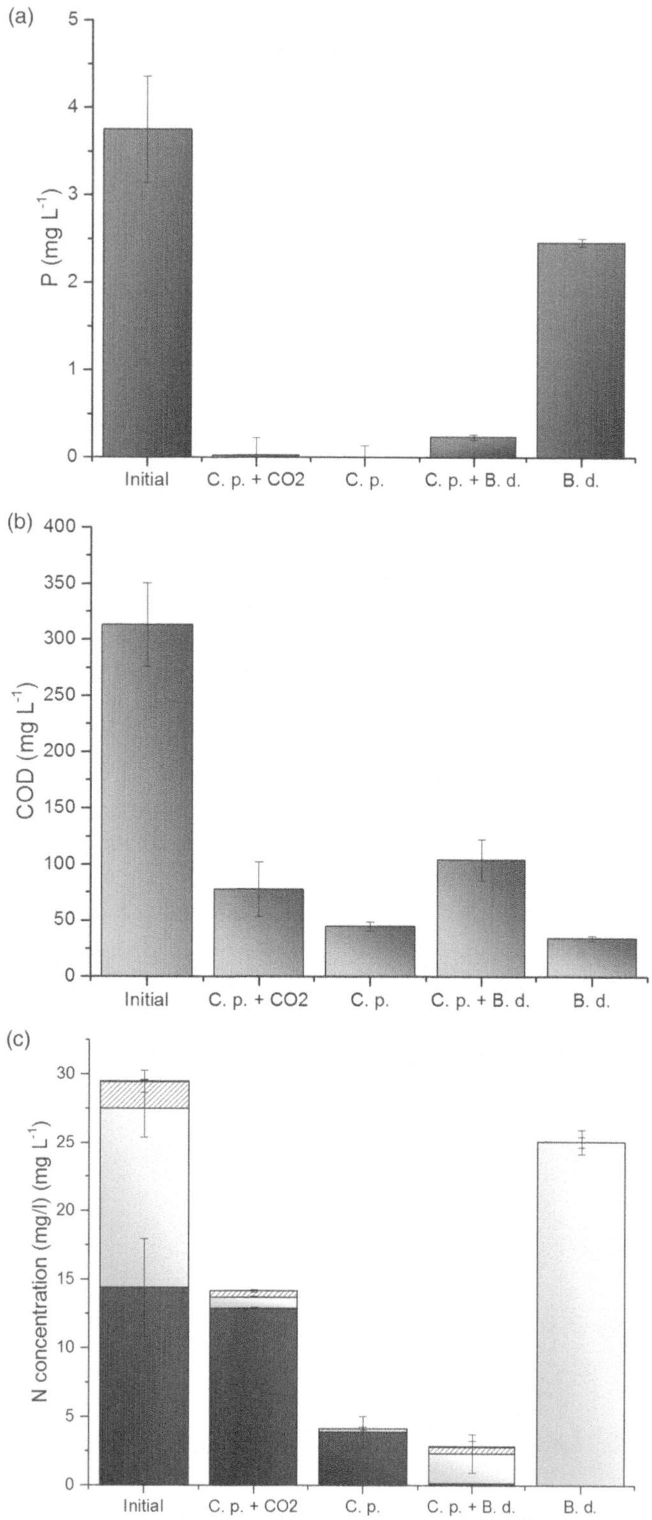

Figure 2 | Initial (first column) and final concentration of phosphorus (a), COD (b) and nitrogen (c) of growth curve in synthetic wastewater of *C. protothecoides* alone without bubbling (C. p.) and with non limiting CO_2 supply (C. p. + CO_2), microalgal–bacterial consortium (C. p. + B. d.) and bacteria alone without bubbling (B. d). In (c), grey portions represent N-NH_3, dark N_{org} and striped N-NO_3.

work, to an extent that depends on CO_2 concentration and light availability (Sforza *et al.* 2012). COD removal in the case of co-cultivation was slightly lower than for *B. diminuta* alone, suggesting that some possible competition for nutrients (including micronutrients), or other kinds of interactions may occur. For instance, as the specific growth rate of the two organisms were quite different, so that it affects the consumption of nutrients in a batch system, in particular if only the final concentration is measured. A decoupled growth of bacteria and microalgae is indeed the possible explanation of this, as confirmed also by the trend of oxygen availability in the mixed growth curve (where the two organisms are cultivated together, see Figure 3 for details). In fact, the dissolved oxygen concentration in the culture strongly decreased in the first day of growth, which corresponds to the exponential phase of bacteria alone (Figure 1). Accordingly, during this phase, bacteria are actively growing, thus consuming all the oxygen available. After that, bacteria reach a stationary phase, but microalgae are still active, as confirmed by the trend of oxygen, with increasing concentration up to 9 mg L^{-1} in the following days. Thus, a sequential growth pattern occurs, with a first phase of bacterial predominance, followed by microalgal growth. Therefore, the kinetics of the two organisms is a key aspect that may influence co-cultivation and nutrient removal, in particular when moving to a real continuous system.

A remarkable result relates to the total nitrogen consumption: as shown in Figure 2(c), it is clear that *B. diminuta*, cultivated axenically, was able to remove all the organic nitrogen and to convert it to ammonium. No nitrification or denitrification were observed, and this species was

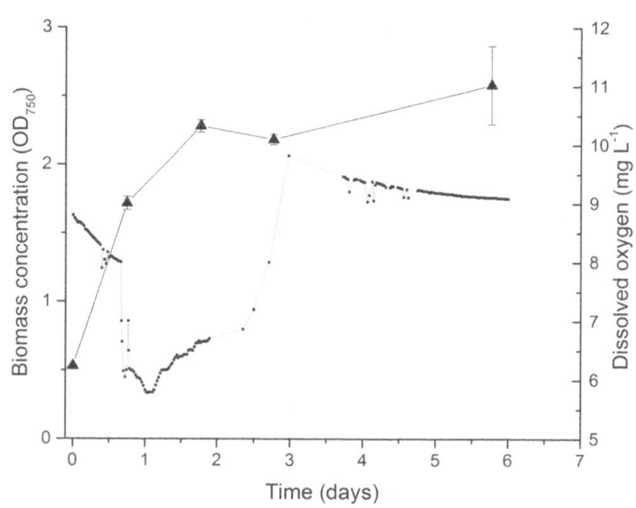

Figure 3 | Dissolved oxygen (dots) in culture medium during the batch experiment of co-cultivation (triangles). Biomass concentration is reported as optical density.

also not able to consume all the ammonia produced. On the other hand, even though *C. protothecoides* was able to mixotrophically consume a fraction of the organic nitrogen, when cultivated alone, its removal capability was not as efficient as the removal of ammonium in the case of co-cultivation. It appeared that one advantage of co-culturing these two species lies in the transformation of nitrogen by *B. diminuta* to a chemical form which is more easily exploited by the microalgal species (Markou *et al.* 2014). As already shown in the literature, the growth of *Chlorella* sp. is promoted by co-cultivation with *B. diminuta* (Park *et al.* 2008), but the interaction is not fully understood. Our results suggest that nutrient exchange may play a role in this mutual interaction, which agrees with current literature (Park *et al.* 2008; Kouzuma & Watanabe 2015; Palacios *et al.* 2016b).

In summary, the co-cultivation of this consortium is very promising for bioremediation purposes, as it resulted in a reduction of 66.7, 69.4, 34.4% of C, N and P respectively, with final concentrations far below the limit for water discharge into surface water bodies (Council of the European Communities (1998) EUR-Lex - 31998L0015).

Continuous experiments: effect of residence time on nutrient removal

As reported in the previous section, *B. diminuta* and *C. protothecoides* are characterized by different specific growth rates. This may actually influence the effectiveness of wastewater treatment in continuous reactors, where biomass is continuously removed from the system, depending on the flow-rate applied. In addition, if the residence time is too low, it may result in the washout of the biomass from the reactor, or an unbalanced microalgae:bacteria ratio. Thus, the growth and nutrient removal performances of the consortium were investigated in a continuous reactor running at steady state, with particular attention to the influence of the residence time.

Results of biomass concentration, in terms of optical density, microalgal cell concentration and dry weight are shown in Figure 4(a) and 4(b).

The overall biomass concentration (Figure 4(b)) increased with the residence time, which is a common result in biological continuous reactors (Sforza *et al.* 2015). On the other hand, as shown in Figure 4(a), at the lower residence time an almost complete removal of microalgal biomass occurred, as the microalgal cell number decreased and the colour of the culture turned from green to white (due to the turbidity of bacteria only). Continuous experiments with axenic culture of *C. protothecoides* were also

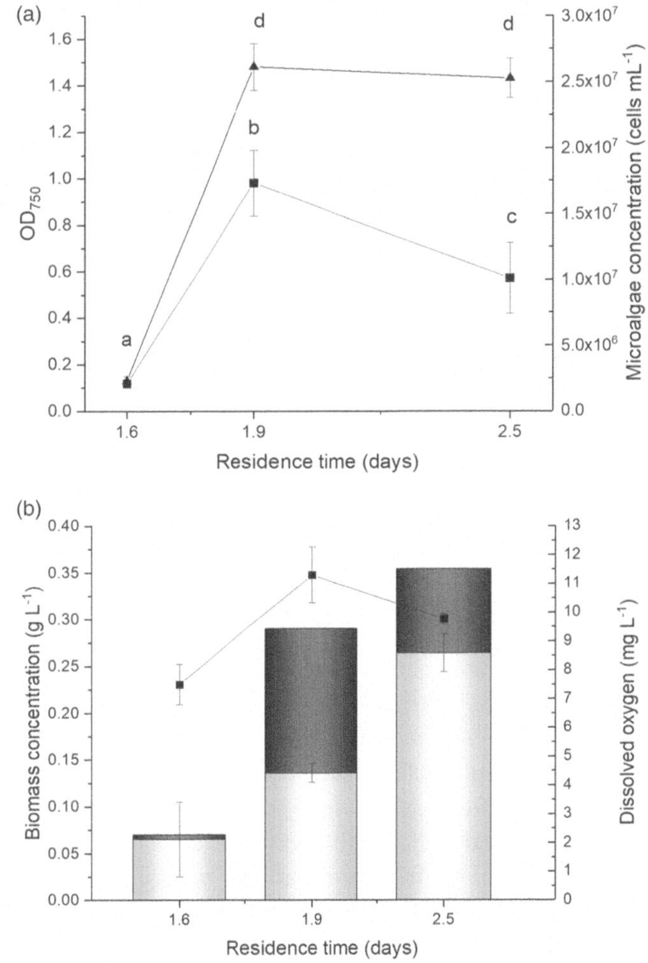

Figure 4 | Biomass concentration data (OD data dark triangles, cell numbers dark squares) at different HRTs tested (a). Biomass concentration expressed as dry weight (bars) and dissolved oxygen concentration (dark squares) (b) at different HRT. In (b), grey portions refer to bacteria concentration, dark to microalgae. Different letters correspond to statistically significant results.

carried out as a control, at the same residence times tested with the consortium: an optical density of 1.13 ± 0.02 and a biomass concentration of $0.18 \pm 0.02\ g\ L^{-1}$ were obtained at 2.5 days, which are lower than for the consortium, while at lower residence times microalgal washout occurred (data not shown). Accordingly, this confirmed that, when *B. diminuta* is present, the growth performance of *C. protothecoides* is enhanced, and the washout occurs at a lower residence time, as the microalgal specific growth rate is enhanced. In the case of the consortium, however, the optical density remained comparable at both 1.9 and 2.5 days, even though the dry weight increased, as a result of the different extinction coefficients of the two organisms, as already clear from Figure 1(a). To better understand how the residence time influenced the population in the reactor, a correlation between dry weight and microalgal cell number was applied (see supplementary material, Figure S1, available with the online version of this paper). As a rough indication, by applying the specific cellular weight to the cell concentration actually counted, it was possible to observe that the residence time also influenced the ratio of the two populations. Thus, if the bacterial concentration increased with residence time, microalgal biomass appeared to reach a maximum at 1.9 days. Of course an effective measure of the biomass concentration of the two distinct populations is challenging, but the correlation of microalgal dry weight may be applied as a rough indication. This trend of microalgal biomass concentration as a function of residence time is confirmed by the trend of the dissolved oxygen concentration measured in the reactor (Figure 4(b)), with a maximum at 1.9 days, where the microalgal:bacteria ratio is higher. This also affected the nutrient removal, as shown in Figure 5.

Remarkable removals were obtained, in particular related to P consumption, which was almost total at higher residence times (about 99%). At 1.6 days, on the other hand, the washout of microalgae caused an inefficient reduction of this nutrient, with a final concentration of about 3.3 mg L^{-1} (removal efficiency of 56%), since the removal of phosphorus is mainly performed by microalgae, as already observed in the case of batch experiments.

COD removal (Figure 5(b)) was satisfactory for all the residence times, with concentrations measured at the outlet being lower than the legal limit value. Removal efficiencies of about 93, 78 and 87% were obtained for 1.6, 1.9 and 2.5 days, respectively. As shown in Figure 5(b), differences in COD outlet concentrations, among the tested HRTs, are statistically significant. This trend, apparently unexplained, may be due to the ratio of the two populations at different residence times (see Figure 4(b)). In fact, it was reported in previous literature that microalgae may also release organic carbon compounds/matter during growth (Watanabe *et al.* 2005; Subashchandrabose *et al.* 2011; Natrah *et al.* 2014), including the molecules produced to sustain the specific interaction with *B. diminuta* (Palacios *et al.* 2016a, 2016b). Thus, *B. diminuta* itself was able to almost completely remove organic carbon alone, at τ = 1.6 days, when microalgae are washed out, while at 1.9 days microalgal concentration prevails, with a possible excretion of organic compounds. At 2.5 days, the ratio of bacteria:microalgae increased, and COD removal was enhanced accordingly. On the other hand, it is noteworthy that, in spite of the differences, the removal of organic carbon was very efficient in all cases.

Finally, it must be pointed out that the presence of microalgae is crucial for nitrogen removal. In fact, as

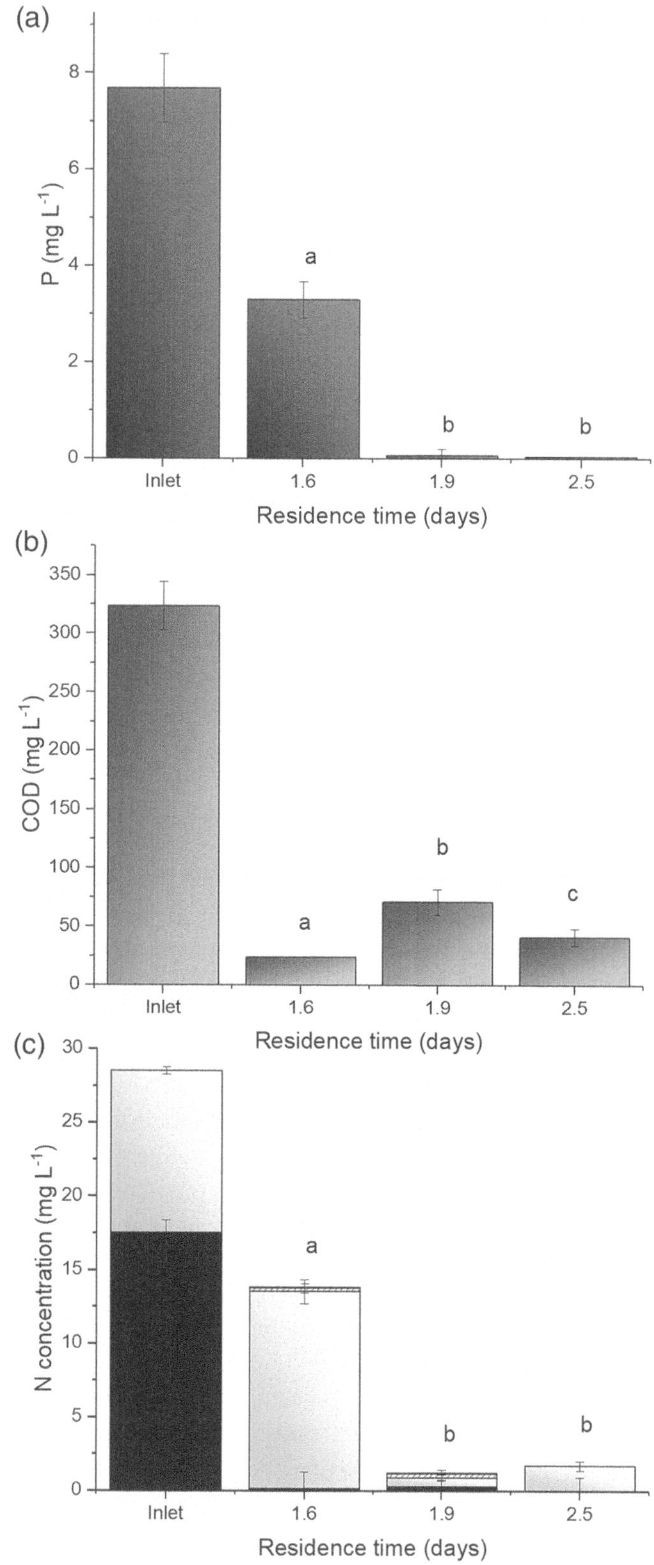

Figure 5 | Inlet (first column) and outlet concentration of phosphorus (a), COD (b) and nitrogen (c) of continuous systems at three retention times. In (c), grey portions represent N-NH_3, dark N_{org} and striped N-NO_3. Different letters correspond to statistically significant results.

shown in Figure 5(c), the washout of microalgae at 1.6 days strongly affected the decrease of nitrogen concentration, which was actually converted from organic to ammonia form, but was not efficiently removed, similar to what was observed in batch experiments with *B. diminuta* cultivated alone. By contrast, at higher residence times, an almost complete removal of nitrogen occurred, suggesting that residence time is the main operating variable affecting the process.

In conclusion, the continuous cultivation of these two species is feasible and efficient in removing pollutants from wastewater, given the proper operating conditions. Clearly, a deeper understanding of the influence of light on microalgal growth, and the subsequent interactions with bacteria, should be pursued in future. In particular, the effect of day–night light cycles on the process should be assessed, with particular attention to the mixotrophic capabilities of microalgae. The capability of the consortium to maintain their interactions in a complex system such as a real WWTP should certainly be investigated, but the preliminary results here reported are promising in terms of an efficient, single-step wastewater treatment. In fact, as COD, nitrogen and phosphorus were simultaneously removed from the wastewater, a possible WWTP consisting of one single biological step may be proposed in the future, without the need for an external oxygen or CO_2 supply, with the consequent abatement of operational costs. A first scaling up should be carried out to validate it in a real system.

CONCLUSIONS

The symbiotic interaction between *B. diminuta* and *C. protothecoides* was assessed with a view to enhancing nutrient removal efficiency from wastewater.

The presence of *B. diminuta* stimulated the growth of the microalgal species, and such an effect is not only due to the gas exchange between the two populations, as *C. protothecoides* in the presence of CO_2 bubbling did not show such a high nutrient removal efficiency. The positive interaction between *C. protothecoides* and *B. diminuta* was probably due to nutrient exchange between the two populations of microorganisms, in particular related to the capability of *B. diminuta* to convert organic nitrogen into ammonium species, which is more easily consumed by microalgae. Even though the two organisms have different specific growth rates, continuous co-cultivation is possible, as the mutual interaction allows an equilibrium to be achieved in the bacteria:microalgae ratio, which was found to be strongly dependent on the residence time applied. Residence times that were too low caused a washout of the microalgae, with a consequent inefficient removal of nitrogen and phosphorus.

ACKNOWLEDGEMENTS

The authors would like to thank Alessandro Spagni for providing instrumentation and valuable discussion.

REFERENCES

Beltràn-Rocha, J. C., Barcèlo-Quintal, I. D., Garcìa-Martinez, M., Osornio-Berthet, L., Saavedra-Villarreal, N., Villarreal-Chiu, J. & Lòpez-Chuken, U. J. 2017 Polishing of municipal secondary effluent using native microalgae consortia. *Water Science and Technology* **75** (7), 1693–1701.

Boelee, N. C., Temmink, H., Janssen, M., Buisman, C. J. N. & Wijffels, R. H. 2014 Balancing the organic load and light supply in symbiotic microalgal-bacterial biofilm reactors treating synthetic municipal wastewater. *Ecological Engineering* **64**, 213–221. [online] http://dx.doi.org/10.1016/j.ecoleng.2013.12.035.

de-Bashan, L. E., Bashan, Y., Moreno, M., Lebsky, V. K. & Bustillos, J. J. 2002 Increased pigment and lipid content, lipid variety, and cell and population size of the microalgae *Chlorella* spp. when co-immobilized in alginate beads with the microalgae-growth-promoting bacterium *Azospirillum brasilense*. *Canadian Journal of Microbiology* **48** (6), 514–521. [online] http://www.nrcresearchpress.com/doi/abs/10.1139/w02-051.

de-Bashan, L. E., Hernandez, J. P., Morey, T. & Bashan, Y. 2004 Microalgae growth-promoting bacteria as 'helpers' for microalgae: a novel approach for removing ammonium and phosphorus from municipal wastewater. *Water Research* **38** (2), 466–474.

de Farias Silva, C. E., Sforza, E. & Bertucco, A. 2017 Stability of carbohydrate production in continuous microalgal cultivation under nitrogen limitation: effect of irradiation regime and intensity on Tetradesmus obliquus. *Journal of Applied Phycology* **30** (1), 261–270.

Doucette, G. J., McGovern, E. R. & Babinchak, J. A. 1999 Algicidal bacteria active against *Gymnodinium breve* (Dinophyceae). I. Bacterial isolation and characterization of killing activity. *Journal of Phycology* **35**, 1447–1454.

Fukami, K., Nishijima, T. & Ishida, Y. 1997 Stimulative and inhibitory effects of bacteria on the growth of microalgae. *Hydrobiologia* **358** (1987), 185–191.

Gonzalez-Bashan, L. E., Lebsky, V. K., Hernandez, J. P., Bustillos, J. J. & Bashan, Y. 2000 Changes in the metabolism of the microalga *Chlorella vulgaris* when coimmobilized in alginate with the nitrogen-fixing *Phyllobacterium myrsinacearum*.

Canadian Journal of Microbiology **46** (7), 653–659. [online] http://www.nrcresearchpress.com/doi/abs/10.1139/w00-041.

Ji, B., Chen, W., Zhu, L. & Yang, K. 2016 Isolation of aluminum-tolerant bacteria capable of nitrogen removal in activated sludge. *Marine Pollution Bulletin* **106** (1–2), 31–34. [online] http://dx.doi.org/10.1016/j.marpolbul.2016.03.051.

Kang, Y. H., Kim, J. D., Kim, B. H., Kong, D. S. & Han, M. S. 2005 Isolation and characterization of a bio-agent antagonistic to diatom, *Stephanodiscus hantzschii*. *Journal of Applied Microbiology* **98** (5), 1030–1038.

Kouzuma, A. & Watanabe, K. 2015 Exploring the potential of algae/bacteria interactions. *Current Opinion in Biotechnology* **33**, 125–129.

Le Chevanton, M., Garnier, M., Bougaran, G., Schreiber, N., Lukomska, E., Bérard, J. B., Fouilland, E., Bernard, O. & Cadoret, J. P. 2013 Screening and selection of growth-promoting bacteria for *Dunaliella* cultures. *Algal Research* **2** (3), 212–222.

Markou, G., Vandamme, D. & Muylaert, K. 2014 Microalgal and cyanobacterial cultivation: the supply of nutrients. *Water Research* **65**, 186–202. [online] http://dx.doi.org/10.1016/j.watres.2014.07.025.

Natrah, F. M. I., Defoirdt, T., Sorgeloos, P. & Bossier, P. 2011 Disruption of bacterial cell-to-cell communication by marine organisms and its relevance to aquaculture. *Marine Biotechnology* **13** (2), 109–126.

Natrah, F. M. I., Bossier, P., Sorgeloos, P., Yusoff, F. M. & Defoirdt, T. 2014 Significance of microalgal–bacterial interactions for aquaculture. *Reviews in Aquaculture* **6** (1), 48–61.

Palacios, O. A., Bashan, Y., Schmid, M., Hartmann, A. & de-Bashan, L. E. 2016a Enhancement of thiamine release during synthetic mutualism between *Chlorella sorokiniana* and *Azospirillum brasilense* growing under stress conditions. *Journal of Applied Phycology* **28** (3), 1521–1531. [online] http://dx.doi.org/10.1007/s10811-015-0697-z.

Palacios, O. A., Gomez-Anduro, G., Bashan, Y. & de-Bashan, L. E. 2016b Tryptophan, thiamine and indole-3-acetic acid exchange between *Chlorella sorokiniana* and the plant growth-promoting bacterium *Azospirillum brasilense*. *FEMS Microbiology Ecology* **92** (6), 1–11.

Park, Y., Je, K. W., Lee, K., Jung, S. E. & Choi, T. J. 2008 Growth promotion of *Chlorella ellipsoidea* by co-inoculation with *Brevundimonas* sp. isolated from the microalga. *Hydrobiologia* **598** (1), 219–228.

Parthasarathy, S., Azam, S., Sagar, A. L., Rao, V. N., Gudla, R., Parapatla, H., Yakkala, H., Vemuri, S. G. & Siddavattam, D. 2017 Genome-guided insights reveal organophosphate-degrading *Brevundimonas diminuta* as *Sphingopyxis wildii* and define its versatile metabolic capabilities and environmental adaptations. *Genome Biology and Evolution* **9** (1), 77–81.

Sforza, E., Cipriani, R., Morosinotto, T., Bertucco, A. & Giacometti, G. M. 2012 Excess CO_2 supply inhibits mixotrophic growth of *Chlorella protothecoides* and *Nannochloropsis salina*. *Bioresource Technology* **104**, 523–529. [online] http://dx.doi.org/10.1016/j.biortech.2011.10.025.

Sforza, E., Urbani, S. & Bertucco, A. 2015 Evaluation of maintenance energy requirements in the cultivation of *Scenedesmus obliquus*: effect of light intensity and regime. *Journal of Applied Phycology* **27** (4), 1453–1462.

Singh, N., Marwa, N., Mishra, S. K., Mishra, J., Verma, P. C., Rathaur, S. & Singh, N. 2016 *Brevundimonas diminuta* mediated alleviation of arsenic toxicity and plant growth promotion in *Oryza sativa* L. *Ecotoxicology and Environmental Safety* **125**, 25–34. [online] http://dx.doi.org/10.1016/j.ecoenv.2015.11.020.

Solovchenko, A., Verschoor, A. M., Jablonowski, N. D. & Nedbal, L. 2016 Phosphorus from wastewater to crops: an alternative path involving microalgae. *Biotechnology Advances* **34** (5), 550–564. [online] http://dx.doi.org/10.1016/j.biotechadv.2016.01.002.

Subashchandrabose, S. R., Ramakrishnan, B., Megharaj, M., Venkateswarlu, K. & Naidu, R. 2011 Consortia of cyanobacteria/microalgae and bacteria: biotechnological potential. *Biotechnology Advances* **29** (6), 896–907. [online] http://dx.doi.org/10.1016/j.biotechadv.2011.07.009.

Tricolici, O., Bumbac, C., Patroescu, V. & Postolache, C. 2014 Dairy wastewater treatment using an activated sludge-microalgae system at different light intensities. *Water Science and Technology* **69** (8), 1598–1605.

Unnithan, V. V., Unc, A. & Smith, G. B. 2014 Mini-review: a priori considerations for bacteria-algae interactions in algal biofuel systems receiving municipal wastewaters. *Algal Research* **4** (1), 35–40.

Watanabe, K., Takihana, N., Aoyagi, H., Hanada, S., Watanabe, Y., Ohmura, N., Saiki, H. & Tanaka, H. 2005 Symbiotic association in *Chlorella* culture. *FEMS Microbiology Ecology* **51** (2), 187–196.

First received 21 September 2017; accepted in revised form 24 March 2018. Available online 6 April 2018

Treatment of food waste digestate using microalgae-based systems with low-intensity light-emitting diodes

Andrés Felipe Torres Franco, Scarlet da Encarnação Araújo, Fabiana Passos, Carlos Augusto de Lemos Chernicharo, César Rossas Mota Filho and Cleber Cunha Figueredo

ABSTRACT

Anaerobic digestion of food wastes coupled with digestate post-treatment using microalgae-based systems could recover large amounts of energy and nutrients worldwide. However, the development of full-scale implementations requires overcoming microalgae inhibition by high ammonia concentrations and low light transmittances affecting photosynthesis. This study evaluated the potential of microalgae-based reactors supplied with red light-emitting diodes (LEDs) at low intensity (660 nm and 15 μmol·m^{-2}·s^{-1}) to treat food waste digestate. LED reactors were compared with control reactors exposed to solar radiation. From a range of species in the inoculum, *Chlorella vulgaris* showed high adaptation to both lighting regimes and digestate environmental conditions, characterized by a C:N:P ratio of 74:74:1. Removal efficiencies for control and LED reactors were 84.0% and 95.8% for soluble chemical oxygen demand (COD) and 89.4% and 53.0% for ammonia, respectively. Approximately 50% of ammonia in control reactor and 15% in LED reactor was lost from the systems, whereas 17% and 36% of ammonia was transformed to organic nitrogen in control and LED reactors, respectively. Low-intensity LEDs maintained microalgae growth in levels similar to solar radiation and supported efficient digestate treatment, showing a potential for further application in optimization of full scale reactors at a relatively low energy cost.

Key words | *Chlorella vulgaris*, food waste digestate, light-emitting diodes, microalgal-based treatment

Andrés Felipe Torres Franco
Fabiana Passos
Carlos Augusto de Lemos Chernicharo
César Rossas Mota Filho (corresponding author)
Department of Sanitation and Environmental Engineering,
Federal University of Minas Gerais (DESA-UFMG),
Belo Horizonte, Brazil
E-mail: *cesar@desa.ufmg.br*

Scarlet da Encarnação Araújo
Cleber Cunha Figueredo
Institute of Biological Sciences,
Federal University of Minas Gerais (ICB-UFMG),
Belo Horizonte, Brazil

INTRODUCTION

Interest in anaerobic processes to treat food waste has grown intensely in the last decades due to its multiple advantages, including the possibility to recover energy. However, the liquid effluent from this process (i.e. digestate) usually contains high concentrations of organic matter, ammonia, phosphorus (Möller & Müller 2012) and, sometimes, pathogens (Sheets *et al.* 2015). Nutrients in food waste digestate could be recovered and applied in agriculture. However, direct application of digestate in agriculture is not recommended, as its carbon to nitrogen ratio could be not adequate for plant growth, which could also cause environmental impacts by e.g. ammonia volatilization.

Microalgae-based technologies, such as stabilization ponds, high-rate ponds and photobioreactors, have been applied as a promising alternative to recover nitrogen and phosphorus from food waste digestates (Mayers *et al.* 2017). However, there are many aspects of these processes that require further research and improvement. For instance, residual solids may limit light penetration, high levels of NH_3 in the digestate may cause inhibitory effects to microalgal growth and predators could significantly affect microalgae productivity (Tricolici *et al.* 2014; Sheets *et al.* 2015). Dilution has been used as a strategy to avoid ammonia inhibition and solids interference in light penetration, although it may not be a cost-effective alternative for full-scale implementation since it may demand high amounts of water, making that option unfeasible or unsustainable. High organic carbon and ammonia removals

doi: 10.2166/wst.2018.198

have been achieved at relatively short hydraulic retention time (HRT) with the photo-activated sludge process, where consortia of microalgae and nitrifying bacteria removed ammonium efficiently (van der Steen *et al.* 2015). Photosynthesis in photobioreactors may ease oxygen limitations for bacterial growth, while artificial lighting can provide adequate photosynthetic active radiation (PAR) for microalgae growth. The use of light-emitting diodes (LEDs) as a light source for algal cultures has gained more attention in recent years since they can produce higher levels of radiation with relatively low energy demand, heating, carbon and area footprints (Mohammed *et al.* 2014), so their use may be a strategy to overcome shadowing effects of solids and optimize biomass production. Previous studies suggest that red light (particularly at 660 nm) at moderate to high light intensities (i.e. 50–2,000 $\mu mol \cdot m^{-2} \cdot s^{-1}$) was optimal for microalgae growth in wastewaters and digestates (Zhao *et al.* 2013; Mohammed *et al.* 2014). However, those light intensities may still represent high energy consumption (0.54 $kWh \cdot kgCOD_{removed}^{-1}$) (COD = chemical oxygen demand) and even exceed photoinhibition levels reported for strains like *Chlorella vulgaris* under artificial lighting (Maxwell *et al.* 1995).

In order to enhance photobioreactors for microalgae-based treatment of concentrated digestates and wastewaters, LED lighting can be used to increase photic depth in hybrid ponds and tanks, reducing area requirements but still taking advantage of natural solar radiation, particularly in tropical climates. However, it is required to determine the minimum levels of PAR for microalgae photosynthesis and global treatment efficiencies. Thus, the current work intended to explore the potential of low intensity LEDs to support phototrophic activity in a microalgae-based system to treat food waste digestate under the hypothesis that a minimum level of 15 $\mu mol \cdot m^{-2} \cdot s^{-1}$ can sustain photosynthesis and become a minimum value of reference for artificial lighting in further full-scale implementations.

MATERIAL AND METHODS

Experimental set-up

Laboratory-scale photobioreactors were operated in batch mode treating food waste digestate from an upflow anaerobic sludge blanket (UASB) reactor (Figure 1). The reactors consisted of six transparent acrylic tubes with a working volume of 800 mL, which were arranged vertically outdoors at the Engineering School of the Federal University of Minas

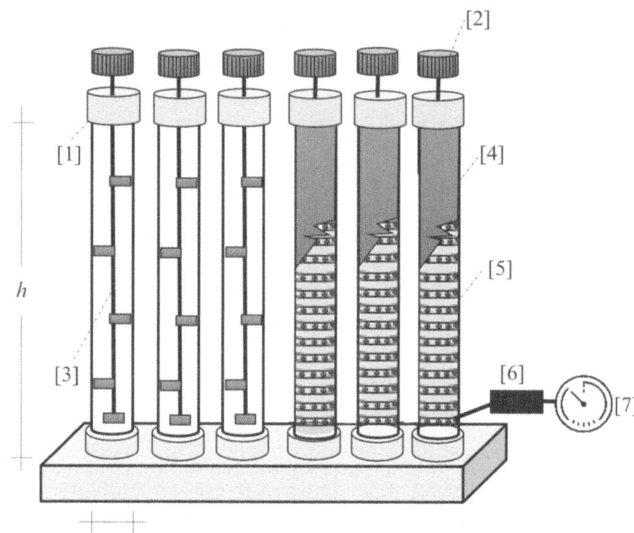

Figure 1 | Experimental set-up for control and LED reactors composed of [1] transparent acrylic tube D = 4 cm; h = 54 cm, [2] motors at 6 rpm, [3] vertical shaft and paddles (2 cm × 1 cm); [4] aluminium foil; [5] LED strips (660 nm), ~5 $W \cdot m^{-1}$, 110 V, [6] wireless dimmers [7] automatic timer.

Gerais (UFMG) (Belo Horizonte, Brazil – 19°52′11.04″S, 43°57′42.92″W). Mixing was achieved by motors operating at 6 rpm, installed at the top of each reactor and welded to a vertical shaft with mechanical paddles. Red LED lights in dimmable strips (wavelength 660 nm, ~25 W) were installed at the wall of the reactors (LED reactor), which were then covered with aluminium foil in order to supply a PAR of 15 $\mu mol \cdot m^{-2} \cdot s^{-1}$. Control reactors were exposed to natural solar radiation with a daily mean PAR of 580 $\mu mol \cdot m^{-2} \cdot s^{-1}$. Photoperiod for LED reactors was fixed for 12 hours, which was approximately the same time between dawn and sunset. Both LED and control reactors were built and operated in triplicate.

Food waste digestate, microalgae and activated sludge inoculum

Food waste digestate was obtained from a treatment system composed of an anaerobic completely stirred tank reactor, followed by a UASB reactor operated in series and fed with 500 kg of food waste per week obtained from the university canteens (UFMG, Belo Horizonte, Brazil).

Microalgae inoculum was obtained from a maturation pond treating municipal wastewater with high relative abundance of *Euglena* sp., *Scenedesmus* sp. and *C. vulgaris.* To increase the microalgae diversity in the initial inoculum, natural samples obtained from an urban reservoir in Belo Horizonte were concentrated with a phytoplankton net and added at a 1:1 ratio. For inoculation, 500 mL of the microalgae previously collected was suspended with 2 L of food waste digestate and 2.5 L of distilled water to avoid ammonia toxicity, which resulted in total ammonia nitrogen (TAN) and total suspended solids (TSS) concentrations of 100 mg·L^{-1} and 500 mg·L^{-1}, respectively. Each reactor was inoculated with 500 mL of this mixture. After an acclimation period of three days (Period I), TSS were measured and activated sludge from a municipal wastewater treatment plant (Belo Horizonte, Brazil) was added after 1 h settling at a ratio of 5:1 (microalgae:activated sludge). Working volume of each reactor was completed to 800 mL with a new pulse of food waste digestate and maintained in operation during 15 days (Period II).

COD, TAN and PO_4-P were determined on filtered samples every 48 h. Operational parameters including pH, dissolved oxygen (DO), temperature (T) and TSS were measured with the same frequency, and at 12 noon. Quantification of total Kjeldahl nitrogen (TKN), nitrate (NO_3-N), coliforms and *Escherichia coli*, as well as microalgae (by microscopy) was performed on the third day, immediately after the addition of activated sludge and digestate, and on the last day of the experiment. Flow cytometry analyses were performed for samples taken from inoculum and reactors at the end of the operational period to establish the ratio of cells fluorescing to red light (670 long pass (LP) filter).

Experimental analyses

Food waste digestate was characterized by measuring total COD, soluble COD (S-COD), TAN, organic nitrogen (org-N), NO_3-N, soluble reactive phosphorus (PO_4-P), total phosphorus (total-P), pH, total alkalinity, TSS and volatile suspended solids (VSS) according to *Standard Methods for the Examination of Water and Wastewater* (Rice *et al.* 2012).

DO and temperature were measured directly in the reactors using a sensor (Hach® HQ30 d) connected to a probe (Hach® LDO101). pH was quantified using a benchtop pH-meter (Denver Instruments® UB-5). Concentrations of NO_3-N were measured by NitraVer® 5 Nitrate Reagent Powder Pillows (HM 8039). Total coliforms (TC) and *E. coli* were measured using the chromogenic substrate (Colilert-18/Quanti-Tray) technique. Microalgae populations were quantified by Utermöhl technique using an inverted microscope (Zeiss, Primovert, Germany). Chlorophyll-*a* (Chl-*a*) content was assessed by ethanol extraction and quantified by measuring absorbances at 645 and 663 nm with a spectrophotometer and using Becker's equation. For flow cytometry, samples were fixed in paraformaldehyde and were evaluated using a FACSCanto™ II cytometer. Side scatter and the 488/530/30 laser and detectors were used. The fraction of cells fluorescence under red light was measured with a 670 LP Filter. Milli-Q water, saline solution and activated sludge were used as negative controls for flow cytometry.

Light intensity in the reactors was measured using a radiometer (Li-cor Li-189, USA). System's performance was assessed in terms of chl-*a* productivity and COD, TAN and PO_4-P removal. chl-*a* productivity was estimated according to Equation (1) from chl-*a* triplicates obtained for the first and last days of the operational time $[chl\text{-}a]_f$, considering A as the illuminated area ($A = \pi Dh$) and t as the number of days of the operational period and V_i and V_f as initial and final volumes in each reactor.

$$chl\text{-}a\ productivity = \frac{V_f\,[Chl\text{-}a]_f - V_i[Chl\text{-}a]_i}{A \cdot t} \tag{1}$$

Removal efficiencies were determined for S-COD, TAN and PO_4-P.

Statistical analysis

Statistical analysis included basic descriptive statistics and a repeated-measures analysis of variance (ANOVA) followed by a Bonferroni test performed in SPSS® for detecting significant differences between control and LED reactors treatment, (between-subject factors) and sampling time (within-subject factor) with a significance level of <0.05. F-values indicating differences are reported for the effects of time within reactor (F calculated for the effect of the interaction *time·reactor*) and between reactors (overall comparison). Normality was tested with a Shapiro–Wilk test. Data which did not follow normality were submitted to a $\log(1+x)$ or a $-1/\ln X$ transformation in order to perform the ANOVA. Homoscedasticity was checked with Mauchly's test of sphericity and Greenhouse–Geisser degrees of freedom correction was used for autocorrelation when necessary.

RESULTS AND DISCUSSION

Food waste digestate characterization

Food waste digestate characterization is summarized in Table 1. The digestate presented high concentrations of organic matter, ammonia, phosphorus and solids. Relatively high pH and total alkalinity are common in food wastes, which are caused by ammonia bicarbonate buffers, CO_2

Table 1 | Physicochemical characterization of the initial condition of the food waste digestate used in the control and in the treatments using LED reactors

Parameter	Value (mean ± std dev.)	Units
Total COD	1,856.7 ± 118.1	$mg{\cdot}L^{-1}$
Filtered COD (S-COD)	808.3 ± 125.2	$mg{\cdot}L^{-1}$
TAN	802.1 ± 20.3	$mg{\cdot}L^{-1}$
TKN	805.1 ± 18.2	$mg{\cdot}L^{-1}$
NO_3-N	12.5 ± 6.46	$mg{\cdot}L^{-1}$
PO_4-P	4.9 ± 7.4	$mg{\cdot}L^{-1}$
Total-P	10.9 ± 2.5	$mg{\cdot}L^{-1}$
pH	7.82	N/A
Total alkalinity	4,116.8 ± 168.0	$mgCaCO_3{\cdot}L^{-1}$
TSS	1,225 ± 125	$mg{\cdot}L^{-1}$
VSS	1,045 ± 55	$mg{\cdot}L^{-1}$

species concentrations of alkali compounds and bicarbonate and carbonate ions formed by successive dissociation reactions of carbonic acid produced during anaerobic digestion (Kang *et al.* 2017). Org-N was very low since the digestate was obtained from a food waste methanization reactor coupled to a UASB reactor and thus all org-N was previously ammonified.

Nutrient concentrations were high enough to support microalgae growth in control and LED reactors. Total alkalinity in food waste digestate presented high concentrations and secured inorganic carbon for autotrophic growth. An experimental C:N ratio based on S-COD of 1:1 and an N:P ratio of 73:1 suggests that carbon and phosphorus could have limited the total assimilation of TAN by autotrophic biomass, which was possibly affected since C:N ratio was relatively low when compared with optimum ratios reported for *Chlorella* sp. cultivated in wastewater with artificial light supply (C:N ratio of 5:1 to 10:1) (Yan *et al.* 2013) and optimum N:P ratio for reaching a maximum biomass production of *Chlorella* and nutrients removal (~10:1) (Choi & Lee 2015).

OPERATIONAL PARAMETERS

Temperature, pH and DO are shown in Figure 2. The control reactors presented a mean temperature of 31.3 ± 3.2°C, whereas the LED reactor temperature was 28.6 ± 2.2°C. No significant differences were found between the reactor's temperatures ($F = 6.713$, $p > 0.05$) or within the reactors, along the operational time ($F = 0.253$, $p > 0.05$). pH ranged between 8.24 and 9.60 for control reactors, and between 8.22 and 9.60 for the reactors with LED. Mean DO values of 9.1 ± 7.8 and 9.0 ± 6.4 mg DO/L were measured for the control and LED reactors, respectively. No significant differences were detected for pH and DO within (pH: $F = 2.709$ $p > 0.05$ and DO: $F = 23.395$ $p > 0.05$) or between reactors (pH: $F = 4.199$ $p > 0.05$ and DO: $F = 4.056$ $p > 0.05$).

While temperature was an independent and not controlled variable, pH and DO were determined by biological processes. Constant pH values were accompanied by alkalinity decreases detected in both reactors. Total alkalinity consumption was 885.4 ± 7.5 mg $CaCO_3$ in the control reactors and 440.9 ± 113.6 mg $CaCO_3$ in the LED reactors. These decreases were attributed to microalgae consumption of dissolved CO_2 during photosynthesis (Park *et al.* 2011), as no precipitate formation or other CO_2 loss pathways occurred in the reactors. In addition, microalgae photosynthesis was probably the major cause of the tendencies observed for DO, while showing that measured concentrations were enough for holding heterotrophic activity. Oxygen consumption was observed after the addition of activated sludge containing bacterial populations, resulting in a slight decrease in DO concentrations. The concentrations of oxygen in the control reactors showed a wider variation when compared to the LED reactors. Control reactors reached maximum DO concentrations around 20 $mg{\cdot}L^{-1}$ due to oversaturation produced by intense photosynthetic activity during hours of maximum sunlight, resulting in a high standard deviation of DO values but no significant differences when compared to LED reactors. Conversely, LED reactors exhibited shorter variations since lighting intensity and consequently the photosynthetic rates were always relatively constant.

MICROALGAE PRODUCTION

Chl-*a* concentrations were evidence that microalgae used the digestate as source of nutrients in both lighting conditions (Figure 3(a)). However, significant differences were detected within ($F = 18.136$, $p < 0.05$) and between reactors ($F = 52.037$, $p < 0.05$), whereby light conditions using LED were considerably better for microalgal biomass production in respect to solar lighting provided in control reactors. Chl-productivity in control and LED reactors at the end of the operational period was 2.9 ± 2.1 and 31.3 ± 6.4 mg Chl-$a{\cdot}m^{-2}{\cdot}d^{-1}$, respectively.

Chlorophyll increased in relation to *C. vulgaris* growth (Figure 3(b)), which was the only species present in both reactors from the beginning of the acclimation (Period I)

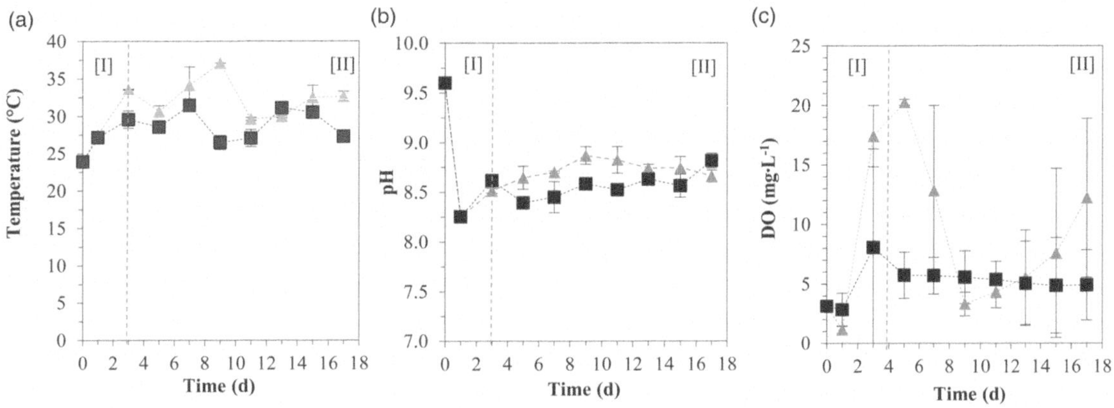

Figure 2 | Temporal series of (a) temperature, (b) pH and (c) DO for [I] acclimation and [II] batch period, for [▲] control and [■] LED reactors.

to the end of the operational period (Period II). The adaptation of *C. vulgaris* to grow on digestate and both lighting conditions was confirmed by the increases in cell counting along the acclimation and operational periods. Regarding microalgae growth, average net value for LED reactors was slightly higher than for control, i.e. $7.83 \cdot 10^6$ and

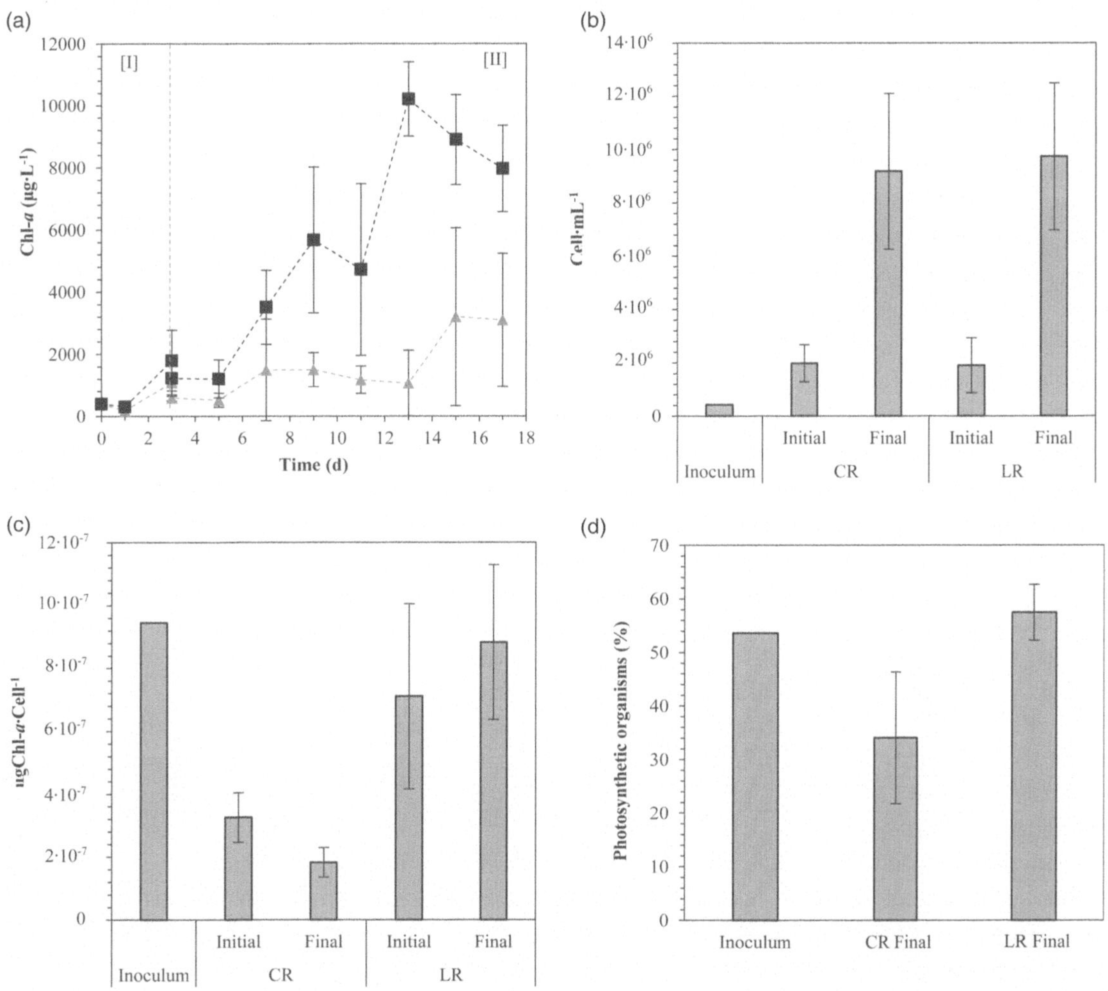

Figure 3 | Temporal series of (a) chl-*a* for [I] acclimation and [II] operational period, for [▲] control and [■] LED reactors, (b) *Chlorella vulgaris* counting of inoculum and initial and final day of operational period (Period II) for control reactors (CR) and LED reactor (LR), (c) chl-*a* content per cell at the initial and final day of operational period and (d) percentage of photosynthetic organisms (fluorescence to red light with a 670 LP filter at FACSCanto™ II Cytometer).

$7.20 \cdot 10^6$ cell·mL^{-1}, respectively and more chl-*a* content was measured in that treatment (Figure 3(c)). These results matched with fluorescence under red light (670 LP filter) in flow cytometry, indicating that the percentage of photosynthetic organisms (Figure 3(d)) increased in the LED reactors (i.e. $57.4 \pm 5.2\%$) and decreased in the control (i.e. $34.2 \pm 12.3\%$). This was possibly a consequence of differences in bacteria and microalgae cell content of chlorophyll. Additionally, heterotrophic bacteria (non-fluorescing cells) seemed to increase their populations in the control, while remaining relatively constant when using LED.

More chl-*a* content per cell in the LED reactors suggests that algae responded to the light condition by altering their photosynthetic pigment content (Wu 2016). It is also possible that LED treatment produced changes in cell sizes since larger cells would have greater amount of chlorophyll for same net biomass quantities, but this was not verified as no measures of cell sizes were made. Cell counting and chl-*a* content were similar to previous values found in outdoor ponds treating wastewater with ammonia concentrations in the range of 800–1,600 mg·L^{-1} and pH 8. However, chl-*a* content in LED reactors in our study was higher than those previously reported (Ayre *et al.* 2017).

LED light at low intensities also seemed to promote a more energy-efficient photosynthesis compared to high PAR intensity sunlight in control reactors, which may have been closed to photoinhibition levels. Probably, the reason for this inefficiency is related to a higher photon flux density, in which the rate of photon absorption by the antenna chlorophylls exceeds the rate at which photosynthesis occurs at maximum, wasting up to 50–80% of absorbed photons (Carvalho *et al.* 2011). Additionally, since each strain and culture may experience photoinhibition at different intensities and culture conditions (Sorokin & Krauss 1958), our results suggest that adaptation of *C. vulgaris* to food waste digestate at outdoor Brazilian temperatures seemed to perform better at low artificial lighting intensities at 660 nm than at high intensities of natural sunlight. Furthermore, low intensities may represent a more cost-efficient solution for wastewater treatment in photobioreactors in respect to higher artificial lighting intensities previously reported (Yan *et al.* 2013; Zhao *et al.* 2013; Mohammed *et al.* 2014).

In respect to species diversity, food waste digestate showed an inhibitory effect, since original inoculum had a higher number of species, composed of *Scenedesmus* spp., *Euglena* spp., *C. vulgaris, Dinobryon* sp., *Monactinus simplex, Botryococcus braunii, Pandorina* sp., *Planktolyngbya limnetica, Aulacoseira granulata, Peridinium* sp., *Desmodesmus* sp., *Trachelomonas volvocina* and *Microcystis aeruginosa*. However, the diversity was quickly lost during the first days of the experiment. This was probably related to high ammonia concentrations in the digestate. In fact, ammonia toxicity is a main constraint on microalgae-based treatments of highly concentrated wastewater, such as observed for *Scenedesmus obliquus* with 534 mg $NH_4Cl \cdot L^{-1}$ at pH 8.4 (Azov & Goldman 1982) or *Neochloris oleoabundans* and *Dunaliella tertiolecta*, inhibited at levels of 2.3 and 3.3 mg $NH_3 \cdot L^{-1}$ at pH 8.0, respectively (Gutierrez *et al.* 2016). However, some species may survive under high ammonia concentrations, such as *Chlorella sorokiniana* and *Nannochloropsis oculata* (Gutierrez *et al.* 2016), which could be a similar response observed for *C. vulgaris* in our experiments.

Digestate treatment performance

S-COD removal

S-COD concentrations significantly decreased along the operational period within both reactors (Figure 4(a)) ($F = 81.68$, $p < 0.05$) with final concentrations of 304.2 ± 77.1 and 84.0 ± 17.4 mg·L^{-1}, in control and LED reactors, respectively ($F = 17.516$, $p < 0.05$). S-COD removal efficiencies increased within reactors along the operational period ($F = 10.529$ $p < 0.05$) (Figure 4(b)) leading to final S-COD removal efficiencies of $84.0 \pm 4.1\%$ in control and significantly higher efficiencies of $95.8 \pm 0.92\%$ in LED reactors ($F = 21.695$, $p < 0.05$).

S-COD removal efficiencies above 80% were obtained in both LED and control reactors after 6 to 8 days of experiment. Higher efficiencies of S-COD removal obtained in the LED reactors suggested that the low irradiances of 15 μmol·m^{-2}·s^{-1} did not limit microalgae-based treatment of food waste digestate and, conversely, exhibited a better performance compared to other studies in which LED lighting at higher irradiances were used (Yan *et al.* 2013; Zhao *et al.* 2013; Mohammed *et al.* 2014; Tricolici *et al.* 2014). Lighting conditions in LED reactors contributed to S-COD removal performance, since microalgae performed photoaeration for heterotrophic activity without increasing pH to maximum values unfavourable for activated sludge bacteria, while slightly higher maximum pH in control reactors may have limited heterotrophic bacterial activity (Mohammed *et al.* 2014).

Nitrogen transformations

Figure 5(a) shows the temporal series for ammonia nitrogen and Figure 5(b) a nitrogen compounds balance for initial

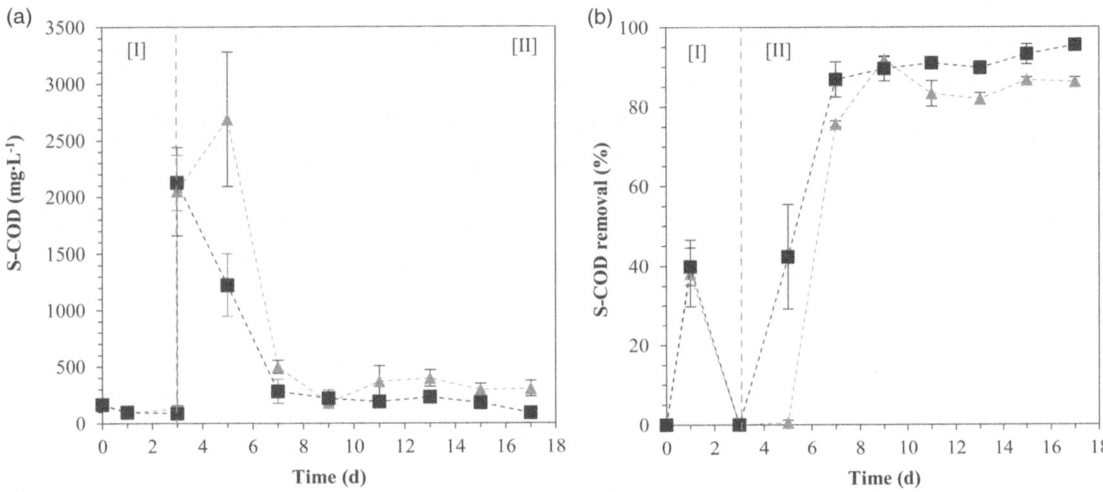

Figure 4 | Temporal series for (a) soluble COD (S-COD) and (b) S-COD removal efficiencies for [I] acclimation and [II] operational period, for [▲] control and [■] LED reactors.

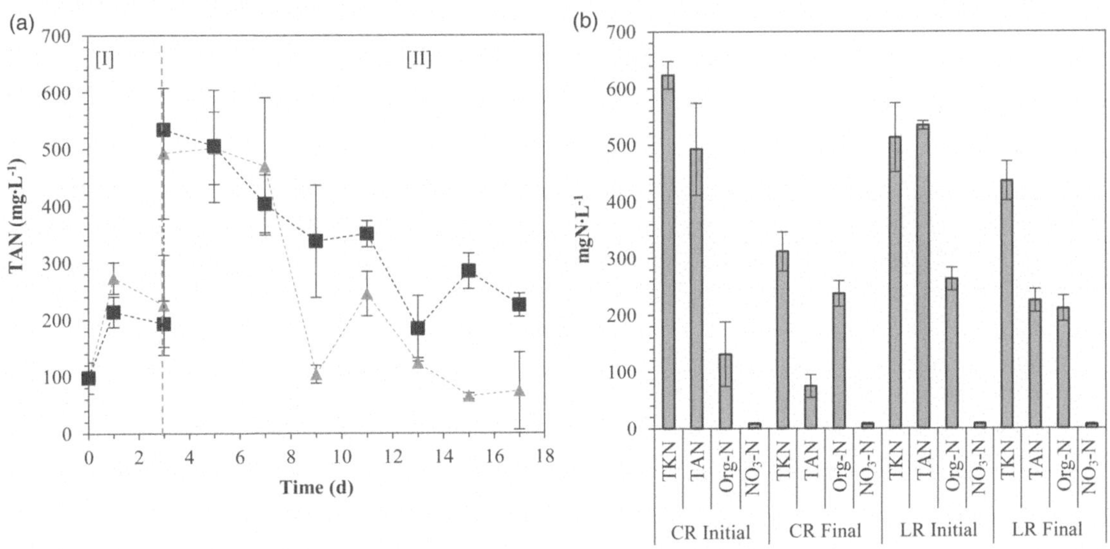

Figure 5 | (a) Temporal series for ammonia nitrogen for [I] acclimation and [II] operational period, for [▲] control and [■] LED reactors and (b) nitrogen balance for initial and final day of the operational period [II].

and final day of operational period. TAN concentrations decreased over time within reactors ($F = 15.402$, $p < 0.05$) and were significantly ($F = 45.302$, $p < 0.05$) lower in control reactors (74.4 ± 48.1 mgTAN·L^{-1}) than in LED reactors (225.0 ± 20.3 mgTAN·L^{-1}), which removed less ammonia (53.0 ± 7.2%) than control reactors (89.6 ± 12.6%).

The balance among different forms of nitrogen seems to be mainly influenced by ammonia volatilization and biological assimilation. Approximately 17% and 36% of TAN was transformed into org-N in control and LED reactors, respectively, whereas around 50% of TAN in control reactors and 15% in LED reactors were lost from the system and unable to be included in the nitrogen balance. The increase in the org-N fraction evidences that high ammonia levels in food waste digestate were not inhibitory for *C. vulgaris*, which was capable of ammonia uptake for biomass production. Higher nitrogen losses in control reactors may be attributed mainly to NH_3 volatilization caused by maximum pH values (>9.0) and temperatures (~35°C), which altered the NH_3/NH_4-N equilibrium to more than 50% of ammonia being present as the neutral and volatile form (NH_3) and potentially volatilized (Emerson *et al.* 1975). Other mechanisms may include nitrate denitrification during night-time when DO concentrations were low ($<1\ mg{\cdot}L^{-1}$) allowing the establishment of anoxic conditions (Park & Craggs 2011). However, nitrification was low in both treatments since a short production of NO_3-N was observed, meaning that nitrifiers were possibly inhibited by pH conditions above the optimum range (6.2–8.0) (Gujer 2010).

TAN removal accounted for $89.4 \pm 12.6\%$ in control reactors and $50.3 \pm 7.2\%$ in LED reactors. The efficiencies for nitrogen removal were relatively low when compared with other studies using or not LED lighting (e.g. Tricolici *et al.* 2014), but they are still results valuable since initial concentrations at food waste digestate were high and dilution was short (~30%). From a sustainable reuse perspective, TAN levels in filtered effluent and org-N in algal biomass can be positive if soil application is planned as a reuse alternative for the system's effluent and/or energetic exploitation of biomass.

Phosphorus and TSS

Statistical analyses of PO_4-P concentrations (Figure 6(a)) revealed no significant differences either within reactors along time ($F = 0.188$, $P > 0.05$) or between reactors ($F = 0.032$, $p > 0.05$). PO_4-P consumption may have been masked by re-mineralization of total-P from organic matter in digestate or even of the algal biomass, especially in control reactors where more losses may have occurred due to temperature and lighting variations. Anyway, removal efficiencies broadly coincide with other experiences with *C. vulgaris*, reporting low values of about 20% for N:P ratios between 61 and 70 (Choi & Lee 2015) or not a clear tendency at all (Tricolici *et al.* 2014). As mentioned for TAN, phosphorus concentrations in effluent may give an agriculture value to the treated digestate, even though higher TAN removal efficiencies are required to obtain an effluent with an N:P ratio of 1:5, which would have a more interesting ratio for e.g. beans fertirrigation.

Figure 6(b) presents time variations for TSS concentrations in both reactors, which remained constant along the operational period within ($F = 3.041$, $p > 0.05$) and between reactors ($F = 1.476$, $p > 0.05$). Considering the high concentration of TSS in food waste digestate, solids were not associated with algal/bacterial biomass, even considering that biomass is a fraction of solids. No significant solids removal was expected in the microalgae-based reactors, since microalgae show low settling rates so an external mechanism like centrifugation or flocculation would be required to obtain higher solids removal efficiencies (Park *et al.* 2011).

Finally, both control and LED reactors inactivated TC and *E. coli* at similar magnitudes (Table 2). Even if final dilutions made for determinations could have been smaller, at a dilution level of 10^2 no presence of coliforms or *E. coli* was found by the end of the operational period, so reactors' effluent would accomplish the WHO's (2006) standard of $<10^3$ colony forming unit/100 mL for reuse in agriculture. Mechanisms of TC and *E. coli* inactivation in maturation ponds and algal systems have been reported as associated with high penetration of UV radiation, elevated pH (especially $pH > 10$), elevated DO concentration (favouring photooxidation) (von Sperling 2002). Since pH was always lower than 10 in both control and LED reactor and no UV radiation was expected in the LED reactor, the

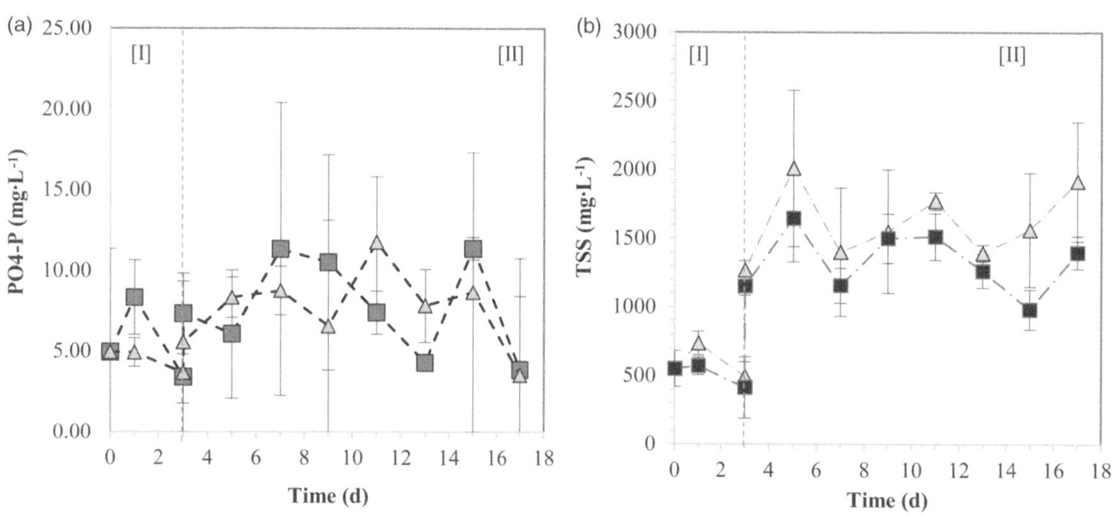

Figure 6 | Temporal series for (a) PO_4-P and (b) TSS for [I] acclimation and [II] operational period, for [▲] control and [■] LED reactors.

Table 2 | Performance of control and LED reactors for total coliforms and *E. coli* inactivation (Ci: initial concentration; Cf: final concentration; CFU: colony forming unit)

	Total coliforms		***E. coli***	
Parameter	**Control**	**LED**	**Control**	**LED**
C_o	$2.8 \cdot 10^5 \pm 1.7 \cdot 10^5$	$4.6 \cdot 10^5 \pm 3.4 \cdot 10^5$	$1.5 \cdot 10^3 \pm 1.0 \cdot 10^3$	$8.8 \cdot 10^3 \pm 6.0 \cdot 10^3$
C_f	$<10^2$	$<10^2$	$<10^2$	$<10^2$
Removal (log-units)	5.3 ± 0.3	5.6 ± 0.3	2.9 ± 0.4	3.8 ± 0.3

predominant mechanism must have been an oxidative stress produced by high DO concentrations.

Sustainability of microalgae-based technology application to food waste digestate

Global results of this work provide a perspective for microalgae-based reactor design and operation for increasing sustainability in treatment of sewage containing high concentrations of organic matter and nutrients, as those observed in food waste digestate. It was evidenced that microalgal productivity and economy of energy can be improved by combining solar and artificial LED lighting. In this work, PAR intensities of $15\,\mu mol \cdot m^{-2} \cdot s^{-1}$ were obtained with dimmed LED strips consuming ~12.5 W during HRT of 15 d and a photoperiod of 12 h, which was equivalent to an energy daily consumption of 0.15 kWh. The overall consumption was of ~0.48 kWh·kg $COD_{removed}^{-1}$ (or 1,875 $kWh \cdot m^{-3}$), which is considerably less than consumptions of activated sludge and/or artificial lighting at higher LED irradiances (e.g. Mohammed *et al.* 2014).

Furthermore, the area required for application of microalgae-based technologies could potentially be reduced by increasing reactor depth using artificial LED lighting applied under the level of the euphotic zone, which would remain illuminated by sunlight. Area requirement limit current applications of microalgae-based technologies in urban and peri-urban areas of developed and developing countries, which are the typical localization of biggest food waste sources. Thus, reducing area requirements is a decisive factor in full-scale implementation of microalgae-based technologies for sustainable treatment of food waste digestate. An example could be as follows: if a typical high rate algal pond (HRAP, reactor height h = 0.30 m) treating 1 m^3 per day of food waste digestate may require an area of ~33 m^2, an optimized HRAP of h = 1.2 m could reduce that requirement to 8 m^2 (~75%) using LED lighting at low irradiance in a depth of 0.90 m, provided by inner vertical panels with a PAR source producing around $400\,\mu mol \cdot m^{-2} \cdot s^{-1}$, which can easily secure minimum values around $15\,\mu mol \cdot m^{-2} \cdot s^{-1}$ in a horizontal distance of 15 cm (Mohammed *et al.* 2013), overcoming lighting blocking and scattering by solids. According to our results, energy consumption in such a reactor will be close to 2,000 $kWh \cdot m^{-3} \cdot month^{-1}$. Current (2018) electric energy price in Belo Horizonte is US\$0.18 kWh^{-1} so ~US\$360 $month^{-1}$ would be required per m^3 of food waste treated. The cost per m^2 of area in Belo Horizonte is about US\$2,000; thus diminishing an area of 33 m^2 to 8 m^2 implies savings of around US\$50,000, comparable to energy costs of about 12 years of operation with LED artificial lighting, which seems an economically and environmentally sustainable scenario for food waste digestate treatment, even without considering the economical values of the reuse of subproducts.

CONCLUSIONS

Food waste digestate with high ammonia concentrations and a C:N:P ratio of 74:74:1 showed to have an inhibitory effect on a high range of microalgae species. However, *C. vulgaris* was able to survive and perform photoaeration and assimilation of nutrients when illuminated with solar lighting in control reactors or LED lights (660 nm, 15 $\mu mol \cdot m^{-2} \cdot s^{-1}$) in the LED reactors. LED reactors exhibited a better performance in S-COD removal, with efficiencies of 95.81 ± 0.92%, higher than efficiencies of 84.02 ± 4.05% obtained in the control reactors. TAN efficiencies of LED reactors were less than those obtained in the control reactors as a consequence of higher ammonia volatilization at slightly higher pH values. No significant differences between control and LED reactors were found for PO_4-P removal efficiencies. Nutrients and microbiological composition of treated food waste digestate suggest that either energetic or agricultural reuses are opportunities for increasing sustainability of food waste management. LED lighting at 660 nm and $15\,\mu mol \cdot m^{-2} \cdot s^{-1}$ is a cost-effective solution to optimize bioreactors for microalgae-based treatment of highly concentrated digestates and other wastewaters.

ACKNOWLEDGEMENTS

Authors acknowledge the Minas Gerais State Agency for Research and Development - (FAPEMIG) for funding support through project 694 - FAPEMIG - APQ-02752-14 and the National Council for Scientific and Technological Development (CNPq) for the scholarship 141428/2016-3.

REFERENCES

Ayre, J. M., Moheimani, N. R. & Borowitzka, M. A. 2017 Growth of microalgae on undiluted anaerobic digestate of piggery effluent with high ammonium concentrations. *Algal Research* **24**, 218–226.

Azov, Y. & Goldman, J. C. 1982 Free ammonia inhibition of algal photosynthesis in intensive cultures. *Applied and Environmental Microbiology* **43** (4), 735–739.

Carvalho, A. P., Silva, S. O., Baptista, J. M. & Malcata, F. X. 2011 Light requirements in microalgal photobioreactors: an overview of biophotonic aspects. *Applied Microbiology and Biotechnology* **89** (5), 1275–1288.

Choi, H. J. & Lee, S. M. 2015 Effect of the N/P ratio on biomass productivity and nutrient removal from municipal wastewater. *Bioprocess and Biosystems Engineering* **38** (4), 761–766.

Emerson, K., Russo, R. C., Lund, R. E., Thurston, R. V. & Zarka, A. 1975 Aqueous ammonia equilibrium calculations: effects of pH and temperature. *Journal of the Fisheries Research Board of Canada* **32**, 2379–2383.

Gujer, W. 2010 Nitrification and me – A subjective review. *Water Research* **44** (1), 1–19.

Gutierrez, J., Kwan, T. A., Zimmerman, J. B. & Peccia, J. 2016 Ammonia inhibition in oleaginous microalgae. *Algal Research* **19**, 123–127.

Kang, J., Kwon, G., Nam, J. H., Kim, Y. O. & Jahng, D. 2017 Carbon dioxide stripping from anaerobic digestate of food waste using two types of aerators. *International Journal of Environmental Science and Technology* **14** (7), 1397–1408.

Maxwell, D. P., Falk, S. & Huner, N. P. 1995 Photosystem II excitation pressure and development of resistance to photoinhibition (I. Light-harvesting complex II abundance and zeaxanthin content in *Chlorella vulgaris*). *Plant Physiology* **107** (3), 687–694.

Mayers, J. J., Nilsson, A. E., Albers, E. & Flynn, K. J. 2017 Nutrients from anaerobic digestion effluents for cultivation of the microalga *Nannochloropsis* sp. – Impact on growth, biochemical composition and the potential for cost and environmental impact savings. *Algal Research* **26**, 275–286.

Mohammed, K., Ahammad, Z. S., Sallis, P. J. & Mota, C. R. 2013 Optimisation of red light-emitting diodes irradiance for illuminating mixed microalgal culture to treat municipal wastewater. *WIT Transactions on Ecology and the Environment* **178**, 263–270.

Mohammed, K., Ahammad, S. Z., Sallis, P. J. & Mota, C. R. 2014 Energy-efficient stirred-tank photobioreactors for simultaneous carbon capture and municipal wastewater treatment. *Water Science and Technology* **69** (10), 2106–2112.

Möller, K. & Müller, T. 2012 Effects of anaerobic digestion on digestate nutrient availability and crop growth: a review. *Engineering in Life Sciences* **12** (3), 242–257.

Park, J. B. K. & Craggs, R. J. 2011 Nutrient removal in wastewater treatment high rate algal ponds with carbon dioxide addition. *Water Science and Technology* **63** (8), 1758–1764.

Park, J. B. K., Craggs, R. J. & Shilton, A. N. 2011 Wastewater treatment high rate algal ponds for biofuel production. *Bioresource Technology* **102** (1), 35–42.

Rice, E. W., Baird, R. B., Eaton, A. D. & Clesceri, L. S. 2012 *Standard Methods for the Examination of Water and Wastewater*. American Public Health Association, American Water Works Association, and Water Environment Federation, Washington, DC.

Sheets, J. P., Yang, L., Ge, X., Wang, Z. & Li, Y. 2015 Beyond land application: emerging technologies for the treatment and reuse of anaerobically digested agricultural and food waste. *Waste Management* **44**, 94–115.

Sorokin, C. & Krauss, R. W. 1958 The effects of light intensity on the growth rates of green algae. *Plant Physiology* **33** (2), 109.

Tricolici, O., Bumbac, C., Patroescu, V. & Postolache, C. 2014 Dairy wastewater treatment using an activated sludge–microalgae system at different light intensities. *Water Science and Technology* **69** (8), 1598–1605.

van der Steen, P., Rahsilawati, K., Rada-Ariza, A. M., Lopez-Vazquez, C. M. & Lens, P. N. 2015 A new photo-activated sludge system for nitrification by an algal-bacterial consortium in a photo-bioreactor with biomass recycle. *Water Science and Technology* **72** (3), 443–450.

Von Sperling, M. 2002 Lagoas de estabilização. In: *Lagoas de estabilização*. DESA/UFMG.

WHO 2006 Guidelines for the Safe Use of Wastewater, Excreta and Greywater (Vol. 1). World Health Organization. Geneva, Switzerland.

Wu, H. 2016 Effect of different light qualities on growth, pigment content, chlorophyll fluorescence, and antioxidant enzyme activity in the red alga *Pyropia haitanensis* (Bangiales, Rhodophyta). *BioMed Research International* **2016**, Article ID 7383918, 8 pages.

Yan, C., Zhang, L., Luo, X. & Zheng, Z. 2013 Effects of various LED light wavelengths and intensities on the performance of purifying synthetic domestic sewage by microalgae at different influent C/N ratios. *Ecological Engineering* **51**, 24–32.

Zhao, Y., Wang, J., Zhang, H., Yan, C. & Zhang, Y. 2013 Effects of various LED light wavelengths and intensities on microalgae-based simultaneous biogas upgrading and digestate nutrient reduction process. *Bioresource Technology* **136**, 461–468.

First received 1 November 2017; accepted in revised form 18 April 2018. Available online 2 May 2018

Synthesis, characterization and application of green seaweed mediated silver nanoparticles (AgNPs) as antibacterial agents for water disinfection

D. Dixit, D. Gangadharan, K. M. Popat, C. R. K. Reddy, M. Trivedi and D. K. Gadhavi

ABSTRACT

A simple and eco-friendly method for the synthesis of hybrid bead silver nanoparticles (AgNPs) employing the aqueous extract derived from natural and renewable source namely tropical benthic green seaweed *Ulva flexuosa* was developed. This route involves the reduction of Ag^+ ions anchored onto macro porous methacrylic acid copolymer beads to AgNPs for employing them as antibacterial agents for *in vitro* water disinfection. The seaweed extract itself acts as a reducing and stabilizing agent and requires no additional surfactant or capping agent for forming the AgNPs.

The nanoparticles were analyzed using high-resolution transmission electron microscopy, UV–Vis spectroscopy, Fourier transform infrared spectroscopy, scanning electron microscopy, energy dispersive X-ray analysis and inductively coupled plasma optical emission spectroscopy. The study elucidates that such biologically synthesized AgNPs exhibit potential antibacterial activity against two Gram positive (*Bacillus subtilis, Staphylococcus aureus*) and two Gram-negative (*Escherichia coli, Pseudomonas aeruginosa*) bacterial strains tested. The bacterial count in treated water was reduced to zero for all the strains. Atomic force microscopy was performed to confirm the pre- and post-state of the bacteria with reference to their treatment with AgNPs. Attributes like facile environment-friendly procedure, stability and high antibacterial potency propel the consideration of these AgNPs as promising antibacterial entities.

Key words | antibacterial, environment, green, silver nanoparticles, *Ulva flexuosa*

D. Dixit (corresponding author)
M. Trivedi
Department of Earth and Environmental Science, K.S.K.V. Kachchh University, Near Changleshwar Mahadev Temple, University Road, Bhuj 370001, Kachchh-Gujarat, India
E-mail: *dharadixit2008@gmail.com*

D. Gangadharan
Department of Sciences, Amrita Vishwavidyapeetham University, Amritanagar, Ettimadai, Coimbatore, Tamil Nadu 641112, India

K. M. Popat
Membrane Science and Separation Technology Division, CSIR Central Salt and Marine Chemicals Research Institute, G.B. Marg, Bhavnagar 364002, Gujarat, India

C. R. K. Reddy
Division of Biotechnology and Phycology, CSIR-Central Salt and Marine Chemicals Research Institute, G.B. Marg, Bhavnagar 364002, Gujarat, India

D. K. Gadhavi
Kutch Ecological Research Centre – The Corbett Foundation, Khatau Makanji Bungalow, P.O. Tera, Taluka Abdasa, District Kachchh 370660, Gujarat, India

ABBREVIATIONS

AgNPs	Silver nanoparticles
LB	Luria Bertani
SEM	Scanning electron microscopy
EDX	Energy dispersive X-ray spectroscopy
HRTEM	High-resolution transmission electron microscopy
FTIR	Fourier transform infrared spectroscopy
UV	Ultraviolet spectroscopy
ICP-OES	Inductively coupled plasma optical emission spectroscopy
AFM	Atomic force microscopy
ATCC	American type culture collection

doi: 10.2166/wst.2018.292

INTRODUCTION

Nanotechnology holds immense potential in advancing water and wastewater treatment to improve treatment efficiency, quality and augment clean water supply through the safe use of unconventional water resources (Qu *et al.* 2013) with least possible environmental disturbances. Microbial contamination of water resources is a serious concern (Ashbolt 2004) in water disinfection systems. A variety of bacterial species are present in raw water as well as in wastewater. Nanoparticles are receiving worldwide attention due to their extremely minute size and high surface-to-volume ratio, leading to the chemical as well as physical developments in their properties as compared to their bulk counterparts (Iravani 2011; Veisi *et al.* 2018). Silver nanoparticle applications have been widely studied as disinfectants in medical institutions, and an increasing amount of research has been carried out on their applications in drinking water treatment and distribution systems (Rodriguez *et al.* 2008; Kumar & Raza 2009; Zhao *et al.* 2010; Alsabagh *et al.* 2015; Deng *et al.* 2017). Silver-based nanomaterial has a mechanism to disrupt bacterial metabolic processes (Cui *et al.* 2013), increase the cytoplasmic membrane permeability (Morones *et al.* 2005) and interact with DNA (Li *et al.* 2011) etc. and ultimately cause lysis of the cell wall. Among silver associated materials, silver nanoparticles are the most important because they have attracted much attention owing to their unique shape-dependent optical, electrical, chemical as well as antibacterial properties. Many protocols have been established for the physical as well as the chemical synthesis of silver nanoparticles (Donga *et al.* 2014; Xu *et al.* 2014; Lu *et al.* 2017; Okafor *et al.* 2017). However, these methods are energy as well as capital intensive (Dahoumane *et al.* 2016). Moreover, they employ toxic chemicals and solvents in the synthesis procedure and later on synthetic additives or capping agents, thus impeding their applications in clinical as well as biomedical fields (Akhtar *et al.* 2013). The need for the advancement of a clean, green, eco-friendly, efficient and cost-effective route to synthesize nanoparticles therefore attracts researchers globally to turn towards 'green' chemistry and bioprocesses (Jain *et al.* 2011; Dahoumane *et al.* 2017).

Several attempts have been made to synthesize different types of nanoparticles using algae as a biological route (Liu *et al.* 2005; Singaravelu *et al.* 2007; Xie *et al.* 2007a, 2007b; Abboud *et al.* 2014; El-Kassas *et al.* 2016). Researchers have also attempted synthesizing silver nanoparticles (AgNPs) utilizing different marine macro algae (Govindaraju *et al.* 2009; Azizi *et al.* 2013; Aboelfetoh *et al.* 2017; Ballesteros *et al.* 2017; Minhas *et al.* 2018). However, the utilization of this untapped marine resource in the synthesis of AgNPs for practical applications is yet to be pursued meticulously. Studies for the exploration of various natural sources need to be taken further in order to progress the prospects for greener synthesis and to develop an improved protocol for the synthesis and practical application of such products (Wang *et al.* 2014). A majority of the reports on the synthesis of AgNPs with a biological origin are in liquid-based/suspension systems and not on anchored AgNPs stabilized using a solid support. Therefore, synthesizing AgNPs through an environmentally friendly, non-toxic and hazard-free route using marine macro alga, stabilizing them by anchoring them on solid macro porous resin beads and applying them as an effective water disinfectant is our novel approach in this study. Immobilization of the AgNPs onto a solid support will ensure their stability for effective disinfection of water. Immobilized nanoparticles could turn out to be promising tactics for practical antibacterial applications like disinfection.

Macro algal species covered under *Ulvales* (Chlorophyta) are commonly encountered worldwide. To date, this biomass has represented quite low-value addition and its utility has been limited to composting (Maze *et al.* 1993; Badescu *et al.* 2017), food consumption (Dubigeon *et al.* 1997; Pirian *et al.* 2018) and methane production (Brand & Morand 1997; Costa *et al.* 2012). Therefore, the utilization of this algal biomass with respect to its biological assets, abundance and most importantly its renewability for the purpose of creating nanoparticles that can act as water disinfectants has been attempted.

EXPERIMENTAL

Materials

All the chemicals, solvents and reagents used were of analytical reagent grade or better and were used as received. Luria Bertani (LB) Agar and LB Broth were purchased from HiMedia (Mumbai, India). All the glass containers were rinsed with deionized water obtained from a Q-GARD, Milli-Q system Millipore, Germany (resistivity ¢18 MV cm) and dried in an oven prior to use.

Bacterial cultures

All the bacterial strains used were of the American Type Culture Collection (ATCC) procured from HiMedia (Mumbai, India). The strains used were *Escherichia coli* (ATCC 8739), *Pseudomonas aeruginosa* (ATCC 9027), *Staphylococcus aureus* (ATCC 25923) and *Bacillus subtilis* (ATCC 6051).

Synthesis of cross-linked methacrylic acid copolymer beads

The synthesis of methacrylic acid copolymer beads was done by following the method reported by Gangadharan *et al.* (2010). Briefly, a monomer mixture containing divinyl benzene methacrylic acid, toluene and benzoyl peroxide (1% w/w) was prepared and added to the preheated suspension medium, maintained at a temperature of 70 °C, in a three-neck round-bottom flask. The complete reaction mixture was slowly heated to 80 °C under reflux condition for further 5 h with constant stirring. The copolymer beads separated out in the form of white opaque beads.

Preparation of aqueous macro algal extract

Green macro alga (*Ulva flexuosa*) was collected from Pingleshwar coast, (N 23° 3′ 53.8″, E 68° 48′ 3.08″), Kachchh District, Gulf of Kachchh, India. Briefly, the seaweed was cleaned thoroughly using surface sterilization with autoclaved seawater followed by deionized water in order to remove the extraneous materials adhered onto the surface. The material was shade dried for a period of 2 weeks and powdered with mixer grinder and meshed. To 1 g of this powder, 100 mL MQ water was added and heated at 80 °C for 30 minutes. The extract was further subjected to cooling and then filtered using 0.45 μm membrane. Freshly prepared extract in this manner was then used for all the studies.

Synthesis of silver nanoparticles bound copolymer beads and its stability check

The copolymer beads were equilibrated with 0.01 M $AgNO_3$ for 16 h on a mechanical shaker. The equilibrated beads were washed with Milli-Q water to remove excess silver ions adhering to the copolymer beads. These copolymer beads were subjected to reduction by freshly prepared aqueous seaweed extract (1% conc.) under sonication at an ambient condition with a high-density ultrasonic 6 mm probe (Labsonic/Ultrasonic Processor, Sonopros-10,000MP) immersed directly in the mixture of copolymer beads and seaweed extract. The reaction time during the ultrasonic irradiation was 10 minutes. This step was undertaken as ultrasound which is considered as a green extraction procedure (Chemat *et al.* 2012; Rombaut *et al.* 2014) that would facilitate the breakdown of glycosidic linkages of the seaweed polysaccharides. The mixed solution was incubated for 24 h at 120 rpm at 25 °C. After 24 h, the beads turned brown indicating the formation of silver nanoparticles that was also confirmed by high-resolution transmission electron microscopy (HRTEM) analysis. The stability of these AgNPs was tested by placing them in aqueous condition at neutral pH for a period of 7 days (168 h). The absorbance spectra of the hybrid bead AgNPs did not show much variation thereby indicating the stability of these particles. These beads were then dried at room temperature (25 °C) and subjected to further characterization.

INSTRUMENTAL CHARACTERIZATION

The AgNPs bound copolymer beads were investigated by field emission scanning electron microscopy (FE-SEM). These images were obtained using a JEOL JSM- 7100F at an acceleration voltage of 5 kV, and energy dispersive X-ray (EDX) analysis was performed using INCA energy (Oxford Instruments Analytical Ltd, UK) coupled with an EDX facility. HRTEM (JEOL JEM 2100, Japan) was used to characterize the microstructure of the nanoparticles. Fourier transform infrared (FTIR) spectroscopy was performed using a PerkinElmer Spectrum GX FT-IR System. UV–Vis absorption spectrum of the sample was carried out using a Shimadzu UV–VIS–NIR Spectrophotometer (Model UV-3600). Atomic force microscopy (AFM) analysis and imaging were performed using NT-MDT Ntegra Aura model (Moscow, Russia). Bacterial sample preparation for AFM analysis and imaging was performed according to Suresh *et al.* (2010). Stereomicroscopy was performed to check the AgNPs surface for any crack or rupture. The detail is given in the supplementary document (Figure S1), available with the online version of this paper.

BACTERICIDAL EFFECT OF AgNPS

Spread plate test

The spread plate test was used to measure the antibacterial effect of the copolymer beads containing silver nanoparticles against the selected bacterial strains. The bacteria were cultured in LB broth at 30 °C. 100 μL of 24 h old bacterial culture was spread onto the LB agar plates. Copolymer beads containing silver nanoparticles were placed over the LB agar plates. The copolymer beads without silver nanoparticles were used as the blank. All the plates were incubated at 37 °C for 24 h to measure the zone of inhibition.

Test tube test

All the sterile test tubes were filled with 20 mL of autoclaved water. Bacterial cultures, namely (*E. coli*: 24×10^6 CFU/mL, *P. aeruginosa*: 19×10^6 CFU/mL, *B. subtilis*: 20×10^6 CFU/mL and *S. aureus*: 18×10^6 CFU/mL) were added to these test tubes. 100 mg (in the dry state) of copolymer beads containing silver nanoparticles were added to the test tubes containing *E. coli*, *P. aeruginosa*, *B. subtilis* and *S. aureus* bacterial suspensions separately and closed with a cotton stopper in order to avoid contamination. Similarly, 200 mg and 300 mg (in the dry state) of copolymer beads loaded with silver nanoparticles were tested to quantify the number of copolymer beads required to inhibit the bacterial growth completely. All the tubes were kept on the shaker for studying the growth inhibitory effect on the bacterial cultures. 100 μL of supernatant fractions were withdrawn after 2 h, 4 h, 6 h and 8 h of time intervals in order to investigate the influence of contact time in killing the bacterial cells. The supernatant fractions withdrawn were plated directly without any further dilution to obtain the bacterial colony count. Copolymer beads without AgNPs were used as blank in the experiment.

Kinetic study of silver release

The quantity of Ag^+ released in water was studied as a function of time (Taglietti *et al.* 2014). The experimental set up contained a series of 25 mL flasks containing 0.1 g of hybrid bead AgNPs with 30 mL of deionized (DI) water. The flasks were further subjected to shaking at room temperature. Sample aliquots were withdrawn at different time intervals, namely 0 h, 6 h, 12 h, 18 h, 24 h, 48 h, 72 h and 96 h for the analysis of silver ions. The concentration of silver ions in the filtrates was assessed using inductively coupled plasma optical emission spectroscopy (ICP-OES). Silver nanoparticle coated beads were ground into fine powder and 0.2 g of this powder was dissolved in a mixture of HF and HNO_3. After the complete dissolution of the powder, it was filtered. The samples containing silver ions eluted from the polymeric beads were diluted up to 50 mL with Milli-Q water prior to quantification. The final concentration was quantified using a set of single-element external calibration standards to enable measurements in the 0–1 ppm concentration range.

RESULTS AND DISCUSSION

Field emission scanning electron microscopy/energy dispersive X-ray analysis

The morphology of copolymer beads bound with AgNPs and their elemental composition were examined with Field emission scanning electron microscopy (FE-SEM) and EDX analysis, respectively (Figure 1). The copolymer beads were examined for their surface characteristics and no cracking was observed on the surface of the copolymer beads (Figure 1(a) and 1(b)) at a scale of 100 μm with a magnification of 180× and 95× respectively. The AgNPs distributed on the surface of the copolymer beads were observed to have uneven size and shape distribution at a scale of 1 μm with a magnification of 5,000× (Figure 1(c)). The distribution of the AgNPs onto the surface of the beads is clearly evident as previously reported by Dahoumane *et al.* (2012). The EDX analysis performed confirmed the presence of silver and chloride on the copolymer beads (Figure 1(d)). The intensity of the peak of silver indicates the formation of AgNPs. The signal of Cl is also observed which may be due to the adherence of seaweed extract used for the reduction.

High-resolution transmission electron microscopy

The morphology, size and the distribution of AgNPs were analyzed by HRTEM. The average size of the silver nanoparticles was observed to be 4.93–6.70 nm. The smaller particle size of AgNPs bound onto the copolymer beads render its close interaction with the microbial membranes, proving them to be good biocidal material. The HRTEM image containing AgNPs is shown in Figure 2 and in supplementary information Figure S5 (available with the online version of this paper).

Fourier transform infrared

The FTIR spectrum of the seaweed extract as well as the copolymer beads without AgNPs (blank) and seaweed mediated AgNPs were obtained. The IR spectrum of the seaweed extract (Figure 3(a)) showed major peaks in the lower frequency regions at 621, 764, 962, 1,120, 1,390, 1,459, 1,543 and 1,652 cm^{-1} which were assigned for C–Cl alkyl halide, alkenes, C–N stretch, C=C stretch and N=O bend, respectively. The IR spectrum of the macro algal extract showed that the major peaks were observed in the lower frequency regions at 1,543 and 1,652 cm^{-1} which were assigned for carbonyl stretching vibrations of amide I and amide II linkages. This clearly indicates the presence of protein/peptide group that serves as a stabilizing as well as reducing agent for the formation of silver nanoparticles (Gopinath *et al.* 2013). The peak at 1,390 cm^{-1} implies the role of C=C in the reduction of silver ion (Stalin Dhas *et al.* 2014). The infrared band detected at 2,358 cm^{-1} may be assigned to combination bands, namely strongly hydrogen bonded water molecules and/or hydroxyls (Frost *et al.*

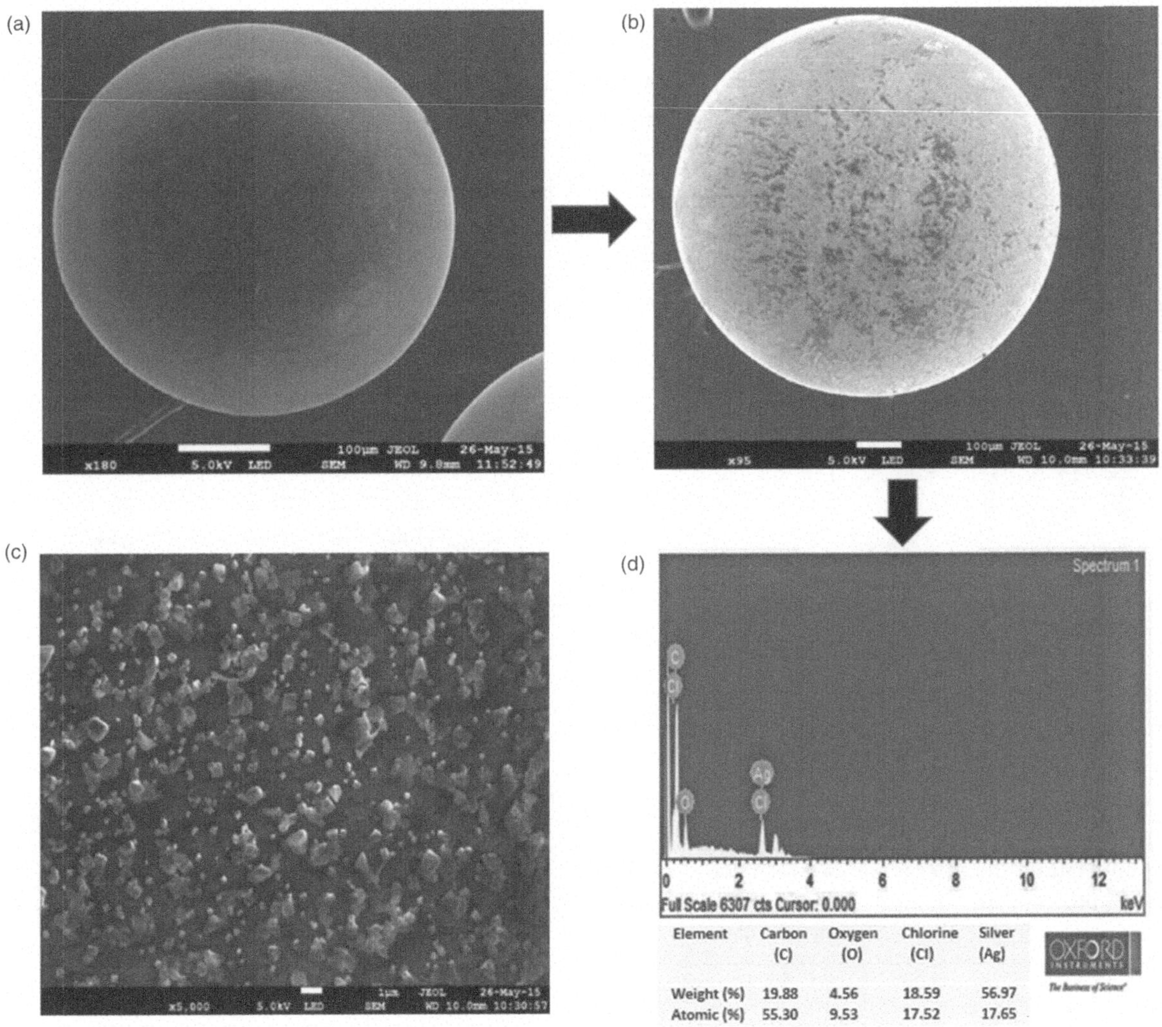

Figure 1 | (a) SEM image of copolymer bead without AgNPs (Blank); (b) SEM image of copolymer bead with AgNPs; (c) enlarged image of AgNPs anchored on the copolymer bead surface; (d) EDX analysis of silver nanoparticles (AgNPs) bound copolymer beads.

2013). It may also correspond to the –PH in phosphine stretch (Afsheen *et al.* 2018) or represent a characteristic peak for CO_2 (Khaleel & Nawaz 2015). The role of the asymmetric –CH bending represented by the peak at 2,927 cm^{-1} designates its role in the reduction of silver ions (Stalin Dhas *et al.* 2014). The absorbance at 3,417 cm^{-1} corresponded to N–H stretch as well as to O–H stretch which is characteristic of the hydroxyl functional group in alcohol and phenolic compounds. Figure 3(b) indicates the IR spectra of the blank. The FTIR spectrum of AgNPs (Figure 3(c)) showed distinct peaks at 514, 1,192, 1,378, 1,539, 1,646, 2,358, 2,936 and 3,427 cm^{-1}. The comparison of resultant spectrum between the green macro algal extract and the copolymer bound AgNPs showed notable shift changes in the peak positions as well as the absorption bands except at 2,358 cm^{-1}. From these results, it was concluded that halide, alkenes, amine and nitro groups which were present in the seaweed extract possessed a strong affinity to bind to the silver ions, thereby attributing their reduction to AgNPs.

UV–Vis spectrum

Figure 4(a) shows a broad peak at 480 nm. The surface plasmon resonance (SPR) peak at $\lambda_{max} = 480$ nm was observed due to the active biomolecules present in the seaweed extract which interacted with silver ions, thereby forming AgNPs. The spectrum in the range of 420 nm to 480 nm indicates the formation of silver nanoparticles through evolvement of characteristic surface plasmon extinction (Jes *et al.* 2015). The polydispersity of the nanoparticles on the copolymer beads is indicated by the broadening of peak (Ghaedi *et al.* 2012). The UV–Vis spectrum of the

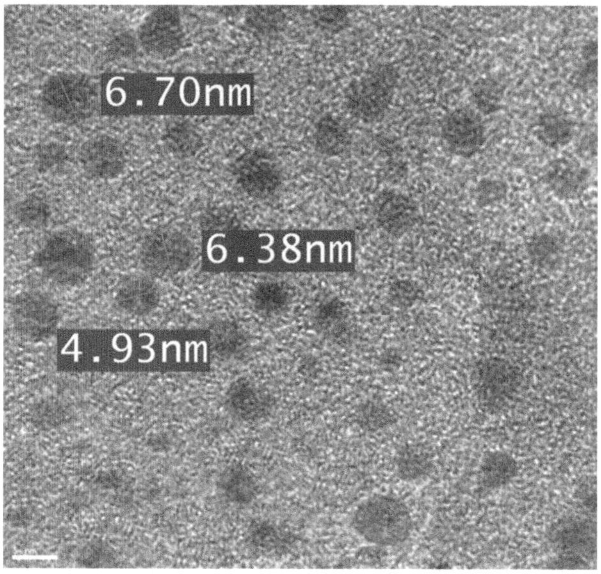

Figure 2 | High-resolution transmission electron microscopy image of the AgNPs formed using *U. flexuosa* extract along with their size measurement.

methacrylic copolymer beads without silver nanoparticles (blank) is shown in Figure 4(b).

Atomic force microscopy

In order to further understand the interaction between the selected bacterial strains and the hybrid bead AgNPs, AFM imaging experiments of the bacteria treated with blank and AgNPs coated beads were carried out (Figure 5). AFM is a convenient tool for examining the changes in cell membrane morphology and surface structure (Suresh *et al.* 2010) and can aid in elucidating morphological changes induced by AgNPs on bacterial cells. AFM images of blank and silver nanoparticle-treated bacterial samples showed notable differences in their outer surface morphology (Shukla *et al.* 2012). AFM images of all the four bacterial strains treated with blank (Figure 5(a1), 5(b1), 5(c1), 5(d1)) and with AgNPs (Figure 5(a2), 5(b2), 5(c2), 5(d2)) showed a notable change in the morphology. The bacteria showed intact structure when treated with copolymeric beads without AgNPs coating (blank). The cell wall was not damaged and the shape of the cell, as well as its integrity, were maintained. While when treated with hybrid bead AgNPs, the cell surface either developed a hole or was completely ruptured. The silver nanoparticles do appear to cause significant structural changes thereby affecting the cell integrity leading to cell death. However, the precise mode of the bacterial-nanoparticle interaction is still unclear.

Microbiological results

A soil as well as water contaminant, *P. aeruginosa* (ATCC 9027) (Prochnow *et al.* 2015), *E. coli* (ATCC 8739) as a

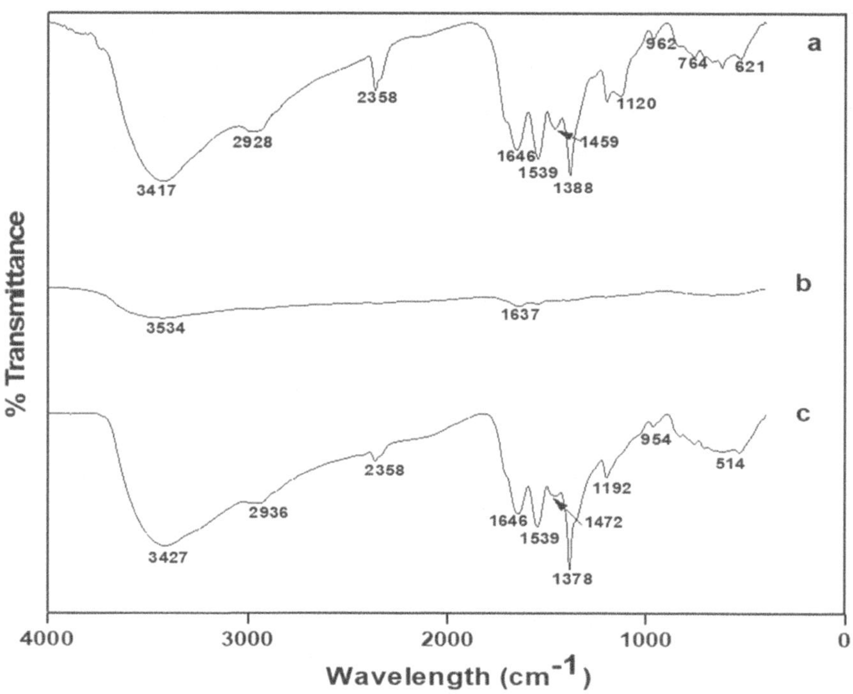

Figure 3 | FTIR spectra. (a) Seaweed extract only; (b) copolymer beads without AgNPs (Blank); (c) hybrid bead AgNPs synthesized using seaweed extract.

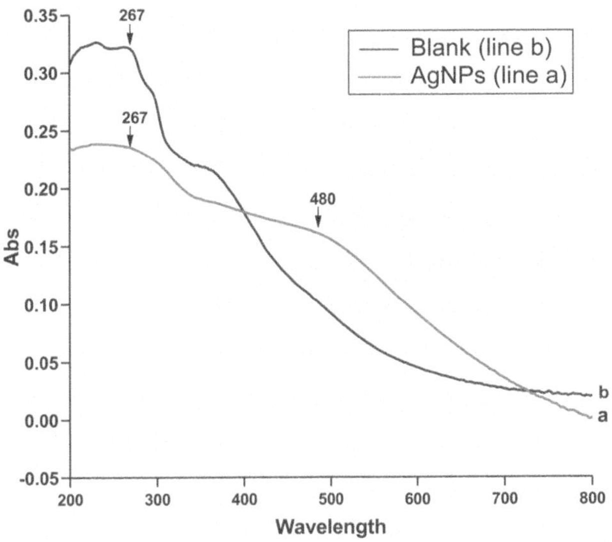

Figure 4 | UV–Vis absorption spectra. Line (a) shows the AgNPs bound copolymer beads (indicated in red online); line (b) shows copolymer beads without AgNPs/Blank (indicated in blue online). Please refer to the online version of this paper to see this figure in colour: http://dx.doi.org/10.2166/wst.2018.292.

faecal contamination indicator (How *et al.* 2013), *S. aureus* (ATCC 25923) (Basso *et al.* 2014) and *B. subtilis* (ATCC 6051) (Dodd & Grace 1989) as human influence indicators of water were selected for the study. The bacterial cultures were grown aerobically in LB broth at 37 °C for 24 h and maintained as stock cultures on LB agar slants at 4 °C.

Spread plate test

A preliminary experiment was conducted with AgNPs-bound copolymer beads. The zones of inhibition were measured for all the four bacterial cultures tested and represented in supplementary information (Figure S2), available online. In case of Gram-negative organisms like *P. aeruginosa*, the zone diameter was found to be 24 mm (Figure S2a) while for *E. coli* it was 15 mm (Figure S2b). The zone diameter against Gram-positive bacterial species like *B. subtilis* was around 18 mm (Figure S2c), while for *S. aureus* it was 13 mm (Figure S2d). The results indicate that AgNPs prepared using the aqueous extract of *U. flexuosa* exhibited good antibacterial activity against all the four selected bacterial strains.

Test tube test

The copolymer beads loaded with AgNPs reduced the bacterial count to zero with the increase in the contact time as well as the quantity of the copolymer beads. Viable cells of *E. coli* (Figure 6(a)) as well as *P. aeruginosa*, (Figure 6(b)) were reduced to zero within 6 h of contact time

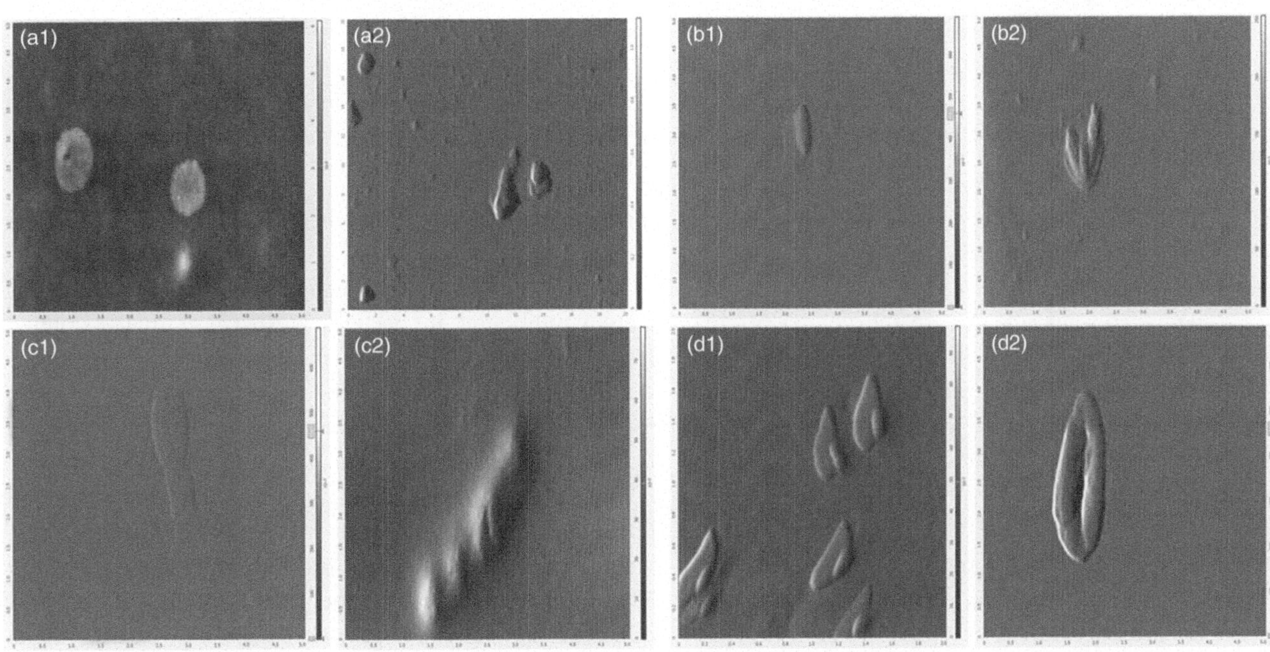

Figure 5 | AFM images showing the interaction of Gram-positive *B. subtilis* and *S. aureus* as well as Gram-negative organisms *E. coli* and *P. aeruginosa* with copolymer beads having AgNPs anchored onto the surface and with copolymer beads without AgNPs (Control/Blank); (a1) *S. aureus* culture with copolymer beads without AgNPs (Blank/Control); (a2) *S. aureus* culture treated with AgNPs (Test); (b1) *B. subtilis* culture with copolymer beads without AgNPs (Blank/Control); (b2) *B. subtilis* culture treated with AgNPs (Test); (c1) *E. coli* culture with copolymer beads without AgNPs (Blank/Control); (c2) *E. coli* culture treated with AgNPs (Test); (d1) *P. aeruginosa* culture with copolymer beads without AgNPs (Blank/Control); (d2) *P. aeruginosa* culture treated with AgNPs (Test).

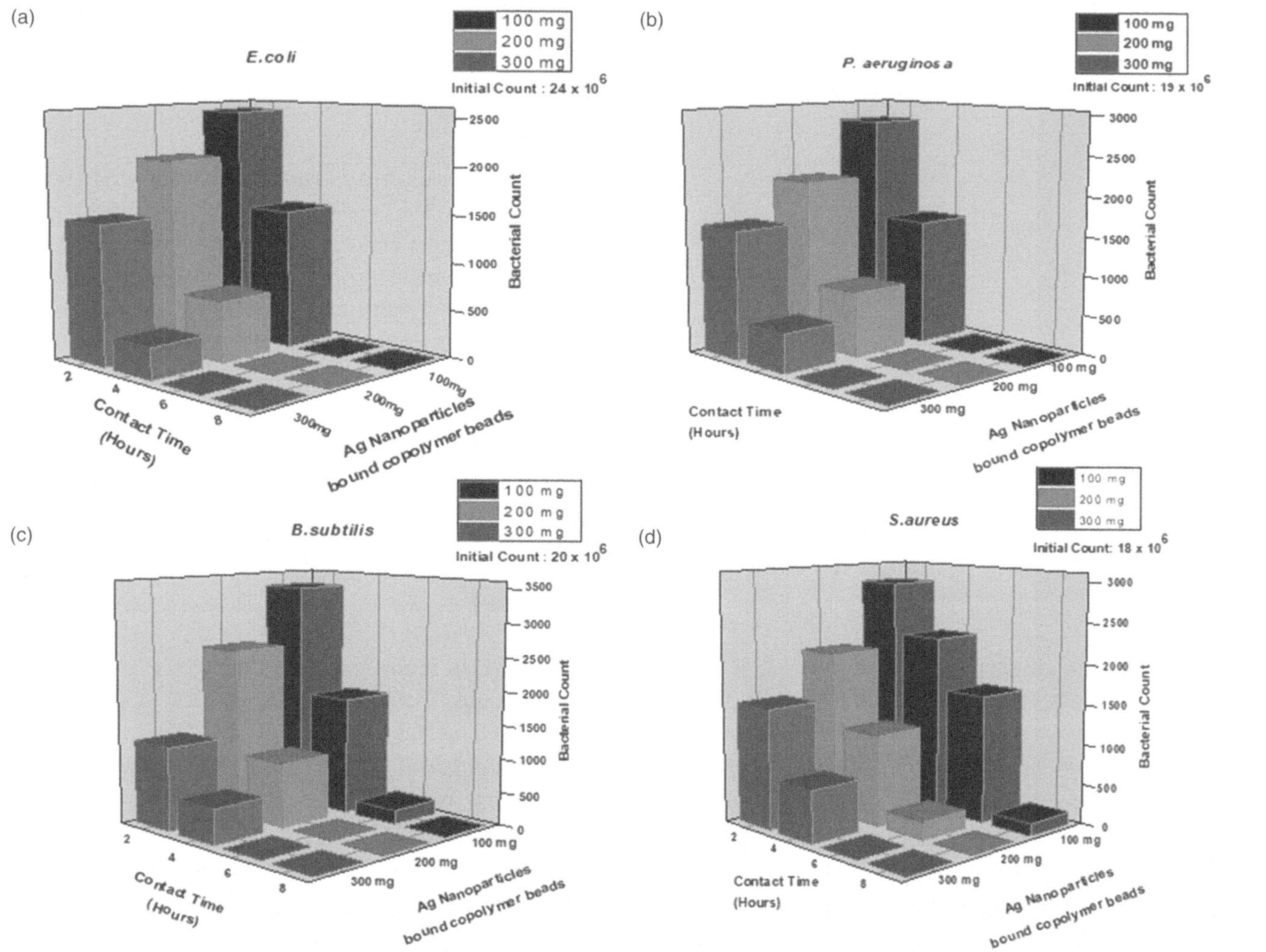

Figure 6 | (a) Graph showing the biocidal activity of the AgNPs against *E.coli*; (b) graph showing the biocidal activity of the AgNPs against *P. aeruginosa*; (c) graph showing the biocidal activity of the AgNPs against *B. subtilis*; (d) graph showing the biocidal activity of the AgNPs against *S. aureus*.

with 100 mg of AgNPs bound copolymer beads, while in the case of Gram-positive bacteria *B. subtilis* (Figure 6(c)) it reaches zero after 6 h of contact time but at a concentration of 200 mg. This may be due to the complexity of the outer cell wall of *B. subtilis*. In the case of *S. aureus* (Figure 6(d)), the count becomes zero after 6 h of contact time for 300 mg of AgNPs bound copolymer beads. This difference in bacterial reduction among the Gram-negative organisms is due to the higher permeable nature of *E. coli* cell wall.

When 200 mg of AgNPs was used, 100% reduction in the bacterial count was observed within 6 h of contact time for all the tested bacteria except *S. aureus*. It took nearly 8 h for the complete reduction of *S. aureus* when treated with 200 mg of AgNPs (Figure 6(d)). Studies suggest that both gene duplication and lateral gene transfer are important in the maintenance of the resistance mechanisms to toxic drugs or heavy metal in *Staphylococcus* (Chan *et al.* 2011).

When 300 mg of copolymer beads containing AgNPs were equilibrated in bacterial suspension they exhibited higher activity compared to the 100 mg copolymer beads within 2 h as obvious from the results. The bacterial count for all the tested strains was reduced to half or more than half with the usage of 300 mg of AgNPs coated copolymeric beads within 2 h of contact time. Complete reduction of both the Gram-negative as well as the Gram-positive viable cells was achieved within 6 h of contact time with 300 mg of AgNPs bound copolymer beads (Figure 6). This indicated that 300 mg copolymer beads with 6 h of contact time could be considered as the minimum inhibitory concentration necessary to render complete disinfection. The plate results of the test tube test are given in supplementary Figure S3 (available online).

Release of silver into the filtrate

The amount of silver loaded onto the polymer beads was 9 mg/g. In this study, the silver concentration in the filtrate was below the detectable range up to 12 h. The silver release gradually increases with the increase in contact time and complete leaching was observed after 96 h (Table 1). The increase in release rate at the initial phase of contact may be due to the liberation of silver nanoparticles. Up to 12 h, the leaching of silver was below the permissible level, thereby proving these hybrid bead AgNP entities to be potential antibacterial agents.

The emergence of protocols involving biological routes for the synthesis of nanoparticles constitutes an evolving facet of nanotechnology. The green synthesis approach of silver nanoparticles that defines the development of eco-friendly process without the usage of toxic chemicals is in vogue (Ramkumar *et al.* 2017; Kim *et al.* 2018; Pugazhendhi *et al.* 2018). This approach offers many advantages over the conventional chemical as well as physical methods. The physical approaches used for the nanoparticle production are quite laborious and expensive, and the chemical methods involve the usage of hazardous chemicals that may adhere to the nanoparticle surface making them unsuitable for biological applications (Chung *et al.* 2016). Chemical processes also involve high energy consumption, are cumbersome for large scale production and generate toxic by-products that pose a grave threat to the environment as well as humans. However, the biological methods have proven to be environmentally safe, rapid and energy-efficient. The selection of the environmentally friendly solvent as well as non-toxic reducing and capping agents marks the most crucial aspect of green chemistry involved in nanoparticle formation (Iravani *et al.* 2014). Silver nanoparticles synthesized through green method have been reported to have biomedical applications (Rasheed *et al.* 2017) as well as in controlling the pathogenic microbes (Ferreira *et al.* 2017). In addition, due to their potent antimicrobial activity, AgNPs have also been used in cosmetics (Kale & Jagtap 2018), food preservation (Kumar *et al.* 2017) and clothing (Wigger 2017). This environmentally friendly and biocompatible approach of using the marine macroalga *U. flexuosa* for the formation of hybrid bead AgNPs was therefore adopted for our study. In addition to this, the other advantages of using marine macroalgal species include their renewability, easy availability and the presence of a variety of primary as well as secondary metabolites that aid in silver reduction.

Table 1 | The release of silver from AgNPs bound to copolymer beads as a function of time

Sample	Time interval for withdrawal of sample (hours)	Volume of fraction collected (mL)	Total silver concentration (mg/L)
1	0	5	Nd
2	6	5	Nd
3	12	5	Nd
4	18	5	0.02
5	24	5	0.034
6	48	5	0.058
7	72	5	0.066
8	96	5	0.069

Nd represents not detected.

CONCLUSION

The present study reveals that hybrid bead bound AgNPs can be formed successfully using *U. flexuosa* via a facile single step green procedure onto the copolymer beads without causing any physical damage to the surface of beads. The stable immobilized AgNPs at a concentration of 300 mg demonstrated prominent antibacterial traits which resulted in 100% bacterial reduction within 6 h of contact time for all the bacterial strains. Complete inhibition was achieved with an increase in the contact time as well as with the increase in the amount of the AgNPs bound copolymer beads. This environmentally friendly procedure employing abundant and naturally available marine benthic algae can, therefore, be considered for antibacterial treatment of water. Such features, in a nutshell, prove these hybrid bead AgNPs coated copolymer beads, synthesized using macro algal extract, as a probable future candidate for water disinfection. However, further studies need to be carried out to assure a considerable performance of the AgNPs in more complex physiological environments.

ACKNOWLEDGEMENT

The author, Dhara Dixit, deeply acknowledges the support rendered by the Director, CSIR-CSMCRI, for providing the infrastructure to conduct this study as well as the Analytical Division (CSIR-CSMCRI) for carrying out the instrumental analysis. I also thank KERC (A Division of The Corbett Foundation) for providing the field assistance during sampling. I thank Dr M. G. Thakkar (Head, Department of Earth & Environmental Science, K. S. K. V. Kachchh University) for providing me the administrative support required for conducting this study.

CONFLICT OF INTEREST

The authors have declared no conflict of interest.

REFERENCES

Abboud, Y., Saffaj, T. & Chagraoui, A. 2014 Biosynthesis, characterization and antimicrobial activity of copper oxide nanoparticles (CONPs) produced using brown alga extract (*Bifurcaria bifurcata*). *Applied Nanoscience* **4** (5), 571–576.

Aboelfetoh, E. F., El-Shenody, R. A. & Ghobara, M. M. 2017 Eco-friendly synthesis of silver nanoparticles using green algae (*Caulerpa serrulata*): reaction optimization, catalytic and antibacterial activities. *Environ. Monit. Assess.* **189**, 349. https://doi.org/10.1007/s10661-017-6033-0.

Afsheen, S., Tahir, M. B., Iqbal, T., Liaqat, A. & Abrar, M. 2018 Green synthesis and characterization of novel iron particles by using different extracts. *Journal of Alloys and Compounds* **732**, 935–944.

Akhtar, M. S., Panwar, J. & Yun, Y. S. 2013 Biogenic synthesis of metallic nanoparticles by plant extracts. *Sustainable Chemistry and Engineering* **1**, 591–602.

Alsabagh, A. M., Fathy, M. & Morsi, R. E. 2015 Preparation and characterization of chitosan/silver nanoparticle/copper nanoparticle/carbon nanotube multifunctional nano-composite for water treatment: heavy metals removal; kinetics, isotherms and competitive studies. *RSC Advances* **5**, 55774–55783.

Ashbolt, N. J. 2004 Microbial contamination of drinking water and disease outcomes in developing regions. *Toxicology.* **198**, 229–238. doi: https://doi.org/10.1016/j.tox.2004.01.030.

Azizi, S., Namvar, F., Mahdavi, M., Bin, A. M. & Mohamad, R. 2013 Biosynthesis of silver nanoparticles using brown marine macro alga, *Sargassum muticum* aqueous extract. *Materials* **6**, 5942–5950.

Badescu, I. S., Bulgariu, D. & Bulgariu, L. 2017 Alternative utilization of algal biomass (*Ulva* sp.) loaded with Zn(II) ions for improving of soil quality. *Journal of Applied Phycology* **29**, 1069.

Ballesteros, N. G., Rodriguez, J. B. G., Arguelles, M. C. R. & Lastra, M. 2017 New application of two Antarctic macroalgae *Palmaria decipiens* and *Desmarestia menziesii* in the synthesis of gold and silver nanoparticles. *Polar Science* doi: 10.1016/j.polar.2017.10.004.

Basso, A. P., Martins, P. D., Nachtingall, G., Sand, S. V. D., De Moura, T. M. & Frazzon, A. P. G. 2014 Antibiotic resistance and enterotoxin genes in Staphylococcus sp. isolates from polluted water in Southern Brazil. *Annals of the Brazilian Academy of Sciences* **86** (4), 1813–1820.

Brand, X. & Morand, P. 1997 Anaerobic digestion of Ulva sp. 1. Relationship between Ulva composition and methanisation. *Journal of Applied Phycology* **9** (6), 511–524.

Chan, C. X., Beiko, R. G. & Ragan, M. A. 2011 Lateral transfer of genes and gene fragments in *Staphylococcus* extends beyond mobile elements. *Journal of Bacteriology* **193** (15), 3964–3977.

Chemat, F., Vian, M. A. & Cravotto, G. 2012 Green extraction of natural products: concept and principles. *International Journal of Molecular Science* **13** (7), 8615–8627.

Chung, I. M., Park, I., Hyun, K. S., Thiruvengadam, M. & Rajakumar, G. 2016 Plant-mediated synthesis of silver nanoparticles: their characteristic properties and therapeutic applications. *Nanoscale Research Letters* **11**, 40.

Costa, J., Gonçalves, P., Nobre, A. & Alves, M. 2012 Biomethanation potential of macroalgae Ulva spp. and Gracilaria spp. and in co-digestion with waste activated sludge. *Bioresource Technology* **114**, 320–326.

Cui, L., Chen, P., Chen, S., Yuan, Z., Yu, C. & Ren, B. 2013 *In situ* study of the antibacterial activity and mechanism of action of silver nanoparticles by surface-enhanced Raman spectroscopy. *Analytical Chemistry* **85**, 5436–5443.

Dahoumane, S. A., Djediat, C., Yepremian, C., Coute, A., Fievet, F., Coradin, T. & Brayner, R. 2012 Species selection for the design of gold nanobioreactor by photosynthetic organisms. *Journal of Nanoparticle Research* **14**, 883.

Dahoumane, S. A., Wujcik, E. K. & Jeffryes, C. 2016 Noble metal, oxide and chalcogenide-based nanomaterials from scalable phototrophic culture systems. *Enzyme and Microbial Technology* **95**, 13–27.

Dahoumane, S. A., Mechouet, M., Wijesekera, K., Filipe, C. D. M., Sicard, C., Bazylinsky, D. A. & Jeffryes, C. 2017 Algae-mediated biosynthesis of inorganic nanomaterials as a promising route in nanobiotechnology – a review. *Green Chemistry* **19**, 552–587.

Deng, C. H., Gong, J. L., Zhang, P., Zeng, G. M., Song, B. & Liu, H. Y. 2017 Preparation of melamine sponge decorated with silver nanoparticles-modified graphene for water disinfection. *Journal of Colloid and Interface Science* **488**, 26–38.

Dodd, V. A. & Grace, P. M. 1989 *Agricultural Engineering: Proceedings of the 11th International Congress, Dublin*, 4–8 September 1989. CRC Press.

Donga, Y. Y., Fu, D., Jin-Jin, Z., Jing, H., Ming-Guo, M., Feng, X. & Run-Cang, S. 2014 Environmentally friendly ultrasound synthesis and antibacterial activity of cellulose/Ag/AgCl hybrids. *Carbohydrate Polymers* **99**, 166–172.

Dubigeon, B. C., Lahaye, M., Guillon, F., Barry, J. L. & Gallant, D. 1997 Factors limiting the biodegradation of Ulva spp. cell-wall polysaccharides. *Journal of the Science of Food and Agriculture* **75** (3), 341–351.

El-Kassas, H. Y., Aly-Eldeen, M. A. & Gharib, S. M. 2016 Green synthesis of iron oxide (fe3o4) nanoparticles using two selected brown seaweeds: characterization and application for lead bioremediation. *Acta Oceanologica Sinica* **35**, 89.

Ferreira, V. D. S., ConzFerreira, M. E., Lima, L. M. T. R., Frases, S., De Souza, W. & SantAnna, C. 2017 Green production of microalgae-based silver chloride nanoparticles with antimicrobial activity against pathogenic bacteria. *Enzyme and Microbial Technology* **97**, 114–121.

Frost, R. L., Palmer, S. J., Xi, Y., Cejka, J., Sejkora, J. & Plasil, J. 2013 Raman spectroscopic study of the hydroxy-phosphate mineral plumbogummite $PbAl_3(PO_4)_2(OH,H_2O)_6$. *Spectrochimica Acta Part A: Molecular and Biomolecular Spectroscopy* **103**, 431–434.

Gangadharan, D., Kumar, H., Gnanasekar, G., Dixit, D., Popat, K. M. & Anand, P. S. 2010 Polymeric microspheres containing silver nanoparticles as a bactericidal agent for water disinfection. *Water Research* **44**, 5481–5487.

Ghaedi, M., Sadeghiana, B., Pebdania, A. A., Sahraeib, R., Daneshfarb, A. & Duranc, C. 2012 Kinetics, thermodynamics and equilibrium evaluation of direct yellow 12 removal by adsorption onto silver nanoparticles loaded activated carbon. *Chemical Engineering Journal* **187**, 133–141.

Gopinath, V., Priyadarshini, S., Meera Priyadharsshini, N., Pandian, K. & Velusamy, P. 2013 Biogenic synthesis of antibacterial silver chloride nanoparticles using leaf extracts of *Cissus quadrangularis* Linn. *Materials Letters* **91**, 224–227.

Govindaraju, K., Kiruthiga, V., Ganesh, K. V. & Singaravelu, G. 2009 Extracellular synthesis of silver nanoparticles by a marine alga, *Sargassum wightii Grevilli* and their antibacterial effects. *Journal of Nanoscience and Nanotechnology* **9**, 5497–5501.

How, J. A., Lim, J. Z. R., Goh, D. J. W., Ng, W. C., Oon, J. S. H., Lee, K. C., Lee, C. H. & Ling, M. H. T. 2013 Adaptation of *Escherichia coli* ATCC 8739 to 11% NaCl. *Dataset Papers in Biology* Volume 2013, Article ID 219095.

Iravani, S. 2011 Green synthesis of metal nanoparticles using plants. *Green Chemistry* **13**, 2638–2650. doi: 10.1039/c1gc15386b.

Iravani, S., Korbekandi, H., Mirmohammadi, S. V. & Zolfaghari, B. 2014 Synthesis of silver nanoparticles: chemical, physical and biological methods. *Research in Pharmaceutical Sciences* **9** (6), 385–406.

Jain, N., Bhargava, A., Majumdar, S. & Panwar, J. 2011 Extracellular biosynthesis and characterization of silver nanoparticles using *Aspergillus flavus NJP08*: a mechanism prospective. *Nanoscale* **3** (2), 635–641.

Jes, E. H., Marta, E. P., Nicolai, H., Klaus, B. M. & Katrin, K. 2015 Green preparation and spectroscopic characterization of plasmonic silver nanoparticles using fruits as reducing agents. *Beilstein Journal of Nanotechnology* **6**, 293–299.

Kale, R. D. & Jagtap, P. 2018 Biogenic Synthesis of Silver Nanoparticles Using Citrus Limon Leaves and Its Structural Investigation. In: *Advances in Health and Environment Safety* (N. Siddiqui, S. Tauseef & K. Bansal, eds). Springer Nature Singapore, Springer Transactions in Civil and Environmental Engineering.

Khaleel, A. & Nawaz, M. 2015 Enhanced catalytic complete oxidation of 1,2-dichloroethane over mesoporous transition metal-doped γ-Al_2O_3. *Journal of Environmental Sciences* **29**, 199–209.

Kim, D. Y., Saratale, R. G., Shinde, S., Syed, A., Ameen, F. & Ghodake, G. 2018 Green synthesis of silver nanoparticles using *Laminaria japonica* extract: characterization and seedling growth assessment. *Journal of Cleaner Production* **172**, 2910–2918.

Kumar, M., Bala, R., Gondil, V. S., Jain, D. V. S., Chhibber, S., Sharma, R. K. & Wangoo, N. 2017 Efficient, green and one Pot synthesis of sodium acetate functionalized silver nanoparticles and their potential application as food preservative. *Journal of Materials Science* **52**, 8568–8575.

Kumar, R. V. & Raza, G. 2009 Photocatalytic disinfection of water with Ag-TiO_2 nanocrystalline composite. *Ionics* **15** (5), e579–587.

Li, W. R., Xie, X. B., Shi, Q. S., Duan, S. S., Ouyang, Y. S. & Chen, Y. B. 2011 Antibacterial effect of silver nanoparticles on *staphylococcus aureus*. *Biometals* **24**, 135–141.

Liu, B., Xie, J., Lee, J. Y., Ting, Y. P. & Chen, J. P. 2005 Optimization of high-yield biological synthesis of single-crystalline gold nanoplates. *Journal of Physical Chemistry B* **109**, 15256–15263.

Lu, Q., Wang, H., Liu, Y., Hou, Y., Li, H. & Zhang, Y. 2017 Graphitic carbon nitride nanodots: as reductant for the synthesis of silver nanoparticles and its biothiols biosensing application. *Biosensors and Bioelectronics* **89** (Part 1), 411–416.

Maze, J., Morand, P. & Potoky, P. 1993 Stabilization of green tides Ulva by method of composting with a view to pollution limitation. *Journal of Applied Phycology* **5**, 183–190.

Minhas, F. T., Arslan, G., Gubbuk, I. H., Akkoz, C., Ozturk, B. Y., Asikkutlu, B., Arslan, U. & Ersoz, M. 2018 Evaluation of antibacterial properties on polysulfone composite membranes using synthesized biogenic silver nanoparticles with *Ulva compressa* (L.) Kütz. and *Cladophora glomerata* (L.) Kütz. Extracts. *International Journal of Biological Macromolecules* **107**, 157–165.

Morones, J. R., Elechiguerra, J. L., Camacho, A., Holt, K., Kouri, J. B. & Ramirez, J. T. 2005 The bactericidal effect of silver nanoparticles. *Nanotechnology* **16**, 2346–2353.

Okafor, O., Weilhard, A., Fernandes, J. A., Karjalainen, E., Goodridge, R. & Sans, V. 2017 Advanced reactor engineering with 3D printing for the continuous-flow synthesis of silver nanoparticles. *Reaction Chemistry & Engineering* **2**, 129–136.

Pirian, K., Piri, K., Sohrabipour, J. & Blomster, J. 2018 Three species of Ulva (Ulvophyceae) from the Persian Gulf as potential sources of protein, essential amino acids and fatty acids. *Phycological Research*. doi:10.1111/pre.12212.

Prochnow, A. M., Bradbury, M. & Murphy, A. B. 2015 Draft genome sequence of *Pseudomonas aeruginosa* ATCC 9027 (DSM 1128), an important rhamnolipid surfactant producer and sterility testing strain. *Genome Announcements* **3** (5), e01259-15.

Pugazhendhi, A., Prabakar, D., Jacob, J. M., Karuppusamy, I. & Saratale, R. G. 2018 Synthesis and characterization of silver nanoparticles using *Gelidium amansii* and its antimicrobial property against various pathogenic bacteria. *Microbial Pathogenesis* **114**, 41–45.

Qu, X., Alvarez, P. J. J. & Li, Q. 2013 Applications of nanotechnology in water and wastewater treatment. *Water Research*. **47**, 3931–3946.

Ramkumar, V. S., Pugazhendhi, A., Gopalakrishnan, K., Sivagurunathan, P., Saratale, G. D., Dung, T. N. B. & Kannapiran, E. 2017 Biofabrication and characterization of silver nanoparticles using aqueous extract of seaweed

Enteromorpha compressa and its biomedical properties. *Biotechnology Reports* **14**, 1–7.

Rasheed, T., Bilal, M., Iqbal, H. M. N. & Li, C. 2017 Green biosynthesis of silver nanoparticles using leaves extract of *Artemisia vulgaris* and their potential biomedical applications. *Colloids and Surfaces B: Biointerfaces* **158**, 408–415.

Rodriguez, S. N., Bright, K. R., Slack, D. C., Uhlmann, D. R. & Gerba, C. P. 2008 Silver as a residual disinfectant to prevent biofilm formation in water distribution systems. *Applied and Environmental Microbiology* **74** (5), e1639–1641.

Rombaut, N., Tixier, A. S., Bily, A. & Chemat, F. 2014 Green extraction processes of natural products as tools for bio refinery. *Biofuels, Bioproducts and Biorefining* **8** (4), 530.

Shukla, M. K., Singh, R. P., Reddy, C. R. K. & Jha, B. 2012 Synthesis and characterization of agar-based silver nanoparticles and nanocomposite film with antibacterial applications. *Bioresource Technology* **107**, 295–300.

Singaravelu, G., Arockiamary, J. S., Kumar, V. G. & Govindaraju, K. 2007 A novel extracellular synthesis of monodisperse gold nanoparticles using marine alga, *Sargassum wightii* Greville. *Colloids and Surfaces B: Biointerfaces* **57** (1), 97–101.

Stalin Dhas, T., Ganesh Kumar, V., Karthick, V., Jini Angel, K. & Govindaraju, K. 2014 Facile synthesis of silver chloride nanoparticles using marine alga and its antibacterial efficacy. *Spectrochimica Acta Part A: Molecular and Biomolecular Spectroscopy* **120**, 416–420.

Suresh, A. K., Pelletier, D., Ji, W., Baohuagu, W., Mortensen, N., Allison, D. P., Joy, D. C., Phelps, T. J. & Doktycz, M. J. 2010 Silver nanocrystallites: biofabrication using shewanella oneidensis, and an evaluation of their comparative toxicity on gram-negative and gram-positive bacteria. *Environmental Science and Technology* **44**, 5210–5215.

Taglietti, A., Arciola, C. R., D'Agostino, A., Dacarro, G., Montanaro, L., Campoccia, D., Cucca, L., Vercellino, M., Poggi, A., Pallavicini, P. & Visai, L. 2014 Antibiofilm activity of a monolayer of silver nanoparticles anchored to an amino-silanized glass surface. *Biomaterials* **35**, e1779–1788.

Veisi, H., Azizi, S. & Mohammadi, P. 2018 Green synthesis of the silver nanoparticles mediated by *Thymbra spicata* extract and its application as a heterogeneous and recyclable nanocatalyst for catalytic reduction of a variety of dyes in water. *Journal of Cleaner Production* **170** (1), 1536–1543.

Wang, T., Jin, X., Chen, Z., Megharaj, M. & Naidu, R. 2014 Green synthesis of Fe nanoparticles using eucalyptus leaf extracts for treatment of eutrophic wastewater. *Science of the Total Environment* **466–467**, 210–213.

Wigger, H. (ed.) 2017 Case study: Clothing textiles with incorporated silver nanoparticles. In: *Environmental Release of and Exposure to Iron Oxide and Silver Nanoparticles*. Springer Vieweg, Wiesbaden.

Xie, J., Lee, J. Y., Wang, D. I. C. & Ting, Y. P. 2007a Identification of active biomolecules in the high-yield synthesis of single-crystalline gold nanoplates in algal solutions. *Small* **3**, 672–682.

Xie, J., Lee, J. Y., Wang, D. I. C. & Ting, Y. P. 2007b Silver nanoplates: from biological to biomimetic synthesis. *ACS Nano* **1**, 429–439.

Xu, Z., Lei, H., Peng, H. & Shaojun, D. 2014 Facile synthesis of small Ag@AgCl nanoparticles via a vapour diffusion strategy and their highly efficient visible-light-driven photocatalytic performance. *Catalysis Science & Technology* **4**, 3615.

Zhao, Y. K., Sung, W. P., Tsai, T. T. & Wang, H. J. 2010 Application of nanoscale silver-doped titanium dioxide as photocatalyst for indoor airborne bacteria control: a feasibility study in medical nursing institutions. *Journal of the Air & Waste Management Association* **60** (3), e337–345.

First received 22 September 2017; accepted in revised form 19 June 2018. Available online 2 July 2018

IWA Publishing's authorised EU representative for General Product Safety Regulations is Diane D'Arras, 15 rue Duret, 75116 Paris, France, e-mail: safety@iwap.co.uk.

Printed and bound by CPI Group (UK) Ltd, Croydon, CR0 4YY
12/05/2026
02108874-0001